Ecological and Practical Applications for Sustainable Agriculture

Kuldeep Bauddh • Sanjeev Kumar •
Rana Pratap Singh • John Korstad
Editors

Ecological and Practical Applications for Sustainable Agriculture

Springer

Editors
Kuldeep Bauddh
Department of Environmental Science
Central University of Jharkhand
Ranchi, Jharkhand, India

Sanjeev Kumar
Department of Environmental Science
Central University of Jharkhand
Ranchi, Jharkhand, India

Rana Pratap Singh
Department of Environmental Science
Babasaheb Bhimrao Ambedkar University
Lucknow, Uttar Pradesh, India

John Korstad
Department of Biology and Renewable Energy
Oral Roberts University
Tulsa, OK, USA

ISBN 978-981-15-3371-6 ISBN 978-981-15-3372-3 (eBook)
https://doi.org/10.1007/978-981-15-3372-3

This Springer imprint is published by the registered company Springer Nature Singapore Pte Ltd.
The registered company address is: 152 Beach Road, #21-01/04 Gateway East, Singapore 189721, Singapore

Dedicated to my beautiful daughter Ms. Aashwita Kuldeep.

—Kuldeep Bauddh

Dedicated to my beloved parents and family.

—Sanjeev Kumar

I dedicate the book to all who are making efforts for sustainable agriculture and safe food security throughout the world.

—Rana Pratap Singh

Dedicated to my wonderful and blessed wife of nearly 48 years, and our 4 beautiful daughters, 4 son-in-laws, and 10 amazing grandchildren. Proverbs 31:10-31 and Psalm 127:3-5.

—John Korstad

Acknowledgment

The authors made this challenging assignment a reality by their kind contribution and cooperation. I would like to acknowledge and thank each one of the authors for their contributions whose work, research, and support helped formulate this book. I express my deep sense of gratitude to my Ph.D. supervisor Prof. Rana Pratap Singh and other two fellow editors of the book Dr. Sanjeev Kumar and Prof. John Korstad for their vision and foresight which inspired me to complete this book along with them.

Words are not enough to express my grateful appreciation for my parents (Mr. Suresh Baboo and Mrs. Shyama Devi), younger brother (Jagdeep Kumar), and all of my relatives. Their dreams and ambition made me strong to complete the target.

I am thankful to my loving wife (Sweta) for her continuous help, for her encouragement, and most importantly, for her cooperating nature.

I take this opportunity to acknowledge the entire team of Springer Publishing and everyone who collaborated in producing this book.

(Kuldeep Bauddh)

This edited book is the outcome of long and dedicated efforts of many individuals, who supported me during the compilation and publication of this valuable edition. I thank all the national and international contributing authors for their valued submissions, cooperation, and providing up-to-date information on the diverse aspects of the subject.

I avail this opportunity to express my profound sense of sincere and deep gratitude to my Ph.D. Supervisor Prof. Rana Pratap Singh (Coeditor) and other coeditors, Dr. Kuldeep Bauddh and Prof. John Korstad, for their continuous guidance, cooperation, and assistance that helped me in bringing this work to light. I express my gratitude to all the faculty members of the Department of Environmental Science, Central University of Jharkhand, Ranchi, Jharkhand, for their wholehearted support.

My sincere appreciation and deep gratitude to my parents (Dr. Mahesh Prasad and Mrs. Parwati Devi), brother (Dr. Rajeev), sisters (Dr. Shashi and Dr. Sudha), and my wife (Dr. Ritu) for their unending support and helpful encouragement to complete this exciting task.

I also acknowledge the production team members for guiding us throughout the process of publication of this work in an efficient and professional manner.

(Sanjeev Kumar)

We acknowledge input from all the authors who have taken pain to develop their chapters in spite of their busy schedule. Our family members and students spared us for this useful input at the cost of their own share and hence we express our regard and gratefulness to them. The publisher Springer has facilitated us at each step with care and patience which deserve a high appreciation and thanks.

(Rana Pratap Singh)

This book was a "labor of love" from the beginning—meaning that it was a challenging assignment in terms of time expenditure for each of us when we were already busy with our regular jobs, but we were (and still are) passionate about the need to explain and encourage best practices for sustainable farming globally. I am especially encouraged by the contribution of some of my students as chapter authors in this book who are just beginning their academic careers. They have a bright future ahead!

(John Korstad)

Contents

About the Editors

Kuldeep Bauddh is currently an Assistant Professor at the Department of Environmental Science, Central University of Jharkhand, Ranchi, India. He received his PhD in Environmental Science from Babasaheb Bhimrao Ambedkar University, Lucknow, India. Dr. Bauddh is currently teaching thrust areas of environmental sciences like environmental pollution and management and environmental toxicology. His main research areas are phytoremediation and sustainable agriculture. He has published three books (Edited), one with Elsevier on *Phytomanagement of Polluted Sites*, and two with Springer on *Phytoremediation Potential of Bioenergy Plants* and *Algae and Environmental Sustainability*. He has published 32 research papers in national and international journals.

Sanjeev Kumar is an Assistant Professor, Department of Environmental Science, Central University of Jharkhand, Ranchi, India. He obtained MSc and PhD in Environmental Sciences from Babasaheb Bhimrao Ambedkar University, Lucknow, India. He has worked in the field of eco-friendly farming practices, nutrient leaching, and gaseous emissions from agricultural fields. He has published 21 research articles, 15 book chapters, and 2 edited books with Springer and presented research in many national and international conferences and seminars. He was awarded "Fellow of Society of Plant Research" and received the recognition award of "Significant contribution in academic activities" by Babasaheb Bhimrao Ambedkar University.

Rana Pratap Singh is Dean Academic Affairs and Professor of Environmental Science at Babasaheb Bhimrao Ambedkar University, Lucknow, India. He has made significant contribution in the understanding of N metabolism, ecological nutrition, and heavy metal toxicity in plants under nonstress and environmental stress. His group has developed carrier-based climate-resilient microbial consortia for sustainable agriculture. He has published over 150 papers and17 books and has about 3900 citations with H-index 35 and i10-79; he has 33 years of PG teaching, 38 years of research, and 14 years of administrative experience, has 8 awards and 4 fellowships, and is president of two academic societies. He is Chief Editor of PMBP (www.springer.com/journals/12298).

John Korstad is Professor of Biology Oral Roberts University, Tulsa, Oklahoma. His educational background includes an MS in Environmental Science from Cal. State Hayward and an MS and PhD in Zoology from the University of Michigan. John teaches courses in zoology, ecology, limnology, marine ecology, research, environmental science, and sustainability. He has vast experience in aquaculture, including research at SINTEF in Trondheim, Norway, supported by a Fulbright Fellowship. His current research focuses on lake management, algal biofuels, phytoremediation, and sustainable development. John is on the editorial board of 5 journals, has over 40 professional publications, and is coeditor of 3 books.

Sustainable Agricultural Approaches for Enhanced Crop Productivity, Better Soil Health, and Improved Ecosystem Services

1

Lala Saha and Kuldeep Bauddh

Contents

Abstract

Agriculture is an important sector that provides food, fiber, and fuel, and other vital commodities which possibly sustains life on Earth. In recent time, the

L. Saha · K. Bauddh (✉)
Department of Environmental Science, Central University of Jharkhand, Ranchi, India
e-mail: kuldeep.bauddh@cuj.ac.in

K. Bauddh et al. (eds.), *Ecological and Practical Applications for Sustainable Agriculture*, https://doi.org/10.1007/978-981-15-3372-3_1

growing human population demands a large amount of agriculture commodities to fulfill the need. Therefore, agriculture has been escalating rapidly, which introduced various modern practices and technologies that affect the environment in many ways. The excessive use of chemical fertilizers and pesticides contaminates the air, water, and soil. Although the use of synthetic fertilizers and pesticides enhances crop productivity, it also deteriorates the soil health. There is a need to explore the economically sound and ecologically viable alternatives which can address these concerns. Numerous sustainable cropping practices like the application of biofertilizer, slow-release fertilizers, biochar, vermicompost, zero or low tillage, etc., have been investigated and found substantially effective. In the present chapter, a thorough discussion about these technologies has been made. Moreover, how these technologies can be incorporated with modern/corporate agricultural tools has also been explored.

Keywords

Climate change · Organic farming · Soil pollution · Sustainable development · Traditional agriculture

1.1 Introduction

Agriculture system plays a pivotal role in sustaining life on Earth. Worldwide people cultivate a variety of food crops, medicinal and industrial plants to meet their demand. In the present scenario, an increase in food demand by almost 3.8% $year^{-1}$ for the booming population, which is further predicted to grow to nearly 9–10 billion by 2050, causes a lot of pressure on the agricultural sector. However, the production of food grows only by 1.2% $year^{-1}$ (FAO 2014). The present agricultural practices are based on various synthetic fertilizers, pesticides, large amount of water, energy, hybrid seeds, and capital investment in the form of agricultural machinery to maximize production to meet the demand. The modern agricultural practices evolve with direct and indirect environmental impacts such as the soil erosion, loss of soil fertility, groundwater depletion, greenhouse gas emissions, the degradation of freshwater ecosystems, antibiotic resistance, losing traditional crop varieties, and apparently losing biodiversity (Maikhuri et al. 1997; Gundi et al. 2005; Maddela and Venkateswarlu 2018; Gowdy and Baveye 2019; Taiwo 2019). Earlier people practiced agriculture in a more sustainable manner by cultivating the indigenous variety of crops, rotating the crops, mixed cropping, and application of farmyard manure etc. There are a number of traditional agricultural practices which have been proven ecologically sound like cultivating the specific crops which have high productivity and low pest attack, intercropping, crops supporting more parasitoids, and the leguminous crops which enrich nitrogen in soil (Drinkwater et al. 1998; Leung et al. 2003; Tscharntke et al. 2005; Singh et al. 2017; Sánchez et al. 2019).

Since the green revolution (of the 1960s), farmers have shifted from traditional agricultural practices and adapted the modern techniques which comprise the

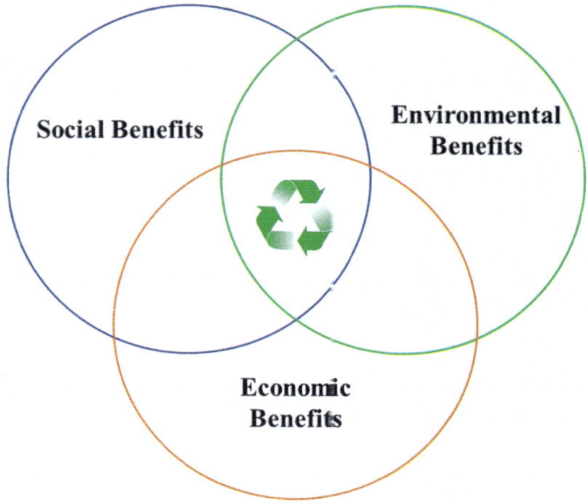

Fig. 1.1 Model of sustainable agricultural practices

application of a variety of chemical fertilizers and pesticides, maximum plowing, monocropping, etc. These practices pose significant adverse impacts on the agroecosystems (Pimentel et al. 1992; Henry et al. 2011; Duran et al. 2015; Maddela and Venkateswarlu 2018; Taiwo 2019).

Both traditional and modern agricultural practices may be integrated to overcome food security issues along with agricultural sustainability. Due to an increase in diseases associated with agricultural practices, the people are again looking toward eco-friendly agricultural practices like switching to organic farming with the application of vermicompost and other organic manure, increasing the crop production, and using recycled water for irrigation purposes. These practices lead to sustainable agriculture which is inclusive of social, economic, and environmental benefits (Fig. 1.1). High market demand and profit attract the private corporation and is rapidly entering into agriculture commonly popularized as agribusiness. The corporate agribusiness bought or leased the agricultural land/wastelands from people and the government as well, where they cultivate the crops of their own interest. In this way, the agriculture sector is increasing and introducing the new practices and crops, for example, introduction of the genetically modified crops, invasive crop, focus on monocropping, crop repetition, etc., which has both positive and negative aspects.

Corporate agriculture can be a good option for utilizing the wasteland for crop production if it can manage in sustainable ways. In India, there are many corporate agricultural practices running in the state of Gujarat, Goa, Karnataka, Madhya Pradesh, Maharashtra, Punjab, and Tamil Nadu (Singh 2006). The agriculture context is changing rapidly; the United Nations General Assembly adopted 17 sustainable development goals (SDGs) in the year 2015 for the year 2030. In that, six goals are directly linked with the agricultural practices as shown in Fig. 1.2.

The present climate change is also a great threat to modern agriculture ecosystems as most of the modern crop's varieties are climate-sensitive.

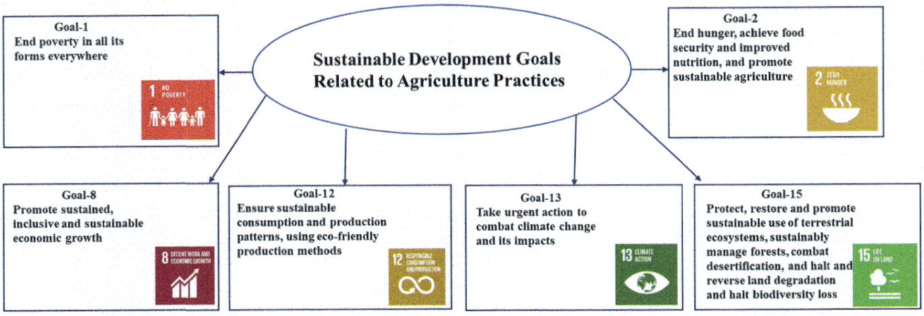

Fig. 1.2 SDGs related to agriculture system (Source: United Nations Summit on Sustainable Development 2015)

1.2 Crop Productivity Along with Soil Health

Soil is an important resource that provides more than 90% of food for human beings (Brevik 2010). It is our responsibility to maintain the soil health. Soil health can be defined as the capacity of soil to function within the natural and man-made ecosystem boundaries, sustain the living organism's productivity, maintain the quality of air and water, and promote the animals, plants, and human health (Brevik 2010; Hoorman et al. 2011). In general, the soil quality and health status can be indicated by its physical, chemical, and biological indicators (Fig. 1.3). However, the properties of a healthy soil may vary from place to place and also depend on many other conditions.

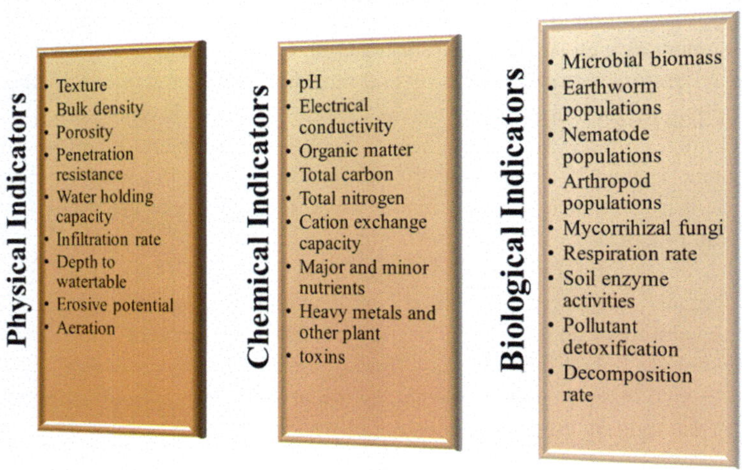

Fig. 1.3 Indicators of soil health (Brevik 2010)

In healthy soil, the plant can grow more comfortably, which results in higher productivity. Sometimes, the available nutrient in the soil determines the suitability of crops for cultivation. For example, if the iron availability is higher in the soil, the soil will be acidic where the blueberries (*Vaccinium corymbosum*) will thrive successfully; however, corn (*Zea mays*) will not give good productivity.

The most important parameter is the organic matter content of the soil, because it provides a profound influence on the physical, chemical and biological properties. Some of the modern agricultural practices have a significant impact on the soil organic matters. For instance, the intense machine-driven cultivation practices like heavy mechanical plowing introduce oxygen up to the deep layer soil that speeds up decomposition rate of organic matters. Mechanical tillage chops the crop residues into smaller pieces, which are immediately available for decomposition and incorporated into the soil. The soil microorganism also easily participates in the decomposition process. The modern agriculture soil manipulation practices improve the crop productivity for a short period of time but lead to the loss of the soil organic matter in the long term, which put the sustainable yield in peril.

To increase the crop productivity, pesticides and fertilizers are being used in a large amount (Zhang et al. 2011; IFA 2015; Bitew and Alemayehu 2017). The crop production inputs include mineral fertilizers like ammonium nitrate, calcium sulfates, urea, and phosphate; pesticides include insecticides, nematicides, fungicides, herbicides, and soil fumigants, and various organic fertilizers and products include livestock manure, biosolid and compost, microbial inoculants, and humic acids (Bunemann et al. 2006; Schreck et al. 2008). The synthetic fertilizers and pesticides have some direct and indirect influences on the soil health and on the beneficial microorganism. Studies have shown that the application of chemical fertilizers affects the soil by forming hydrochloric and sulfuric acid that leads to acidic soil. These acids dissolve the soil crumbs, which are responsible for binding the rock particles together (Bitew and Alemayehu 2017). The mineral-rich soil crumbs play a crucial role in soil drainage and help in air circulation in the soil.

1.3 High-Yielding Varieties (HYVs) and Genetically Modified Varieties (GMVs)

With the beginning of modern agricultural practices (after the green revolution), the cultivation of high-yielding varieties was started rapidly. The HYVs may be the same as the natural crops as they are investigated through screening methods or are developed by cross-pollinating different related cultivars. These varieties can successfully cultivate over the generations, because they may not share the desired traits, which are responsible for high productivity and healthy seeds. Indeed, the HYV varieties were more productive, although they require more fertilizer and water to grow. With time, these varieties are vulnerable to pest and diseases (Sarkar and Vanloon 2015). On the other hand, genetically modified varieties are developed in the laboratory using biotechnological interferences like gene splicing, gene silencing, etc. The new varieties were developed by crossing two different biological organisms such as plant and bacteria, for example, Bt corn and cotton which were developed by crossing the genetic material of bacteria called *Bacillus thuringiensis.* These crops are pest resistant, and when the pest eats the plant, it dies.

Several genetically modified crops have been proven to give higher yield and also require less or no pesticides. Albeit there is fear that Bt crops that are consumed by the livestock including human beings may work as pesticides, still researchers and scientists are working on safety issues of genetically modified crops. A number of studies have found worrying results on GMVs (Carman et al. 2013; Séralini et al. 2014). However, another study suggests that GMVs such as soybeans and maize are found safe and nutritious (Domingo and Bordonaba 2011; Hilbeck et al. 2015). Several issues which are associated with GMVs are adverse impacts on microbial diversity of soil, transgene flow, impacts on nontarget fauna and flora, etc. The GMVs should be tested thoroughly at laboratory scale and also at field level so that these concerns are resolved.

1.4 Modern Agricultural Practices and Their Impacts on Ecosystem Services (ES)

In modern agriculture, application of inorganic fertilizer and pesticides has become standard practice to enhance the crop productivity. The intensive modern agricultural practices without considering the ecological consequences deteriorate the soil quality, deplete groundwater, and decline the agrobiodiversity (Kesavan and Swaminathan 2007; Sarkar and Vanloon 2015). These intensive agricultural practices may lead to desertification. Further, to manage the irrigational systems, a number of dams have been constructed, and many of them lost their water-holding capacity due to the silting and disturbances in the drainage system associated with landscape changes. These landscape changes result in flood, increased soil salinity, and crop loss (Hazell 2009; GOI 2013) (Fig. 1.4).

A recent study on ecosystem services of both traditional and modern agricultural practices shows that Chinese traditional agriculture proves more sustainable with high ecosystem services. However, the flow of ecosystem services (ES) and ecosystem disservices (EDS) directly relies on the management of agroecosystems (Zhang et al. 2012). For example, a cost-benefit analysis (CBA) of traditional rice-fish cultivation system (TRFC) and modern rice monoculture shows in terms of the economic value of ES and EDS that the net monetary value of TRFA was 3.31×10^4 CNY.ha^{-1}, and for the rice monoculture, it was 1.99×10^4 CNY.ha^{-1} (Zhang et al. 2012). Further, the same study shows, when the CBA for monetary value for the environmental loss for the traditional vs. modern agriculture, it shows that net earnings of rice-fish cultivation were 1.94×10^4 CNY.ha^{-1} which is higher than rice monoculture. However, the cost-benefit ratio of rice-fish farming is less than that of rice monoculture. So, the traditional farming practice is not priority choice for the farmers.

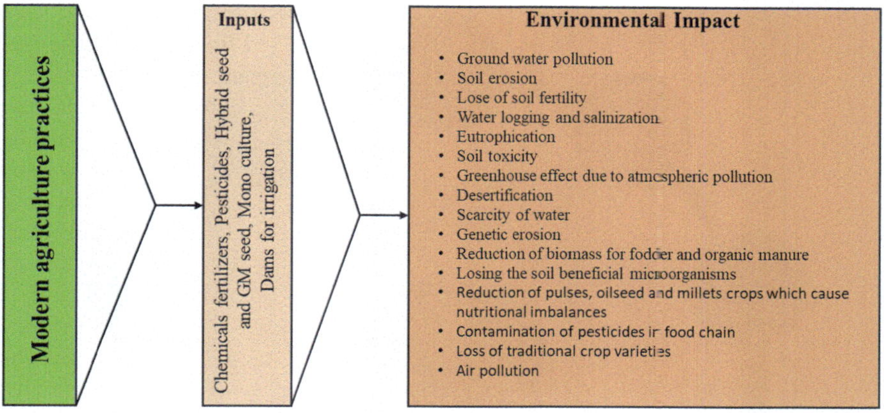

Fig. 1.4 Modern agricultural practices and their direct and indirect environment impact (Source: Shiva and Singh 2015 with modification)

1.5 Environmentally Friendly Agricultural Practices and Ecosystem Services

Environment friendly agricultural practices (EFAPs) are helpful to mitigate the nonpoint sources of agricultural pollution with enhanced ecosystem services like nutrient cycling, hydrological cycles, maintenance of soil health, etc. To ameliorate the agricultural practices, many traditional agriculture systems are being re-evaluated from time to time, and further intensive research is being conducted to develop the new eco-friendly agriculture methods, such as application of recommended amount of fertilizer after soil test, no or minimum tillage, controlled and slow release of fertilizers, and leaving the crop residues in the field for manure (Zheng 2010). The advantages of no or minimum tillage result in conserving the soil moisture, improving the soil structure, and saving fuel and labor (Stevenson et al. 2014). Another method is the application of fertilizer in a controlled manner which releases the nutrients slowly that leads to utilizing the whole nutrients by plant and results in reduction of the fertilizer application in the field (Zhang and Zhang 2005). This practice is also helpful in reduction of leaching of nutrients into the groundwater as well as their emission into the atmosphere in the form of gases. Testing the soil and application of the recommended amount of fertilizer result in balance of soil fertility (Luo et al. 2016). Agricultural practices like application of the crop straw into the field without burning it, lowering the air pollution, and also improving soil fertility. Furthermore, studies also conducted in these EFAPs found that it decreases the nitrate leaching and emission of N_2O from the agriculture fields (Qin et al. 2015). Overall recommended amount of fertilizer application after the soil proved to be more helpful to mitigate the environmental pollution followed by the application of organic manure and minimum tillage.

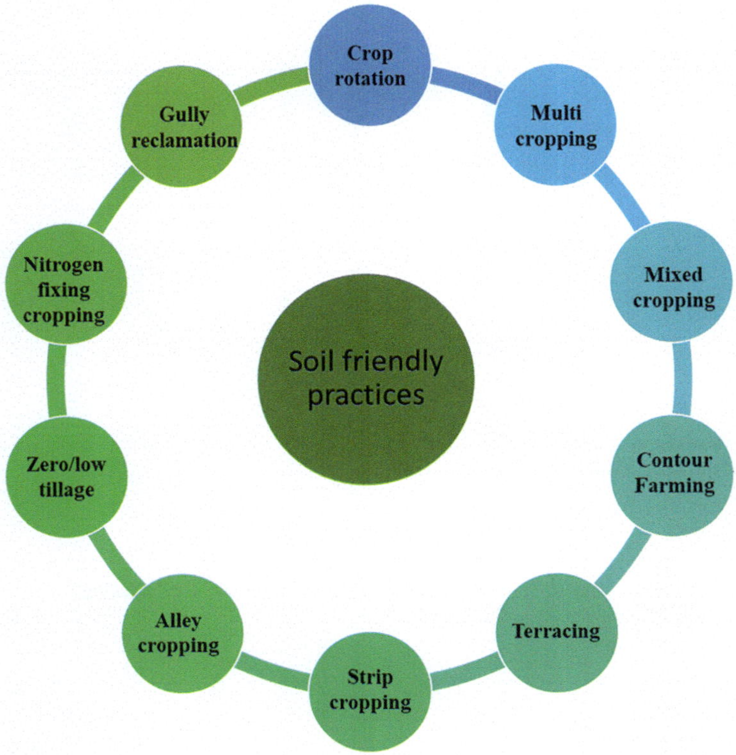

Fig. 1.5 Different soil-friendly agricultural practices

The eco-friendly agricultural practices overall aim to manage the resources for the rural people's welfare and to conserve/improve the soil health. These eco-friendly agricultural practices will ensure that the agribusiness activity is consistent without disturbing the ecosystem functioning. For example, the soil nutrient cycle and the structure of biodiversity will be intact and give a perfect agriculture system with self-pest-resistant capacity and self-regulating soil nutrients (Mishra 2013).

The main concept in agriculture sustainability includes the protection of resources, decrease of negative externalities, and encouragement of positive externalities. The main issues include soil pollution and erosion, water quantity and quality, air pollution, protection of biodiversity and landscape protection, protection of living organisms, and food safety. To overcome these problems, practicing the green methods and using soil-friendly pesticides and fertilizers are highly helpful. Agricultural practices such as crop rotation, multi-cropping, mix cropping, counter farming, strip farming, etc., have proven better over modern farming systems (Fig. 1.5). Application of soil-friendly fertilizers and pesticides derived from the natural sources like biopesticides, biofertilizers, green manures,

Fig. 1.6 Soil-friendly fertilizers and pesticides for environmentally friendly agricultural practices

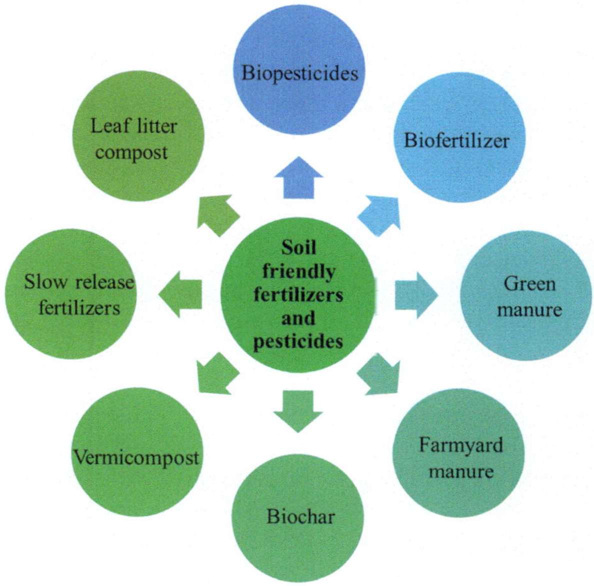

farmyard manure, and vermicompost has no adverse impact on the soil ecosystem (Fig. 1.6).

It has been shown that the major cause of modern agricultural practices is the flexibility in cropping pattern. All earlier methods of cultivation patterns were nature-friendly. For example, the rotational and mixed cropping systems are planned in such a way that successively disturbs the insect's life cycle. Biopesticides and push-pull cropping can be used to control the residual pest activity (Shiva and Singh 2015). Push-pull cropping system refers to the system where pest repellent plants are planted in between the crop to distract from the crop and planting those plants in the boundary which attract the pests. In this way, the pest repellent plants push the pest away and the boundary planted plant pulls them.

1.6 Soil-Friendly Crops

Plant species play a vital role in maintaining soil health by enriching beneficial microbes with the help of root exudates. Different crops have different constituents of exudates, which vary by nature, and the corresponding microbes use the resources and develop its community in the rhizosphere (Tilak et al. 2005). There are varieties of bacteria available in the soil. Bacteria that are living directly in the soil are known as free-living bacteria as they do not rely on the organic constituents of the exudates to survive. The rhizospheric bacteria utilize the root exudates (Barraquio et al. 2000). These bacteria are largely available in the leguminous crop fields which are beneficial to plant growth, known as plant growth-promoting bacteria (PGPB). PGPB are responsible for fixing the atmospheric nitrogen, solubilize minerals like phosphorous, produce siderophores that solubilize and sequester iron, and also produce hormones known as plant growth regulators (Tilak et al. 2005). Crops such as soybean (*Glycine max*), chickpea (*Cicer arietinum*), cowpea (*Vigna sinensis*), lentil (*Lens esculenta*), berseem (*Trifolium alexandrinum*), dhaincha (*Sesbania aculeata*), guar (*Cyamopsis tetragonoloba*), and Indian clover (*Medicago parviflora*) are host to several rhizobium species which are helpful in fixing nitrogen in saline and alkaline stress soils which improves the soil health and promote plant growth (Rehman and Nautiyal 2002; Tilak et al. 2005; Mia and Shamsuddin 2010; De Meyer et al. 2018). The leguminous crops are recognized as the soil-building crops. It helps to improve soil properties through its effects on soil physical, chemical, and biological characteristics.

Arbuscular mycorrhizal fungi (AMF) are also found in the rhizospheric region of most of the plant species which form a mutual relationship with the host plant root. These fungi are helpful in plant growth and increase plant nutrients. Worldwide about 8% of plant families are associated with AMF (Sadhana 2014). AMF controls the diffusion of different ions such as P, Cu, and Zn in soil, because this plant utilized the nutrients immediately from the surroundings. The mycorrhizal fungal hyphae enlarge its network toward the nutrient depletion areas and capture the immobile elements. AMF provides direct benefits such as increasing the phosphate uptake by plants, in some cases zinc and copper which increase the yield and also provide indirect benefits such as enhancing the soil structure and aggregation (Leifheit et al. 2014; Sadhana 2014; Augé et al. 2015). In addition, to enhance nutrient supply, AMFs help plants to sustain drought and salinity (Porcel et al. 2011; Berruti et al. 2016). A recent study demonstrated that AMF also contribute to reducing the emission of N_2O which is an important greenhouse gas; thus, it is believed that it may play a critical role in the mitigation of climate change (Berruti et al. 2016). A number of studies confirm that AMF colonize in various vegetable crops such as carrot (*Daucus carota*) and tomato (*Solanum lycopersicum*), in tea plantation soil, in black pepper-growing area, and in the field of pigeon pea (*Cajanus cajan*), pearl millet (*Pennisetum glaucum*), maize (*Zea mays*), and chickpea (*Cicer arietinum*) (Rajeshkumar and Selvaraj 2006; Sadhana 2014; Baum et al. 2015). Apart from this, there are several host plants which can be useful to increase the AMF in crop field such as sorghum (*Sorghum vulgare*), onion (*Allium cepa*), clover (*Trifolium subterraneum*), bahiagrass (*Paspalum notatum*), cenchrus grass (*Cenchrus ciliaris*), barley (*Hordeum vulgare*), strawberry (*Fragaria sp.*), coleus (*Coleus sp.*), etc. (Sadhana 2014; Baum et al. 2015).

1.7 Soil-Friendly Fertilizers and Pesticides

The soil-friendly fertilizers and pesticides are derived from natural sources/materials such as from plant, animal, and microbes. By 2013, there were more than twelve hundred registered biopesticide products available in the market (Raja 2013). Some of the soil-friendly fertilizers and pesticides have been discussed in Fig. 1.6.

1.7.1 Biofertilizer

Biofertilizers are cost-effective for the plant nutrient supplements which reduce the application of chemical fertilizers. Biofertilizers contain living microorganisms such as bacteria, cyanobacteria, fungi, etc., which are helpful to the plants, nutrient mobilization, and phosphorous solubilization and therefore increase the productivity of the soil. Biofertilizers such as *Azotobacter, Bacillus, Pseudomonas, Rhizobium,* and *Azospirillum* have been used widely due to its environmentally friendly nature and cost-effectiveness. Biofertilizers can be classified into some categories based on the functional attributes such as nitrogen-fixing, plant growth-promoting rhizobacteria and potassium-, phosphate-, and micronutrient-solubilizing biofertilizers (Raimi et al. 2017). Nitrogen-fixing biofertilizers contain oxygen-sensitive enzyme complex which is composed of dinitrogenase and dinitrogenase reductase which are responsible for the conversion of atmospheric N_2 into reactive ammonia (Dighe et al. 2010). Bacteria such as diazotrophs can present both symbiotically and non-symbiotically with the plants. Diazotrophs can fix the nitrogen gas in nitrogen-deficient soil via the biological nitrogen fixation process with the help of the nitrogenase enzyme. *Bradyrhizobium, Sinorhizobium,* and *Rhizobium* inhabit symbiotically in the root nodules of many leguminous crops such as cowpea (*Vigna unguiculata*), soybean (*Glycine max*), mung bean (*Vigna radiata*), and groundnuts (*Arachis hypogaea*) (Oldroyd et al. 2011; Raimi et al. 2017).

In soil, phosphorous and potassium are coupled with aluminum, iron, and calcium to form a stable compound which is not easily available for plant uptake (Mohammadi 2012). In this situation, microbes play a crucial role to solubilize and mobilize the nutrients like phosphate. P is solubilized mostly by certain bacteria and fungi such as *Aspergillus, Bacillus, Penicillium,* and *Pseudomonas.* These microorganisms produce organic P by secreting the phosphatase enzymes followed by enzymatic hydrolysis process. They save up to 30–50 kg of phosphate fertilizers per hectare (Raimi et al. 2017). Phosphate-solubilizing biofertilizers (PSB) such as *Pseudomonas putida, P. chlororaphis, Rhizobium leguminosarum, Bacillus circulans*, and *Azotobacter chroococcum* have been applied on several crops such as in pulse, potato, wheat, and tomato (Malusà and Ciesielska 2014). Potassium-solubilizing biofertilizers (KBS) include *Acidithiobacillus, Burkholderia, Bacillus, Pseudomonas,* and *Paenibacillus* genera that solubilize K from compounds such as biotite, illite, mica, orthoclase, and muscovite by producing hydroxyl anions and organic ligands (Shanware et al. 2014; Bahadur et al. 2014; Raimi et al. 2017).

Plant growth-promoting rhizobacteria (PGPR) are composed of heterogeneous groups of rhizospheric microbes which include the genera *Azospirillum, Azotobacter, Burkholderia, Bacillus, Pseudomonas, Serratia*, and *Erwinia*. PGPR stimulates plant growth directly and indirectly through various mechanisms (Soltani et al. 2010). They also control the diseases indirectly by generating antimicrobial metabolites such as tensin, hydrogen cyanide, and phenazines (Raimi et al. 2017). Studies reported that KSB and PSB such as *Pseudomonas, Burkholderia,* and *Bacillus* increase the plant growth and yield of wheat, pepper, and cucumber (Shanware et al. 2014; García-Fraile et al. 2015). Numbers of studies suggested

Table 1.1 Different biofertilizer applications and their benefits

Biofertilizer	Crop	Remarks	References
Bacillus subtilis, Lactobacillus spp., Saccharomyces spp., and *Pseudomonas spp.*	Maize (*Zea mays* L.)	Biofertilizer application leads to an increase in leaf area and enhances growth	Kumar et al. (2019)
Biofertilizer (EM-1)	Three varieties of wheat (*Triticum aestivum*) (Abu-Ghraib3, Buhooth22, and Ibaa99)	Frequent spraying of EM-1 during the vegetative stages results in the highest yield by Ibaa99	Al-Naqeeb et al. (2018)
Azotobacter	Wheat (variety Gautam)	The study suggests that *Azotobacter* can be applied as a biofertilizer for higher yield	Mahato and Kafle (2018)
Bacillus subtilis OSU-142 and *Bacillus* in T8	Sour cherry (*Prunus cerasus* L.)	Application of these bacteria separately and combined increases the leaf area, shoot length, and yield per hectare	Arikan and Pirlak (2016)
Oscillatoria angustissima and *Nostoc entophytum*	Pea (*Pisum sativum*)	The result shows that inoculation of soil with individual spp. or combined significantly increases the germination rate and stimulates growth parameters and green pigment	Osman et al. (2010)

that the inoculation of different biofertilizers results in many benefits as mentioned in Table 1.1.

1.7.2 Biopesticides

Biopesticides are environmentally friendly pesticides derived from the natural resources especially from microbes and plants. It includes living creatures (natural enemies) or their products such as microorganisms, phytochemicals, or by-products used for controlling the pests in the crop field (Kandpal 2014). Commonly living organism is used as biopesticides which acts as a pathogen for the pest such as bioinsecticides (*Bacillus thuringiensis*), biofungicides (*Trichoderma*), and bioherbicides (*Phytophthora*).

There are a number of plant products used as biopesticides such as extract of neem. Its oil and leaf extract are used to control different chewing and sucking insects. Neem product is known as the most effective and eco-friendly in nature, which effectively controls 350 arthropod species, 15 fungi species, 12 nematode species, three viruses, one crustacean, and two snail species (Kandpal 2014). Linal-ool and limonene plant products are applied to control the aphids, fleas, mites, fire ants, flies, crickets, and paper wasps. Rotenone plant product is used to control leaf-feeding insects such as beetles, aphids, and caterpillar. Sabadilla is used to control leafhoppers, caterpillars, harlequin bugs, squash bugs, stink bugs, and thrips. Plant extract such as andrographis and sida kashayam is effective to control borers and aphids in lady finger and brinjal crop. Plant extract from adathoda, pudina, triphala, and garlic is useful to regulate bacterial leaf blight, leaf folder, and helminthosporium leaf spot disease (Kandpal 2014).

1.7.3 Green Manure

Green manure (GM) is derived from plant or plant products especially fast-growing crops like azolla (*Azolla microphylla*), clitoria (*Clitoria ternatea* L.), desmanthus (*Desmanthus virgatus* L.), sunn hemp (*Crotalaria juncea* L.), soybean (*Glycine max* L.), etc. After decomposition, GM releases nitrogen slowly which is utilized better by the plant compared to the synthetic fertilizers and therefore increases the crop productivity with lower nutrient leaching (Cline and Silvernail 2002; Cherr et al. 2006). GM improves soil organic matter and microbial biomass for longer duration (Biederbeck et al. 1998; Goyal et al. 1999; Cherr et al. 2006). Studies suggested that GM may also work as habitat and resources for the beneficial organisms (Nicholls and Altieri 2001).

There are a number of plant species available which can be used as green manure such as azolla (*Azolla microphylla*), clitoria (*Clitoria ternatea* L.), pigeon pea (*Cajanus cajan* L.), desmanthus (*Desmanthus virgatus* L.), sunn hemp (*Crotalaria juncea* L.), soybean (*Glycine max* L.), indigo (*Indigofera tinctoria* L.), velvet bean (*Mucuna pruriens* L.), Japanese millet (*Echinochloa crus-galli* L.), lablab (*Lablab purpureus* L.), buckwheat (*Fagopyrum esculentum*), Rhodes grass (*Chloris gayana*), sesbania (*Sesbania macrantha*), black gram (*Vigna mungo* L.), black lentil (*Lens culinaris*), mung bean (*Vigna radiata* L.), alfalfa (*Medicago sativa* L.), cowpea (*Vigna unguiculata* L.), berseem clover (*Trifolium alexandrinum* L.), and oilseed radish (*Raphanus sativus* L.) (Cherr et al. 2006; Zandvakili et al. 2017; Couëdel et al. 2018; Arora and Kaur 2019).

1.7.4 Farmyard Manure (FYM)

Farmyard manure consists of waste materials, typically dung, urine, litter, and leftover roughages or fodder of the farm animals. The waste products contain a high amount of major nutrients like nitrogen and micronutrients required for plant growth. Application of farmyard manure improves the soil's biological, chemical, and physical properties. It also improves the soil structure which creates a suitable condition for root development. Studies show that the application of FYM enhances the water-holding capacity and balances the nutrient level in the soil which leads to an increase in the crop productivity (Tadesse et al. 2013). Another study shows that the application of farmyard manure in weathered upland soil increases the phosphate fertilizer use efficiency in rice (Andriamananjara et al. 2018). One major constraint associated with FYM is its bulk requirements and availability. The application of FYM in combination with reduced dose of synthetic fertilizers or other organic manures may overcome this concern.

1.7.5 Biochar

Biochar is a solid material generated by burning of natural biomass (i.e., crop waste, wood chips, etc.). Biochar application in the soil influences the soil structure, texture, density, porosity, and particle size distribution (Atkinson et al. 2010). In the upland soil, application of biochar enhances the soil quality by beneficial changes in physical, chemical, and biological characteristics which leads to improved crop productivity (Palansooriya et al. 2019). Biochar can regulate the unsuited soil acidity, salinity, and alkalinity and cleans the contaminated soil while improving the soil organic carbon retention and water-holding capacity.

Most biochar has a key physical feature like their large surface area and highly porous structure which provides shelter for the beneficial soil microorganism such as bacteria and mycorrhizae which influence the nutrient cations and anions (Atkinson et al. 2010; Yang et al. 2017). Biochar also enhances the soil macronutrients such as nitrogen and phosphorous and improves the soil pH, cation exchange capacity, and electrical conductivity.

1.7.6 Vermicompost

The term vermicomposting refers to composting the organic residues using earthworms. The casting from the earthworms is rich in major nutrients such as nitrogen (N), phosphorous (P), potassium (K), magnesium (Mg), and calcium (Ca). Earthworms create passage in the soil which is suitable condition for the growth of bacteria and actinomycetes. Actinomycetes are six times higher in the worm casts compared to the original soil (Misra et al. 2003). Vermicompost contains humus which helps to increase water-holding capacity which decreases the water requirement for irrigation by 30–40% (Sinha et al. 2010). It also increases the plant growth, germination, and crop yield. A recent study by Blouin et al. (2019) on the vermicompost effect on plant growth shows that it increases the average commercial yield by 26%, total biomass by 13%, shoot biomass by 78%, and root biomass by 57%.

Vermicompost influences the microbial activity, enhances oxygen availability, improves soil porosity, and maintains the soil temperature (Rekha et al. 2018). Vermicompost application on (*Capsicum annuum* L.) Hepper improves the exo-morphological features such as shoot length, number of leaves, branches, and length of internode (Rekha et al. 2018). Doan et al. (2015) found that the vermicompost application in maize can resist water stress.

1.7.7 Slow-Release Fertilizers

Slow-release fertilizers (SRFs) release nutrients such as N, P, and K in a regulated manner, where plants utilized the maximum amount of fertilizer without loss (Singh et al. 2010; Kumar et al. 2012, 2013a, b, 2014a, b, c, 2015; Rai et al. 2017). The application and suitability of SRFs have been extensively studied in several crops like wheat, rice, *Rauwolfia*, etc. In this process, microbes decomposed nitrogen products. SRFs can be classified into two types on the basis of source, i.e., natural and artificial SRFs. Natural SRFs include plant matters, cover crops, animal manures, compost, etc. (Liu et al. 2014). Natural SRFs are organic in nature; it takes a long time to decompose by the microorganisms. On the other hand, synthetic SRFs are water-soluble and release the nutrient in a controlled manner. Application of SRFs reduces the nutrient inputs, lowers the environmental contamination, and decreases the agriculture production cost (Wang et al. 2011; Tian et al. 2016). SRFs lower the fertilizer-related risks such as leaf burning, eutrophication, reducing runoff, and leaching and water contamination problems (Liu et al. 2014; Singh et al. 2010). Studies confirm that SRF application in turmeric (*Curcuma longa* L.) field increases the wet rhizome yield and maintains the soil nutrients during all stages (Jagadeeswaran et al. 2007). A study conducted by Li et al. (2017) found that slow-release nitrogen fertilizers increase the yield and fruit quality of tomato.

1.8 Soil-Friendly Farming Practices

1.8.1 Crop Rotation

Crop rotation is the practice of farming different crop varieties in recurrent succession in a particular time interval in the same land. It is an ancient practice which was practicing by the Romans who used to sow the cereal and leguminous crops in alteration (Sumner 2018). Modern crop rotation practices are established slowly when farmers know the benefits of rotating the crops. Studies reported that it is a sustainable practice to gain high yield and profit with improved soil health, for example, 2 years of rotation of maize with soybean gives 5–20% more yields compared to continuous maize cultivation (Bullock 1992). Studies also suggested that crop rotation controls the weeds and increases production. Crop rotation of small grain crops such as wheat, barley, oats, and rye containing allelochemicals and their residues controls the weeds which alternatively leads to high yield (Bullock 1992). It also improves soil physical properties and organic matter content. Venter et al. (2016) show that crop rotation provides higher microbial diversity and richness. Rotation of normal crops with leguminous species enhances soil fertility especially bioavailable form of nitrogen in the soil.

1.8.2 Multi-cropping

Multi-cropping refers to cultivating more than one crop in a single season on the same land (Petrie and Bates 2017). Multi-cropping has many benefits such as improving soil health; controlling pests, diseases, and weeds; and increasing production. In this practice, there will be various interactions that occur between plant and soil such as pest and disease attenuation, interspecific competition, microbial composition, structure and nutrient cycling, and structural dynamics of soil. The interspecific competition between different plants successfully increases the nature and length of root networks that leads to efficient use of resources in time and space. The increase in competitive ability and allelopathic interactions of the crop can reduce disease, pest, and weeds (Ehrmann and Ritz 2014).

1.8.3 Mixed Cropping

Mixed cropping is the growing of two or more crops in the same field and at the same time. Mixed cropping has multiple benefits such as balancing soil nutrients, controlling the insect's pest and disease, resistance to climate extremes, managing the scarce resource, and increasing the yield (Picasso et al. 2008; Hiddink et al. 2010). There are mixed cropping practices of major grain crops with other species for risk management (Brooker et al. 2015). Worldwide mixed cropping of cereals and legumes is practiced very frequently because the cereal crop gets benefited by utilizing the nitrogen which is fixed by the nitrogen-fixing bacteria associated with roots of leguminous species. A study shows that mixed cropping of dryland and wetland crop has the potential to overcome the field stress (Iijima et al. 2016).

1.8.4 Strip Cropping

Strip cropping is a type of agricultural practice in which different crops are planted in the alternate strip, for example, the row of crop soybeans or corn planted in a wide strip and the next strip planted with the cover crop such as alfalfa which covers the soil and reduces the soil erosion as well. The row and cover crop reduced the water runoff and prevented the spread of disease and pests from one strip to another. It may also help in keeping soil fertility intact by availing nitrogen from the nitrogen-rich crops if they are planted in some strips. A study done by Ma et al. (2007) found that strip cropping of wheat and alfalfa biologically controls the wheat aphid (*Macrosiphum avenae*) by the mite (*Allothrombium ovatum*).

1.8.5 Zero/Low Tillage

No-tillage or medium-tillage system is a system in which crops are directly planted through mulch without causing much disturbance to the soil. In the minimum-tillage process special tiller breaks, drills, and loosens the surface soil without interrupting the topsoil, cover vegetation, and previous crop residue. In case of no-tillage process special seed planting machines inject the seeds directly to the soil. No-tillage and minimum-tillage methods lower the labor charges and soil erosion and save the fuel which is used for plowing the field. The integration of the no-tillage system and cover crops conserves the soil moisture for the crop which can be a suitable agricultural practice (Nascente et al. 2013). Research conducted by Balota et al. (2003) found that no-tillage system increases the soil microbial biomass and nitrogen phosphorous percentage at 0–5 cm depth. It also increases the carbon to N, sulfur (S), and P ratio.

1.8.6 Nitrogen-Fixing Cropping

Planting the nitrogen-fixing crops is a good option to maintain the soil nutrient. The crops such as chickpea, bean, pea, alfalfa, clover, and lentil produce nitrogen through fixing the atmospheric nitrogen with the help of bacteria present in their root's nodules. These plants work as a renewable source of nitrogen for the crop field. It has shown that the cultivation of legume crops in a crop sequence improves the yield of the following crops (Peoples et al. 2009). Apart from legume crop nitrogen-fixing capability, authors have found that other nonlegume crops can fix the nitrogen symbiotically by forming para-nodules associated with *Azospirillum* (Saikia et al. 2012). If this is possible with all crop varieties, it boosts crop production and maintains the soil health in a sustainable way. The mycorrhizal association in the crops improves plant water and nutrient uptake. The legume crops break the pest cycles and control the disease spread on the nonlegume crops too through the increase in the microbial activity and diversity (Lupwayi et al. 2011).

1.9 Conclusion

Sustainable agriculture can play a significant role in solving multiple current and future concerns related to agroecosystems. By adopting environment friendly agricultural practices, farmers can produce crops with improved soil health. The consequences of modern agricultural practices are soil erosion, loss of soil fertility, loss of nutrient-rich indigenous crop varieties, and groundwater depletion that can be managed by adoption of traditional agricultural practices along with soil friendly modern agriculture and associated technologies. Soil is the prime component in the agricultural ecosystem, and its health is a crucial component for achieving better productivity. There are various organic fertilizers which not only enhance the crop productivity but also improve soil physicochemical and biological properties along with other ecosystem services. Agribusiness companies can follow the same procedure to increase crop production without damaging the soil and environment.

Acknowledgment The authors are thankful to the Science and Engineering Research Board (SERB), New Delhi, India, for the award of research grant (EEQ/2017/000476).

References

Al-Naqeeb MAR, Al-Hilfy IHH, Hamza JH, Al-Zubade ASM, Al-Abodi HMK (2018) Biofertilizer (EM-1) effect on growth and yield of three bread wheat cultivars. J Cent Eur Agric 19 (3):530–543

Andriamananjara A, Rakotoson T, Razanakoto OR, Razafimanantsoa MP, Rabeharisoa L, Smolders E (2018) Farmyard manure application in weathered upland soils of Madagascar sharply increase phosphate fertilizer use efficiency for upland rice. Field Crops Res 222:94–100

Arikan Ş, Pirlak L (2016) Effects of plant growth promoting rhizobacteria (PGPR) on growth, yield and fruit quality of sour cherry (*Prunus cerasus* L.). Erwerbs-obstbau 58(4):221–226

Arora M, Kaur A (2019) *Azolla pinnata*, *Aspergillus terreus*, and *Eisenia fetida* for faster recycling of nutrients from wheat straw. Environ Sci Pollut Res 26(31):1–12

Atkinson CJ, Fitzgerald JD, Hipps NA (2010) Potential mechanisms for achieving agricultural benefits from biochar application to temperate soils: a review. Plant Soil 337(1–2):1–18

Augé RM, Toler HD, Saxton AM (2015) Arbuscular mycorrhizal symbiosis alters stomatal conductance of host plants more under drought than under amply watered conditions: a meta-analysis. Mycorrhiza 25:13–24

Bahadur I, Meena VS, Kumar S (2014) Importance and application of potassic biofertilizer in Indian agriculture. Int J Biol Sci 3(12):80–85

Balota EL, Colozzi-Filho A, Andrade DS, Dick RP (2003) Microbial biomass in soils under different tillage and crop rotation systems. Biol Fertil Soils 38(1):15–20

Barraquio WL, Segubre EM, Gonzalez MS, Verma SC, James EK, Ladha JK, Tripathi AK (2000) In the quest for nitrogen fixation in rice. IRRI, Los Banos, pp 93–118

Baum C, El-Tohamy W, Gruda N (2015) Increasing the productivity and product quality of vegetable crops using arbuscular mycorrhizal fungi: a review. Sci Hortic 187:131–141

Berruti A, Lumini E, Balestrini R, Bianciotto V (2016) Arbuscular mycorrhizal fungi as natural biofertilizers: let's benefit from past successes. Front Microbiol 6:1–6

Biederbeck VO, Campbell CA, Rasiah V, Zentner RP, Wen G (1998) Soil quality attributes as influenced by annual legumes used as green manure. Soil Biol Biochem 30:1177–1185

Bitew Y, Alemayehu M (2017) Impact of crop production inputs on soil health: a review. Asian J Plant Sci 16(3):109–131

Blouin M, Barrere J, Meyer N, Lartigue S, Barot S, Mathieu J (2019) Vermicompost significantly affects plant growth. A meta-analysis. Agron Sustain Dev 39(4):1–15

Brevik EC (2010) Soil health and productivity. In: Soils, plant growth and crop protection, vol 1. UNESCO, p 106

Brooker RW, Bennett AE, Cong WF, Daniell TJ, George TS, Hallett PD et al (2015) Improving intercropping: a synthesis of research in agronomy, plant physiology and ecology. New Phytol 206(1):107–117

Bullock DG (1992) Crop rotation. Crit Rev Plant Sci 11(4):309–326

Bunemann EK, Schwenke GD, Zwieten LV (2006) Impact of agricultural inputs on soil organisms - a review. Aust J Soil Res 44:379–406

Carman JA, Vlieger HR, Ver Steeg LJ, Sneller VE, Robinson GW, Clinch-Jones CA et al (2013) A long-term toxicology study on pigs fed a combined genetically modified (GM) soy and GM maize diet. J Org Syst 8(1):38–54

Cherr CM, Scholberg JMS, McSorley R (2006) Green manure approaches to crop production. Agron J 98(2):302–319

Cline GR, Silvernail AF (2002) Effects of cover crops, nitrogen, and tillage on sweet corn. Hortic Technol 12:118–125

Couëdel A, Alletto L, Tribouillois H, Justes É (2018) Cover crop crucifer-legume mixtures provide effective nitrate catch crop and nitrogen green manure ecosystem services. Agric Ecosyst Environ 254:50–59

De Meyer SE, Ruthrof KX, Edwards T, Hopkins AJ, Hardy G, O'Hara G, Howieson J (2018) Diversity of endemic rhizobia on Christmas Island: implications for agriculture following phosphate mining. Syst Appl Microbiol 41(6):641–649

Dighe NS, Shukla D, Kalkotwar RS, Laware RB, Bhawar SB, Gaikwad RW (2010) Nitrogenase enzyme: a review. Der Pharmacia Sinica 1(2):77–84

Doan TT, Henry-des-Tureaux T, Rumpel C, Janeau JL, Jouquet P (2015) Impact of compost, vermicompost and biochar on soil fertility, maize yield and soil erosion in northern Vietnam: a three year mesocosm experiment. Sci Total Environ 514:147–154

Domingo JL, Bordonaba JG (2011) A literature review on the safety assessment of genetically modified plants. Environ Int 37:734–742

Drinkwater LE, Wagoner P, Sarrantonio M (1998) Legume-based cropping systems have reduced carbon and nitrogen losses. Nature 396:262–265

Duran RE, Kilic S, Coskun Y (2015) Response of maize (*Zea mays* L. *saccharata* Sturt) to different concentration treatments of deltamethrin. Pest Biochem Physiol 124:15–20

Ehrmann J, Ritz K (2014) Plant: soil interactions in temperate multi-cropping production systems. Plant Soil 376(1–2):1–29

FAO (2014) Ethiopia: El Nino-Southern Oscillation (ENSO) and the main Kiremt rainy season an assessment using FAO's Agricultural Stress Index System (ASIS). http://www.fao.org/es/GIEWS/english/shortnews/enso27062014.htm

García-Fraile P, Menéndez E, Rivas R (2015) Role of bacterial biofertilizers in agriculture and forestry. AIMS Bioeng 2(3):183–205

Government of India (2013) State of Indian agriculture 2012–13. Ministry of Agriculture, Department of Agriculture and Cooperation, Directorate of Economics and Statistics, New Delhi

Gowdy J, Baveye P (2019) An evolutionary perspective on industrial and sustainable agriculture. In: Lemaire G, Carvalho PCF, Kronberg S, Recous S (eds) Agroecosystem diversity. Academic, Cambridge, MA, pp 425–433

Goyal S, Chandler K, Mundra MC, Kapoor KK (1999) Influence of inorganic fertilizers and organic amendments on soil organic matter and soil microbial properties under tropical conditions. Biol Fertil Soils 29:196–200

Gundi VA, Narasimha G, Reddy BR (2005) Interaction effects of insecticides on microbial populations and dehydrogenase activity in a black clay soil. J Environ Sci Health 40(2):69–283

Hazell PBR (2009) The Asian green revolution. IFPRI discussion paper 00911. The International Food Policy Research Institute (IFPRI). https://core.ac.uk/download/pdf/6257689.pdf

Henry RS, Johnson WG, Wise KA (2011) The impact of a fungicide and an insecticide on soybean growth, yield, and profitability. Crop Prot 30(12):1629–1634

Hiddink GA, Termorshuizen AJ, Van BAH (2010) Mixed cropping and suppression of soilborne diseases. In: Lichtfouse E (ed) Genetic engineering, biofertilisation, soil quality and organic farming. Springer, Dordrecht, pp 119–146

Hilbeck A, Binimelis R, Defarge N, Steinbrecher R et al (2015) No scientific consensus on GMO safety. Environ Sci Eur 27(1):1–6

Hoorman J, Aziz I, Reeder R, Sundermeier A, Islam R (2011) Soil terminology and definitions. Agric Nat Res Fact Sheet SAG-19-11:1–8

IFA (2015) IFADATA. International Fertilizer Association. http://ifadata.fertilizer.org/ucSearch.aspx

Iijima M, Awala SK, Watanabe Y, Kawato Y et al (2016) Mixed cropping has the potential to enhance flood tolerance of drought-adapted grain crops. J Plant Physiol 192:21–25

Jagadeeswaran R, Murugappan V, Govindaswamy M, Kumar PS (2007) Influence of slow release fertilizers on soil nutrient availability under turmeric (*Curcuma longa* L.). Asian J Agric Res 1 (3):105–111

Kandpal V (2014) Biopesticides. J Environ Res Develop 4(2):191–196

Kesavan PC, Swaminathan MS (2007) Strategies and models for agricultural sustainability in developing Asian countries. Philos Trans R Soc Lond Ser B Biol Sci 363(1492):877–891

Kumar M, Bauddh K, Sainger M, Sainger PA, Singh JS, Singh RP (2012) Increase in growth, productivity and nutritional status of rice (*Oryza sativa* L. c.v. basmati) and enrichment in soil fertility applied with an organic matrix entrapped urea. J Crop Sci Biotechnol 15(2):137–144

Kumar M, Bauddh K, Kumar S, Sainger M, Sainger PA, Singh RP (2013a) Increase in growth, productivity and nutritional status of wheat (*Triticum aestivum* L. C.v. Wh-711) and enrichment in soil fertility applied with organic matrix entrapped urea. J Environ Biol 34:1–9

Kumar S, Bauddh K, Barman SC, Singh RP (2013b) Evaluation of conventional and organic matrix entrapped urea and diammonium phosphate for growth and productivity of *Triticum aestivum* L. and mobilization of NO_3^-, NO_2^-, NH_4^+ and PO_4^{-3} from soil to plant leaves. Int J Agron Plant Prod 4(6):1357–1368

Kumar S, Bauddh K, Barman SC, Singh RP (2014a) Amendments of microbial biofertilizers and organic substances reduces requirement of urea and DAP with enhanced nutrient availability and productivity of wheat (*Triticum aestivum* L.). Ecol Eng 71:432–437

Kumar S, Bauddh K, Barman SC, Singh RP (2014b) Organic matrix entrapped bio-fertilizers increase growth, productivity and yield of *Triticum aestivum* L. and mobilization of NO_3^-, NO_2^-, NH_4^+ and PO_4^{-3} from soil to plant leaves. J Agric Sci Tech 16(2):315–329

Kumar M, Bauddh K, Sainger M, Sainger PA, Singh RP (2014c) Increase in growth, productivity and nutritional status of wheat (*Triticum aestivum* L.) and enrichment in soil microbial population applied with biofertilizers entrapped with organic matrix. J Plant Nutr 38:260–276

Kumar M, Bauddh K, Kumar S, Sainger M, Sainger PA, Singh RP (2015) Enhancing efficacy of Azotobacter and Bacillus entrapping in organic matrix for rice cultivation. Agroecol Sust Food Syst 39:907–923

Kumar AS, Wafula WN, Korir NK (2019) Effect of biofertilizer on growth and yield characteristics of *Zea mays* L. in different ecological zones in Kenya. Asian J Soil Sci Plant Nutr 4(3):1–7

Leifheit EF, Veresoglou SD, Lehmann A, Morris EK, Rillig MC (2014) Multiple factors influence the role of arbuscular mycorrhizal fungi in soil aggregation—a meta-analysis. Plant Soil 374:523–537

Leung H, Zhu Y, Revilla-Molina I, Fan JX et al (2003) Using genetic diversity to achieve sustainable rice disease management. Plant Dis 87(10):1156–1169

Li Y, Sun Y, Liao S, Zou G, Zhao T et al (2017) Effects of two slow-release nitrogen fertilizers and irrigation on yield, quality, and water-fertilizer productivity of greenhouse tomato. Agric Water Manag 186:139–146

Liu G, Zotarelli L, Li Y, Dinkins D, Wang Q et al (2014) Controlled-release and slow-release fertilizers as nutrient management tools. Hort Sci Dept HS1255:1–7

Luo L, Qin L, Wang Y, Wang Q (2016) Environmentally-friendly agricultural practices and their acceptance by smallholder farmers in China—a case study in Xinxiang County, Henan Province. Sci Total Environ 571:737–743

Lupwayi NZ, Kennedy AC, Chirwa RM (2011) Grain legume impacts on soil biological processes in sub-Saharan Africa. African J Plant Sci 5(1):1–7

Ma KZ, Hao SG, Zhao HY, Kang L (2007) Strip cropping wheat and alfalfa to improve the biological control of the wheat aphid *Macrosiphum avenae* by the mite *Allothrombium ovatum*. Agric Ecosyst Environ 119(1–2):49–52

Maddela NR, Venkateswarlu K (2018) Impact of Acephate and Buprofezin on soil amylases. In: Insecticides soil microbiota interactions. Springer, Cham, pp 41–48

Mahato S, Kafle A (2018) Comparative study of Azotobacter with or without other fertilizers on growth and yield of wheat in Western hills of Nepal. Ann Agrar Sci 16(3):250–256

Maikhuri RK, Semwal RL, Rao KS, Nautiyal S, Saxena KG (1997) Eroding traditional crop diversity imperils the sustainability of agricultural systems in central Himalaya. Curr Sci 73 (9):777–782

Malusà E, Ciesielska J (2014) Biofertilisers: a resource for sustainable plant nutrition. Fertil Technol 1(1):282–319

Mia MB, Shamsuddin ZH (2010) Rhizobium as a crop enhancer and biofertilizer for increased cereal production. Afr J Biotechnol 9(37):6001–6009

Mishra M (2013) Role of eco-friendly agricultural practices in Indian agriculture development. Int J Agric Food Sci Tech 4(2):11–15

Misra RV, Roy RN, Hiraoka H (2003) On-farm composting methods. Food and Agriculture Organization of the United Nations, Rome, pp 28–29

Mohammadi K (2012) Phosphorus solubilising bacteria: occurrence, mechanisms and their role in crop production. Resour Environ 2(1):80–85

Nascente AS, Crusciol CAC, Cobucci T (2013) The no-tillage system and cover crops—alternatives to increase upland rice yields. Eur J Agron 45:124–131

Nicholls CI, Altieri MA (2001) Manipulating plant biodiversity to enhance biological control of insect pests: a case study of a northern California vineyard. In: Gliessman SR (ed) Agroecosystem sustainability: developing practical strategies. CRC, Boca Raton, FL, pp 29–50

Oldroyd GE, Murray JD, Poole PS, Downie JA (2011) The rules of engagement in the legume-rhizobial symbiosis. Annu Rev Genet 45:119–144

Osman MEH, El-Sheekh MM, El-Naggar AH, Gheda SF (2010) Effect of two species of cyanobacteria as biofertilizers on some metabolic activities, growth, and yield of pea plant. Biol Fertil Soils 46(8):861–875

Palansooriya KN, Ok YS, Awad YM, Lee SS et al (2019) Impacts of biochar application on upland agriculture: a review. J Environ Manag 234:52–64

Peoples MB, Brockwell J, Herridge DF, Rochester IJ et al (2009) The contributions of nitrogen-fixing crop legumes to the productivity of agricultural systems. Symbiosis 48(1–3):1–17

Petrie CA, Bates J (2017) 'Multi-cropping', intercropping and adaptation to variable environments in Indus South Asia. J World Prehist 30(2):81–130

Picasso VD, Brummer EC, Liebman M, Dixon PM, Wilsey BJ (2008) Crop species diversity affects productivity and weed suppression in perennial polycultures under two management strategies. Crop Sci 48(1):331–342

Pimentel D, Acquay H, Biltonen M, Rice P et al (1992) Environmental and economic costs of pesticide use. Bioscience 42(10):750–760

Porcel R, Aroca R, Ruiz-Lozano JM (2011) Salinity stress alleviation using arbuscular mycorrhizal fungi. A review. Agron Sustain Dev 32:181–200

Qin LH, Wang Y, Wu YF, Wang Q, Luo LG (2015) Assessment of nitrate leakage and N_2O emission from five environmental-friendly agricultural practices using fuzzy logic method and empirical formula. Environ Monit Assess 187:1–12

Rai A, Kumar S, Bauddh K, Singh N, Singh RP (2017) Improvement in growth and alkaloid content of *Rauwolfia serpentina* on application of organic matrix entrapped biofertilizers (*Azotobacter chroococcum, Azospirillum brasilense* and *Pseudomonas putida*). J Plant Nut 40 (16):2237–2247

Raimi A, Adeleke R, Roopnarain A (2017) Soil fertility challenges and Biofertiliser as a viable alternative for increasing smallholder farmer crop productivity in sub-Saharan Africa. Cogent Food Agric 3(1):1–26

Raja N (2013) Biopesticides and biofertilizers: ecofriendly sources for sustainable agriculture. J Biofertil Biopestici 4(1):1–2

Rajeshkumar S, Selvaraj T (2006) Influence of native arbuscular mycorrhizal fungi on the growth, nutrition and biomass production of tea var., UPASI-9. Indian J Appl Pure Biol 21(1):31–38

Rehman A, Nautiyal CS (2002) Effect of drought on the growth and survival of the stress-tolerant bacterium rhizobium sp. NBRI 2505 Sesbania and its drought-sensitive transposon Tn5 mutant. Curr Microbiol 45:368–377

Rekha GS, Kaleena PK, Elumalai D, Srikumaran MP, Maheswari VN (2018) Effects of vermicompost and plant growth enhancers on the exo-morphological features of *Capsicum annum* (Linn.) Hepper. Int J Recycl Org Waste Agric 7(1):83–88

Sadhana B (2014) Arbuscular Mycorrhizal Fungi (AMF) as a biofertilizer-a review. Int J Curr Microbiol App Sci 3(4):384–400

Saikia SP, Bora D, Goswami A, Mudoi KD, Gogoi A (2012) A review on the role of Azospirillum in the yield improvement of non-leguminous crops. Afr J Microbiol Res 6(6):1085–1102

Sánchez NV, Zornoza R, Faz Á, Fernández JA (2019) Comparing legumes for use in multiple cropping to enhance soil organic carbon, soil fertility, aggregates stability and vegetables yields under semi-arid conditions. Sci Hortic 246:835–841

Sarkar A, vanLoon GW (2015) Modern agriculture and food and nutrition insecurity: paradox in India. Public Health 129(9):1291–1293

Schreck E, Geret F, Gontier L, Treilhou M (2008) Neurotoxic effect and metabolic responses induced by a mixture of six pesticides on the earthworm *Aporrectodea caliginosa nocturna*. Chemosphere 71:1832–1839

Séralini GE, Clair E, Mesnage R, Gress S, Defarge N, Malatesta M et al (2014) Republished study: long-term toxicity of a roundup herbicide and a roundup-tolerant genetically modified maize. Environ Sci Eur 26(1):1–17

Shanware AS, Kalkar SA, Trivedi MM (2014) Potassium Solubilizers: occurrence, mechanism and their role as competent biofertilizers. Int J Curr Microbiol Appl Sci 3(9):622–629

Shiva V, Singh V (2015) Wealth per acre. Natraj, New Delhi

Singh S (2006) Corporate farming in India: is it must for agricultural development? W.P. no. 2006-11-06, IIM Ahmedabad

Singh RP, Sainger M, Bauddh K, Senger RS, Jaiwal PK (2010) Sustained nutrient supply reduced nutrient loss and high plant productivity with slow release fertilizers. In: Senger RS, Sharma AK (eds) Stable food production and sustainable agriculture. Studium Press, Lanham, pp 62–79

Singh A, Weisser WW, Hanna R, Houmgny R, Zytynska SE (2017) Reduce pests, enhance production: benefits of intercropping at high densities for okra farmers in Cameroon. Pest Manag Sci 73(10):2017–2027

Sinha RK, Agarwal S, Chauhan K, Valani D (2010) The wonders of earthworms and its vermicompost in farm production: Charles Darwin's 'friends of farmers', with potential to replace destructive chemical fertilizers from agriculture. Agric Sci 1(2):6–94

Soltani AA, Khavazi K, Asadi-Rahmani H, Omidvari M, Dahaji PA, Mirhoseyni H (2010) Plant growth promoting characteristics in some *Flavobacterium* spp. isolated from soils of Iran. J Agric Sci 2(4):106–115

Stevenson JR, Serraj R, Cassman KG (2014) Evaluating conservation agriculture for small-scale farmers in sub-Saharan Africa and South Asia. Agric Ecosyst Environ 187:1–10

Sumner DR (2018) Crop rotation and plant productivity. In: Handbook of agricultural productivity. CRC, Boca Raton, pp 273–314

Tadesse T, Dechassa N, Bayu W, Gebeyehu S (2013) Effects of farmyard manure and inorganic fertilizer application on soil physico-chemical properties and nutrient balance in rain-fed lowland rice ecosystem. Am J Plant Sci 1(4):275–301

Taiwo AM (2019) A review of environmental and health effects of organochlorine pesticide residues in Africa. Chemosphere 220:1126–1140

Tian C, Zhou X, Liu Q, Peng JW, Wang WM et al (2016) Effects of a controlled-release fertilizer on yield, nutrient uptake, and fertilizer usage efficiency in early ripening rapeseed (*Brassica napus* L.). J Zhejiang Univ Sci B 17(10):775–786

Tilak KVBR, Ranganayaki N, Pal KK, De R, Saxena AK et al (2005) Diversity of plant growth and soil health supporting bacteria. Curr Sci 89(1):136–150

Tscharntke T, Klein AM, Kruess A, Steffan-Dewenter I, Thies C (2005) Landscape perspectives on agricultural intensification and biodiversity–ecosystem service management. Ecol Lett 8 (8):857–874

United Nations Summit on Sustainable Development, New York (2015) https://sustainabledevelopment.un.org/sdgs

Venter ZS, Jacobs K, Hawkins HJ (2016) The impact of crop rotation on soil microbial diversity: a meta-analysis. Pedobiologia 59(4):215–223

Wang RF, An DG, Hu CS et al (2011) Relationship between nitrogen uptake and use efficiency of winter wheat grown in the North China plain. Crop Pasture Sci 62(6) 504–514

Yang DING, Yunguo LIU, Shaobo LIU, HUANG X, Zhongwu LI et al (2017) Potential benefits of biochar in agricultural soils: a review. Pedosphere 27(4):645–661

Zandvakili OR, Ebrahimi E, Hashemi M, Barker AV, Akbari P (2017) The potential of green manure mixtures to provide nutrients to a subsequent lettuce crop. Commun Soil Sci Plant Anal 48(19):2246–2255

Zhang Q, Zhang CH (2005) Why do slow- and controlled-fertilizer release fertilizers develop slowly? Chin Rural Sci Tech 3:28–29

Zhang W, Jiang F, Ou J (2011) Global pesticide consumption and pollution: with China as a focus. Proc Int Acad Ecol Environ Sci 1:125–144

Zhang D, Min Q, Liu M, Cheng S (2012) Ecosystem service tradeoff between traditional and modern agriculture: a case study in Congjiang County, Guizhou Province, China. Front Environ Sci Eng 6(5):743–752

Zheng X (2010) Analysis of the influencing factors on the farmers' use of manures in Danjiangkou reservoir area. J Hunan Agric Univ (Soc Sci) 1:11–15

Ecologically Sound and Practical Applications for Sustainable Agriculture

2

Anna Christine Taylor and John Korstad

Contents

Abstract

Rampant soil erosion of as much as 20 tons per acre per year doesn't merely affect farmers. The nitrogen and phosphorus contained in the soil affect aquatic life and marine ecosystems, creating the dead zones and cultural eutrophication in aquatic ecosystems. Losing topsoil is like losing time—one cannot get it back. Many fear the depletion in topsoil will eventually affect food availability. The authors examine how no-till farming, crossbreeding and domestication of perennial plants, and a purposeful shift toward sustainable intensification and polyculture farming will positively impact future generations. Author Anna Christine Taylor also includes her interviews with professionals who practice sustainable farming or sustainable living. As a millennial, she presents her hope that a focus on

A. C. Taylor
Oral Roberts University, Tulsa, OK, USA

J. Korstad (✉)
Department of Biology and Renewable Energy, Oral Roberts University, Tulsa, OK, USA
e-mail: jkorstad@oru.edu

K. Bauddh et al. (eds.), *Ecological and Practical Applications for Sustainable Agriculture*, https://doi.org/10.1007/978-981-15-3372-3_2

communication, education, and economic viability will pave the way for the sustainable future.

Keywords

Sustainable agriculture · No-till farming · Polycultures · Soil erosion · Sustainable intensification

...We treat agricultural lands with three million metric tons of pesticides per year; we fix more chemical nitrogen fertilizer, using natural gas via the Haber-Bosch process, than all natural processes combined; we have already lost 20% of the world's top-soil; 20% of agricultural land is now so degraded that it is no longer able to support food production; and species extinctions are three orders of magnitude higher than the geologic baseline. (Dornbos paraphrase of Raven, Persp Sci Christian Faith 64:51–61, 2012)

2.1 Introduction

Dr. David Dornbos, retired global head of Syngenta Seeds' genetic research development and current chair of the Biology Department at Calvin College in Grand Rapids, Michigan, USA, summarizes the sobering prospects agriculture faces worldwide. Water and wind erosion claim 20 tons of soil per acre in 1 year (Mann 2008). Rain washes excess nitrogen and phosphorus fertilizer from fields into bodies of water, triggering algal blooms and starting the eutrophication cascade. As algae and other plant life die, bacteria decompose the organic matter, stripping the water of oxygen and creating zones of hypoxia (Ramos et al. 2018). Researchers estimate that over 400 hypoxic zones exist globally, including the Gulf of Mexico (Paine 2012; Crews et al. 2018). Low oxygen levels have been traced to debilitating genetic mutations in fish, altering the metabolism and immune system, and killing benthic animals such as shrimp and other crustaceans that fishermen rely on for their livelihood (Dasgupta et al. 2015). Monoculture farming causes this ecological discrepancy (Huggins and Reganold 2008). Successful monocultures, fields of only one crop such as corn, offer efficiency, producing a maximum yield for a minimal mechanically driven effort; however, monocultures lack effectiveness. "For example, industrialized food production requires ten kilocalories of fossil fuel energy to produce one kilocalorie of supermarket food, much of which is processed corn and soybean" (Dornbos 2012). Monoculture farming ranks as the top threat to biodiversity, pollution of waterways, and water consumption (Pretty et al. 2018a, b). However, one of the greatest side effects of monoculture farming is soil erosion (Dornbos personal interview 2018; Crews et al. 2018).

Continuous tilling kills weeds but exposes deeper soil to sunlight. Without a root system, the soil has nothing to secure it or shield it from the elements (Huggins and Reganold 2008). Conventional tillage leaves soil exposed for months, leading to excessive erosion (Crews et al. 2018; Dornbos personal interview 2018). The global median rate of soil erosion at 1.52 mm per year outpaces soil genesis at 0.004 mm per year (Dornbos personal interview 2018; Crews et al. 2018).

Today, scientists and farmers strive to remedy and reverse topsoil loss through methods such as no-till farming, crossbreeding and domestication of perennial plants, and a shift toward sustainable intensification and polyculture farming.

2.2 The Need for Topsoil

Soil itself is not lifeless, merely consisting of particles, gravel, and dirt. Soil functions similarly to the human body, requiring essential nutrients, structure, and organic matter. These components in unison give soil the ability to sustain plant, invertebrate, fungal, and microbial life.

Plants require 17 essential elements, macronutrients including C, Mg, N, Ca, O, P, H, K, and S and limited quantities of micronutrients such as Mn, Zn, Cu, Cl, Mo, Fe, Ni, and B to sustain life (Welch and Shuman 1995). Utilizing large quantities of macronutrients, plants synthesize organic molecules such as lipids, carbohydrates, proteins, and nucleic acids (Tully and Ryals 2017). Amino acids, the building blocks of proteins and enzymes, which make up every cell in organisms, contain nitrogen. Phosphorus is a critical element in forming DNA, RNA, and ATP (which provides the energy to power any reaction in living tissue). Phosphorus also makes up the structure of the phospholipid bilayer of cell membranes, which regulate every substance that enters the cell (Tully and Ryals 2017; Stewart personal interview 2016). Positively charged macronutrients, such as phosphorus and nitrogen, ionically bond to negatively charged soil particles; thus, rain and erosion sweep away both soil and the vital nutrients that are essential for plant survival (Dornbos personal interview 2018; Adegboyega 2019).

Less than half of soil consists of pore spaces for water and air. The other half contains a heterogeneous mixture of sand, silt, clay, and soil organic matter (SOM), which consists of organic substrates from decomposing organisms (Crews et al. 2018). For instance, glomalin, a protein secreted by mycorrhizal fungi, binds minuscule clumps of soil particles, which contain organic matter, into aggregates. Repeated tilling of the soil fragments the aggregates, allowing bacteria to consume the released organic matter. Thus, every soil disturbance reinitiates this process and reduces SOM (Crews et al. 2018).

Organisms such as bacteria, fungi, and earthworms benefit soil, recycling wastes back into inorganic nutrients (Reganold et al. 1990). A wealth of microorganisms leads to a great biodiversity of species in the soil and allows for mutualistic relationships between root systems and fungal hyphae (Dornbos personal interview 2018; Reganold et al. 1990). The hyphae's wide horizontal reach allows it to "forage" and capture nutrients such as phosphorus, which benefit the plant, while the fungus benefits from extracting carbohydrates from the root system (Tully and Ryals 2017; Crews et al. 2018).

2.3 No-Till

Using no-till farming methods, farmers essentially leave the soil alone, leaving harvested crop residue atop to decompose into additional nutrients. This layer of organic matter prevents rapid wind and water erosion, protecting the soil beneath and reducing surface runoff, saving water, and preventing downstream pollution from pesticides (Huggins and Reganold 2008). No-till practices retain the SOM within the soil aggregates and reduce the global median rate of soil erosion per year from 1.52 to 0.065 mm. Crews et al. state, "These rates of soil loss are 360 and 16 times the rate of soil formation" (Crews et al. 2018; Montgomery 2007).

No-till farming reduces fuel usage but requires expensive specialized machinery designed to slit open crop residue-covered earth. Stagnant moisture promotes rampant fungal diseases, which can devastate crops (Huggins and Reganold 2008). To prevent weeds and pests from overwhelming crops, farmers apply more than the usual amount of herbicides and pesticides. Unfortunately, these chemicals can build up in the soil and bioaccumulate up the food chain, threatening animal and human health (Huggins and Reganold 2008).

2.4 Crossbreeding and Genetics

Co-founder of the Land Institute, geneticist Wes Jackson dreams of combining low-maintenance perennial plants with food-bearing annual plants through crossbreeding and domestication. Annual plants occupy 80% of global agriculture. With their minuscule root systems and constant need for tillage, annual plants promote rampant soil erosion (Glover et al. 2007).

The massive root systems of perennial plants, often 2 m deep, anchor soil and, thus, provide structure for the soil and for microorganism inhabitants (Glover et al. 2007). The carbon contained in the root systems also adds to SOM and reduces carbon dioxide levels in the air (Crews et al. 2018; Huggins and Reganold 2008).

Annual plants require herbicides and pesticides, whereas some perennials naturally repel pests and diseases (Glover et al. 2010). Some estimate a reduction in cost up to eight times compared to the previous amount of herbicides used for annuals (Glover et al. 2007).

Thus, perennials reduce planting time, herbicide use, tillage, and, therefore, fuel consumption (Glover et al. 2007). Since their root systems and leaves are already established, unlike annuals, perennial plants maximize sunlight during the time annuals are developing underground growth (Glover et al. 2010).

Universities and research centers around the world in Canada, China, Argentina, Sweden, and Australia, including Jackson's Kansas-based Land Institute, research progeny species (Glover et al. 2007, 2010). Using biotechnology, researchers select genotypes that will code for certain phenotypes, physical manifestations of the desired genetic trait (Glover et al. 2007). However, "Of the 13 most widely grown grain and oilseed crops, 10 are capable of hybridization with perennial relatives...When it reaches the reproductive stage... the hybrid genetic anomalies frequently manifest as an inability to produce seed" (Glover et al. 2007).

Often, progeny are sterile or only female due to parent chromosomes misaligning during meiosis; thus, researchers backcross the plant, breeding the progeny with their parent to produce a fertile offspring (Glover et al. 2007). Researchers strive toward creating a stable, fertile hybrid high in nutrients such as wheat, maize, rice, pigeon peas, sunflower, flax, and sorghum (Glover et al. 2010).

The Land Institute reached success with Kernza® grain, domesticated perennial intermediate wheatgrass (*Thinopyrum intermedium*) crossed with annual wheat. Its roots reach down to 3.3 m, anchoring the soil. However, the plants' small seeds and lesser grain quality need improvement (The Land Institute 2019).

Genetic manipulation takes time, and changing an annual into a perennial plant will take more than adding a few genes (Glover et al. 2007). However, researchers estimate perennial crops may dominate the market within 20 years (Glover et al. 2010).

2.5 Case Studies Around the World

According to a global analysis of agriculture led by Jules Pretty of England's University of Essex and conservation agriculturist expert John Reganold of Washington State University in the USA, one-third of agriculture employs sustainable practices (Pretty et al. 2018a, b). Sustainable agriculture (redefined by Pretty as "sustainable intensification") strives to reduce pesticide use and erosion while improving caloric qualities, biodiversity, and human health (Pretty et al. 2018a, b).

The focus of sustainable intensification (SI) is efficiency, substitution, and redesign (Pretty et al. 2018a, b). "Efficiency" includes precision farming, which uses satellite-equipped machinery that assesses the soil's need for fertilizer through a combination of information provided from Global Information Systems (GIS) and Global Positioning Systems (GPS) (Pretty et al. 2018a, b; GPS 2018). GIS uses GPS to interpret signals from satellites into a spatial plane, creating a map. Although the two systems sound similar, GPS relays locational information to a vehicle's receiver, while GIS creates maps laden with information that can be used for geographical, economic, political, and agricultural purposes (GPS 2018). Computer software analyzes the information accumulatively, allowing a user access to the data, whether in the tractor, home, or office (Pretty et al. 2018a, b; GPS 2018). Using this geospatial information, farmers apply only the needed amount of fertilizer, insecticides, and any other needed treatments to specific areas and, thus, improve accuracy, cost-effectiveness, environmentally safe methods, and overall farmer understanding (GPS 2018).

"Substitution" consists of planting alternative crops that are genetically resistant to drought or pests. "Redesign" physically alters the environment to reflect the checks and balances of a natural ecosystem. Pretty and fellow researchers champion seven aspects of redesign, "(1) integrated pest management, (2) conservation agriculture, (3) integrated crop and biodiversity, (4) pasture and forage, (5) trees in agricultural systems, (6) irrigation water management, and (7) intensive small and patch systems" that produce positive effects (Pretty et al. 2018a, b).

Using integrated pest management (IPM), a farmer strategically places a plant, such as *Desmodium* (tick clover), throughout a plot, which acts as a biocontrol, deterring pests and perhaps attracting natural predators of those pests. Placing certain plants in a border around a plot repels other pests. This attracting and repelling strategy is christened "push-pull." Pretty writes, "It is estimated that 132,000 farmers have adopted push-pull in Kenya, Uganda, Tanzania, and Ethiopia" (Pretty et al. 2018b). Farmer Field Schools (FFS) teach practical ways to incorporate agroecology for IPM and provide a network of like-minded people in over 90 countries "with some 19M farmer graduates, 20,000 of whom are running FFS for other farmers as expert trainers" (Pretty et al. 2018a, b).

Farms in Australia, New Zealand, New Guinea, and South America have adopted conservation agriculture, which emphasizes soil coverage and healthy microbe-rich soils. Countries such as New Zeeland, China, India, Nepal, Indonesia, and Vietnam utilize redesign for their pastureland or rice paddies. Although many of these same countries, including Turkey, Pakistan, Sri Lanka, and the Philippines, committed to water management, they have failed to implement it. Farmers in China and East Africa have utilized raised beds and other efficient designs for gardening. In India, the states Kerala and Sikkim committed to only farming organically (Pretty et al. 2018a, b).

Pretty et al. (2018a, b) and other sources document more of these global efforts and success stories for effective sustainable (intensive) farming. We are hopeful and supportive of these ecologically sound and practical applications for sustainable agriculture and look forward to continued growth in these farming reforms.

Anna Christine Taylor's phone interview with Dr. David Dornbos, retired global head of Syngenta Seeds genetic research development, department chair of Calvin College, and professor at Au Sable Institute in Northern Michigan. Taylor took field biology with Dr. Dornbos at Au Sable Institute. (https://www.ausable.org/dave-dornbos)

Taylor: What are the major discrepancies with monoculture farming and how to improve it?

Dornbos: The two biggest problems are soil erosion and biodiversity loss in my opinion. The reliance on chemical fertilizers and pesticides incurs significant greenhouse gas costs but also reduces biodiversity.

Soil erosion [is a major issue] because much of the soil surface is uncovered during a significant part of the calendar year. . . . These soils are exposed to wind and rain all winter long as well as the part of the spring. . . .Soil particles [will] wash into watersheds and cause this eutrophication problem. The minerals that plants need, some of which were applied as fertilizer, [are] attached to those soil particles. You're not just losing the soil; you're also losing your fertility.

The rates that soil forms are a small fraction of one centimeter per year. Soil conservation reduces soil erosion by 75–80%. [However, soil erosion] still happens consistently and soil continues to be lost at an unreplaceable rate even with minimum tillage. So there's still a tenfold difference. . .a tenfold higher loss than soil regeneration. Obviously, to be fully sustainable, rates of soil genesis need to be equal to the rate of erosion.

Polycultures are the best overall solution, utilizing the developing practices of agroecology, to mitigate soil erosion. Polycultures ensure a more efficient use of solar radiation and water, and they cycle nutrients more efficiently, especially nitrogen. If organized well, polycultures promote diversity of soil microbes which make food plants more efficient, and they can attract insect predators, which, in turn, reduce the reliance on inorganic pesticides.

Think of healthy ecosystems like a forest or a tallgrass prairie—these ecosystems utilize virtually all the available photosynthetically active sunlight year-round. Because corn is a semitropical crop, it ends up getting planted typically in May in temperate areas of North America. Up until the time corn plants canopy the soil surface in June, rays of sunlight or photon energy of light that don't hit any leaves hit the ground or weeds. They go straight from the sun through our atmosphere to the ground. That's lost potential energy. Healthy natural ecosystems have plants that start growing very, very early. Think of the ephemeral plants in the forest understory. Growth begins in February or March. These plants are already utilizing early spring sunlight. A monoculture field typically only has plants growing out there for only three months, and they are only canopying the floor of the field for about a month and a half, so there's a big loss in potential solar energy.

Polyculture systems or forests and prairies have much lower rates of soil loss than they do rates of new soil formation; they've got fully established permanent root systems year-round. Some natural tillage can occur and be okay but very minimal. Hooves of bison or burrowing of mammals are natural forms of tillage.

Now I want to circle back to biodiversity. As we've talked a little bit about polycultures, we've talked about the optimal use of nutrients, light, and water, but [we need to address] pesticides.

I think that when a lot of people think of biodiversity, they are thinking about polycultures—such as apple trees, raspberries, beans, corn, squash, lettuces. There are other biological kingdoms that turn out to be really, really important. Bacterial and fungal species living in the soil help plants grow in mutualistic relationships. There's microbiological diversity in soil that plants rely on, mycorrhizal fungi, for example.

Taylor: Yes!

Dornbos: These crazy fungi will literally penetrate the roots of particular plants, their symbiont, and they derive carbon nutrition from the plant because the plant has plenty of that. A plant can do photosynthesis while the fungus can't. At the same time, the fungal mycelium acts as a very extensive system to get mineral nutrients for the plant. So whether you're talking about fungal or bacterial organisms, roughly 95–98% of these organisms live in the soil. They are often species-specific that can help these crop plants grow much more efficiently.

A monoculture favors only one kind of microbiological symbiote, which means most of the others cannot persist in that place because they never see their host plant for, let's say, 7 years if you're on a 7-year rotation. Imagine in one tablespoon of soil there are like 3 billion bacteria, many different kinds, and they all have different kinds of relationships with each other and the plants that are growing there. If you're only growing one kind of plant, you're only going to favor a narrow range of those kinds of soil microorganisms.

And the thing is, if you start throwing a bunch of pesticides and nutrients in the soil, who knows what effect that variety of pesticides will have on them? Or if you're applying lots of nitrogen, you're creating these huge pH swings. Imagine going from a pH of 12 to 3 and back to 6 in the weeks following of anhydrous ammonia as a nitrogen fertilizer. Broad pH swings are probably really hard on those microbes, too.

Aboveground, I think about the variety of insects. None of us in our class appreciated mosquitoes all that much, but mosquitoes are important, feeding a whole lot of birds, aquatic macroinvertebrates, and fish.

Taylor: And the dragonflies.

Dornbos: And the dragonflies are also going to eat mosquitoes and some other pests that are on your crop plants. . . . The chipping sparrows, like many of the other sparrows and warbling birds, are little insectivores. They eat half of their body weight in insects every day. If you start thinking about all the other kinds of insects, the parasitic wasps that are out there, dozens and dozens of kinds of parasitic wasps—many with very specific host targets—and dragonflies and whatever other kinds of critters that are floating around out there, these can serve as your pesticides in a biodiverse production system. If you set up a year-round polyculture sensitive to these insects' life cycles, it turns out you will seldom if ever need pesticides. All these other little bugs and bacteria will be doing your bidding for you. I almost never have to apply pesticides in my garden. I don't need to because I never get insect problems. I've got birds and other things hitting about everything except cabbage loopers. Because many different crops grow in my garden, the more devastating specialized insect pests never can build up to troublesome populations.

Taylor: How would you set up polyculture farming? Would you introduce species such as birds and insects or let them naturally come? How would you introduce new species?

Dornbos: Generally, you can simply set up a biodiverse system and let them naturally come and build their own populations once suitable habitat is available to them. I envision extensive localized crop rotations, far fewer cereal grains (which implies far less grain grown for feed and fuel). Imagine vegetable gardens or CSAs on steroids. Forests are naturally laid out in that sort of way, too. At Au Sable, we looked at plant communities with beech trees and oak trees arranged in kind of a mosaic. You typically don't have all beeches in one area and another area with all red oaks and another area of all sugar maples. They are all interspersed, suggesting some sort of relationship. The Native Americans in the story of the "three sisters" crops were clearly creating an environment like that. Do you know the three sisters? Does that ring a bell?

Taylor: No, but it makes me think of the three Fates from Greek and Norse mythology. Shakespeare references them in *A Midsummer Night's Dream*. They are basically the same goddesses that would sit at the tree of life and weave everybody's fate. That's what I thought.

Dornbos: [laughs] Okay, so this is a little different. It was squash, beans, and corn—those were the three sisters. Native Americans, the Iroquois, particularly—and others picked up on it, too—would plant corn and beans in the spring together at the same time. The beans were like pole beans, so they climb, not the typical bush beans you might think of in a garden. The corn grows early, and the beans climb it. The beans are legumes, so they fix nitrogen that the corn uses as they are both growing together, creating a bit of a tangle, but you end up with a strong corn plant because the corn plant has access to the nitrogen that would otherwise limit its growth. When the bean and corn plants were about half mature, that's when the Native Americans would plant squash late June or early or mid-July. Squash typically starts growing later in the year anyway, so you're in the normal life cycle of these plants. The idea is you go ahead and harvest corn and harvest your beans. By the time the beans and corn are dead, the squash grows really effectively in the cold months of the fall. The squash fruits don't grow until August, September, and October. The idea is that in one area you plant an intermix of those three crops that you can optimize the use of sunlight, water, nitrogen, and vertical space because they're climbing on each other.

Imagine a four-dimensional matrix of perennial plants like berry bushes and fruit trees more or less permanently in place with a variety of vegetables and grains rotated over time. With a variety of plant species and minimal use of pesticides and fertilizers, insects and birds and soil microbiota will come naturally and proliferate in the more favorable environment if they are fed year-round. Food animals could certainly be integrated into such a system.

Taylor: How would a polyculture be operated? Spraying nitrogen by hand?

Dornbos: I would think that if engineers would put effort into smaller scale implements for organic and CSA sorts of farms, something close to the scale they do now for big industrialized farming equipment, I think much of what causes us to labor today can be reduced. I am certainly not opposed personally to using technology, but this would need to be re-envisioned to provide alternatives to the industrialized monoculture model we now almost exclusively employ. Even how we grow crops can change by narrowing rows, for example, competition for weeds can be reduced immensely.

Taylor: Do you think that a polyculture system would be possible in that people would be open to polycultures?

Dornbos: Some will, certainly not all. Most in our society take food for granted and many don't want to lift a finger.... I would argue some level of work for food follows biblical instruction, creates awareness of and respect for nature, and will produce food of the types consistent with good human health. The success of organic farming and CSAs in many parts of the USA suggests to me that people are open to a different food production system than we currently have. Generationally, as soil, water, and fossil fuel deplete and energy costs limit the production of nitrogen fertilizer and pesticides, agroecologically, savvy systems will grow, I believe. We are on the cusp of a revolution in food systems now, I think and hope.

Taylor: How do we know that no-till is working?

Dornbos: We know no-till works because soil erosion rates have been and are going down. We need to continue these sorts of practices as no soil equals no food. The calculus here is pretty simple. Topsoil does regenerate quickly, and while soil quality can be improved over time, we are speaking of geologic time and not generational time, certainly not the short-term requirements of profit-loss statements. Somehow, we need to financially account for the "loss" of resources that we depend upon and don't seem to value until they are gone. The challenge of cost/benefit in food consumers' minds is problematic when much food of all kinds is inexpensive and available in the local grocery. This apparent largess promotes the perception of low value. Most Americans don't even save for their own retirement, so getting those Americans to take action on soil and biodiversity improvement will take patience, persistence, and time.

Taylor: What is being done in the government now to encourage conservative tillage?

Dornbos: The government does encourage sustainable agriculture but almost exclusively within the context of industrialized monocultures—so-called intensive systems. Of course, this is largely the result of big lobbies (corn growers, beef growers, dairy producers, and big companies). Very, very, very little taxpayer money is devoted via the farm bill, for example, to sustainably intensified food production system development. This should change and needs to change, but I don't see our government having any interest on this topic. This is going to have to be promoted and supported from the grassroots level.

Taylor: What are your thoughts on what a "sustainable future" in farming looks like, or what practices will lead to sustainable farming in different countries and regions of the world?

Dornbos: First, I would hope that we don't continue to export unsustainable food production methods to developing countries.

Second, we need to move away from monocultures. We need to identify more sustainable systems. I fully realize that we cannot move completely from monocultures for feed crops, but there are better alternatives. And, if we as a culture would eat *much* less meat, we would be healthier, and the environment would be healthier. Anything that can be done to increase no-till use and thereby conserve soils should be done. Anything that we can do to increase functional diversity should be done.

Farmers are sort of like serfs in a system; they must largely act at the whim of input suppliers and costs, a public that demands cheap food, and a globalized system that drives cost efficiency. If we are able to internalize the plethora of externalized costs and think in a timeframe of decades and centuries (not monthly balance sheets or daily stock market returns), we would realize that the real costs of producing food in the way we do are extremely expensive, not at all as cheap as it appears. We should value the inputs and products of the process, and the process itself, much more than we do.

Anna Christine Taylor's interview with Nathan Feller and Anna Mueller, Oral Roberts University students who were part of the Zimbabwe Healing Team: the team partnered with a local organization to impact the local economy through wicking beds, a form of sustainable agriculture (https://oru.edu/news/oru_news/20180726-oru-zimbabwe-healing-team.php).

Taylor: Before this project, how much did you know about sustainable farming?

Feller: I had the pleasure of being exposed to the sustainable farming world before this project took place, but this project taught me a great deal more than I ever could have imagined.

Mueller: I honestly had only heard about sustainable farming in classes like environmental science (EVR 250). I have learned so much about the practical applications of sustainable farming through the research and on-the-ground work of this project. For example, Nathan Pickard taught me much about permaculture during the research phase of this project.

Taylor: For someone who's never heard of wicking beds, how would you explain how they work?

Feller: Essentially, a wicking bed is a closed system that takes advantage of a natural biological process known as capillary action. Water is polar, meaning that one end of the molecule has a positive charge and the other end has a negative charge. This molecular interaction is a fundamental process that enables plants to continually replace the evaporated water molecules from the leaves with water that is in the soil via a long chain of water molecules that flows up the plant. The fineness of a capillary tube will determine its surface tension relative to gravity. The finer the tube, the higher the water will rise. Capillary action or wicking is at the heart of a wicking bed. The bed can be broken down into a few components for simplicity. If you were to cut right down the middle of the bed, you would quickly find that it has two main sections: the water reservoir and the soil. These sections are approximately equal in size and are separated by geotextile fabric so as not to allow the soil into the water reservoir. After selecting an area for the bed, a plastic liner is placed in a 6 × 6 foot area that is 2 feet deep. An L-shaped tube is placed flush with the bottom of the bed with one end shooting up and out of the bed above the soil layer. This is where the water will be added to the system. A coarse media is then placed over the tube, thus creating the water reservoir. Next, the geotextile fabric will be placed on top of the coarse media and the soil, which is topsoil, compost, and manure mix. Mulch should be placed on top of the soil to reduce evaporation. Initial wetting of the soil should be performed. After this, water should only be placed into the tube, filling the reservoir from the bottom up.

Mueller: The only other thing that I would say is that this system allows for more water retention and conservation and also much greater yields in the beds than a normal garden due to the design.

Taylor: Since one of the Healing Team's goals was to empower the local people and their economy through sustainable farming, how do you foresee the wicking beds impacting the communities' future?

Feller: One of the greatest things about this project is that it was born and bred in Zimbabwe. We had the honor of working with the Foundation for Farming (FFF), which was located a few miles from the community we were working in, and it was FFF that came up with the wicking bed design that we implemented. I believe that this empowered the local people by showing them a design that was created for their people by their people. We were merely vectors for communication.

Mueller: Our hope is that, like the Latrine project from the previous year's Healing Team, the community members would be inspired to take the initiative to build their own wicking beds. We established a loose system of mentorship with the community leaders where they were required to train three members of the community on how to build a wicking bed during/after we had helped them build their own bed. We also encouraged them to help contribute some funds for others to build beds. If this project takes off as we had hoped (I have not heard about how it is doing now), this project has the potential to greatly increase the crop production and economic success of many members of the community.

Taylor: Did the wicking beds change the community and the culture? If so, how did seeing that change affect you?

Feller: Due to the nature of short-term missions work, I was not able to see the change take place. I have heard that the beds were successful and that families were able to benefit economically from the produce. Hearing about the success brought me immense joy and has shown me how important empowering local people truly is.

Mueller: As Nathan said, it has only been a few months since we launched the pilot, and I have not heard much news about the success in the community of Hatcliffe. For me, it has been amazing to have a part in this project and participate in something that could have a lasting impact on a community.

Taylor: How did you ensure people would plant a diversity of crops in their gardens so they wouldn't saturate the market with one crop? Does everyone pool their revenue and draw from that? How do they organize the process from planting to selling?

Feller: Diversity was a topic that we discussed and strongly recommended; however, it was ultimately up to the individual to choose what they would plant. The economic gain from a type of kale was the most promising for their ROI, and, thus, the majority of the wicking bed owners choose to plant that in their beds. The revenue is on a personal level. Planting and selling are also on a personal level, and it is up to the individual to pick a price and market for the goods that were produced.

Mueller: When talking with the business students on our team, we were mindful of flooding the market, but when it came down to planting the gardens for the community leaders, they all chose between two options because of seedling price, market price, and personal preference. Only time will tell how this impacts the local economy.

As Dr. Korstad mentioned, we did discuss the beginning of a plan to coordinate between the community members a more business-oriented plan to prevent things like flooding the market. As of now, I think this would take a higher level of coordination than we have at the moment, but it could be a development in the future.

We talked about families having many wicking beds—some for their own consumption with a crop rotation model and a few for profit. This would pay for the materials and make excess income within a matter of months. While we were there, they simply planted one bed with the intention of making a profit.

Anna Christine Taylor's Interview with Dr. David Unander, former professor at the Eastern University and current professor at the Au Sable Institute Costa Rica campus. Dr. Unander was also an assistant professor of horticulture at the University of Puerto Rico (https://www.ausable.org/dave-unander).

Taylor: What farming practices are working in Costa Rica? Do they employ polycultures?

Unander: Costa Rica and New Jersey are close to the same land area—and even have two coasts. And there the similarities end! Costa Rica has 9–10 active volcanoes and 100 s of inactive or extinct ones. The geology is mostly igneous, with sedimentary rocks close to the coast (such as youngish coral limestone layers). However, the highest and oldest mountains are uplifts, not volcanic—but they get up above the freeze line, so you even have some tropical alpine ecosystems (a relatively small area) but fairly sizeable cool tropical areas.

[There are lots of] valleys and watersheds, as well as two quite different coasts. A general rule for the tropics is prevailing winds from the East until one is close to the equator. Costa Rica falls between 8 and 11 degrees North, with the ocean on both sides, one of which is like a partially confined, warm bathtub (Caribbean) and the other the largest ocean on Earth, with huge, complicated circulations, affected by cyclic patterns.

There are wet seasons and dry seasons, although the dates of these vary a little with the location. Sometimes winds come from the West, sometimes from the East. Because of the mountains, there's often a rain shadow effect.

To really understand each location, you'd also have to learn the seasonality of the rainfall: does it rain all year consistently, or is there a dry-wet cycle every year? You'd also have to add in what the temperatures are like from elevation. And then consider the soil type and—definitely unlike Pennsylvania, New Jersey, or Oklahoma—effects from volcano emissions, which are acidic, sometimes dramatically so, but also add P, K, and S to the soil as fertilizer.

So, that permits a remarkable diversity of crops grown in Costa Rica, and the dominant crop one sees (or few or no crops at all) can change quickly as one drives around this relatively small country.

Absent would be temperate staples like wheat and barley. I also can't think of ever seeing potatoes, although I'm sure there are places in Costa Rica where they would thrive.

The farming practices vary enormously as does the economic organization. For example, Costa Rica has a long history of agricultural cooperatives, some of which have been very successful, including their international marketing. Other crops are dominated by large corporate plantations and corporate exports and still others by small landholdings with a diversity of plants and animals, plus usually one or two major market products. There's a wide range of scale and practices one encounters. Costa Rica is usually classed as a First World country. That prosperity extends to the level of capital accessible to many farmers and corporations. Their purchases include heavy use of synthetic fertilizers and pesticides, such that the watersheds in the major agricultural areas are heavily contaminated. The use of fertilizers and pesticides is comparable to the USA. At the same time, Costa Ricans are generally literate and educated, and many are concerned about their environment, so there is an increasing market for plants and animals produced on organic-certified farms. There's a lot of small producers and active farmers markets everywhere.

Bananas, pineapples, sugar, and African oil palm are mostly grown in monoculture plantations, sometimes enormous acreages for export. Roads to the port of Limon are always clogged with 18 wheelers with containers loaded for North America and Europe. Bananas and pineapples eaten in Costa Rica are often grown on small farms, some organic and some not, but most of them grow in more of a polyculture. Some small farmers are successfully competing with Dole, Chiquita, Del Monte, etc., by organic production of bananas and cacao, sold through marketing cooperatives. My Au Sable class visits several successful coops.

Coffee, of course, is extremely important in the Costa Rica economy and is historically the traditional family farm crop. With the exception of a plantation Starbucks recently has opened, coffee seems to be produced on family or small (Crican) corporate farms but marketed through agricultural coops. Quality is strictly adhered to by farmers and by the coops. Due to government regulations, one can't even grow the lower quality but higher-yielding coffee species and qualify it as Costa Rican coffee. This is how they've held their edge in a highly competitive global market.

Ecologically, some farms do shade coffee, which is better for the environment. Many do sun coffee, which produces a higher yield though it is more environmentally degrading. Many use synthetic fertilizers and pest control. Some coffee farms are organic. High-quality coffee requires a particular ecosystem, tropical but cool and wet, yet well drained, so in those fortunate locations, coffee is almost a monoculture, although there's often some acreage for a polyculture of some fruits and vegetables the farmer sells as a sideline. Little family coffee shops/bakeries are also a common sideline in coffee country.

Cattle, pigs, and chickens are really important in Costa Rica. They are a major Central American milk and cheese producer (one source of immigrants in the 1800s came from dairy regions of Europe, like southern Germany, Switzerland, etc.). The area around Vara Blanca looks a little like Wisconsin, with all the Holstein dairy cattle grazing on the rolling hills. The Wisconsin image is broken by the smoking volcanoes looming behind the cows, huge tree ferns, and occasional tapirs that come out of the cloud forest to graze with the cows. Dairy farms range from small family farms (say, 10–20 cows) to huge ranches with hundreds or more. They all are still a major source of eutrophication in the watersheds. Sadly, I learned that was true even for an organic yogurt and ice cream factory we visited: the milk was organic, but so was the waste they just washed downstream.

The northwest coastal plain probably is the closest to Oklahoma you would see, a seasonal dry forest/savanna region. Historically, the agriculture has been, and still is, beef cattle. Costa Rica raises and exports a lot of beef. Dan Janzen, a tropical biologist from U/PA, thinks the cattle and horses the Spanish brought have subtly returned the dry forest toward the ecosystem that was there before humans arrived. The first Indian tribes seem to have hunted all the largest grazing mammals to extinction. Some of the dominant trees that depended on seeds passing through those animals then became less common. Since the 1500s, pollen records show those trees slowly returning to dominance, which Janzen attributes to horses and cattle eating pods from those particular trees, thus aiding germination! Anyway, that agricultural system has been in place for centuries: the rainforest wasn't really cleared for the cattle; it was always savanna, probably reminiscent to the Spanish of savannas/chaparral in southern Spain where cattle ranches also have existed for centuries. In contrast, the highland dairy cows are in a rainforest and cloud forest that was/are clear-cut for them.

Chickens and pigs are raised on every scale you would see in the USA, from backyards (very common for chickens) to huge factory-like operations. Due to coyotes and other predators, free-range chickens have to be inside some sort of protective enclosure.

Trout is worth mentioning. Apparently, wherever European men went, they sent for their trout sooner or later. The Spanish and other immigrants introduced trout all over Central and South America, where they weren't native. The USA now has a mix of introduced European and native North American trout species. Anyway, small trout farms are a small business common in high mountain areas with cold, clean streams.

There are many tropical and some temperate fruits, often in smallish plantations, occasional larger plantations (such as for mangos or papaya or, at cool elevations, avocados), sometimes a mix of many different kinds. All the tropical staple roots are grown: yuca/cassava, yautia, malanga, sweet potatoes/batatas, and ñame. I think those are all for local consumption or maybe some export to Panama or Nicaragua. Everyone eats them in Central America. A lot of vegetables and small fruit species are grown, again on many different scales of size and organic/not organic. Vara Blanca, where the Au Sable course is based, is famous for its dairy farms and its strawberries. Roadside vendors there sell boxes of strawberries with a local sweet cream dip to eat them with!

There is no single way to characterize Costa Rican agriculture.

Taylor: When you go into a community, what are your first steps? Do you blend the Gospel with creation care? How do you impact the economy through agriculture?

Unander: (1) I stress language learning as essential, wherever one goes. It just can't be a shortcut to be effective at anything personally. Everything else someone does needs to be combined with continuous language learning. It clearly communicates that one honestly values that community.

(2) I'm a fan of being "naturally supernatural," working with what one senses God desires to do at that moment for a particular situation and a particular person. So, yes, it's often very appropriate. It's nice to be able to introduce Jesus into a conversation when it naturally fits the topic. I enjoy it when someone asks me, "Are you some kind of missionary?" and I can say, "No, I'm a biologist."

(3) Sustainability in agriculture has to fit with the farmer feeding his/her family. Any way we help—as outsiders—to make this happen accelerates its acceptance. Plant with Purpose, the Christian service mission that promotes reforestation, learned decades ago, they had to include trees with marketable wood or fruits, to make up for the lost income of planting trees in some sloping land rather than crops.

The most successful agricultural innovations fit the local ecosystem and (usually) existing farming techniques and existing types of foods. World Vision successfully introduced carrots to the Quechua farmers in highland Peru, a region where most of the staple crops in the diet are roots (potatoes, sweet potatoes, cassava, yacón, arracacha, mashua, oca, ulluco, etc.). That added an additional rich vitamin A source to their diet, and basically the agricultural techniques were the same, and also carrots are cooked the same way as most any other root. Carrots also keep better than sweet potatoes (the other good vitamin A source they had); most people immediately like the taste of carrots, so they quickly became a market vegetable to sell as well as eat.

Taylor: What are your thoughts on what a "sustainable future" in farming looks like, and what practices will lead to sustainable farming in different countries and regions of the world?

Unander: See (3) above. Ecologically, the agriculture has to fit a given region's ecology. For example, cattle in Oklahoma fill the niche of buffalo (as still do, of course, native buffalo themselves). The prairie there has always had a large hoofed herbivore, and the grasses actually grow better being clipped by them, at a rational stocking level. The grasses are potentially sustainable versus a feedlot, which depends on bringing in corn or hay grown elsewhere. Sometimes a feedlot makes economic sense in the short run, but the entire capital expense is greater and the environmental impact from large animals crowded in one place.

Ask as many questions of local farmers and gardeners as possible, over time, and among different individuals, and take careful notes; maybe take photos as well (but always ask first!). This can fit very well with the extremely important task of learning or improving skill in the local language that one should always be thinking of. Since I first made a serious commitment to learn Spanish 36 years ago, I still try to "program" some Spanish into every week, so that I keep progressing—reading the Bible in Spanish, listening to local or Internet Spanish radio stations, talking to Spanish neighbors, etc. Almost always, I learn some new words and current pop songs, which people make allusions to, just like people do in English.

I observe that many North American students—Americans and Canadians both, but I think Americans are a little worse—are intimidated by foreign languages. That was me, too! One thing that really helped me one day was suddenly remembering parts of my childhood when my grandfather lived with us for several years. He was an immigrant; in fact, he came as an illegal immigrant, although eventually he became a citizen. He learned English all on his own, beginning from ground zero by trying to talk to English speakers on the construction sites where he worked as a young man, to people riding streetcars with him on the way to work, etc., all around him in Chicago. He became quite fluent, although he never lost a thick foreign accent even after 60 years of speaking English. In spite of his accent, he only rarely mangled English sentence structure in ways that made him unintelligible. No one could ever mistake him for a native English speaker—but so what? He had become part of the community, a nice, hardworking man who just happened to have a foreign accent. There is no shortage of people with accents in any major city of the world in our day.

Perhaps we Americans set our sights too high, unconsciously thinking we either reach native fluency in a second language or it's wasted effort. Back when I was fearfully contemplating learning Spanish (a job offer in agricultural research in Puerto Rico had just come), I suddenly realized one day that, just like my grandfather in English, I'd never fool people that I'm a native Spanish speaker—but so what? The truth is, I'm not! But I've plugged away over the years to keep improving competency. It's such a blessing now to be able to travel and work in completely Spanish-speaking places. Most importantly, it seems perfectly natural now to talk with people about Jesus in Spanish—and pray and worship in Spanish. And it really does communicate, without directly saying so, that one cares about people and their culture, to take the trouble to speak their language.

Except when a farmer has a secret technique that's an "edge" over other farmers, most enjoy taking the role of a teacher to describe their farm to a visitor. Farmers and gardeners are usually proud of being asked what they grow and how and why they do it this way and not that way. They all spend a lot of time alone with their plants, thinking about what's the best use of their time to make a living for one more year. Even on short-term trips in an area, even just a few hours on someone's farm, I try to meander around with a farmer and ask some open-ended questions and just be a listener and writer. With an older farmer, it's often really interesting to ask how farming has changed in their lifetime. So, before trying to do anything much in technical realms, when entering a community, I'd want to learn as much as possible, within whatever time one has (1 day? 1 week? 1 semester? 2 years?), what people in that place believe is the best way to farm and garden and why they think that.

If it's a tropical farm, it will definitely use different approaches for some things than a temperate farm, even though the plant biochemistry is essentially the same. Sometimes farmer traditions are very wise, and sometimes they are quite wrong—just like agricultural extension recommendations can be either in the USA! As outsiders, we may be able to bring a new perspective and solutions to a long-standing problem or, we may be clueless about what to do, but the traditional farmers have a workable, local solution. Or, together, we may discover a new solution neither known there or in the USA.

I hope this is useful!

In Christ,

Dr. U.

Anna Christine Taylor's transcribed phone interview with Nathan Pickard, entrepreneur and founder of the nonprofit company Restoration Collective. Pickard spoke at Oral Roberts University on sustainability, sharing the way he chooses to live and how he strives to impact his community for the better (https://www.restorationcollectivetulsa.org/).

Taylor: I remember your talk about the food forest along the highway, and I just read your articles and watched your video interview. Would you consider your food forest a polyculture?

Pickard: Definitely. The goal is to have different things and diversity in every direction possible, like the root depths—the plants are feeding at different levels. The plant heights and fruiting times are different. You can have spring ephemeral plants that can provide seed/feed before the trees get their leaves and shade out other plants. One of my goals is putting together a good database of our native plants and the native animals that they support and really trying to get it down to where you can say, "OK we've got this pest problem; what are the predators of that pest so we can have a natural check?"

Taylor: Yes. Biocontrol.

Pickard: Yeah. What habitat does the predator need to survive? What attracts it? Whatever habitat that predator needs, let's plant those plants and create that habitat. It gets pretty complex when you start getting into the polyculture, but it's fun, too. The joy that I get from it is to see how amazing God is. To see his design. It was perfect. He created all this amazing diversity, and it all keeps itself in check and protects itself.

Taylor: Yes, that's really neat. Did you discover this database, or did you put it together?

Pickard: No, unfortunately, I couldn't find anyone who had really done it to a large extent. There was a school; I think in Virginia. It was actually an elementary school, and they had put together the best thing I could find. But then their school system did this major update on their website, and so all that information was lost. I was able to find it on an archive, and I paid one of my employees to transcribe all that information onto a database that we have. The information was so great because it included all the predator-prey relations.

Taylor: Like all the different species?

Pickard: Yeah, and the habitat. So if you have a chigger problem, which we do around here.

Taylor: Ew.

Pickard: Yeah, you can look up the ten things that actually naturally eat chigger, and it's like, "Oh awesome. I want those ten things." But what do those ten things need to thrive and to come to this area? I found the database actually because of the chigger problem!

Taylor: Was that the book you mentioned in your lecture that pushed you to live sustainably?

Picard: No, that was *Pollution and the Death of Man* by Francis Schaeffer. It's written from a theological perspective—why Christians should care about God's creation and stewarding it and be like the biggest environmentalists there are.

Taylor: That was my next question—what inspired you? I remember you talking about living off of things you had grown in your backyard and the different herbs and raising chickens. Was that book the main thing that changed your point of view?

Pickard: Yeah, it's been a lot of things, but that was the best from a theological perspective of why I, as a Christian, should be doing that. That's probably the deepest motivation and change that I had.

Taylor: Was there anything else other than that?

Pickard: I just read all kinds of books. The more you read about everything and think about people and God's creation, I think you just have to end up there, like to really think through my actions and to love my neighbors. Being this huge American consumer is harming my neighbors all over the world, and it's easy to live ignorantly and consume and consume and consume, but when you're willing actually to see the effects of your choices, then that motives you to change.

Taylor: Could you elaborate on your plan for the school? I remember you worked with the city board, and you wanted to teach children about agriculture, ecosystems, and the species native to Oklahoma?

Pickard: With the school, one of the books I read was *Last Child in the Woods*, and it addresses so many of the problems unique to today that children are facing, like this huge massive attention-deficit disorder. The author really connects back to the fact that children are not out in nature like they used to be and not able to take risks and be out in nature alone. Everything they do is so structured and guided. Their parents are scared to let them be outside by themselves.

Taylor: I think you talked about that at ORU—was that the story you told of kids being scared to go into the woods because it's an unknown? That's what we used to do—everybody would just go play in the woods.

Pickard: Yeah, so I started doing some research, and I found some interesting schools that actually allow that. . . . So I got pretty inspired by that, especially since we live in a low-income area just where kids can't travel out to parks; they don't have the transportation abilities to go see nature. So, how we bring that to them? This school had 20 acres. It was all grass, with no trees at all. It was like a totally blank canvas. We wanted to bring back what was originally here. That's taken a lot of study, so I hope it can be a history lesson for the kids. They'll end up seeing what all this area looked like before.

Taylor: Are they going to incorporate taking care of the trees and berry bushes into the curriculum?

Pickard: That's the goal. So the cool part is they are switching to a public Montessori school, and that adds too much more flexibility for the kids. Conventional education doesn't leave a lot of time in a day. With the Montessori model, if they work hard, kids can finish their work at the beginning of the week, so that they can have all of Friday to do whatever they want. Then they get to do the things that interest them. My goal is to give them 20 acres of interest where they can go out and explore. That was the idea behind the database. We made it so that it's very user-friendly. I think it would be so fun for the kids to say, "I really want to see this type of butterfly" or whatever type of wildlife they want to see. Then they come up with a plan to create the habitat, and then they watch and see if they can find that wildlife.

Taylor: That's very neat. And similar to what my parents let me do because I was homeschooled.

Pickard: So was I.

Taylor: Oh, cool! Yeah, my mom was a teacher for years, and she has her master's in education, so she decided to teach me.

Pickard: Wow.

Taylor: That's what I loved about homeschooling. After we were done with my classes, she would let me go outside and play. I would have tea parties and pretend to make medicines out of plants in my own lab. Being outside in the fresh air helped shape who I am and my interest in ecology and biology.

Pickard: Yup!

Taylor: I really like your idea of incorporating the outdoors with education. That's really cool.

Pickard: Yeah, I'm excited about it.

Taylor: Dr. Korstad actually wanted me to ask what you think of having college students like Anna Mueller, Jake Lanferman, and Nathan Feller intern with you. Do you have any thoughts on that?

Pickard: Yeah, it's been awesome. They went on that mission's trip, and they learned the skills like how to make wicking beds, and it was just so cool to bring that back from Africa. It's like Africa's giving us this gift instead of us thinking we can only give them gifts. Through the students, we built the wicking beds. We filmed the making of it, and I've got a guy who works for Channel 6 News here in town. He'll create a promotional video for it, so we can raise funds to make 12 more of them for the school. That way, every class will have its own wicking bed. The vegetable bed is a part of the Montessori curriculum, and the wicking bed will be right outside their classroom, so they can daily care for that.

I don't have the bandwidth to do all that, so I can't be more excited to have ORU students help. It's so crazy! Because it's really hard to raise funds like grants—all those things take a lot of time. Even a small grant takes so much time to write, so much information. And that's what I have a limit of—time. To be given this gift of time from students has been a huge help.

We're using the wicking beds' video for the Go Fund Me campaign. Hopefully, the parents can take ownership and fund the Go Fund Me for the rest of the materials. The video covers all the information about what's needed for every piece and how to do it. All of that should empower the parents to both fund the materials and then actually be involved in making them for their kids. That's the goal. I love doing stuff that is grassroots, where the people who will be affected take ownership of a project. I've seen the opposite happen. The Tulsa Health Department had all these beds made from a big grant, and there are all these raised beds just sitting there. They had me come speak because they had no involvement. No one cares about these beds. And it's just like, "Well, you kinda started off all wrong."

Taylor: Yes, I can see that. I learned that in my business classes. My professor, Dr. Jeff Paul, would always say, "Get people involved, and then they'll care more because their personal time and effort are at risk."

Pickard: Yup, definitely.

Taylor: Dr. Korstad's second question is: What do you think practical permaculture might look like in different regions of the world, like in the USA or Brazil or Europe or China or Africa?

Pickard: It usually doesn't take going very far back to find the people who practiced it without it being called permaculture, you know? In American history, you go back and find the natives who totally practiced permaculture. I think you can do that in any of these countries. There's been a permaculture leader named Geoff Lawton. He's got a lot of videos of the projects he's done. It's interesting to see the methods he's used in the Jordan and Palestine area where it's almost complete desert. He starts with one tree that can handle the conditions, and that provides enough shade that you can start planting other things and start changing the soil. Then there's one of my heroes on the permaculture side. He lives on a mountain in Austria. He created all these microclimates using ponds, and he's able to grow lemon trees up on this freezing mountain. It is genius. The principles of permaculture are everywhere.

Taylor: Do you have more thoughts on permaculture?

I think when people see permaculture projects, they think, "That's cute, but how can that feed the world?" I thought that, too, until I read the book *Restoration Agriculture*. The book asks if a 1-acre cornfield is more productive than a polyculture. The author has a large-scale 100-acre polyculture farm. He concludes that if you consider a broader view of calories, a food forest has as many calories per acre as does corn. That was my major inspiration for the food forest. The book throws that assumption on its head, saying, "You can actually feed the world with a food forest. You're building soil and cleaning the air and doing all these things—feeding the world actually healthy calories because that corn isn't."

Taylor: So inflammatory!

Pickard: Yeah, a negative calorie!

Taylor: I remember in your talk you addressed the economic part of sustainability and how you run the accounting part of your organization. Remember talking about your peach tree—how you spend this much on it, and it keeps giving and giving?

Pickard: Yeah, I have a degree in accounting. I started teaching a sustainable personal finance curriculum, which I created that asks the question, "If I consume less, I'll have more time, and if I have more time, what can I start producing for myself?" I think it's the way out of the market system that says, "I need to specialize in one thing and then pay for everything else."

Most people think of saving their money and only getting interest in return. Instead of saving your money, I love thinking about investing it to give multiple different yields, and that is a permaculture principle. For example, I bought the building next door. It's given us some financial returns because we rent out its commercial kitchen to a nonprofit. We took out the parking lot and planted a forest, which provides us food. I could have put that $80,000 into the stock market and gotten some interest off of it, but it's provided so many yields for the community and for us. It's crazy the difference when it comes to investing money. Even for people who don't have a lot to invest, you can buy a fruit tree for $10 instead of spending that at McDonald's. We have probably harvested 1000 pounds of fruit from one of our peach trees, and we paid $5 for it. That kind of return from a financial perspective is incredible—to invest $5 and to get $3000 back in a 5-year period is pretty much unheard of in the stock market. Investing in living things provides a financial return but so much more, like purifying our air and beautifying our surroundings.

Taylor: Thank you so much for taking the time to call! I know you are super, hectic busy. Dr. Korstad suggested interviewing you, and I'm so glad it worked out.

Pickard: You're welcome. Of course!

2.6 Conclusion

Why are people so afraid of sustainability? But, of course. Despite its noble character, sustainability sounds like the agenda of "environmental wackos," convinced that polar bears are more important than people, that sunlight and wind could make more energy than geological resources like coal and gas, and that global warming is more deadly than terrorism or nuclear war. Any proposition with "the environment" in the title reads as "regulations" and "bad press" (Hume 2011). Companies avoid it like a money-sucking leech.

However, Jib Ellison, CEO of Blu Skye sustainable consulting agency, encouraged Wal-Mart executives to be a sustainable company. He defines sustainability and Blu Skye's mission statement, saying:

> Sustainability is the greatest untapped source of competitive business advantage in the twenty-first century. It is the dynamic interplay of convening the right players to develop the right answers but also doing so in such a way that they feel a great sense of ownership that they actually built the trust, wherewithal and the will and the courage and the agreements to actually act in concert to capture the value. (Blu Sky 2020)

Ellison's statement dovetails with his epiphany, his revelation that sustainability could drive a company forward as the main engine, not as a tacked on PR generator aimed to please environmental wackos and to distract a prying public as most CEOs saw it. Focusing on how to do business effectively and efficiently, targeting waste—whether excess water use, cardboard, or plastic—changed the entire perspective and strategy (Humes 2011).

The heart of sustainability is "people, planet, and profit," not only planet, letting the people die and businesses go to waste, as how some environmental wackos put it; not only planet and people, as how some environmentalists put it; and not only profit and people, as how some businesses put it. Sustainability is how Dr. John Korstad puts it—a balanced understanding of doing things that protect all people and don't harm but restore the environment in profitable ways. Economic viability is crucial for people; a healthy environment is crucial for people; healthy bodies are crucial for people. All three parts should work together as one.

Thus, how should we then approach farming sustainability? Like the definition of sustainability itself, the answer is threefold—communication, economic viability, and education.

2.7 Communication

Often when experts in their field expound on their favorite subject, it sounds like this:

> Eru Ilúvatar created Eä through the song of the Ainur, who are the Valar and the Maiar.

When people talk about biology, ecology, agroecosystems, conservative agriculture, chemistry, or sustainability, it's as if they are speaking Quenya, or Mandarin, or Greek. Recently, when I explained to a friend how to remember that corals, hydra, and sea anemone belong to phylum Cnidaria, while sea cucumbers, starfish, sand dollars and sea urchins belong in phylum Echinodermata, in all seriousness, a fellow student turned to me and asked if I were speaking English.

Taking time to explain acronyms, the process and effects of eutrophication, and the effects of soil erosion invites people to understand one's language, to glimpse the inner complexity of one's field, and to share one's intended meaning. I have learned that listening and relating topics to other fields, whether biology, marketing, or industries, I've studied increases the probability that the expert and I will understand each other. Listening and understanding leads one closer to communicating meaning, and, therefore, one step closer to brainstorming a viable solution.

2.8 Economic Viability

Some ideas, like prions, infectious proteins, can be deadly, restructuring neural networks and webs of connections, reducing the brain to a liquid sponge. Such ideas are exemplified in the socialistic "Green New Deal," a deliberate restructuring of the foundation of the economy in the name of sustainability. Sustainability without economic viability is an idealistic prion.

Newly elected socialist Congresswoman Alexandria Ocasic-Cortez proposes that the "Green New Deal" will eliminate cars, planes, coal, and gas and require stripping buildings of their conventional energy generators and installing green ones all over the country. Her deal would cost "$4.6 trillion at minimum" and completely tank our economy (Pruden 2019). Writer Steven Moore observes that renewable energy, such as wind and solar, accounts for only 8%, costing $150 billion of taxpayer's money. He concludes that to implement her plan, taxes "would exceed $2 trillion while displacing some 10 million Americans in high-paying oil and gas industries from their jobs" and goes on to say:

> Initially, the Green New Deal planned to abolish nuclear, and natural gas from the energy mix—which is absurd given that shale gas has contributed to a major reduction in carbon emissions and nuclear plants emit no greenhouse gases whatsoever. Along with hydropower, these are the cleanest forms of energy today.

Ironically, the passenger next to me on a recent plane ride turned out to be a biofuel engineer. When I asked him his thoughts on the Green New Deal, he laughed a long time and told me it would never work.

That's the great thing about America. America does not have a dictator, a collectivist culture, or a rigged legal system. Americans are not denied access to free speech, free or nearly free media, a fair trial, legal recognition of property, or self-defense. If someone spoke out for social equality or just about anything in some totalitarian nations, he'd be in jail or dead. Here, he'd get "likes" on Twitter.

Americans are free to start companies, to engage in joint ventures, and to buy and trade all over the world. Our Founding Fathers constructed a nation where economic, social, and political opportunities are possible. Our economy thrives. Unemployment has plunged, giving companies the room to explore sustainability. In fact, the market inspires entrepreneurship and breakthroughs in technology, even embracing biomass energy in a quest for economic sustainability (Backer 2019). Benjamin Backer, president of the American Conservative Coalition, says:

> Today, for the first time ever, transportation emissions have surpassed electric power emissions in the United States, thanks to a dramatic drop in energy emissions via clean energy innovation through the free market. In conservative Texas, through deregulation, the state is rapidly transitioning to renewables unlike anywhere in the country.

As Dr. Korstad says, businesses will drive sustainability in the market place. Backer confirms it, citing that the USA leads the emission reduction movement. However, webs of government regulations ensnare US companies, who like Bilbo Baggins search for a much-needed sword, to cut the spiderwebs. Even the Harvard Business Review admits the spiderwebs sucked $70 billion from the solar industry. Unfortunately, the highly regulated socialist governments of India and Portugal keep their countries bound like the hapless cocooned dwarves in Mirkwood, which emitted 10 million metric tons of carbon dioxide, while Sweden's free market has led to their idyllic environmentally friendly star status (Backer 2019).

Forcing green energy through socialism would devastate the economy, people's jobs, our military capacity for safety, and our way of life. Zizou Corder paints this reality in her masterpiece, *The Lionboy* trilogy. Everyone's life at the Corporacy is perfect, timely, and controlled. Everyone goes to work. Everyone smiles—fake, *dead* smiles. They don't remember how they got there. They don't remember who they are. They don't know the Corporacy makes its money experimenting on animals and *humans* and selling their discoveries. They are drugged pawns in someone else's game, having given up their voice, controlled by a bloated government.

Fiction mirrors reality as a warning—no toilet paper, food, clean water, or enough medicine. It sounds like the backdrop for an apocalyptic thriller, except it's the reality of Venezuela.

Maybe my friend Tosca Lee, New York Time's best-selling author and queen of psychological thrillers, will write a sequel to *The Line Between*, which offers a cure for the prion infection devastating her fictitious world inflicting madness and death…

Or was the prion really green socialism all along?

It's still a free market after all.

2.9 Education

"Show, don't tell," my friend author Daniel Schwabauer continuously reminded me through the process of drafting my first novel. If you show the consequences of something, you never have to say anything. In one of my classes, I presented a case study on plastics polluting the ocean. I played a video clip of a scientist pulling over 200 pieces of plastic out of the stomach of a dead bird, piling the fragments high, flies swarming over the mess of macroplastics and gastric juices. I showed them how plastics never biodegrade or organically break down. My class sat with their mouths open. Students on the front row looked like they were going to vomit. And these were biology students—animal dissections are a way of life. After class, a fellow student approached me. He told me, "I never knew that. I'm going to recycle now!" After showing the presentation to my parents, my mom began filling our recycling bin daily, and my dad donated to an ocean cleanup organization. Even after my own presentation, I think about how every water bottle cap could end up in a sea turtle and a bird. Two bags filled with recyclables, waiting to be taken to the Mr. Murph, sit in my dorm room as I write this.

That is the power of showing.

My high school biology teacher, Dr. Christopher Oglivie, began every single class with the question, "What did you have for breakfast?" After listening to the list of pop-tarts, bananas, orange juice, milk, Frosted Flakes, Nutella, and toast, he would announce, "That is all sugar." Then he would describe in detail the digestive process of those sugar molecules from the mouth all the way through the digestive tract. Listening to his description of the inflammation milk causes in your body, how human enzymes that break down milk degenerate after the human body turns 5 years old, and how broccoli actually has more calcium than milk, I stopped drinking milk, and I was the queen of milk.

Again, the power of showing.

Through Dr. Oglivie, I learned that calories don't matter in the long run. What matters is what you are putting in your body. Was it grown in the ground? On a tree? Will it cause inflammation, like tiny fires burning throughout your body, in your cells?

Next, we would discuss how we could improve our breakfast—add protein, to slow down absorption of sugars to prevent an overload of glucose in the blood so the pancreas won't have to secrete insulin over time; fiber, to help clean out any toxins stored in the gall bladder; water with lemon to give the water molecules carrier proteins to ride on and help hydrate your body.

Education and nutrition dovetail. As my friends Dr. Dornbos and Mr. Pickard agree, corn is empty calories. I think most Americans are sadly misinformed regarding nutrition. Of course, an occasional indulgence, chips, pizza, and soda, is fine. But what chemicals were added to market and sell that snack as one hundred calories? Fruit is great, but it is still natural sugar. Fat isn't all bad. The brain and muscles require protein and healthy fats to build myofilaments and myelinate neuro-connections. The power of showing, not telling, will illuminate the needed changes in any discipline from ocean plastic and nutrition to agriculture.

2.10 Final Thoughts

After researching sustainability, I thought that Jib Ellison's use of appreciative inquiry, focusing on the positives rather than the negatives, was genius. My parents and my professors taught me this principle. Your thoughts and words like seeds create the environment around you. Dr. Jeff Paul, my business professor, once said, "You can think your way down the drain and can think your way up the drain." It depends on you. How you direct your thoughts, your workforce, and your business all influences your success or failure.

Sustainable farming should be approached in this manner. Showing—not telling, demeaning, or commanding—will incite people to change. Communicating between industries that are involved will improve efficiency and cost-effectiveness. Wal-Mart tackled its dairy industry this way, bringing together everyone involved from the farmers to the business executives (Humes 2011). Letting the market drive sustainability through private companies will innovate processes and provide checks and balances, cutting short inefficiency.

Problems can be mountains. I used to look at the mountain of impassable stress blockading the book I wanted to write. My thoughts raged, "What if the story is wrong? What if the whole feel is wrong? What version do I use?" At 2 AM, February 16th, God handed me the idea, the perfect synthesis of everything I'd been trying to write for years. Looking back on my notes that night, I had written out a list of similar crushing questions and ended the page with the words, "Okay, God will send it. It will come." Just when I trusted, the perfect ideas poured down like rain.

Isn't that what life is about? Trusting in Someone other than yourself to come in and pick you up when you can't do it anymore? It's the principle of appreciative inquiry. Instead of looking at the missing pieces of a story, of an industry, of sustainability and saying, "I don't know what fits," why don't we say, "Well, all of this has worked. What else will work?"

References

Adegboyega ER (2019) The impact of soil erosion on agricultural land and productivity in Efon Alaaye, Ekiti state. Int J Agric Policy Res 7(2):32–40

Backer B (2019, February 7) Marrying socialism and climate change: the shortsighted folly of Alexandria Ocasio-Cortez. The Washington Times. https://m.washingtontimes.com/news/2019/feb/7/marrying-socialism-and-climate-change/. Accessed 16 Feb 2019

Blu Sky (2020) https://blusky.me/. Accessed 25 Apr 2020

Crews T, Carton W, Olsson L (2018) Is the future of agriculture perennial? Imperatives and opportunities to reinvent agriculture by shifting from annual monocultures to perennial polycultures. Globa Sustain 1(11):1–18

Dasgupta S, Huang IJ, McElroy AE (2015) Hypoxia enhances the toxicity of Corexit EC9500A and chemically dispersed southern Louisiana sweet crude oil (MC-242) to Sheepshead minnow (*Cyprinodon variegatus*) larvae. PLoS One 10(6):e0128939

Dornbos D (2012) How should Christians promote sustainable agriculture in agrarian systems? A normative evaluation. Persp Sci Christ Faith 64:51–61

Dornbos D (2018) Personal interview

Glover J, Reganold J, Cox C (2007) Future farming: a return to roots. Large scale agriculture would become more sustainable if major crop plants lived for years and built deep root systems. Sci Am 297(2):82–89

Glover J, Reganold J, Bell L, Borevitz J, Brummer E, Buckler E, Cox C, Cox T, Crews T, Culman S, DeHaan L, Eriksson D, Gill B, Holland J, Hu F, Hulke B, Ibrahim A, Jackson W, Jones S, Murray S, Paterson A, Ploschuk E, Sacks E, Snapp S, Tao D, Van Tassel D, Wade L, Wyse D, Xu Y (2010) Increased food and ecosystem security via perennial grains. Science 328 (5986):1638–1639

GPS: National Coordination Office for Space-Based Positioning, Navigation, and Timing (2018). https://www.gps.gov/applications/agriculture/. Accessed 7 Jan 2019

Huggins D, Reganold J (2008) No-till: the quiet revolution. Sci Am 299(1):70–77

Humes E (2011) Force of nature. Harper Collins, New York, pp 14–23, 87, 160–180

Mann C (2008) Our good earth. Natl Geogr Sep 2008:84–107

Montgomery DR (2007) Soil erosion and agricultural sustainability. PNAS 104(33):13268–13272

Paine V (2012) What causes ocean "dead zones?" https://www.scientificamerican.com/article/ocean-dead-zones/. Accessed 2019 June 11

Pretty J, Benton T, Bharucha Z, Dicks L, Flora C, Godfray C, Goulson D, Hartley S, Lampkin N, Morris C, Pierzynski G, Prasad V, Reganold J, Rockström SP, Thorne P, Wratten S (2018a) Global assessment of agricultural system redesign for sustainable intensification. Nat Sustain 1:441–446

Pretty J, Benton T, Bharucha Z, Dicks L, Flora C, Godfray C, Goulson D, Hartley S, Lampkin N, Morris C, Pierzynski G, Prasad V, Reganold J, Rockström SP, Thorne P, Wratten S (2018b) Global assessment of agricultural system redesign for sustainable intensification: supplementary information. Nat Sustain 1:441–446

Pruden W (2019) Second thoughts in the Unicorn Caucus. Jewish World Review. http://www.jewishworldreview.com/cols/pruden021219.php3. Accessed 16 Feb 2019

Ramos TB, Darouich H, Gonçalves M, Brito D, Castelo Branco M, Marins J, Fernades M, Pires F, Morals M, Neves R (2018) An integrated analysis of the eutrophication process in the Enxoé reservoir within the DPSIR framework. Water 10(11):1576

Reganold J, Robert P, Parr J (1990) Sustainable agriculture: traditional conservation minded methods combined the modern technology can reduce farmers' dependence on possible dangerous chemicals. The rewards are both environmental and financial. Sci Am 262(6):112–114

Stewart R (2016) Personal interview

The Land Institute (2019) Kernza® Grain: toward a perennial agriculture. https://landinstitute.org/our-work/perennial-crops/kernza/. Accessed 7 Jan 2019

Tully K, Ryals R (2017) Nutrient cycling in agroecosystems: balancing food and environmental objectives. Agroecol Sustain Food Sys 41(7):761–798

Welch R, Shuman L (1995) Micronutrient nutrition of plants. Crit Rev Plant Sci 14(1):49–82

Destruction of Soil Health and Risk of Food Contamination by Application of Chemical Fertilizer

3

Lekeanju Nguatem Tayoh

Contents

Abstract

Chemical fertilizers are used to increase crop quantity and quality and to influence the taste of many food crops. On the other hand, there are increasing environmental concerns and food-related illnesses connected to the use of these chemicals in order to boost food production and meet up with existing food demand. The problems have been that of inappropriate application and misuse of chemical fertilizers at different stages in farm management. The major concern of fertilizer contamination to food is that of nitrate pollution leading to nitrate poisoning common to both man and livestock. The tendency for food contamination is high for short-cycle crops like vegetables and maize, root crops like carrots and tubers like yams and cassava. This depends largely on the type of fertilizer application, the quantity and rate of application. Many studies have revealed that above 5% of nitrogen nutrient from fertilizer application is lost to water during the cropping season. Two-thirds of it is due to inappropriate application practices.

L. N. Tayoh (✉)
Sustainable Agricultural Centre for Rural and Community Development (SACRCoD), Buea, Cameroon

K. Bauddh et al. (eds.), *Ecological and Practical Applications for Sustainable Agriculture*, https://doi.org/10.1007/978-981-15-3372-3_3

Most farmers carry out mix application of agrochemicals. This is rampant with liquid chemical application on food crops, where liquid pesticides are often mixed with either powder or liquid chemical fertilizers during application or vice versa. This makes identification and characterization of the risks of inorganic fertilizers on soils and food crops a difficult task and needs more research and investigations.

Keywords

Chemical fertilizers · Fertilizer application · Soil nutrients · Soil contamination · Food contamination

3.1 Mineral or Chemical Fertilizers

By functioning and need, mineral, inorganic or chemical fertilizers can be defined as elements, either natural or manufactured, containing nutrients essential for the growth and development of plants (Isherwood 2000). Three of such nutrients are most used in large quantities, nitrogen (N), phosphorus (P) and potassium (K), known as NPK fertilizers. They constitute the main essential nutrients for plant growth and are called macro plant nutrients. Other nutrients like sulphur, calcium and magnesium are also required in substantial amounts. According to Adiaha and Agba (2016) and Isherwood (2000), these nutrients are constituents of many plant components such as proteins, nucleic acids and chlorophyll and are essential for processes such as energy transfer, maintenance of internal pressure and enzyme actions. Nine other elements are required in very small quantities compared to other nutrients and are referred to as "micronutrients" or "trace elements". A deficiency of any one of these nutrients can compromise the development of the plants but cannot limit plant growth.

Fertilizers are used in order to:

- Meet the demand of crops by maintaining soil nutrient supply
- Produce economically substantial yields and maintain a sustainable high yield potential
- Improve unfavourable or to maintain good soil conditions for plant growth and development

3.2 Application of Chemical Fertilizers

Chemical fertilizers exist as solid granules and as concentrated powders and are applied either as solid granules or as soluble solutions to plants. Application largely depends on the method of application, types of plants, field management, timing and existing soil fertility. Application also varies from one geographical zone to another. The application and use of chemical fertilizers in Europe differ from the application and use in sub-Saharan Africa, due to differences in crop varieties, soil types,

Table 3.1 Types of fertilizer application

Direct application		Indirect application	
Surface feeding	Foliage feeding	Farm irrigation	Precipitations
Broadcasting	Spraying on leaves	Irrigation rivers	Rainfall
Ring placement		Irrigation wells	Atmospheric dew
Side or row placement			

geo-climatic differences, availability, know-how and follow-up (Lekeanju et al. 2016; Mateo-Sagasta et al. 2017).

The application of chemical fertilizers to plants can either be direct application, via intention farming like irrigation or indirect via atmospheric processes from rain pollution. In direct application, chemical fertilizers are applied to crops using two main methods: surface feeding and foliar feeding (Table 3.1).

3.2.1 Direct Application

Direct application of inorganic fertilizers to the soils is mainly to increase soil nutrients for plant uptake, and this is done using different methods of application, which can either be directly to soils or on plant leaves depending on the stage of crop development and farmers' expectations.

3.2.1.1 Surface Feeding

There are different types of surface feeding, which include broadcasting, ring or roll application and hole application (Adiaha and Agba 2016; Hoyt 2018; Finneran 2012).

Broadcasting involves application of dry granular fertilizer over large space, which is the most common method of placement done by using either hand or a variety of mechanical fertilizer spreaders. Broadcasting is best when leaves are dry so that wet leaves will not dissolve the fertilizers causing them to stick to the leaf surface and cause burned spots. When watered into the soil, it is called top dressing (Hoyt 2018; Finneran 2012).

Fertilizers can also be hand-applied on top soil to specific plants or plant rows in a ring around each plant in proportion to its size and rooting area that will provide nutrients in a more concentrated area. This method helps in reducing waste between rows and plants (Hoyt 2018). Another surface-feeding method is the hole method. The fertilizer is placed in the area of planting days before planting in order for the nutrients to filter into the soil (Hoyt 2018; Finneran 2012). This method is common with the preparation of seed bed for nursery development of vegetables and short-cycle fruits like water melon (Fig. 3.1).

3.2.1.2 Foliage Feeding

Soluble fertilizer products are sold either in liquid form or solid concentrate. Once mixed with water at the manufacturer's recommended rate or by farmer's discretion,

Fig. 3.1 Types of surface spreading: (**a**) broadcasting and (**b**) focus spreading

the solution is applied using a spray or drench over the leaves where it will be absorbed through the tissues of plant leaves (Finneran 2012). Nitrogen, phosphorous, potassium and a number of trace elements are readily absorbed through the leaves of some plants than others. According to Finneran (2012), absorption and plant response tends to be fairly quick and more effective. Working with local farmers producing vegetables and other short-cycle crops in Buea, Cameroon, reveals that soluble fertilizers are mostly used to promote leaf development, flowering, fruit development and ripening of fruits for crops like tomatoes. The challenge is that most fields lost 2–20% of liquid fertilizers sprayed on crops via evapotranspiration (Savci 2012). This forms the bulk of indirect agrochemicals that comes in contact with adjacent fields through precipitation and air drift processes.

3.2.1.3 Indirect Application

Indirect application of chemical nutrients to plants is common in areas where large agricultural plantations crisscross rivers that supply a large amount of farms downstream. Excess fertilizers in soils of plantation area are often drained, are leached or flow into the running river that is used for irrigation purpose downstream (Mateo-Sagasta et al. 2017). As a result, uncalculated soluble nutrients are supplied to crops. This can lead to crop failures, when the flow is from tree crop plantations to food crop farming and accumulation of other heavy metals in the food crops. Other unintended forms of soluble fertilizer may comes from wind drifts and atmospheric moisture of close by farms and plantations under fertilizer application.

3.3 Contamination of Soil Health

The exact effects of mineral fertilizers on soils and agricultural production are debatable (Savci 2012; Isherwood 2000; Mateo-Sagasta et al. 2017), but millions of field trials carried out throughout the world clearly reveal the great influence of chemical fertilizers on crop yields (Isherwood 2000; Adiaha and Agba 2016; Savci

2012). Soils have strong buffering power, which determines the rate of nutrient uptake (Savci 2012; McCauley et al. 2017). Thus, the risk in affecting soil health can be minimal in fields where production is short and irregular. Continuous cultivation of land over time can lead to nutrient depletion, soil degradation, loss of soil fertility and deterioration of soil nutrient balance (Lekeanju et al. 2016 Isherwood 2000). This has become the motivating factor in the use and application of chemical fertilizers in crop land in order to increase the corresponding soil nutrients. The risk of soil contamination and destruction becomes inevitable when application becomes too much and not corresponding to soil needs.

3.3.1 Soil Health

Before assigning specific problems that can result from excess use of chemical fertilizers, it is better to define soil health and understand the different components of a healthy soil. Soil health is defined by Doran et al. (1994), Brevik (2010), CRI (2016) and Laishram et al. (2012) as the capacity of a soil to function, within natural or managed ecosystem boundaries, to sustain biological productivity, maintain environmental quality and promote plant, animal and human health. Soil health is established through sustainable interactions of soil's physical, chemical and biological soil properties and soil air (Brevik 2010; Al-Kaisi 2017; CRI 2016), which together determines soil fertility in agro-ecosystems. Thus, soil fertility is the capacity of any soil to adequately provide nutrients required by plants for growth and development (Brevik 2010; CRI 2016; Lekeanju et al. 2016). This implies that a healthy soil is a fertile soil. To better understand how poor soil management can destroy soils, a brief description of soil's physical, chemical and biological properties with reference to Brevik (2010), Al-Kaisi (2017) and CRI (2016) is important because they define the interactions within healthy soils:

- Physical properties comprise of soil texture, a physical measurement of the percentage of sand, silt and clay.
- Soil structure, the arrangement of individual soil particles including sand, silt and clay into aggregates or "clumps".
- Chemical properties of a soil measure the nutrient-carrying capacity and pH (acidity) in soil and soil solution.
- Biological properties constitute the community of soil organisms (principally bacteria, fungi and actinomycetes and other soil animals).

In addition to the above properties are indicators for healthy soils, which include good tilt, mineralization ability, good rooting medium, toxicity level, resistance to adverse conditions such as drought and others (Table 3.2).

When soil properties and indicators are tampered with, by human activities in agro-ecosystems, the tendency for a destructive soil health can be inevitable. Although chemical fertilizers boost crop production, their overuse and increased application have caused different problems to soil health for different crops. These

Table 3.2 Indicators of a healthy soil: Adapted from Brevik (2010), CRI (2016), Al-Kaisi (2017) and Laishram et al. (2012)

Indicators	Characteristics
Mineralization	This reflects the ability of the soil to transform nutrients (macro and micro) and a moderate pH (~6.0–7.0) that allows the nutrients to be both held in the soil and available to plants and soil microbes as needed
Good tilt	This includes a good structure that resists degradation (e.g. erosion and compaction) and provides adequate aeration and sustainable water infiltration that accepts, holds and releases water to plants and groundwater
Rooting medium	Soil that promotes good root growth and maintains good biotic habitat that sustains high and diverse populations of beneficial organisms and low populations of pests and pathogens
Soil air	Its primary role is to provide oxygen to fuel the aerobic (oxygen-requiring) activities of microorganisms and plant roots. For example, soil bacteria that associate with roots of legumes such a beans and peas use the nitrogen component of soil air to "fix" nitrogen in a form that plant roots can assimilate
Organic matter (OM)	Carbon-containing substances, including living organisms, plant biomass and the carbonaceous remains of organisms and plants
Soil water and solutions	Soil water or the soil solution carries dissolved nutrients that flow to and are actively intercepted by plant roots. Thus, the soil solution is the vehicle for nutrients to "flow" into plants and, along with the products of photosynthesis, "grow" the plant
Toxicity	Sustainable soils have low salinity levels and low levels of potentially toxic elements (e.g. boron, manganese and aluminium)
Resistance	Good soils have high resilience and are able to withstand deleterious events, such as drought and flooding

include soil hardening, decreased fertility and strengthened pesticides (Isherwood 2000; Savci 2012; Lekeanju et al. 2016). Lekeanju et al. (2016) and Isherwood (2000) indicated that continuous use of chemical fertilizers depletes essential soil nutrients and minerals that naturally exist in fertile soil. Some like phosphorous fertilizers do not dissolve in water, and its overuse may cause soil hardening. Likewise, overapplication of alkaline fertilizers like sodium nitrate increases soil alkalinity and reduces its fertility and makes it barren (Savci 2012; Isherwood 2000). Soil fertility and crop development depend much on the balanced supply of essential soil nutrients and minerals. As such, overuse and/or application of specific nutrients may cause nutrient imbalance. This can result in soil degradation and the loss of soil nutrient equilibrium in a stable soil (Savci 2012).

Chemical fertilizers are also known to cause soil acidification and root or fertilizer burns (Rahman and Zhang 2018). Their overuse has also been associated with the killing of soil-friendly microorganisms (Savci 2012; McCauley et al. 2017). Some studies (Wu et al. 2012; Ngo et al. 2012) have revealed that earthworms increase soluble organic matter (OM) transformation, microbial activity and colony-forming units of microorganisms in soil during their feeding activity, thus posing a positive influence on the number of microorganisms like the *Rhizobium spp.* for the fixing of mineral N in soils. In excess use of chemical fertilizers, the activities of earthworms to transform soluble OM and increase microbial functions can be affected.

Particularly high levels of sodium- and potassium-containing fertilizers have negative impact and influence on soil. This may lead to increase in soil pH and deterioration of soil structure (Savci 2012). Continuous use of acid-forming nitrogen fertilizers can also cause a decrease in soil pH. This can lead to a sudden drop in the yields and quality of food crops (Rahman and Zhang 2018; Savci 2012) and consequently to soil pollution by accumulation (Savci 2012).

Savci (2012) indicated an increase in soil pH in Nevsehir over the last 25 years as a result of nitrogen fertilization of potatoes. The soil pH has fallen to two, and *Rhizobium* sp. activities, such as symbiotic nitrogen fixing by microorganisms, are negatively affected. In addition, more nitrogenous fertilizers limit the activities of nitrifying bacteria.

Excessive nutrients can also cause adverse effects on plant growth and increase the potential for environmental contamination due to leaching of nutrients from farmlands. In particular, above-optimum nitrogen and phosphorus levels have been the cause of excessive plant and algal growth in many waterways: degraded drinking water reservoirs, fisheries and recreational areas (Allen and Mangan 2015; Mateo-Sagasta et al. 2017).

3.4 Chemical Fertilizers and the Risk of Food Contamination

The application of chemical fertilizers is aimed at increasing plant growth and improves health of plant tissues. This is best at optimum nutrient levels at which plant growth and development is maximized. Above-optimum nutrient level plant growth is halted and it is a waste of resources (Allen and Mangan 2015). In most cases worldwide, crop lands have often witnessed excessive application of chemical fertilizers, as a result of either excessive human activities or natural factors like excess rainfall leading to over-leaching. The resulting consequences of overapplication to food crop and food need deep investigations and examinations.

Chemical fertilizers have become a major need for plant growth and development, and the risk of food contamination from excess application depends on many factors:

- The rate of and quantity of application
- Time between application and harvest
- The rates of nutrient uptake and breakdown by plant tissues
- The agro-biodiversity of the crops

Food contamination will vary from short-cycle food crops to mid-cycle, annual, root crops and perennial crops. Perennial crops have well-established root systems and patterns, and uptake of nutrient is slower over a longer period of time. They can effectively absorb nutrients for the development of fruits, nuts and leaves over prolonged periods of time.

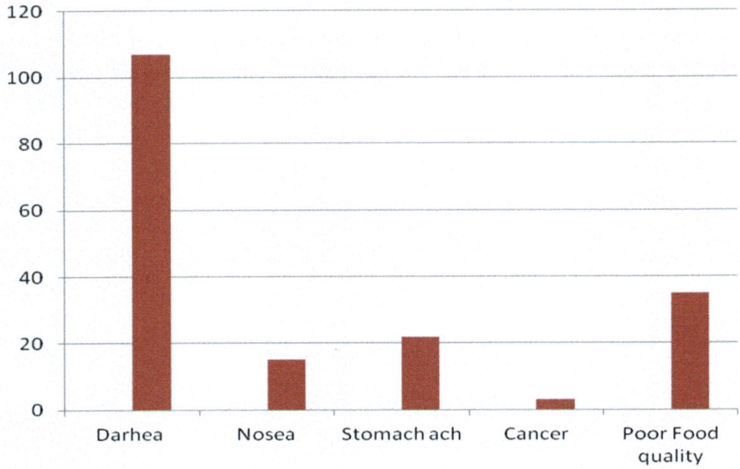

Fig. 3.2 Effects of mineral fertilizers to humans and food quality: Buea, Cameroon; Lekeanju et al. (2016)

Generally, plant nutrient uptake is highest during the vegetative growth stage of plants (Haun 2015). This implies for short-cycle crops; uptake of fertilizer nutrients is higher and shorter than for annual and perennial crops in well-watered soils. For root crops like carrots, yams, cassava and cocoyam, the uptake and accumulation of nutrients from natural and conventional sources can also be faster and higher, because their roots are the major storage of needed food. This indicates that a high amount of chemical fertilizers on short-cycle and root crops can be problematic when considering the determining factors such as the quantity and rate or time between one application and the other. Studies carried out for different types of food crops, after application of chemical fertilizers, reveal different contamination and health problems (Lekeanju et al. 2016; Matt et al. 2011).

The use of chemical fertilizers is common with vegetables than with any other class of food crops (Lekeanju et al. 2016; Whittier 2011). Health problems associated with chemical fertilizers have been most recorded after the consumption of vegetable meals than with other food crops worldwide. The most common types of food contamination-related illness have been diarrhoea and stomach ache (Fig. 3.2).

In addition to the above findings, many studies (Isherwood 2000; Savci 2012; Matt et al. 2011; Wiederholt and Johnson 2005; Silva et al. 2000) have unfolded different diseases caused by nitrates from the application of chemical fertilizers to food crops. Examples are alimentary canal tumours, leukaemia and methaeminoglobinaemia in newborn babies, small children and the elderly. Methaeminoglobinaemia is mostly associated with nitrate-polluted water, and occurrence of this disease is now rare, although, for some reasons, some cases have been recorded in Hungary and Romania (Isherwood 2000). Nitrates from chemical fertilizers are not very harmful in humans, but they are transformed into very toxic nitrites in the body by intestinal microflora to nitrites which are six to ten times more toxic than nitrates in humans and livestock (Matt et al. 2011; Whittier 2011). This is more severe in livestock which feed mostly on plants, a condition known as nitrate poisoning (Glunk et al. 2015; and Whittier 2011). Some plants known as nitrate accumulators absorb and store nitrate than others and when consumed may cause serious health problems. Whittier (2011) and Matt et al. (2011) have recorded some examples including sorghums, small grains (wheat, oats, rye and barley) and millet. Some perennial grasses (fescue and Johnson grass) and weeds (pigweed, mustard, kochia, nightshade and lamb's quarters) can contain dangerous levels.

Food quality problems of food crops have also been recorded. Local farmers and consumers of food crops grown with chemical fertilizers have attested to loss of taste, reduced shelf life and destruction of regenerative seed capacity (Lekeanju et al. 2016; Allen and Mangan 2015). Other studies have attested to the claim that crops grown with "artificial fertilizers" are less healthy than, for example, organically grown or produced crops (Isherwood 2000; Matt et al. 2011). Nitrites from chemical fertilizers can also reduce the nutritional value of food, decreasing assimilation of protein, fat and beta carotene, causing decomposition of the B group of vitamins and reducing the content of vitamin A (Matt et al. 2011). This was confirmed by farming surveys for farmers growing food crops in Fako Division, Cameroon, that crops and vegetables conventionally produced with chemical fertilizers have higher nitrate level (sour and tasteless) when compared with organically produced or unfertilized vegetables, especially the green and root vegetables (Lekeanju et al. 2016).

3.5 Discussion

As the world population keeps increasing, hunger and the need for food continue to be major issues on the world food discussion table; the need for more food to fight hunger is increasing. To meet this demand, the application of chemical fertilizers to crop lands in order to increase food production keeps increasing in many countries of the world. Chemical fertilizers are used to increase crop quantity and quality and to influence the taste of many food crops. On the other hand, there are increasing environmental concerns and food-related illnesses connected to the use of agrochemicals in order to boost food production and meet up with existing food demand. The problems have been that of inappropriate application and misuse of fertilizers at different stages in farm management, which leave behind soil- and food-related risks.

Some major concerns to soil contamination have been that of soil acidification, which affects enzyme activities in different soil types and farm management systems. Continued chemical fertilizer application has led to nutrient imbalance and soil degradation in most farming systems.

The major concern of fertilizer contamination to food is that of nitrate pollution to food crop and water bodies (Allen and Mangan 2015; Mateo-Sagasta et al. 2017). The tendency for food contamination is high for short-cycle crops like vegetables and maize, root crops like carrots and tubers like yams and cassava. This depends largely on the type of fertilizer application, the quantity and rate of application. High dose of surface spread fertilizer might cause problems to soil and reduce the quality of root crops and tubers. This is also true for foliage chemical fertilizers. Studies on food contamination from chemical fertilizers revealed related food problems from vegetables harvested after sprayed with liquid fertilizers shortly before harvest (Lekeanju et al. 2016). This is because excess nutrients for building plant tissues tend to accumulate faster in leaves and roots for short-cycle crops and root crops where growth is focused (Haun 2015). The excess fertilizer has been associated with different illnesses like diarrhoea in humans and livestock after the consumption of vegetables, weeds and foliage plants from conventional farming sites.

The impacts of indirect fertilizer applications are uncommon, and there is a need to define clear-cut lines, risks and problems from different agrochemicals. Conventional farming systems involve the use of chemical fertilizers and pesticides to increase and improve crop production. There are limited studies defining which soil- and food-related problems are tied to only pesticides and/or only chemical fertilizers. Some of these agrochemicals are carriers of similar chemical elements like nitrates, which are problematic to both the environment and humans. Nitrates from farmland are leached to nearby water bodies to cause eutrophication, a network of plants that cover surface of water bodies. These plants consume oxygen needed by fish and other aquatic animals and also block sunlight from reaching the bottom of rivers, lakes and reservoirs. Labelled (nitrogen) N experiment indicates that more than 5% of the nitrogen from fertilizer is lost to water resources during the growing season and two-thirds of it due to bad application practices (Mateo-Sagasta et al. 2017). Excess nitrates that are incorrectly applied are also leached into groundwater and cause groundwater pollution in many cases that when consumed may pose health problems.

In many farming systems, assessing problems from chemical fertilizers is a complex issue due to different methods of application. Most farmers carry out mix application of agrochemicals. This is rampant with liquid chemical application on food crops in Cameroon, where liquid pesticides are often mixed with either powder or liquid chemical fertilizers during application or vice versa. This practice can lead to other reactions and production of other chemical compounds not friendly to the food crops, humans and the environment. In this light, much still needs to be done in connection with the problems of chemical fertilizers to nature and food. There is a need for detailed and specific risk of chemical fertilizers on crops. There is a need to demarcating which contaminations are specifically tied to chemical fertilizers from those caused by pesticides.

3.6 Conclusion

The global need of chemical fertilizers to increase food production and crop quality is inevitable as the world population continues to increase and challenges of food security become a major issue for many underfed nations. The risks and problems from chemical fertilizers are due to excess and untimely application on crop lands.

The major problem from chemical fertilizers is that of nitrate fertilizers. Continuous use of acid-forming nitrogen fertilizers can lead to a decrease in soil pH, which can lead to a sudden drop in crop yield, risk of contamination and reduced crop quality. Together with phosphate fertilizers, they are major sources of nutrient for eutrophication of water bodies, and when leached into groundwater in excess, they may pose health problems. Excess applications of nitrate fertilizers to crops like vegetables and fruits have also been indicated to cause food contamination and subsequent health problems. Excess application can lead to high accumulation in the feed and forage of livestock, causing nitrate poisoning in grass feeders. Worldwide disparity in data makes it difficult to establish major types of environmental and food-related risks caused by the use of chemical fertilizers. Some concludes that excessive accumulation of nutrients may increase adverse health impacts, such as blue baby syndrome, due to high levels of nitrate in drinking water. These conclusions are mostly drawn from testimonies rather than on conclusive laboratory results.

References

Adiaha MS, Agba OA (2016) Influence of different methods of fertilizer application on the growth of maize (*Zea mays* L.) for increase production in South Nigeria. World Sci News 54 (2016):73–86

Al-Kaisi M (2017) Basic field indicators for assessing soil health. Strategies for building healthy soils: soil health conference 02, 2017

Allen T, Mangan F (2015) Over-fertilization of soils: its causes, effects and remediation. US Composting Council field guide to compost use. Fact sheet. Ref. No. SPTTL. 11, March 2015

Brevik EC (2010) Soils, plant growth and crop production. Vol. I: Soil health and productivity. ©Encyclopedia of Life Support Systems (EOLSS)

CRI (2016) Three components of soil health. AgSourc Laboratories

Doran JW, Coleman DC, Bezdicek DF, Stewart BA (eds) (1994) Defining soil quality for a sustainable environment. Special publication #34. American Society of Agronomy, Madison, WI

Finneran R (2012) Fertilizer application methods and placement. Michigan State University Extension, East Lansing, MI

Glunk E, Olson-Rutz K, King M, Wichman D, Jones C (2015) Nitrate toxicity of Montana forages. MT200205AG Revised 3/15

Haun W (2015) Plant nutrient uptake mechanisms. TIGER®TECH: NEWSLETTER

Hoyt R (2018) Four methods for applying fertilizers. https://homeguides.sfgate.com/four-methods-applying-fertilizers-25517.html. Accessed 05 2019

Isherwood KF (2000) Fertilizer use and the environment, revised edition. International Fertilizer Industry Association. IFIA, Paris. isbn: 2-9506299-3-8

Laishram J, Saxena KG, Maikhuri KS, Rao KS (2012) Soil quality and soil health: a review. Int J Ecol Environ Sci 38(1):19–37. © National Institute of Ecology, New Delhi

Lekeanju NT, MIK L, Mbunya FN (2016) Chemical fertilizer application and farmers perception on food safety in Buea, Cameroon. Agric Sci Res J 6(12):287–295. December, 2016: issn: 2026

Mateo-Sagasta J, Zadeh SM, Turral H, Burke J (2017) Water pollution from agriculture: A global review. Executive summary: Food and Agriculture Organization of the United Nations and the International Water Management Institute on behalf of the Water Land and Ecosystems research program Colombo

Matt D, Rembialkowska E, Luik A, Peetsmann E, Pehme S (2011) Quality of organic vs. conventional food and effects on health. Report by Estonian University of Life Sciences

McCauley A, Jones C, Ruts KO (2017) Soil PH and organic matter. Nutrient management module no. 8. Copyright © 2017 Montana State University (MSU) Extension

Ngo PT, Rumpel C, Doan TT, Jouquet P (2012) The effect of earthworms on carbon storage and soil organic matter composition in tropical soil amended with compost and vermicompost. Soil Biol Biochem 50(2012):214e220

Rahman AKM, Zhang D (2018) Effects of fertilizer broadcasting on the excessive use of inorganic fertilizers and environmental sustainability. J Sustain 10(3):759

Savci S (2012) An agricultural pollutant: chemical fertilizers. Int J Environ Sci Dev 3(1):73

Silva JA, Evensen CI, Bowen RL, Kirby R, Tsuji GY, Yost RS (2000) Managing fertilizer nutrients to protect the environment and human health. In: Plant nutrient management in Hawaii's soils. Approaches for tropical and subtropical Agriculture

Whittier JC (2011) Nitrate poisoning. Livestock series/health. Fact sheet no. 1.610

Wiederholt R, Johnson B (2005) Unintended impacts of fertilizer and manure mismanagement on natural resources. North Dakota State University: NDSU Extension service. Jan 2005

Wu F, Wan JHC, Wu S, Wong M (2012) Effects of earthworms and plant growth–promoting rhizobacteria (PGPR) on availability of nitrogen, phosphorus, and potassium in soil. J Plant Nutr Soil Sci 175:423–433. https://doi.org/10.1002/jpln.201100022

Impacts of Synthetic Pesticides on Soil Health and Non-targeted Flora and Fauna

4

Ankit, Lala Saha, Vimal Kishor, and Kuldeep Bauddh

Contents

Abstract

The complications associated with excessive use of synthetic pesticides are well known. Synthetic pesticides are lethal to a host of living organisms, be it insect, weed, rodent, etc. At the same time, they are also accountable for degrading soil quality. The adverse impact of pesticides on non-target organisms is quite

---

Ankit and Lala Saha contributed equally.

Ankit · L. Saha · K. Bauddh (✉)
Department of Environmental Science, Central University of Jharkhand, Ranchi, Jharkhand, India
e-mail: kuldeep.bauddh@cuj.ac.in

V. Kishor
Department of Education, Central University of Jharkhand, Ranchi, Jharkhand, India

K. Bauddh et al. (eds.), *Ecological and Practical Applications for Sustainable Agriculture*, https://doi.org/10.1007/978-981-15-3372-3_4

noticeable because it disrupts the entire natural ecosystems, particularly soil and aquatic ecosystems. Because of this, the application of several pesticides has been banned. Most of the pesticides like organochlorine are significantly resistant to biodegradation, and therefore, they have high risk of entering into the food chain, thus causing adverse impact on non-target species like pollinator insects, birds, fishes, beneficial microorganisms, etc. In this chapter, the emphasis has been laid upon the impacts of synthetic pesticides on the health of soil as well as non-targeted flora and fauna.

Keywords

Pesticides · Toxicity · Soil health · Flora · Fauna

4.1 Introduction

Considering the fact that pesticides are important to control pests, it is essential to assess their ecotoxicity especially on soil health and non-targeted flora and fauna, which is largely unknown. Soil microorganisms are very important players that help to regulate various soil processes like nutrient cycling and decomposition of organic matter and are also responsible to a considerable extent for the function of soil ecosystem (Bowles et al. 2014). Modern molecular means like DGGE (denaturing gradient gel electrophoresis), rDNA (amplified ribosomal DNA), TGGE (temperature gradient gel electrophoresis) and other PCR (polymerase chain reaction) techniques prove to be more efficient in analysing the effects of pesticides on soil microbial community composition. Assessing the effect of pesticide on total nitrogen and carbon mineralisation is a standardised component for testing the effect of pesticides on the non-target organism. A better understanding of its effects on both processes is crucial for understanding pesticide interaction in soil and its role in supporting the growth of plants and the overall health of the ecosystem.

It is an intriguing fact that a minuscule amount of the pesticides, i.e. less than 0.3%, actually falls on the crop, and the rest 99.7% is accumulated in the environment (Pimentel and Burgess 2011) that is detrimental not only to the environment but also to non-targeted beings. It is also important to understand the fate of these pesticides (Fig. 4.1). Approximately 300,000 people worldwide lose their lives every year due to poisoning from pesticides (Sabarwal et al. 2018). The principal route for removal of pesticides from soils is through soil microbes that perform conjugation, hydrolysis, oxidation and reduction (Aislabie and Lloyd-Jones 1995; Bending et al. 2006), thereby contributing to increasing soil's fertility. Therefore, any loss to the microbial community would ultimately lead to a significant loss in soil fertility, which is a prerequisite for the proper growth of plants. Ideally, a pesticide must not affect non-target organisms. However, there have been findings that pesticides adversely affect non-targeted flora and fauna (Singh et al. 2015; Kumar et al. 2018a). Indiscriminate application of synthetic pesticides perturbs the soil environment, thereby

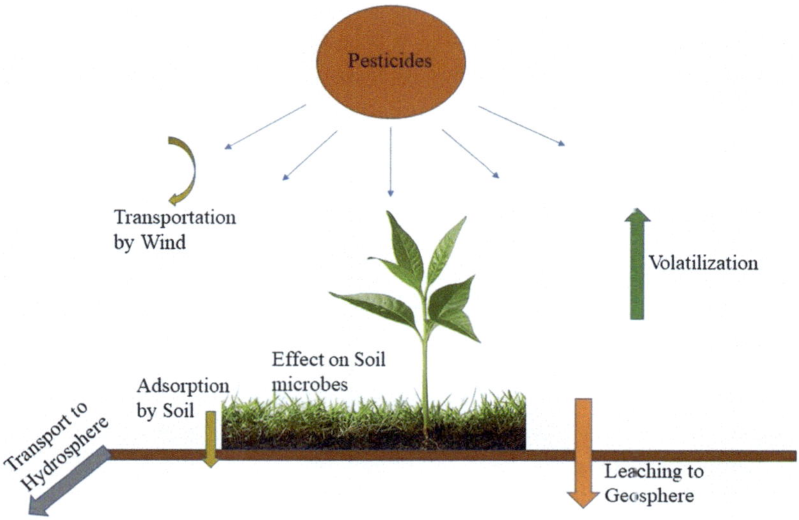

Fig. 4.1 Fate of pesticides on soil, flora, fauna and the environment

affecting the flora, fauna and several physicochemical properties of soil such as alkalinity, salinity and pH leading to poor soil fertility (Bünemann et al. 2006; Kah and Brown 2006; Sarnaik et al. 2006; Chowdhury et al. 2008; Arora et al. 2019).

Organophosphate pesticides are used extensively to check the growth of pests, diseases and weeds of crops. These insecticides tend to have a considerable residence time in the environment, and so they cause serious pollution problems (Mulla et al. 2020). Persistent organic pesticides (POPs) were found in several media of the environment like soil, air and water. They may accumulate in almost all living organisms (Devi 2020). Carbamates, organochlorine and organophosphates have been found to be genotoxic. They have been held responsible for suppressing mitosis in meristematic cells of plants (Ogut 2019).

Soil is a highly complex, living and dynamic ecosystem that supports an exceptionally large variety of macro and micro fauna and flora that in turn has an influence on the soil. It mainly comprises inorganic mineral nutrients, organic matter, water and air. It forms the basis of agriculture. Besides being the principal medium for growth of plants, soil performs several other important functions like maintaining the flow of energy, gaseous exchange, pollutant detoxification, etc. So maintaining the soil quality and health is indispensable to ensure agricultural productions which are sustainable and also to maintain soil microbial diversity. But with imprudent and excessive use of pesticides in agriculture, the non-target effects on flora and fauna are critical to the health of the soil.

Infestation by insects is a serious issue faced by crop plant (Singh and Kaur 2018), but we must resort to sustainable means of agriculture for they are considered as economically feasible, provide safe and nutritious food and meanwhile take care of natural resources along with needs of future generations (Dhakal and Singh 2019). Of late, there has been an increase in awareness among masses regarding the negative effects of pesticides on soil and water bodies (Al-Zaidi et al. 2011; Hou and Wu 2010; Kalia and Gosal 2011; Mustapha et al. 2019).

4.2 Need of Pesticides

Ever-increasing population will put extra pressure on the environment and poses a question regarding food security. The available additional fertile agricultural land is very limited. It must be understood that each and every expansion will take place at the cost of wildlife containing natural habitats and forests. Besides growing agricultural crops, the land will also be used for fibre and biofuel. So we have even less land to produce our food grains with limited water and fertiliser than we use today. So there is a double whammy of producing more crops on less land.

It has been estimated that on an average approximately 35% of the yield is lost because of pests. In order to make agriculture more profitable and productive in times of rising standards and cost of living of humans, a combination of best technologies must be used. Most of the increase in yield per unit area is due to effective control of stress rather than an increase in the potential of yield. The reduction in yield is due to weeds, pathogens and especially pests which pose the biggest hurdle in agricultural production (Oerke and Dehne 2004). The crop protection by the use of pesticides worldwide has increased the yield by 15–20-fold (Oerke 2005). Assisting farmers in decreasing the crop loss due to pests will be a primary factor that will ensure food security because it is the first step towards the economic independence of farmers (FAO 2009).

The positive outcomes of the use of pesticides corroborate the fact that pesticides will continue to play as an important tool in a diverse range of technologies that may improve the living standards of people across the globe. But these positive outcomes fail to justify the high social and environmental cost of the use of pesticides. The spillover effect and negative externalities of pesticides must be addressed as soon as possible (National Research Council 2000). The productivity of crops may be enhanced by using high-yielding varieties, fertilisers, other techniques of cultivation and improved soil and water management. An increased yield is sometimes associated with susceptibility to pest attack which leads to increasing loss rates and absolute losses (Oerke et al. 1994).

Since the 1940s, further enhancement in the production of food was due to the application of chemicals that protect crops. Across the world, the production of pesticides has increased at the rate of 11% per annum from a minuscule 0.2 million tonnes in 1950 to a whopping 5 million tonnes by 2000 (FAO 2017) (Fig. 4.2).

Pesticides directly benefit the health of the public by decreasing the chances of vector-borne diseases, viz. Zika virus and malaria that are carried by mosquitoes to humans, but a not foreseen consequence of checking mosquitoes may be serious harm to honeybees which are chief pollinators, and that will ultimately lead to

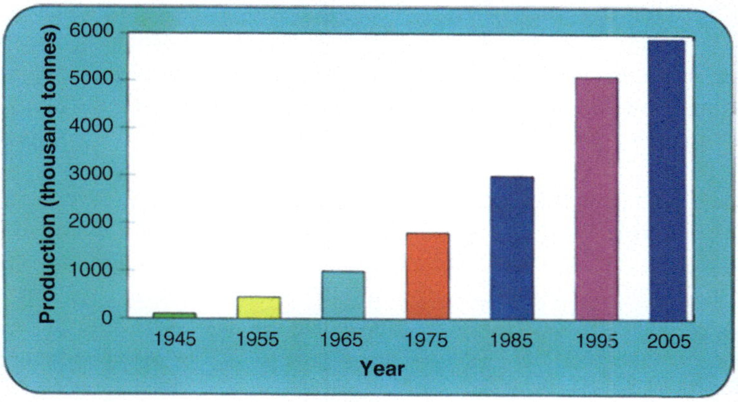

Fig. 4.2 Overall production of synthetic pesticides (based on FAO statistics 2017)

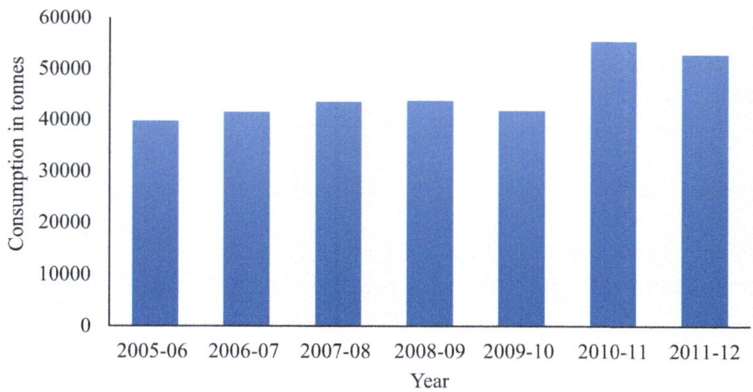

Fig. 4.3 Pesticide consumption in India from 2005–2006 to 2011–2012 (data source: Indiastat.com)

reduction in crop yield (Bonner and Alavanja 2017). The global population of circa 7 billion is estimated to grow at a rapid rate of whopping 70 million per year, and it is expected to increase by 2050 at an approximate rate of 30% to 9.2 billion (United Nations 2015). India also shows an increase in pesticide consumption (Fig. 4.3).

Increasing fear of crop loss due to pests has to be encountered with state-of-the-art crop protection methods. These may be either through the use of chemicals, IPM (integrated pest management), training farmers, etc. There has been a dramatic increase in the consumption of pesticides since 1960, and it coincides with an increase in the average yield of maize, wheat and rice, which form the major source of human nutrition. If pesticides would not have been present, the prices of crops would have soared with diminishing production and increasing prices and farmers would become less competitive in global markets for primary commodities (Popp et al. 2013).

4.3 Synthetic Pesticides

Pesticides are chemicals which are used to destroy, repel, prevent or mitigate any pest ranging from animals, insects, weeds to microorganisms (Grube et al. 2011). There are number of benefits and risks associated with the use of pesticides (Fig. 4.4).

Pesticide is an umbrella term for any substance used to kill pests. Pesticides may be insecticides to kill insects, rodenticides to kill rodents, herbicides to kill weeds, etc. They may be obtained from natural sources, for example, pyrethrin insecticide is obtained from *Chrysanthemum*, azadirachtin from neem tree, etc. They may also be prepared synthetically such as pyrethroids that kill insects (Mitra et al. 2011). Persistent organic pollutants (POPs) are highly toxic chemical substances for human health as well as environment. They are toxic, persistent and bioaccumulative in nature and can be transported to very large distances (Weber et al. 2008). POPs may lead to endocrine, neurologic, behavioural, reproductive and developmental problems. They can cross the placental membrane, as well as can be found in breast milk. People get exposed to these pesticides primarily through food grown on POP-contaminated soil. There are a number of POPs out of which 12 have been identified for causing detrimental effect on the ecosystem and humans, and they can be classified under 3 heads:

1. **Pesticides**: aldrin, chlordane, DDT, dieldrin, endrin, heptachlor, hexachlorobenzene, mirex, toxaphene
2. **Industrial chemicals**: hexachlorobenzene, polychlorinated biphenyls (PCBs)
3. **By-products**: hexachlorobenzene, polychlorinated dibenzo-*p*-dioxins and polychlorinated dibenzofurans (PCDD/PCDF) and PCBs (Stockholm Convention on initial POPs)

Chemical pesticides have been extensively used for controlling phytopathogens of several kinds. So their consumption has been ever increasing. To reduce crop loss due to pests and keep pace with increasing food demands, their consumption has

Fig. 4.4 Benefits and potential risks of pesticides

Benefits

- Protection of Crops
- Preservation of Foods
- Disease Control
- Preservation of Materials

Risks

- Impact on Ecosystems and Environment
- Contamination of food chain
- Bioaccumulation and Biomagnification
- Non-targeted effects

increased by leaps and bounds by the end of the last century (Prashar and Shah 2016).

Soil microflora is one of the main constituents of agriculture ecosystems which plays a crucial role not only in basic soil processes but also in increasing soil fertility, thereby increasing crop productivity. Microbial assays have revealed that pesticides like methamidophos inhibit fungi and increase gram-negative bacteria (Wang et al. 2008). Atradex has been responsible for a change in the structural and functional condition of aquatic microflora (Littlefield-Wyer et al. 2008). A number of studies have been done in the past by using microbial enzymatic assays to know the effects of various pesticides. Chlorpyrifos is responsible for a dose-dependent inhibition in the activity of the enzyme urease, whereas the application of both endosulfan and chlorpyrifos has been observed to cause a decrease in bacterial dehydrogenase and soil cellulase (Kumar 2011). Cypermethrin did not have any significant effect on soil enzymatic activities of urease, dehydrogenase, acid phosphatase or β-galactosidase (Zhuang et al. 2011). However, this methodology has some lacunae because contributions made by all the enzymatic activities from dead microorganisms cannot be ignored. A reduction in the population of bacteria fixing nitrogen and nitrifying bacteria upon exposure to methylpyrimifos and chlorpyrifos has been observed (Martinez-Toledo et al. 1992). The process of nitrification was found to be very susceptible to the application of fenamiphos, and its inhibition enhanced at increased concentrations of fenamiphos (Cáceres et al. 2008). Metabolites of chlorpyrifos, viz. 3,5,6-trichloro-2-methoxypyridine and 3,5,6-trichloro-2-pyridinol, and metabolites of quinalphos such as quinoxaline-2-thiol and 2-hydroxyquinoxaline have been responsible for suppressing the process of bacterial ammonification in loamy sandy soil and sandy loam (Menon et al. 2004).

An intriguing fact is that biopesticides which are considered to be safer have widely been used, but azadirachtin, which is one of the common biopesticides obtained from neem plant, is inhibitory to several soil microbes as well as their enzymatic activity at high doses (Gopal et al. 2007). Cypermethrin and chlorpyrifos are two popular pesticides used in Indian agriculture. It has been found that bacterial diversity is accountable for the maintenance of functional resilience and stability within the community after the disturbance caused by contaminants (Girvan et al. 2005). The ultimate effect of these synthetic pesticides is severe on active communities of bacteria; it is revealed by the abundance of transcripts of 16s rRNA. The key players in maintaining plant and soil health are active microbial population. The number of transcripts of 16s rRNA was found to be profoundly low in almost all pesticide-treated soils. Azadirachtin was also found to be similar in effects as compared to synthetic pesticides which are considered to be environmentally 'safe' (Singh et al. 2015).

By the year 1961, there was a huge surge in the use of pesticides. But after 1962, there was a deceleration in the development of new pesticides as it drew public attention towards environmental degradation caused by them. In 1962, the world witnessed a new revolution against pesticides in the form of a book named *Silent Spring* which was authored by an American scientist named Rachel Carson. This book mainly focused on the imprudent use of DDT (dichlorodiphenyltrichloroethane) which caused the sudden death of non-targeted organisms. *Silent Spring* silenced the process of research and development in the field of pesticides. It opened new avenues in integrated pest management (IPM). It is a method in which parasites or biological predators are used for controlling pests. Overuse of pesticides has repercussions on the aquatic ecosystem as well. It has been found that it has led to a serious threat to aquatic fauna including salmon. They have also been seen to affect macroinvertebrates and primary producers (Macneale et al. 2010).

4.4 Types of Synthetic Pesticides

Pesticides may be classified by several criteria like mode of action, toxicity, functional groups and chemical classes (Fig. 4.5) (Garcia et al. 2012). The most active ingredient of pesticides is either organic or inorganic (lime, sulphur, copper, ferrous sulphate, copper sulphate, etc.) (Gunnell et al. 2007). Chemicals present in organic pesticides are less soluble and more complex than inorganic pesticides (Debost-Legrand et al. 2016). There are several heads under which pesticides may be classified. Some of them have been mentioned in Fig. 4.5 (Table 4.1).

Organic pesticides may be divided into two subgroups—natural, those pesticides that are obtained from naturally occurring substances, and synthetic, those pesticides that are prepared from chemical substances. There are many classes of pesticides, some of which have been discussed.

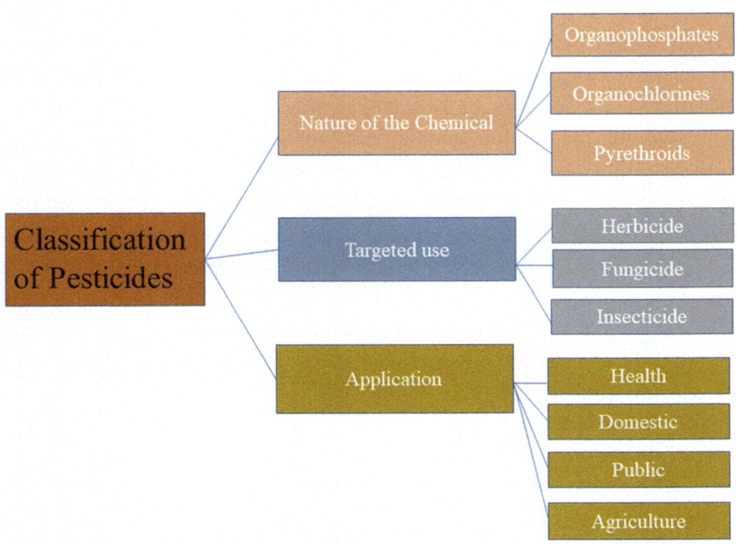

Fig. 4.5 One of the classifications of pesticides (Jayaraj et al. 2016)

Table 4.1 Classification on the basis of toxicity criteria

Type	Level of toxicity	LD_{50} for the rat (mg kg^{-1} body weight)	
		Oral	Dermal
Ia	Extremely hazardous	<5	<50
Ib	Highly hazardous	5–50	50–200
II	Moderately hazardous	0–2000	200–2000
U	Unlikely to present acute hazard	5000 or higher	

Source: WHO (2009)

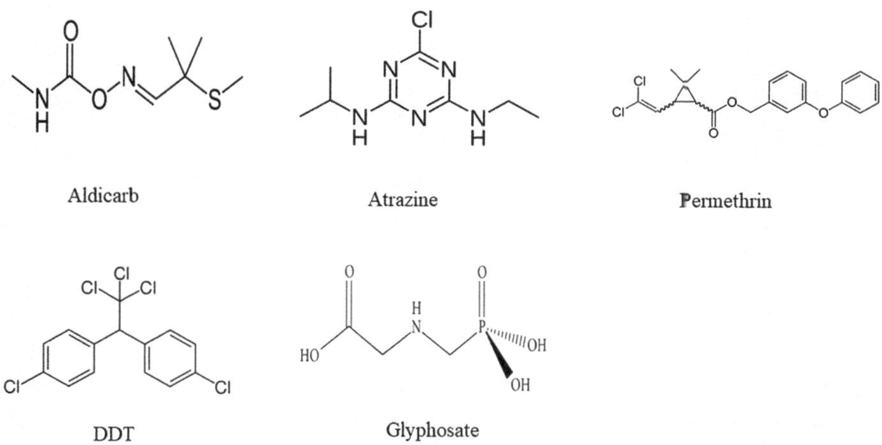

Fig. 4.6 Chemical structure of some popular pesticides (image source: Google)

4.4.1 Organochlorine Pesticides

The most popular organochlorine pesticide is DDT (Fig. 4.6), which is an insecticide; rampant use of this chemical has raised several human and environmental concerns (Alewu and Nosiri 2011; Turusov et al. 2002; van den Berg 2009). Dicofol, dieldrin, heptachlor, endosulfan and methoxychlor are some of the organochlorine pesticides.

It is a ubiquitous chemical that is believed to be present in every living organism, and it is mainly accumulated in the fatty tissues (Turusov et al. 2002; Rodríguez-Alcalá et al. 2015). Both DDT and its metabolite p,p′-dichlorodiphenyldichloroethylene (DDE) also function as endocrine disruptors (Turusov et al. 2002). Both DDT and DDE are also responsible for neurodevelopmental effects in children in utero (Eskenazi et al. 2006). Also, a recent study has found DDE to be responsible for lipid dysfunction of the liver in rats (Rodríguez-Alcalá et al. 2015).

4.4.2 Organophosphorus Pesticides

These pesticides were promoted as an alternative to organochlorines (Jaga and Dharmani 2003). These encompass a wide variety of pesticides; most popular pesticide in this type is glyphosate. Glyphosate is a Group M herbicide manufactured by Monsanto. This chemical helps farmers to get rid of weeds. It is also applied to clear parks, water bodies, railway tracks, etc. In most of the countries, it is used as a preharvest desiccant. It is applied on a standing crop to ease harvesting. Glyphosate kills the plants by blocking the enzyme that assists in the synthesis of essential nutrients and amino acids. But the adverse impacts of glyphosate are acute poisoning, liver and kidney damage, gut microflora alteration, endocrine disruption, cancer, immune system dysfunction, etc. Exposure to this chemical leads to a reduction in soil microbial population. Over and above, this herbicide is also a chelating agent that binds to nutrients, making them unavailable to plants.

Other known pesticides in this class are dimethoate, parathion and malathion (Mnif et al. 2011; McKinlay et al. 2008; Gasnier et al. 2009). A study done recently showed possible linkage of genetically modified (GM) crops, glyphosate and health deterioration in the USA (Nicolopoulou-Stamati et al. 2016).

4.4.3 Carbamates

These pesticides are organic in nature, and they are derivatives of carbamic acid (NH_2COOH). Carbamate esters are the functional group present in these insecticides; their modus operandi is by reversible inactivation of the enzyme acetylcholinesterase (Goel and Aggarwal 2007). Ziram, carbofuran and aldicarb are some of the chemical pesticides falling in this class of pesticides. These are also known to affect endocrine function (Mnif et al. 2011; Goad et al. 2004). In vitro studies have proved them to be cytotoxic and genotoxic on hamster ovarian cell (Soloneski et al. 2015).

4.5 Toxicity of Synthetic Pesticides

The toxicity of pesticides can be seen in lower to higher organism which includes soil microorganisms, plants, insects, fish, birds and other wildlife. The toxicity of pesticides can be known by their ability to harm or cause illness. They can contaminate soil, water, air, vegetation and turf and directly food and produce harmful impact on the living organism. Synthetic pesticides have longer persistence in nature, so their impacts on soil and to the microflora also last longer. Pesticides get accumulated in living organism and transform from one medium to another via different routes (Fig. 4.7). Their toxicity varies depending upon the chemical formula, dosage, the target organism and the chemical composition of the soil. The toxicity of synthetic pesticides also relies on soil biotic and abiotic factors as

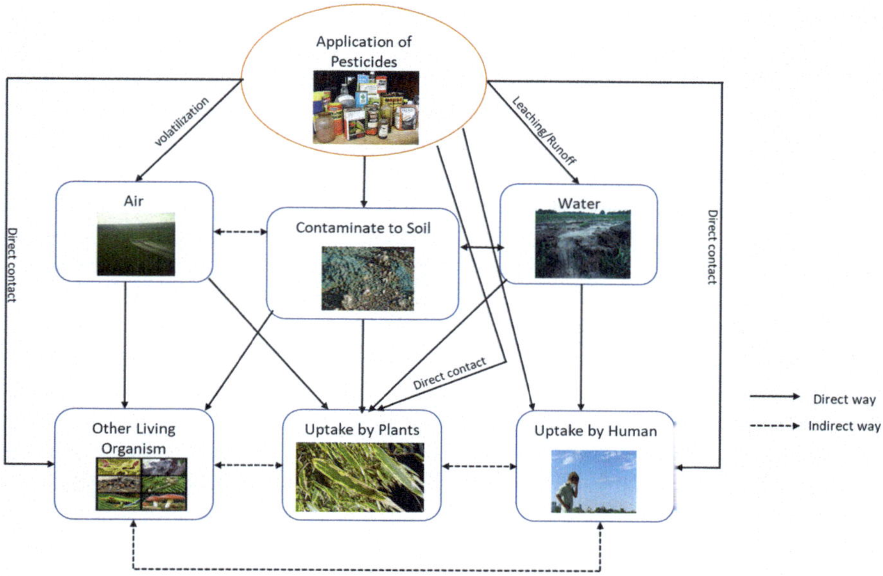

Fig. 4.7 Contamination of pesticides via various routes and their effect on living organisms

well (Prashar and Shah 2016). The dose of pesticide is one of the important factors which determines the toxicity level.

Pesticide toxicity is classified into two categories as acute and chronic. Acute toxic pesticides have immediate effect/effect within the first exposure on the organism. For example, insecticides are generally classified as more acutely toxic pesticide group. Chronic toxic pesticides refer to pesticides which have no immediate effect, but they can affect the organism after repeated application for a long period of time. However, both types of effects depend on dose of exposure.

4.5.1 Toxicity to the Animals

Nowadays application of pesticides is very common in agriculture practices to control the disease and increase the production, so the demand for synthetic pesticides is also increasing steadily. These chemicals enter the ecosystem through direct or indirect way and cause problem to the animals including human beings. Human beings may get hurt by pesticides through poisoning and injuries. The toxic effect of pesticides ranges from minor (like skin and eye irritation) to more severe symptoms such as nausea, strong headache and dizziness. Some organophosphate pesticides may result in neurological disorder, coma and even death (Damalas and Koutroubas 2016). Synthetic organochlorine pesticides like benzene hexachloride (BHC) and dichlorodiphenyldichloroethylene (DDE) cause prickling sensation of the mouth, nausea, headache, lethargy, vomiting, anxiety, cyst in hands, confusion, dizziness, psoriasis, skin rashes, eczema, anaemia, itching, anorexia, leucoderma, hyperexcitability and nervous tension to human beings (Subramaniam and Solomon 2006; Klaassen and Amdur 2013; Jayaraj et al. 2016; Alharbi et al. 2018). Organochlorine pesticides have been detected in human breast milk from many parts of the world, and when infants are exposed to these pesticides through breastfeeding, they cause severe health effects (Lee et al. 2013; Klinčić et al. 2014; Ferronato et al. 2018; Makris et al. 2019). Another study shows that children lose their learning ability when coming in contact with dioxins (Lee et al. 2007). When exposed to higher concentration of p,p′-DDE and PCBs, there is a chance of attention deficit hyperactivity disorder (ADHD) (Sagiv et al. 2010). Exposing parents to DDT, HCB, β-BHC and mirex causes decrease in birth weight of newborn babies (Guo et al. 2014).

In recent years the population of pollinators including bees has declined worldwide (Lebuhn et al. 2013). It has been estimated that more than 70% of the food crops are pollinated by bees, and if the pollinators decline, it leads to a decrease in food production (Kumar et al. 2018b). There are a number of factors which are considered to be responsible for the declining honeybee population, and application of synthetic pesticides is the most important (Van et al. 2013). Several studies have confirmed that honeybee gets affected by different pesticides like prochloraz and deltamethrin which act as an insecticide, even if the used dose is 50 times lesser than recommended (Kumar et al. 2018b; Fulton et al. 2019). Pesticides like clothianidin, neonicotinoids, imidacloprid, etc. are responsible for the reduction of and the disturbed foraging activity of pollinators (Schneider et al. 2012; Gill et al. 2012; Feltham et al. 2014). Pesticides affect the reproduction of fish-eating birds, and it was found that the eggshells become thin which easily break during the nesting period (Cooke 1973; Jayaraj et al. 2016; Orłowski et al. 2016). Another study shows that pesticides like imidacloprid affect the chicken embryo development; lead to failure of retraction of the yolk sac, head enlargement, visceral ectopia and limb defects; and lower the crown-rump length and weight of chick (Hussein and Singh 2016). Pesticides pose adverse impacts to the amphibian like frogs by damaging tail, leading to gonadal abnormalities, decreasing snout-vent length at metamorphosis and taking longer time to metamorphosis (Howe et al. 2004).

4.5.2 Toxicity to Plants

The increasing application of pesticides is affecting the plant kingdom in various ways. Pesticides enter into the plants and alter the physiology such as reducing the net photosynthetic rate, growth and biomass production which ultimately reduced the productivity and yield. It is shown that after the use of methyl parathion, the lettuce photosynthesis rate declines, which results in its reduced diameter and head weight (Henry et al. 2011). In addition, there are a number of studies that have highlighted whether the pesticides negatively affect the metabolic and physiological activities such as pigment content, photosynthesis and modification in reproductive organ in plants (Garcia et al. 2003; Saladin et al. 2003; Petit et al. 2008). Pesticides are also responsible for altering the carbon and nitrogen metabolism (Saladin et al.

2003). Youngman et al. (1990) show that treating cotton crops with methyl parathion leads to reduced boll retention and lint weight. When plants uptake the insecticides, it can lead to inhibition of the electron transport in photosystem II and simultaneously decreased adenosine triphosphate production/impaired stomatal conductance (Toscano et al. 1982; Henry et al. 2011).

Fungicides like epoxiconazole and azoxystrobin increase the antioxidative potential of wheat (*Triticum aestivum*) responsible for protecting the plants from the harmful active oxygen species; as a result, it delays senescence (Wu and Von 2002). Treating the plants with fungicides interrupts the capacity of RuBisCO carboxylation, decreases the ribulose-1.5-bisphosphate reproduction and lowers RuBisCO content, resulting in decreased CO_2 assimilation and attributing to stomatal and nonstomatal effect (Nason et al. 2007; Petit et al. 2008). Further studies show that with the application of fungicides in cucumber (*Cucumis sativus*) and apple (*Malus domestica*), the net CO_2 assimilation decreases with altered intercellular CO_2 concentration and stomatal conductance (Untiedt and Blanke 2004; Xia et al. 2006). Studies also observed that pesticides like benzimidazoles, dithiocarbamate and triazole decreased the maximal quantum efficiency of PSII (Fv/Fm) and quantum yield of PSII (Φ_{PSII}), which causes reduction in photochemical quenching (qP) (Van Iersel and Bugbee 1996; Untiedt and Blanke 2004; Xia et al. 2006). Pesticides like strobilurin fungicides may be responsible for choking the electron transport between PSI and PSII (Nason et al. 2007).

A study conducted by exposing the different relevant dosages (0.01, 0.05, 0.1 and 0.5 ppm) of deltamethrin pesticides on maize seeds shows that all seedling growth, stomatal dimension, stomatal density and pigments like chlorophyll a, chlorophyll b, carotenoids and total chlorophyll were reduced with increasing the deltamethrin concentration. The higher dosage effects the radical length of maize decrease up to 61% ($P < 0.05$) (Duran et al. 2015).

4.5.3 Toxicity to the Soil

The impacts of pesticide application are very serious for soil due to its direct contact with soil. The pesticides are persistent and resistant to degradation, which causes accumulation in soil and sediments and transfers into plant and animals (Taiwo 2019). The organochlorine pesticides travel from a long distance via wind and are deposited in soil and water which later affect the ecosystem (Leadprathom et al. 2009). Soil acidification is a major challenge in the present day. There are a number of factors responsible for the acidification process such as acidic precipitation. In addition, scientist believes that synthetic pesticides are also responsible for the soil acidification which affects soil fertility (Maddela and Venkateswarlu 2018). Long-term study confirms that use of synthetic pesticides affects the soil's physical, chemical and biological properties which disturb the structural and functional aspect of living organisms present in soil (Nicholson and Hirsch 1998; Yang et al. 2000; Gundi et al. 2005; Srinivasulu and Ortiz 2017; Madhavi et al. 2019). Soil is a dynamic ecosystem which contains an enormous number of macro and micro flora and fauna. These living organisms such as bacteria, viruses, nematodes, protozoa, algae, fungi and small insects and rodents are directly affected by the application of pesticides. Studies show that an advisable amount of bayleton, dimethoate and imazetapir application causes negative effect on growth, nitrogen fixation, viability and nif genes of beneficial bacteria like *Azotobacter chroococcum* and *Azotobacter vinelandii* (Khudhur and Askar 2013). It is also confirmed that synthetic pesticides affect the indole acetic acid (IAA) and P solubilisation of soil bacteria which regulate the plant growth (Madhaiyan et al. 2006).

Pesticides lower the pH of topsoil which may affect the microbial activity, lead to a decrease in the legume nodulation and cause nitrogen deficiency (Abd-Alla et al. 2000). The continuous application of specific pesticides brings permanent changes in soil microbial diversity which indirectly affect the other soil-dependent organisms. For example, the application of imidacloprid at high concentration reduced the bacterial population in soil and also changed the dominant bacteria (Moghaddam et al. 2011). There are number of studies which found the differential effect of various pesticides on soil microorganism which has been tabulated in Table 4.2.

Once the pesticides affect the soil microorganism, they directly and indirectly affect the soil flora and fauna which are available in the affected sites. In the case of small organism, for example, beneficial arthropod (spiders and beetles) and annelids (earthworms) get poisoned by the pesticides (Desneux et al. 2007; Korenko et al. 2016; Santos et al. 2017; Yu et al. 2019). Studies confirmed that the application of pesticides reduced the diversity of beneficial insects and worms which are responsible for controlling the pest population and improving the soil quality in sustainable ways (Santos et al. 2017; Jiang et al. 2018; Yu et al. 2019). The effect of synthetic pesticides on different soil organisms has been depicted in Table 4.3.

4.6 Impact of Synthetic Pesticides on Non-targeted Beneficial Insects

Beneficial insects encompass fungi, bacteria, insects, mites and organisms that predate on and parasitise the pest species. Rampant use of pesticides affects these non-targeted beneficial insects. Sometimes the importance of these organisms has been underestimated. Non-targeted beneficial insects play a crucial role in pollination and pest control in a natural way. Both the processes get adversely affected due to the application of pesticides. They reduce the fecundity and alter the suitability of hosts for parasitising; decreased emergence of parasitoids from sprayed host eggs may lead to direct mortality. This has brought serious concern for pollinators and biological control agents. Reduction in the number of natural enemies may lead to even more serious repercussions (Ndakidemi et al. 2016).

In-depth studies have been done on honeybees primarily because of the fact that their biochemical systems are well known. Effect of pesticides on enzymatic activity during and after the application of pesticides has been studied through experiments.

Table 4.2 Effects of different synthetic pesticides on soil microorganism

Pesticides	Dosage	Host	Remarks	References
Kitazin, glyphosate, hexaconazole, metalaxyl, atrazine, quizalofop, monocrotophos, fipronil and imidacloprid	2400, 900, 1800, 1500, 900, 1200, 2100, 1800 and 2700 μg mL^{-1}, respectively	Bacteria (*Azotobacter vinelandii)*	At higher conc. it caused cellular/ structural damage and decreased the cell viability	Shahid et al. (2019)
Benomyl, dithane Z-78, abamectin benzoate and oxydemeton-methyl	10, 25, 50, 75 and 100 μg g^{-1} in soil	*Azospirillum* sp. in groundnut soil	The population of bacteria increased up to conc. 10–25 μg g^{-1} and decrease when the conc. of pesticides was from 50 to 100 μg g^{-1}	Madhavi et al. (2019)
Trifluralin	Concentration were 0, 0.84, 8.4 and 84 mg kg^{-1}	Soil microorganism	Study shows it enhances the bacterial abundance and reduced the fungal abundance. Trifluralin disturbs the community structure of soil bacteria. It inhibits the abundance of AOA, AOB and NO3−-N content	Du et al. (2018)
Hexaconazole	0.6 and 6 mg kg^{-1}	On soil microbes (red and black soil of paddy field)	Half-lives of two doses on red soil were 270–345 days and 122–135 days for black soil. Both dosages did not affect the nitrogen-fixing bacteria and total bacterial diversity but transitorily reduced the total bacterial population in both the soil	Ju et al. (2017)

(continued)

Table 4.2 (continued)

Pesticides	Dosage	Host	Remarks	References
Chlorpyrifos	0.5 kg ha^{-1} in rice field	Soil microbes and nematodes	Within field dose, there is no significant effect on microbes. Nitrifying microbial groups found sensitive to chlorpyrifos	Kumar et al. (2017)
Fomesafen	100 times recommended dose	Soil microbial community	Adversely impact soil microbial activity and reduced respiratory quotient (qCO_2, Q_R). The abundance of nitrogen-fixing genes was decreased with the application of higher fomesafen	Wu et al. (2014)
Bayleton 50, dimethoate and imazetapir	2 g L^{-1} bayleton 50, 100 g L^{-1} imazetapir and 3 g L^{-1} dimethoate. Incubated for 10 days at 28°C	*Azotobacter chroococcum* and *Azotobacter vinelandii* with wheat (*Triticum aestivum*)	Imazetapir shows no effect on bacteria but bayleton 50 and dimethoate have effect on the growth rate and on nif genes	Khudhur and Askar (2013)
Metribuzin, glyphosate, hexaconazole, metalaxyl, kitazin, imidacloprid and thiamethoxam	Recommended at a higher dose	Bacteria (*Mesorhizobium* sp.)	Reduced all plant growth-promoting traits. Application of hexaconazole three times showed maximum stress on siderophore biosynthesis and decreased 2,3-dihydroxy benzoic acid (DHBA) and salicylic acid by 47% and 40%	Ahemad and Khan (2012)
Afugan, Gramoxone, Brominal, Selecron and sumi oil	3.0, 0.75, 0.6, 0.9 and 0.13 mg kg^{-1} in soil, respectively	*Arbuscular mycorrhiza* of legumes cowpea, common bean and lupin	The result shows accumulation of N, P and K was lower in pesticide-treated plants compared to control plants	Abd-Alla et al. (2000)

Table 4.3 Effects of synthetic pesticides on soil flora and fauna

Pesticides	Dosage	Host	Remarks	References
Abamectin, atrazine, chlorpyrifos and lambda-cyhalothrin	Mixtures of all pesticides in different compositions as binary, ternary, quaternary and quinquenary	Earthworm (*Eisenia fetida*)	Two quaternary and one quinquenary mixture record synergistic effects on *E. fetida*	Yu et al. (2019)
Clothianidin	Lethal and sublethal doses	Seven-spotted lady beetle (*Coccinella septempunctata* L.)	It was more toxic to larvae and adults of *C. septempunctata* affect their survival and development	Jiang et al. (2018)
Azadirachtin, thiamethoxam chlorpyrifos, etofenprox, phosmet and imidacloprid	Recommended maximum field concentration by the Brazilian agriculture ministry	Larvae, eggs and adults of lady beetles (*Hippodamia convergens*)	All tested insecticides decrease the larval hatching rates. Chlorpyrifos and phosmet reduced larval and adult survival and slowed the larval growth time. Chlorpyrifos also reduced fertility fecundity and longevity of egg	Santos et al. (2017)
Spinosad and λ-cyhalothrin	Spinosad 0.06% and λ-cyhalothrin 0.01% at different temp. 10, 17, 25 and 31 °C	*Philodromus* spiders	A maximum number of spiders died at 31 °C due to λ-cyhalothrin. The prey capture rate reduced and functional response of spiders changed due to the pesticides at all temperatures	Michalko and Košulič (2016)
Eight different herbicides	Recommended concentrations	Wolf spider (*Pardosa agrestis*)	All herbicides negatively affected the predatory activity of spider	Korenko et al. (2016)
Clothianidin	Lethal and sublethal doses	Monarch butterflies (*Danaus plexippus*)	It affects the larva size and works as stressor to monarch populations	Pecenka and Lundgren (2015)

Cypermethrin (pyrethroid) and fenitrothion (organophosphorus) caused a decrease in acetylcholinesterase (AChE) and Na^+/K^+ ATPase activities (Bendahou et al. 1999). This enzymatic inhibition resulted in downgraded glycaemic disorders.

Na^+/K^+ ATPase is a transmembrane enzyme which releases energy that is important for cell metabolism, and it also maintains cell potential. So inhibiting Na^+/K^+ affects a wide range of cellular functions.

Earthworms provide a host of ecosystem services, be it providing aeration to soil or proper mixing of soil nutrients and making the soil fertile, etc. These services can be severely disrupted by using pesticides. Pesticides severely hamper their reproductive capability and growth, increase individual mortality, disrupt their enzymatic activity and also alter their individual behaviour like feeding rate (Pelosi et al. 2014). Affecting these key organisms will jeopardise our entire agricultural system.

Arbuscular mycorrhizal fungi (AMF) are another group of organisms that are beneficial for the host plant. Application of pesticides may have detrimental effects on the establishment of arbuscular mycorrhiza and other fungal communities in the soil. Pesticides affect AMFs in a dose-dependent fashion (Hage-Ahmed et al. 2019). Application of pesticides like glyphosate is known to reduce AMF spore viability as well as root colonisation. Although glyphosate is rapidly degraded by microorganisms, toxic metabolite of AMF, i.e. AMPA (aminomethylphosphonic acid), is known to decrease the spore viability and its property to colonise roots (Druille et al. 2013)

4.7 Conclusion

The application of pesticides disturbs the overall ecosystem function because it kills several non-targeted organisms that form the food chain and ultimately, the food web. The effect of pesticides on non-targeted flora and fauna has been a concern worldwide. Pesticide affects organisms of all taxonomic groups. Several beneficial arthropods like pollinating insects get affected due to pesticides. Protecting agricultural crops from pests is a must, and it dates back to antiquity when both natural and synthetic pesticides were used to control pests. Current agricultural practices are relying heavily on the use of synthetic pesticides, thus making a shift from traditional methods of agriculture. Current practices in agriculture are responsible for several negative effects on wildlife, natural environment as well as human health. Current agriculture has to face problems such as food security, the health risk from lethal chemical pesticides, population growth, resistance from pesticides, climate change, and degradation of the natural environment. Chronic exposure to synthetic pesticides requires extensive research. The exposure of soil and non-targeted flora and fauna to these synthetic pesticides reveals the real complex nature of the problem. These pesticides are potential carcinogens and endocrine disruptors. These problems emphasise upon the need for a new concept in agriculture as soon as possible.

Acknowledgement Author Kuldeep Bauddh is thankful to the Science and Engineering Research Board (SERB) for the award of Research Grant (EEQ/2017/000476). Ankit is thankful to the University Grant Commission for granting Junior Research Fellow (UGC-JRF).

References

Abd-Alla MH, Omar SA, Karanxha S (2000) The impact of pesticides on arbuscular mycorrhizal and nitrogen-fixing symbioses in legumes. Appl Soil Ecol 14(3):191–200

Ahemad M, Khan MS (2012) Effects of pesticides on plant growth promoting traits of Mesorhizobium strain MRC4. J Saudi Soci Agri Sci 11(1):63–71

Aislabie J, Lloyd-Jones G (1995) Bacterial degradation of pesticides. Aust J Soil Res 33:925–942

Alewu B, Nosiri C (2011) Pesticides and human health. In: Stoytcheva M (ed) Pesticides in the modern world—effects of pesticides exposure. In Tech, Rijeka, pp 231–250

Alharbi OM, Khattab RA, Ali I (2018) Health and environmental effects of persistent organic pollutants. J Mol Liq 263:442–453

Al-Zaidi AA, Elhag EA, Al-Otaibi SH, Baig MB (2011) Negative effects of pesticides on the environment and the farmers awareness in Saudi Arabia: a case study. J Anim Plant Sci 21 (3):605–611

Arora S, Arora S, Sahni D (2019) Pesticides use and its effect on soil bacteria and fungal populations, microbial biomass carbon and enzymatic activity. Curr Sci 116(4):643–649

Bendahou N, Bounias M, Fleche C (1999) Toxicity of cypermethrin and fenitrothion on the hemolymph carbohydrates, head acetylcholinesterase, and thoracic muscle Na+, K+-ATPase of emerging honeybees (*Apis mellifera mellifera. L*). Ecotox Environ Safe 44(2):139–146

Bending GD, Lincoln SD, Edmondson RN (2006) Spatial variation in the degradation rate of the pesticides isoproturon, azoxystrobin and diflufenican in soil and its relationship with chemical and microbial properties. Environ Pollut 139:279–287

Bonner MR, Alavanja MC (2017) Pesticides, human health, and food security. Food Energy Secur 6(3):89–93

Bowles TM, Acosta-Martínez V, Calderón F, Jackson LE (2014) Soil enzyme activities, microbial communities, and carbon and nitrogen availability in organic agroecosystems across an intensively-managed agricultural landscape. Soil Biol Biochem 68:252–262

Bünemann EK, Schwenke GD, Van Zwieten L (2006) Impact of agricultural inputs on soil organisms—a review. Soil Res 44(4):379–406

Cáceres TP, He W, Megharaj M, Naidu R (2008) Effect of insecticide fenamiphos on soil microbial activities in Australian and Ecuadorean soils. J Environ Sci Heal B 44(1):13–17

Chowdhury A, Pradhan S, Saha M, Sanyal N (2008) Impact of pesticides on soil microbiological parameters and possible bioremediation strategies. Indian J Microbiol 48(1):114–127

Cooke AS (1973) Shell thinning in avian eggs by environmental pollutants. Environ Pollut 4 (2):85–152

Pimentel D, Burgess M (2011) Small amounts of pesticides reaching target insects. Environ Dev Sustain 14:1–2

Damalas C, Koutroubas S (2016) Farmers' exposure to pesticides: toxicity types and ways of prevention. Toxics 4(1), 1

Debost-Legrand A, Warembourg C, Massart C (2016) Prenatal exposure to persistent organic pollutants and organophosphate pesticides, and markers of glucose metabolism at birth. Environ Res 146:207–217

Desneux N, Decourtye A, Delpuech JM (2007) The sublethal effects of pesticides on beneficial arthropods. Annu Rev Entomol 52:81–106

Devi NL (2020) Persistent organic pollutants (POPs): environmental risks, toxicological effects, and bioremediation for environmental safety and challenges for future research. In: Saxena G, Bharagava R (eds) Bioremediation of industrial waste for environmental safety. Springer, Singapore, pp 53–76

Dhakal R, Singh DN (2019) Biopesticides: a key to sustainable agriculture. Int J Pure App Biosci 7 (3):391–396

Druille M, Cabello MN, Omacini M, Golluscio RA (2013) Glyphosate reduces spore viability and root colonization of arbuscular mycorrhizal fungi. Appl Soil Ecol 64:99–103

Duran RE, Kilic S, Coskun Y (2015) Response of maize (*Zea mays L. saccharata Sturt*) to different concentration treatments of deltamethrin. Pest Biochem Physiol 124:15–20

Du P, Wu X, Xu J, Dong F, Liu X, Zheng Y (2018) Effects of trifluralin on the soil microbial community and functional groups involved in nitrogen cycling. J Hazard Mater 353:204–213

Eskenazi B, Marks AR, Bradman A, Fenster L, Johnson C, Barr DB (2006) In utero exposure to dichlorodiphenyltrichloroethane (DDT) and dichlorodiphenyldichloroethylene (DDE) and neurodevelopment among young Mexican American children. Pediatrics 118:233–241

FAO (2009) Feeding the world in 2050. World agricultural summit on food security 16–-18 November 2009. Food and Agriculture Organization of the United Nations, Rome

FAO (2017). Available at http://www.fao.org/faostat/en/#home

Feltham H, Park K, Goulson D (2014) Field realistic doses of pesticide imidacloprid reduce bumblebee pollen foraging efficiency. Ecotoxicology 23(3):317–323

Ferronato G, Viera MS, Prestes OD, Adaime MB, Zanella R (2018) Determination of organochlorine pesticides (OCPs) in breast milk from Rio Grande do Sul, Brazil, using a modified QuEChERS method and gas chromatography-negative chemical ionisation-mass spectrometry. Int J Environ Anch 98(11):1005–1016

Fulton CA, Hartz KEH, Fell RD et al (2019) An assessment of pesticide exposures and land use of honey bees in Virginia. Chemosphere 222:489–493

Garcia PC, Rivero RM, Ruiz JM (2003) The role of fungicides in the physiology of higher plants: implications for defense responses. Bot Rev 69(2):162–172

Garcia FP, Ascencio SYC, Oyarzun JCG (2012) Pesticides: classification, uses and toxicity. Measures of exposure and genotoxic risks. J Res Environ Sci Toxicol 1(11):279–293

Gasnier C, Dumont C, Benachour N (2009) Glyphosate-based herbicides are toxic and endocrine disruptors in human cell lines. Toxicology 262:184–191

Gill RJ, Ramos RO, Raine NE (2012) Combined pesticide exposure severely affects individual-and colony-level traits in bees. Nature 491:105–108

Girvan MS, Campbell CD, Killham K, Prosser JI, Glover LA (2005) Bacterial diversity promotes community stability and functional resilience after perturbation. Environ Microbiol 7 (3):301–313

Goad ER, Goad JT, Atieh BH, Gupta RC (2004) Carbofuran-induced endocrine disruption in adult male rats. Toxicol Mech Methods 14:233–239

Goel A, Aggarwal P (2007) Pesticide poisoning. Natl Med J India 20:182–191

Gopal M, Gupta A, Arunachalam V, Magu SP (2007) Impact of azadirachtin, an insecticidal allelochemical from neem on soil microflora, enzyme and respiratory activities. Bioresour Technol 98(16):3154–3158

Grube A, Donaldson D, Kiely T, Wu L (2011) Pesticides industry sales and usage. US EPA, Washington, DC

Gundi VA, Narasimha G, Reddy BR (2005) Interaction effects of insecticides on microbial populations and dehydrogenase activity in a black clay soil. J Environ Sci Heal 40(2):69–283

Gunnell D, Eddleston M, Phillips MR, Konradsen F (2007) The global distribution of fatal pesticide self-poisoning: systematic review. BMC Public Health 7:357–371

Guo H, Jin Y, Cheng Y (2014) Prenatal exposure to organochlorine pesticides and infant birth weight in China. Chemosphere 110:1–7

Hage-Ahmed K, Rosner K, Steinkellner S (2019) Arbuscular mycorrhizal fungi and their response to pesticides. Pest Manag Sci 75(3):583–590

Henry RS, Johnson WG, Wise KA (2011) The impact of a fungicide and an insecticide on soybean growth, yield, and profitability. Crop Prot 30(12):1629–1634

Hou B, Wu L (2010) Safety impact and farmer awareness of pesticide residues. Food Agr Immunol 21(3):191–200

Howe CM, Berrill M, Pauli BD (2004) Toxicity of glyphosate-based pesticides to four North American frog species. Environ Toxicol Chem 23(8):1928–1938

Hussein M, Singh V (2016) Effect on chick embryos development after exposure to neonicotinoid insecticide imidacloprid. J Anat Soc India 65(2):83–89

Jaga K, Dharmani C (2003) Sources of exposure to and public health implications of organophosphate pesticides. Rev Panam Salud Publica 14:171–185

Jayaraj R, Megha P, Sreedev P (2016) Organochlorine pesticides, their toxic effects on living organisms and their fate in the environment. Interdiscip Toxicol 9(3–4):90–100

Jiang J, Zhang Z, Yu X, Ma D, Yu C, Liu F, Wei M (2018) Influence of lethal and sublethal exposure to clothianidin on the seven-spotted lady beetle, *Coccinella septempunctata* L. (Coleoptera: Coccinellidae). Ecotoxicol Environ Saf 161:208–213

Ju C, Xu J, Wu X, Dong F, Liu X, Tian C, Zheng Y (2017) Effects of hexaconazole application on soil microbes community and nitrogen transformations in paddy soils. Sci Total Environ 609:655–663

Kah M, Brown CD (2006) Adsorption of ionisable pesticides in soils. In: Whitacre DM (ed) Reviews of environmental contamination and toxicology. Springer, New York, pp 149–217

Kalia A, Gosal SK (2011) Effect of pesticide application on soil microorganisms. Arch Agron Soil Sci 57(6):569–596

Khudhur M, Askar KA (2013) Effect of some pesticides on growth, nitrogen fixation and nif genes in *Azotobacter chroococcum* and *Azotobacter vinelandii* isolated from soil. J Toxicol Environ Health Sci 5(9):166–171

Klaassen CD, Amdur MO (eds) (2013) Casarett and Doull's toxicology the basic science of poisons, vol 1236. McGraw-Hill, New York

Klinčić D, Romanić SH, Sarić MM (2014) Polychlorinated biphenyls and organochlorine pesticides in human milk samples from two regions in Croatia. Environ Toxicol Pharmacol 37(2):543–552

Korenko S, Niedobová J, Kolářová M (2016) The effect of eight common herbicides on the predatory activity of the agrobiont spider *Pardosa agrestis*. BioControl 61(5):507–517

Kumar S (2011) Fluctuation of soil bacterial dehydrogenase activity in response to the application of endosulfan and chlorpyrifos. Cell Tissue Res 11(2):2847–2851

Kumar U, Berliner J, Adak T, Rath PC, Dey A, Pokhare SS, Jambhulkar NN, Panneerselvam P, Kumar A, Mohapatra SD (2017) Non-target effect of continuous application of chlorpyrifos on soil microbes, nematodes and its persistence under sub-humid tropical rice-rice cropping system. Ecotoxicol Environ Saf 135:225–235

Kumar S, Singh R, Behera M, Kumar V, Sweta RA, Kumar N, Bauddh K (2018a) Restoration of pesticide contaminated sites through plants. In: Pandey VC, Bauddh K (eds) Phytomanagement of polluted sites. Elsevier, Amsterdam, pp 313–327

Kumar S, Joshi PC, Nath P (2018b) Impacts of insecticides on pollinators of different food plants. Entomol Ornithol Herpetol 7(2):1–6

Leadprathom N, Parkpian P, Satayavivad J (2009) Transport and deposition of organochlorine pesticides from farmland to estuary under tropical regime and their potential risk to aquatic biota. J Environ Sci Heal B 44(3):249–261

Lebuhn G, Droege S, Connor EF (2013) Detecting insect pollinator declines on regional and global scales. Conserv Biol 27(1):113–120

Lee DH, Jacobs DR, Porta M (2007) Association of serum concentrations of persistent organic pollutants with the prevalence of learning disability and attention deficit disorder. J Epidemiol Community Health 61(7):591–596

Lee S, Kim S, Lee HK (2013) Contamination of polychlorinated biphenyls and organochlorine pesticides in breast milk in Korea: time-course variation, influencing factors, and exposure assessment. Chemosphere 93(8):1578–1585

Littlefield-Wyer JG, Brooks P, Katouli M (2008) Application of biochemical fingerprinting and fatty acid methyl ester profiling to assess the effect of the pesticide atradex on aquatic microbial communities. Environ Pollut 153:393–400

Macneale KH, Kiffney PM, Scholz NL (2010) Pesticides, aquatic food webs, and the conservation of Pacific salmon. Front Ecol Environ 8:475–482

Maddela NR, Venkateswarlu K (2018) Impact of acephate and buprofezin on soil amylases. In: Maddela NR, Venkateswarlu K (eds) Insecticides soil microbiota interactions. Springer, Cham, pp 41–48

Madhaiyan M, Poonguzhali S, Hari K (2006) Influence of pesticides on the growth rate and plant-growth promoting traits of *Gluconacetobacter diazotrophicus*. Pest Biochem Physiol 84:143–154

Madhavi A, Anuradha B, Rangaswamy V (2019) Influence of pesticides on *Azospirillum sp.* population and its nitrogen fixation in groundnut (*Arachis hypogaea L.*) soils. J Adv Microbiol 15(4):1–12

Makris G, Chrousos GP, Anesiadou S (2019) Serum concentrations and detection rates of selected organochlorine pesticides in a sample of Greek school-aged children with neurodevelopmental disorders. Environ Sci Pollut Res 26(23):23739–23753

Martinez-Toledo MV, Salmeron V, Gonzalez-Lopez J (1992) Effect of the insecticides methylpyrimifos and chlorpyrifos on soil microflora in an agricultural loam. Plant Soil 147 (1):25–30

McKinlay R, Plant JA, Bell JNB, Voulvoulis N (2008) Endocrine disrupting pesticides: implications for risk assessment. Environ Int 34:168–183

Menon P, Gopal M, Prasad R (2004) Influence of two insecticides, chlorpyrifos and quinalphos, on arginine ammonification and mineralizable nitrogen in two tropical soil types. J Agr Food Chem 52(24):7370–7376

Michalko R, Košulič O (2016) Temperature-dependent effect of two neurotoxic insecticides on predatory potential of *Philodromus* spiders. J Pest Sci 89(2):517–527

Mitra A, Chatterjee C, Mandal FB (2011) Synthetic chemical pesticides and their effects on birds. Res J Environ Toxicol 5(2):81–96

Mnif W, Hassine AIH, Bouaziz A (2011) Effect of endocrine disruptor pesticides: a review. Int J Environ Res Public Health 8:2265–2203

Moghaddam NS, Zakaria MP, Omar D (2011) Effects of imidacloprid on the biodiversity of soil microbes in selected soils of Malaysia. In: Proceedings 2nd international conference on environmental science and development IPCBEE, vol 4

Mulla SI et al (2020) Organophosphate pesticides: impact on environment, toxicity, and their degradation. In: Saxena G, Bharagava R (eds) Bioremediation of industrial waste for environmental safety. Springer, Singapore, pp 265–290

Mustapha MU, Halimoon N, Johar WLW, Shukor MYA (2019) An overview on biodegradation of carbamate pesticides by soil bacteria. Pertanika J Sci Technol 27(2):547–563

Nason MA, Farrar J, Bartlett D (2007) Strobilurin fungicides induce changes in photosynthetic gas exchange that do not improve water use efficiency of plants grown under conditions of water stress. Pest Manag Sci 63(12):1191–1200

National Research Council (2000) The future role of pesticides in US agriculture. National Academy Press, Washington, DC

Ndakidemi B, Mtei K, Ndakidemi PA (2016) Impacts of synthetic and botanical pesticides on beneficial insects. Agri Sci 7(06):364

Nicholson PS, Hirsch PR (1998) The effects of pesticides on the diversity of culturable soil bacteria. J Appl Microbiol 84(4):551–558

Nicolopoulou-Stamati P, Maipas S, Kotampasi C, Stamatis P, Hens L (2016) Chemical pesticides and human health: the urgent need for a new concept in agriculture. Front Pub Health 4:148

Oerke EC (2005) Crop losses to pests. J Agr Sci 144:31–43

Oerke EC, Dehne HW (2004) Safeguarding production—losses in major crops and the role of crop protection. Crop Prot 23:275–285

Oerke EC, Dehne HW, Schonbeck F, Weber A (1994) Crop production and crop protection—estimated losses in major food and cash crops. Elsevier, Amsterdam, 808 pp

Ogut S (2019) Genotoxic effects of pesticides. J Environ Ecol 20(1):224–229

Orłowski G, Hałupka L, Klimczuk E (2016) Shell thinning due to embryo development in eggs of a small passerine bird. J Ornithol 157(2):565–572

Pecenka JR, Lundgren JG (2015) Non-target effects of clothianidin on monarch butterflies. Sci Nat 102(3–4)19:1–4

Pelosi C, Barot S, Capowiez Y, Hedde M, Vandenbulcke F (2014) Pesticides and earthworms. A review. Agron Sustain Dev 34(1):199–228

Petit AN, Fontaine F, Clément C (2008) Photosynthesis limitations of grapevine after treatment with the fungicide fludioxonil. J Agric Food Chem 56(15):6761–6767

Popp J, Pető K, Nagy J (2013) Pesticide productivity and food security. A review. Agron Sustain Dev 33(1):243–255

Prashar P, Shah S (2016) Impact of fertilizers and pesticides on soil microflora in agriculture. In: Lichtfouse E (ed) Sustainable agriculture reviews, vol 19. Springer, Cham, pp 331–361

Rodríguez-Alcalá LM, Sá C, Pimentel LL et al (2015) Endocrine disruptor DDE associated with a high-fat diet enhances the impairment of liver fatty acid composition in rats. J Agri Food Chem 63(42):9341–9348

Sabarwal A, Kumar K, Singh RP (2018) Hazardous effects of chemical pesticides on human health-cancer and other associated disorders. Environ Toxicol Phar 63:103–114

Sagiv SK, Thurston SW, Bellinger DC (2010) Prenatal organochlorine exposure and behaviors associated with attention deficit hyperactivity disorder in school-aged children. Am J Epidemiol 171(5):593–601

Saladin G, Magné C, Clément C (2003) Effects of fludioxonil and pyrimethanil, two fungicides used against *Botrytis cinerea*, on carbohydrate physiology in *Vitis vinifera L.* Pest Manag Sci 59 (10):1083–1092

Santos KFA, Zanardi OZ, deMorais MR (2017) The impact of six insecticides commonly used in control of agricultural pests on the generalist predator Hippodamia convergens (Coleoptera: Coccinellidae). Chemosphere 186:218–226

Sarnaik SS, Kanekar PP, Raut VM, Taware SP, Chavan KS, Bhadbhade BJ (2006) Effect of application of different pesticides to soybean on the soil microflora. J Environ Biol 37 (2):423–426

Schneider CW, Tautz J, Grünewald B (2012) RFID tracking of sublethal effects of two neonicotinoid insecticides on the foraging behavior of *Apis mellifera*. PLoS One 7(1):e30023

Shahid M, Zaidi A, Ehtram A (2019) In vitro investigation to explore the toxicity of different groups of pesticides for an agronomically important rhizosphere isolate *Azotobacter vinelandii*. Pest Biochem Physiol 157:33–44

Singh B, Kaur A (2018) Control of insect pests in crop plants and stored food grains using plant saponins: a review. LWT 87:93–101

Singh S, Gupta R, Sharma S (2015) Effects of chemical and biological pesticides on plant growth parameters and rhizospheric bacterial community structure in Vigna radiata. J hazard mater 291:102–110

Soloneski S, Kujawski M, Scuto A, Larramendy ML (2015) Carbamates: a study on genotoxic, cytotoxic, and apoptotic effects induced in Chinese hamster ovary (CHO-K1) cells. Toxicol In Vitro 29(5):834–844

Srinivasulu M, Ortiz DR (2017) Effect of pesticides on bacterial and fungal populations in Ecuadorian tomato cultivated soils. Environ Proc 4(1):93–105

Subramaniam K, Solomon J (2006) Organochlorine pesticides BHC and DDE in human blood in and around Madurai, India. Indian J Clin Biochem 21(2):169–172

Taiwo AM (2019) A review of environmental and health effects of organochlorine pesticide residues in Africa. Chemosphere 220:1126–1140

Toscano NC, Sances FV, Johnson MW, Lapré LF (1982) Effect of various pesticides on lettuce physiology and yield. J Econ Entomol 75(4):738–741

Turusov V, Rakitsky V, Tomatis L (2002) Dichlorodiphenyltrichloroethane (DDT): ubiquity, persistence, and risks. Environ Health Persp 110(2):125–128

United Nations (2015) World population projected to reach 9.7 billion by 2050. Available at http://www.un.org/en/development/desa/news/population/2015-report.html

Untiedt R, Blanke MM (2004) Effects of fungicide and insecticide mixtures on apple tree canopy photosynthesis, dark respiration and carbon economy. Crop Prot 23(10):1001–1006

Van SJP, Simon-Delso N, Goulson D (2013) Neonicotinoids, bee disorders and the sustainability of pollinator services. Curr Opin Environ Sustain 5(3–4):293–305

van den Berg H (2009) Global status of DDT and its alternatives for use in vector control to prevent disease. Environ Health Perspect 117(11):1656–1663

Van Iersel MW, Bugbee B (1996) Phytotoxic effects of benzimidazole fungicides on bedding plants. J Am Soc Hortic Sci 121(6):1095–1102

Wang MC, Liu YH, Wang Q et al (2008) Impacts of methamidophos on the biochemical, catabolic, and genetic characteristics of soil microbial communities. Soil Biol Biochem 40(3):778–788

Weber R, Gaus C, Tysklind M et al (2008) Dioxin and POP-contaminated sites—contemporary and future relevance and challenges overview on background, aims and scope of the series. Environ Sci Pollut Res 15:363–393

WHO (World Health Organization) (2009) The WHO recommended classification of pesticides by hazard

Wu YX, Von TA (2002) Impact of fungicides on active oxygen species and antioxidant enzymes in spring barley (*Hordeum vulgare L.*) exposed to ozone. Environ Pollut 116(1):37–47

Wu X, Xu J, Dong F, Liu X, Zheng Y (2014) Responses of soil microbial community to different concentration of fomesafen. J Hazard Mater 273:155–164

Xia XJ, Huang YY, Wang L (2006) Pesticides-induced depression of photosynthesis was alleviated by 24-epibrassinolide pretreatment in *Cucumis sativus* L. Pest Biochem Physiol 86(1):42–48

Yang YH, Yao J, Hu S (2000) Effects of agricultural chemicals on DNA sequence diversity of soil microbial community: a study with RAPD marker. Microb Ecol 39(1):72–79

Youngman RR, Leigh TF, Kerby TA, Toscano NC, Jackson CE (1990) Pesticides and cotton: effect on photosynthesis, growth, and fruiting. J Econ Entomol 83(4):1549–1557

Yu Y, Li X, Yang G (2019) Joint toxic effects of cadmium and four pesticides on the earthworm (*Eisenia fetida*). Chemosphere 227:489–495

Zhuang R, Chen H, Yao J et al (2011) Impact of beta-cypermethrin on soil microbial community associated with its bioavailability: a combined study by isothermal microcalorimetry and enzyme assay techniques. J Hazard Mater 189(1–2):323–328

Ecological Consequences of Genetically Modified Crops on Soil Biodiversity

5

Aditya Kumar Jha, Sukalyan Chakraborty, Khushbu Kumari, and Kuldeep Bauddh

Contents

Abstract

Uncontrolled population raised an important concern of food security in front of the entire world. To increase the global food productivity, numerous technological interventions have been done, and development of genetically modified organisms (GMOs) especially crops was considered as a novel approach. Genetically modified crops (GMCs) are designed in such a way to fight against both biotic and abiotic stresses and to give a better yield than conventional crops. Several GMCs have been adopted in many countries of the world and many more are under trial. Like many other technologies, use of GMCs in the natural fields is found to have some ecological complications like their impacts on non-target

A. K. Jha · S. Chakraborty (✉)
Department of Civil and Environmental Engineering, Birla Institute of Technology, Mesra, Ranchi, Jharkhand, India
e-mail: sukalyanchakraborty@bitmesra.ac.in

K. Kumari · K. Bauddh
Department of Environmental Science, Central University of Jharkhand, Ranchi, Jharkhand, India

K. Bauddh et al. (eds.), *Ecological and Practical Applications for Sustainable Agriculture*, https://doi.org/10.1007/978-981-15-3372-3_5

organisms, loss of biodiversity, flow of transgene, etc. In this chapter, efforts have been done to explore the concept and role of GMCs in agriculture. Further, adverse impacts of GMCs on the environment, soil biodiversity, and non-target plants and animals have also been discussed thoroughly.

Keywords

Genetically modified crops · Biodiversity · Non-target organisms · Soil organisms · Gene flow

5.1 Introduction

Genetically modified crops (GMCs) are crops produced by alteration of DNA of normal crops to introduce some special properties which are beneficial for crops or farmers or their consumers in some way. Some of these properties include resistance to some pests and diseases, tolerance to herbicides, or increased yield.

Arable lands are decreasing and degrading throughout the world. Desertification is a common phenomenon. Soil erosion and decreasing organic content in soil and changing rain pattern have a great impact on crop yield. Human population reached 7.59 billion in 2018 and is expected to reach 8.5 billion in 2030 and 9.7 billion in 2050 (Goujon 2018). This unchecked growth in human population has a drastic effect on world biodiversity. Expansion in human activities is a major threat with 85% of all species described in red list having a direct negative effect. Urbanization and increasing cultivable land surface requirement lead to deforestation and loss of biodiversity (Lynch et al. 2004). Irreversible damages have been done to ecosystem and biodiversity in a large part of earth. Feeding this increasing population with a limited cultivable area is going to be a major challenge. Herbicide-tolerant crop increased yield by 9% and insect-resistant crop increased yield by 25%. Farmers who have adopted GMCs have 69% higher profits. In developing countries, 14% increase in yield has been recorded. GM maize, cotton, and soybean contribute to \$160 billion per year (Paul et al. 2018). Crop yield must increase to prevent food insecurity in the future. Herbicide and insect resistance, drought tolerance, as well as increase in the number and size of grains and rate of growth of leaves and plant can be achieved by genetic modification of crops (Paul et al. 2018).

Commercialization of GMCs started in 1996, when 1.7 million hectares area was cropped in six countries. In 2011, the number of countries increased to 29. Before the start of 2015, 357 GMCs were approved. In 2016, 180 million hectares of land were cropped with GMCs (Pellegrino et al. 2018). But this created controversies in countries like Japan, Korea, Brazil, Taiwan, Egypt, and the European Union. There are different regulations in different countries. Health risk, environmental hazards, and economic impacts are the main concerns. GMCs can provide higher nutritional content, shelf life, better flavor, texture, color, and processing characteristics.

GMCs are used as fodder for animals, which later give us milk, egg, or meat. Some study indicates that some DNA may not degrade in their body and reach various organs and tissue of animals. This may sometimes be harmful for them and for humans, if consumed. Organic and agricultural wastes from GMCs when they go into water bodies have impact on invertebrates living in the water body (Davies et al. 2003; Benjamin 2008). Leaching from such wastes releases toxins, which are harmful for insects and non-target organisms (Aktar et al. 2009). Reduction of pesticides may sometimes lead to increase in minor pests which are not affected by the proteins present in the GMCs.

5.2 Genetically Modified Crops

GMCs have their genetic material (DNA) altered in a way that cannot occur naturally. It was in 1944 that scientists discovered about the transfer of genetic material from one species to another. First commercialized in 1996, just in 20 years in 2016, 13% of world's arable land was cropped with GMCs. Soybean, maize, cotton, canola, eggplant, alfalfa, sugar beet, papaya, squash, etc. are some of the major commercialized GMCs.

The gene(s) for certain trait is inserted into the cell of crop by direct transfer of DNA or indirect transfer by bacterial vehicle or direct edition of genomic DNA. Proteins of interest are identified and isolated in the genes. They are then transferred into the host through recombinant DNA technology (Lusser and Davies 2013). They are then regenerated through tissue culture. Genetic modification can be achieved through non-genetic engineering and genetic engineering. Genetic engineering methods comprise targeted and non-targeted manipulation. Targeted manipulation of genetic material is done by methods such as cloning, viral vectors, electroporation, microinjection, or microprojectile bombardment. Non-targeted manipulation of genetic material is done by methods such as random DNA mutations induced through either physical or chemical mutagens. Biological delivery methods depend on the use of plasmids which are frequently used (Actis et al. 1999; Zamana et al. 2018). Non-biological-based delivery methods such as particle bombardment/biolistics, microinjection, chemical transduction, electroporation, silicon carbide fiber, and liposome-mediated transfer.

Before commercialization of any GMC, comparison of its compositional, phenotypic, molecular, and agronomic trait is done with its near isoline crop. Nutritional content, toxicity, and allergic effect need to be investigated for any adverse effect on the health of humans or animals (Key et al. 2008; Kramkowska et al. 2013). In case the composition of the food or feed produced from GMC changes, or there is an indication of such, study on experimental animals is necessary before its commercial release. Dry matter intake, average daily weight gain, feed conversion, and reproduction are generally studied on experimental animal. The aim of monitoring plan is to study if there can be adverse effect of the new GMC before its release on the environment or human and/or animal health and monitor and identify any unintended effect after release. If any food crop is produced with altered nutritional value or composition or any health claim, post-market monitoring is required (Dawkar et al. 2018).

5.3 Role of Genetically Modified Crops in Agriculture

Crop yield must increase to prevent food insecurity in the future. Herbicide- and pest-resistant crops, drought-resistant crops, as well as increased number and size of grains and rate of growth of leaves and plants/crops can be achieved by genetic modification of crops. There is a need to produce more food on limited land area. GMCs are necessary for the sufficient production of food for ever-increasing human population, feed for animals, and fuel for the world in the future. This brings economic benefits to farmers and reduces the chances of failure of crops. GMCs have proved beneficial to small- and large-scale farmers and consumers. The benefits may be economical, agronomical, or/and environmental in nature depending upon the trait introduced. The changes in trait increase profit of farmers by less expenditure on herbicides and pesticides and also improved yield. This helps the government as the amount spent on subsidies decreases and this can be reassigned for targeted environmental measures. Reduction in the consumption of insecticides benefits both economically and environmentally. At present 99% of GMCs are designed for HT or/and IR. No high-yield crops have been commercialized with complex trait. Present modifications have no to very little impact on increasing the yield of crops.

Genetic modification of crops is the fastest adopted technology of modern times. Genes from soil bacterium *Bacillus thuringiensis* are introduced to the crops to act against pests. They are resistant to disease from insects and actively control pest population. Such crops cover over 100 million hectare of cropped area by 2017 as per estimation (ISAAA 2017). They also guard the crops against fungal infection, as fungi grow in area damaged by pests. This reduces the mycotoxins produced by some species of fungi in crops, which is an added advantage. A benefit of $167.8 billion was analyzed due to GM technology from 1996 to 2015. This was achieved by 72% increase in yield and grain production and 28% from reduced cost of production. Literature-based analysis revealed an average decrease of 37% pesticide use and 39% reduction in pesticide use. Production of Bt maize increased by 0.25–0.75 tons per hectare over its non-transgenic line in 1998–1999. Use of Bt cotton increased profit of farmers by 50% in India. Risks associated with GMCs have been proven from low to negligible.

Water is the most important abiotic factor which limits the productivity of crops. Similar to herbicide-tolerant and pest-resistant crops, drought-tolerant crops have been designed using genetic modification. For this transgene is introduced into the crop, which directly affects its water use. Few researches have been published on drought-tolerant crops. DroughtGard® is the only drought-tolerant corn available in the market. Other similar crops include Verdeca's HB4 soybean, which is in the process of approval, and PT Perkebunan Nusantara Xi'sNXI-4T sugarcane, which is being cultivated in Indonesia. But the lack of commercial products does not mean lack of research and development. The main problem is the failure of such crops in commercial production. They do not exhibit much advantage over their non-GM counterparts. The developed GMCs are efficient in lab conditions, but field conditions change throughout the growing season. The duration and intensity of irrigating a crop vary from place to place, which reduces its efficiency from lab condition. But such crops are more tolerant to extended periods of water deficiency. Traits such as drought-tolerant crops do not always prove economical as they neither reduce input cost nor increase the yield. Higher price of seeds is one of its drawbacks. But it proves advantageous in years of drought. It may also help farmers grow such crops in areas with lower or irregular rainfall then generally required by non-GM variety. Biotech companies attempt to recover their investment made in product development by increasing the cost of seed. Thus, GMCs with small increase in yield do not prove beneficial to farmers. Price of GM crops is generally lower than their non-GM counterpart except for a few cases where the product is more nutritious or provides some medicinal values. There are many restrictions for the export of GM crops to various countries.

Changes in nutritional profile are desirable to consumers. Increase in healthy components and decrease in unhealthy components can increase demand of a product. Such crops are beneficial for both farmers and producers as consumers are ready to pay higher price for better products. Increase in omega-3 in oils, removal of harmful proteins, changes in vitamin and fat content, etc. can make a crop healthier. Such crops can help eradicate malnutrition in poor countries or serve as alternate to some medicinal supplements. Addition of medicinal properties into crop can help improve health condition of consumers. Such crops may serve as alternative to many medicines.

5.4 Environmental Consequences of GM Crops

GMCs can increase nutrition, texture, flavor, color, and yield. GMCs have the potential to bring environmental benefits. They can grow in difficult environmental condition and reduce insecticide consumption and other resources, thus proving economical. This is necessary to feed growing human population and farm animals. A meta-analysis suggests that GM technology has increased the yield by 22% and reduced pesticide use by 37%. Average profit of farmers cultivating GMCs increased by 68% globally. The area with GMCs in developing countries is greater than that of developed countries. In 2016, GMCs covered around 13% of total cropland. 18 million farmers from 26 countries were involved in GMC farming. Most of these countries are located in North America, South America, and Asia. The United States has the largest of GM-cropped area (39%), followed by Brazil (27%), Argentina (13%), India (6%), and a number of countries. Europe and Africa are least involved in GMCs due to limited public acceptance. GMCs have benefits for biodiversity and ecosystem. Reduced pesticide use has led to reduction in farmer pesticide poisoning. With increased productivity from GMCs, around 25 million hectare of cropland would have to be cultivated globally to maintain production, thus preventing deforestation.

Many traits in GMCs have been developed and tested, but most of them have not been commercialized due to lengthy regulatory policy and cautious policy attitudes.

But there are GMCs which are unidentifiable and undetectable. Simple genetic analysis cannot determine their trait. Event or elemental specific methods are necessary for their analysis. Such crops include unauthorized GMCs, GMCs produced by unintended cross-pollination between wild and GMC, unintended cross between two GMCs, and genuine crops. As these crops are unintended, no study on their nature is done. They can have varying properties. At times they can have bad effects on humans or animals consuming it, soil biodiversity, other plant species, and other organisms. They can have lower yield or lower-quality yield than their normal variety. They may be illegal to grow and put the farmer under problem. This overall can lead to economic damage to the farmer (de Vendomois et al. 2009). With growing GMC cultivation, there is a fear that it will lead to loss of genetic diversity of crops (Ammann 2005).

5.4.1 Impact on Human Health

Proteins are the main source of allergens in food and may lead to skin reactions, respiratory disorder, and problems in the circulatory system and can together lead to anaphylactic shock in human (Bernstein et al. 2003; Ladics et al. 2011; Kramkowska et al. 2013). GMCs have allergenic potential when the protein sequence is similar to an allergen sequence. Around 2% of adults and 6% of children suffer food allergies (D'Agnolo 2005; Bernstein et al. 2003; Kramkowska et al. 2013). Attention must be given to allergic reactions caused by the use of transgenic foods. The modified maize plant (StarLink) has a transgene, resistant to pesticides. This entered the market exclusively as animal feed, but was found in human food after some time. This led to food allergies in humans such as headache, nausea, and vomiting (Bernstein et al. 2003; Domingo 2007; Batista and Oliveira 2009; Kramkowska et al. 2013). Another case was reported in GM soy enriched in methionine, by using a gene from Brazil nut. An amino acid was obtained by synthesis of a gene that was isolated from Brazil nuts, which have been known to cause allergy in humans. In such a case, there is danger to the people who have allergy from such nuts and consume GM soy (Key et al. 2008; Kramkowska et al. 2013). It was reported that GM sunflower seeds also caused similar allergy (Kramkowska et al. 2013). Cellular synthesis of toxic substance in GMCs can increase case of neoplasm. One such case was reported in rapeseed oil derived from GMCs which entered the market in Spain in 1983, which led to death of many people. In the United States, transgenic tryptophan led to pain in muscles and joints and in some cases resulted in death in 1989. The health effects of GM foods have been investigated in different animals. MON810 maize which is pest resistant affected the liver, kidneys, intestine, and pancreas in rodents (Kramkowska et al. 2013). GM maize MON810 and MON863 which are resistant to insects and NK603 which is herbicide tolerant were toxic to the liver and kidneys of rats and could lead to palpable tumors with extended use (de Vendomois et al. 2009; Seralini et al. 2014). GM potatoes contain agglutinin and lectin which are toxic to the growth and development of mammals. This can cause deregulated cell division within gastric mucosal cells (Domingo 2007).

The use of GM crops as feed increased in cattle, pig, and poultry. Effects of GM crops have been studied on dairy, beef cattle, pig, and fish that include safety aspects.

No significant change in fat, protein, lactose, milk solids, non-fat, and total solid contents was noticed in the milk of dairy cows that were fed GM feeds. No proteins (Bt proteins Cry1Ac, BG-II cotton seed, Bt cotton seed, or Bt corn silage) were detected in the milk of such cows. Similarly herbicide-tolerant corn silage has no effect on milk yield or its composition in cow milk. Bt corn hybrids reported no impact on the yield of beef (Sidhu et al. 2000; Bohme et al. 2001; Folmer et al. 2002; Singh et al. 2003; Onkin et al. 2003; Lutz et al. 2005; Calsamiglia et al. 2007; Mohanta et al. 2010).

GM-based diet increased the performance significantly in case of pigs (Hyun et al. 2004; Broll et al. 2005). GM crops are insect protected, preventing them from fungal damage proving economical as feed. Bt cotton-feeding lambs reported no change in growth, hemato-biochemical profile, and histopathology. No specific DNA fragments were detected in tissues of rabbits fed with GM soybean meal (Tudisco et al. 2006; Tripathi et al. 2011).

GM corn and soybean, wheat, and canola are fed to poultry. Various feeding trials have been conducted on chicken. No effect on growth, feed efficiency, performance of bird, or presence of toxicity due to use of GM feed was reported in study by Scheideler et al. (2008) and Mejia et al. (2010). Use of GM feed on fish has increased. The results of feeding trials showed no interference of GMCs on economic traits of fish (Halver and Hardy 2002; Brown et al. 2003; Glencross et al. 2003; Sanden et al. 2005; Chainark et al. 2008).

5.4.2 Impact on Plants

Most GMCs in the market are first-generation crops, which have agronomic traits such as insect resistance and herbicide tolerance to improve crop production. First-generation GMCs are similar to non-GMC counterparts as they do not interfere with the metabolism of the plant cell. First-generation GM Bt maize contained lower level of mycotoxin (EFSA 2008; de Vos and Swanenburg 2018; Pellegrino et al. 2018). Second-generation GMCs can increase output traits like nutritional properties, with the aim of improving human and animal nutrition and health. This is achieved by increasing the level of desired substances like amino acids and fatty acids or by decreasing the level of undesirable substances like anti-nutrients such as phytate (D'Agnolo 2005; Batista and Oliveira 2009). Complex genetic modifications can lead to substantial changes in plant metabolism and composition. Introduction of nutrient precursor-like β-carotene, alteration of concentration of specific nutrients like amino or fatty acids, or fiber content, concentration of a nutrient enhancer like enzymes, increase in medicinal substances like prebiotics (insulin), and reduction of anti-nutritional factor like phytate or toxic substances like mycotoxins can be achieved in GMCs. Previous studies indicate better biological control of pests in GMCs (de Santis et al. 2018).

GM maize grain contains lower concentration of fumonisin (31%), mycotoxins (29%), and thricotecens (37%) than its non-GM counterpart (Pellegrino et al. 2018). This may be due to less attack of insect attack, as it is 59.6% less damaged compared to its non-GM isolines or near isolines. Fungal colonization on such crops is also very low (Pellegrino et al. 2018). Mycotoxins are toxic and carcinogens, and their high concentration causes market rejection of such grains or reduction in their market price. By contrast, there is lower mycotoxin content in GE maize grain.

Due to rapid climate change, salt concentration in soil, and reduction in rainfall, crops are more prone to disease and death caused by viruses, fungi, and bacteria. GMCs can tolerate and grow better in such situations. Management of weed and insects becomes a lot easier. It also reduces labor, fertilizers, herbicides, pesticides, and cost input in the crop and increases yield and decreases environmental pollution. GMCs can reduce the stress on irrigation system (Dawkar et al. 2018). GM rice has increased the yield of rice by 35%. Such increases were also observed in other crops. Manyfold increase in crop yield since the Green Revolution has saved millions of people from death during famine due to starvation. Increasing the quality, nutrients, taste, and storage period increased the usability and economics of the crop. Amflora potato was made by GM to produce useful composition of starches (Key et al. 2008; Kramkowska et al. 2013; Domingo 2007). GM canola and soybean can produce more healthy oil, and a GM rice strain produces grain with higher iron, zinc, and vitamins (de Vos and Swanenburg 2018). Pollens from GMCs can reach forest area and produce crossbreeds and superweeds. This may lead to decrease in nutrients or increase in toxins and allergens. This could also lead to wipeout of native breeds of plants available at a place and death of insects and microorganisms useful for other crops. At present, there is no proof, but microorganism could become more tolerant to antibiotics and damage non-GM crops at higher pace. Swapping one pest for another is possible.

5.4.3 Impacts on Soil Biodiversity

Soil biodiversity plays a major role in the growth of a crop. Nature of soil, amount of organic content, climate, rain, and biological activities in the soil determine if a crop can grow and thrive in that soil. Soil biota has a great role in the functioning of ecosystem by facilitating decomposition, nutrient recycling, and energy flow (Ammann 2005).

Farming requires a lot of water, and a lot of waste water is generated in the process. This leads to surface water and groundwater pollution. Unplanned cropping can destroy soil biodiversity and air and water quality. Agricultural activity is a major source of greenhouse gases. Use of GMCs has modified cropping pattern in many areas, due to higher income. Cultivation of one GMC repeatedly without mixing it with other crops has been reported. This tends to take certain nutrients from the soil which is not good for soil biodiversity. As GMCs can grow in difficult conditions, more land is being cropped with them. In some countries, natural areas and forests are slowly being replaced with such crops, leading to deforestation, soil erosion, habitat destruction, loss of topsoil cover, and wetland destruction. No direct consequences of GMCs have been reported on soil health. Some authors believe that they have good effects due to less use of chemical fertilizers and pesticides, less till, less irrigation costs, and less use of power. Such data have been reported on GM soybean, cotton, maize, etc. However, others believe that they have more negative effects due to continuous cropping, decrease in soil fertility, deforestation, impact on non-target organisms, crossbreeding from wild plants, increased toxicity, etc. (Phelinas and Choumert 2017). Use of GMC was reported to have low pressure on microorganisms, fauna, and biofunction by a group of experts (Ismail et al. 2012). GMCs do not have much direct impact on weed, but change in cropping pattern and techniques leads to reduction in weed concentration.

Decomposition of plant tissue determines the nutrients and amount of organic matter in soil. Plant biomass decomposition is determined by quality and quantity of lignin (Pellegrino et al. 2018). Study on GM maize indicates no change in lignin concentration in leaves and stalks from their isolines but significant differences during the loss of total biomass which includes all crop residues. This could be due to differences in proportion and composition of GMC and their isolines. A significant reduction was found in the number of amoebae, earthworms, flagellates, ciliates, and nematodes. Various GM crops have various effects on soil biodiversity. The range varies from no to small impact.

Potential interaction between GMCs and soil microbial community is not completely known. These interactions have the potential to change microbial biodiversity and affect the functioning of the ecosystem. There is a possibility of transfer of transgenes from GMCs to soil microorganisms through horizontal gene transfer. Cry proteins may be released from GMCs into the soil, whose degraded products would affect soil biodiversity. The composition of microbial community can be altered through their selective growth. Minor variation in their composition can affect the soil fertility and ecosystem functioning. Death of beneficial bacteria can reduce the nutrients in soil and affect the growth of crops.

By DNAse-sensitive process, competent bacteria are able to take up free DNA. Some of them can integrate into genome of bacteria, while others can form autonomous replicating element. In nature everything goes everywhere and gene flows from one organism to another. This has the potential to bring mutation in various organisms (Giovannetti et al. 2005).

There is a risk of plant invasiveness or dispersal of the plant itself in native area, gene transfer through pollen transfer, or horizontal flow to microorganisms. Elimination of herbicides and pesticides can help in the growth of superweed and pests, which may be harmful to other crops. Cry proteins from GMCs are eventually released into the soil ecosystem. Decomposition of litter eventually release them into soil ecosystem. Their interaction with soil community depends upon tillage. When there is no tillage, these proteins remain concentrated on the surface of the soil and get a chance to interact with low numbers of microorganisms. If there is conventional tillage, these proteins get distributed in a low concentration throughout the topsoil and get a chance to interact with a large number of microorganisms.

If the transgenes are introduced in a crop to reduce the impact of bacteria and fungus, by producing enzymes or compounds to kill them, such transgenes can lead to the formation of antibiotic-resistant bacteria (Ammann 2005).

The most important question is the way these proteins can interact with microorganisms and their effects on microbial biodiversity. Stable productivity in ecosystem depends upon stable biodiversity. A gram of fertile soil contains more than one billion bacteria, and this becomes a real risk. Released free DNA stay bounded to clay, available for take-up by bacteria. Life of free DNA in soil depends upon various biotic and abiotic factors. But the possibility of transfer of such DNA to soil microorganisms is very low and is restricted to microhabitats.

Sometimes the microbial community for one site was entirely different from the other with same crop. This suggests that the environment plays a great role in determining the impact of GMCs in field conditions. Results of a small-scale study at a greenhouse vary considerably with field observation (Dunfield and Germida 2004). Study shows that rhizospheric community of first transgenic line of potato was different from second transgenic line of potato and both were different from non-transgenic line. This may be due to different DNA structures in all three varieties and different cry proteins expressed by the transgenic line of potatoes.

Cry proteins of Bt corn which are insecticidal in nature remain active for at least 180 days. Thus, it can affect target and non-target soil organisms (Giovannetti et al. 2005). GM potatoes producing concanavalin A and *Galanthus nivalis* agglutinin led to the reduction of microbial activity by 10%. GM potatoes having cysteine proteinase inhibitors led to the reduction of microbial and fungal abundance by 23% at the end of the second year. Chitinase expressing rice led to extreme reduction in intraradical fungi population and tenfold increase in bacterial population (Giovannetti et al. 2005). Many GMCs have been designed to resist growth of non-selective herbicides, for weed control in field. Glyphosate- and glufosinate-resistant *Brassica napus* and corn did not alter the permanent composition as well as the diversity of the soil microorganisms. Cry1Ab gene in Bt corn led to higher lignin content in its stem. Such alteration in composition of crop can change the carbon, nitrogen, and phosphorus content in the soil after degradation. Arbuscular mycorrhizal fungi are found in the root zone of soil and produce biofertilizers for crops. Transgenes from GM potatoes spread up to 2 meters during the growing phase (Giovannetti et al. 2005). GMCs have the potential to inhibit growth of such fungi, ultimately leading to reduction in nutrients in soil. There are chances for escape of transgenes through pollens. There is a high risk of fertilization of non-GMCs with these pollens. They may produce unintended and unwanted GMCs with no information of their effects and risks. This may lead to the formation of superweed, or high growth of invasive/exotic species, affecting the growth of other species and disturb soil biodiversity. Gene transfer between GMCs and fungi has been observed in both co-culture and in planta system. This can be a risk to fungal pathogens as well as symbionts living in plant cell (Giovannetti et al. 2005).

Collectively it can be said that different GMCs affect the biodiversity in soil. But with comparison to environmental factors, these effects are minor and localized. GMCs may lead to accumulation of large amounts of compounds in the soil in the long term (Dunfield and Germida 2004). This can affect soil macroflora and microflora. Presence of cry proteins is directly harmful to many microorganisms and invertebrates. Their take-up by bacteria has much more risk. The GM bacteria and fungi may be harmful for crops as well as other organisms (Giovannetti et al. 2005). Cry proteins of decaying crops or modified microorganisms can travel from soil to water and cause widespread damages. Removal or addition of bacteria species such as rhizobacteria, pathogenic organisms, or key organisms performing nutrient cycling process can have a major impact on ecosystem sustainability. There is a need to address long-term effects on ecosystem and its sustainability (Dunfield and Germida 2004).

Soil is a complex system. Interaction of soil and various microorganisms living in it is difficult to monitor in short-term research. Precautionary approach must be taken before adopting a new GMC, and detailed lab-based and field study must be done before the commercialization of the product. Agricultural practices such as crop rotation, tillage, biopesticides, herbicide usage, and irrigation are capable of increasing the productivity and are environmental friendly (Giovannetti et al. 2005).

5.4.4 Impact on Non-targeted Insects

Food production is a pivotal ecosystem service among many others that depends on several biotic and abiotic components associated with it. Microorganisms and different species of insects are two important elements of an agro-ecosystem. Microorganisms are considered a highly dynamic factor (Welbaum et al. 2004; Yadav et al. 2018), and their presence or absence directly influences the crop productivity. Microorganisms perform several utmost required functions to the agro-ecosystems like enhancing plant nutrient bioavailability, soil organic matter (thereby soil carbon sequestration), mineralization and demineralization of soil organic matter, and other components and many more functions (Altieri 1999; Neher 1999; Shennan 2008; Ramesh et al. 2019). The functions of beneficial insects to the agri-ecosystems cannot be overlooked. Beneficial insects are considered as natural pollinators, enemies for pest, important component of a food chain, nutrient recycler, etc. (Fig. 5.1).

Although GMCs are being designed to overcome biotic and abiotic stresses and express the resistance toward pests, associated risk assessments must be performed at various stages with laboratory and field experiments which ensure no ecological harm, especially to the non-target organisms (Groot and Dicke 2002). Numerous beneficial insects are dependent on the plant species for water as well as nutrients, and if the toxins are released from the GMCs, the host species will be affected. Several studies observed the possible toxic effects of GMCs on non-target species, including scavengers, herbivores, soil fauna, etc. (Schuler 2000; Hilbeck 2002; O'Callaghan et al. 2005; Romeis et al. 2014; Lazebnik et al. 2017); however, the adverse impacts depend on the toxin released and duration of exposure. This leads to disbalances in the entire food chain and subsequently food web, especially of agricultural ecosystems. A non-target species is any species that the GMO was not planned to control, and therefore all the species whether plants or animals that are affected by a GMO are considered as non-target species. In agro-ecosystems the life of these species may also be affected by the cultivation of GMCs which ultimately

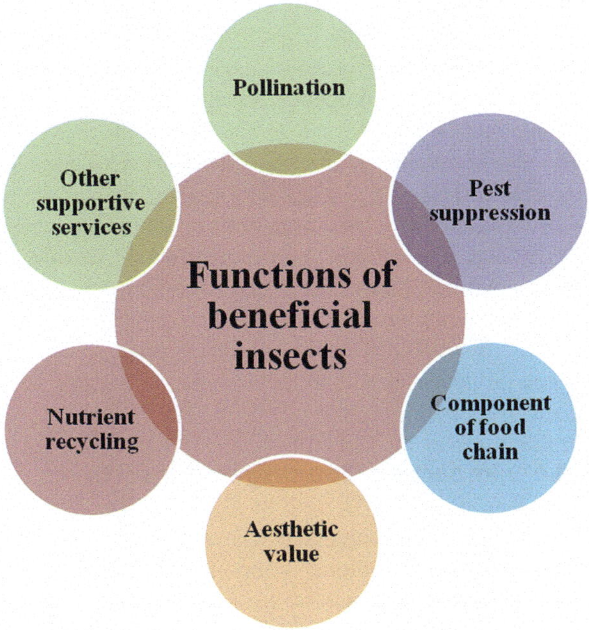

Fig. 5.1 Functions of beneficial insects in an ecosystem (Irtwange 2006; Nicholls and Altieri 2013; Macfadyen et al. 2015; Ndakidemi et al. 2016)

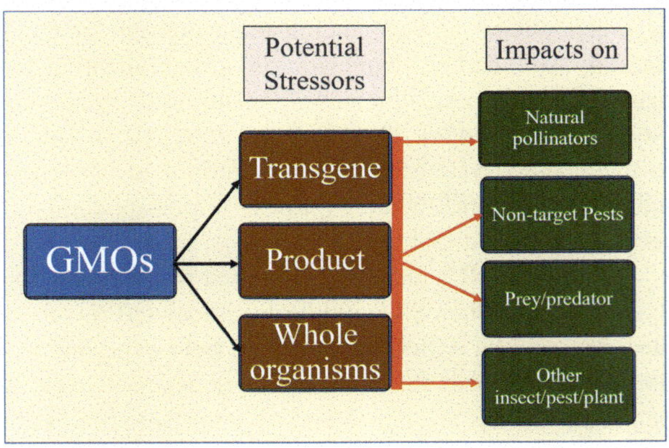

Fig. 5.2 Potential stressors and their impacts

disturb all ecosystem functions in soil (Dunfield and Germida 2004; Giovannetti et al. 2005). Several studies revealed that the toxin released from GMCs has Bt potentially insecticidal and may persist in the soil up to 180 days (Saxena and Stotzky 2000; Saxena and Stotzky 2001; Saxena et al. 2002). Most of the experimental studies conducted to assess the impacts of GMCs on other organism are laboratory based and need further and thorough investigations at field levels. The potential stressors which are considered as threats for a natural ecosystems are transgene, product, or entire genetically modified organisms (Fig. 5.2).

5.5 Risk of Gene Flow

Gene flow in the ecosystems is a natural process and commonly found in the traditional cropping systems. Gene flow may become an important type of environmental pollution in the near future, if GMCs will not be produced in a well-managed and well-controlled manner. Transgene releases from the GMCs are very specific and sometimes are fit into different ecosystems. Albeit the genetic modification in the agricultural crops is investigated as an approach to enhance growth and productivity, it has several environmental concerns, and gene flow of transgene in the ecosystems is a major drawback of this biotechnological intervention (Johannessen et al. 2006; Warwick et al. 2009; Moon et al. 2011; Verma 2013; Zhang et al. 2018). Genetically modified plants may affect the conventional cropping systems (Rieben et al. 2011; Chandler and Stevenson 2014; Herman et al. 2019). In the genetically modified cropping systems, the introduced gene may transfer to the traditional crops through the wind and pollinator insects (Giddings 2000; Jialin et al. 2013; Pu et al. 2014; Yan et al. 2015; Herman et al. 2019). A transgene released from genetically modified crops may contaminate other plant species through several ways like direct crop to crop flow, crop of weed gene flow, and crop to wild gene flow (Fig. 5.3).

Gene flow from GM crops can force a change in the natural environment due to enhanced genetic diversity (Gustafson et al. 2005) and that may lead to invasiveness, a pivotal environmental risk, due to its efficiency to change both the structure and function an ecosystem (Clark 2005). A gene flow from wheat to jointed *Aegilops cylindrica* (goat grass) was observed by Wang et al. (2001). It was reported by Martinez-Ghersa et al. (2003) that development of resistance of Roundup in two soybean weeds *Amaranthus rudis* and *Abutilon theophrasti* is an example of futuristic risk of gene flow through genetic technology. A review done by Gealy et al. (2003) observed the gene flow from GM rice to weedy rice; however, the hybridization rate of crop to weed was very low (approximately 0.01–1.0%).

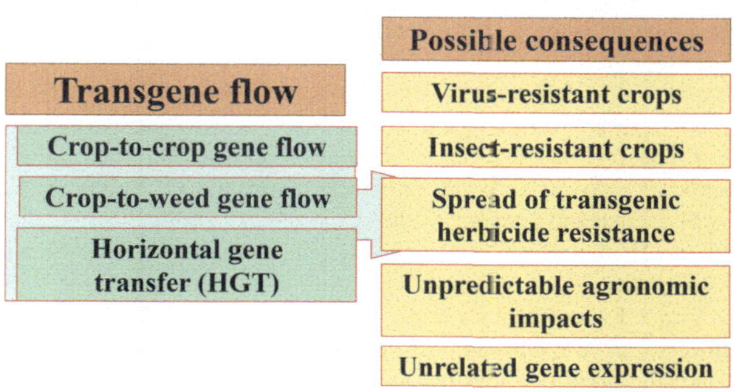

Fig. 5.3 Ways of flow of a transgene and its possible consequence (Bevan et al. 2001; Martinez-Ghersa et al. 2003; Lu and Snow 2005; Lu and Yang 2009)

5.6 Conclusion

Food security for the increasing population forced the farmers to use excessive amount of synthetic fertilizers and pesticides which caused potential damages to the soil health and a decrease in soil biodiversity and raised many other concerns. Development of genetically modified crops and their wise cultivation have been proven to reduce the pesticide as well as synthetic fertilizer use with enhanced crop productivity. Several stress-tolerant and pest-resistant varieties have been introduced and are found to give substantially better productivity than conventional varieties. However, thorough investigations of GMCs must be done during preparation and before and after cultivation to avoid current health issue and future environmental consequences. The threats related to GMCs which include adverse impact on tradition varieties, loss of agricultural biodiversity, transgene flow, adverse impact on non-targeted beneficial flora and fauna, etc. are supposed, and some of these have been experimentally proven. These problems associated with GMCs may be avoided or managed by doing proper research and investigation at field level.

Acknowledgment Kuldeep Bauddh is thankful to Science and Engineering Research Board (SERB) Research Grant No. EEQ/2017/000476 for financial support.

References

Actis LA, Tolmasky ME, Crosa JH (1999) Bacterial plasmids: replication of extra-chromosomal genetic elements encoding resistance to antimicrobial compounds. Front Biosci 4:43–62

Aktar W, Sengupta D, Chowdhury A (2009) Impact of pesticides use in agriculture: their benefits and hazards. Interdiscip Toxicol 2(1):1–12

Altieri MA (1999) The ecological role of biodiversity in agro-ecosystems. Agric Ecosyst Environ 74:19–31

Ammann K (2005) Effects of biotechnology on biodiversity: herbicide tolerant and insect resistant GM crops. Trends Biotechnol 23(8):388–394

Batista R, Oliveira MM (2009) Facts and fiction of genetically engineered food. Trends Biotechnol 27(5):277–286

Benjamin A (2008) Pesticides: Germany bans chemicals linked to honeybee devastation. The Guardian

Bernstein JA, Bernstein IL, Bucchini L, Goldman LR, Hamilton RG, Lehrer S, Rubin C, Sampson HA (2003) Clinical and laboratory investigation of allergy to genetically modified foods. Environ Health Perspect 111(8):1114–1121

Bevan M, Mayer K, White O, Eisen JA, Preuss D, Bureau T, Salzberg SL, Mewes HW (2001) Sequence and analysis of the *Arabidopsis* genome. Curr Opin Plant Biol 4:105–110

Bohme H, Aulrich K, Daenicke R, Flachowsky G (2001) Genetically modified feeds in animal nutrition. 2nd communication: glufosinate tolerant sugar beets (roots and silage) and maize grains for ruminants and pigs. Arch Tierernahr 54(3):197–207

Broll H, Zagon J, Butschke A, Leffka A, Spiegelberg A (2005) The fate of DNA of transgenic inulin synthesizing potatoes in pigs. J Anim Feed Sci 14(Suppl. 1):337–340

Brown PB, Wilson KA, Jonker Y, Nickson TE (2003) Glyphosate tolerant canola meal is equivalent to the parental line in diets fed to rainbow trout. J Agric Food Chem 51(15):4268–4272

Calsamiglia S, Hernandez B, Hartnell GF, Phipps R (2007) Effects of corn silage derived from a genetically modified variety containing two transgenes on feed intake, milk production, and composition, and the absence of detectable transgenic deoxyribonucleic acid in milk in Holstein dairy cows. J Dairy Sci 90(10):4718–4723

Chainark P, Satoh S, Hirono SI, Aoki T, Endo M (2008) Availability of genetically modified feed ingredient: investigations of ingested foreign DNA in rainbow trout *Oncorhynchus mykiss*. Fish Sci 74(2):380–390

Chandler SF, Stevenson TW (2014) Gene flow and risk assessment in genetically modified crops. In: Pratap A, Kumar J (eds) Alien gene transfer in crop plants, vol 1. Springer, New York, pp 247–265

Clark EA (2005) Environmental risks of genetic engineering. Euphytica 148:47–60

D'Agnolo G (2005) GMO: human health risk assessment. Vet Res Commun 29(Suppl. 2):7–11

Davies J, Honegger JL, Tencalla FG, Meregalli G, Brain P, Newman JR, Pitchford HF (2003) Herbicide risk assessment for non-target aquatic plants: sulfosulfuron—a case study. Pest Manag Sci 59(2):231–237

Dawkar VV, Chougale AD, Barvkar V, Tanpure RS, Giri AP (2018) Genetically engineered crops: opportunities, constraints, and food security at a glance of human health, environmental impact, and food quality. In: Holban AM, Grumezescu AM (eds) Genetically engineered foods. Academic, New York, pp 311–334

de Santis B, Stockhofe, Wal JM, Weesendorp E, Lalles JP, Dijk JV, Kok E, De Giacomo M, Einspanier R, Onori R, Brera C, Bikker P, Meulen JVD, Kleter G (2018) Case studies on genetically modified organisms (GMOs): Potential risk scenarios and associated health indicators. Food Chem Toxicol 117:36–65

de Vendomois JS, Roullier F, Cellier D, Seralini GE (2009) A comparison of the effects of three GM corn varieties on mammalian health. Int J Biol Sci 5:706–726

de Vos CJ, Swanenburg M (2018) Health effects of feeding genetically modified (GM) crops to livestock animals: a review. Food Chem Toxicol 117:3–12

Domingo JL (2007) Toxicity studies of genetically modified plants: a review of the published literature. Crit Rev Food Sci Nutr 47(8):721–733

Dunfield KE, Germida JJ (2004) Impact of genetically modified crops on soil- and plant-associated microbial communities. J Environ Qual 33(3):806–815

EFSA Panel on Genetically Modified Organisms (GMO) (2008) Guidance on the environmental risk assessment of genetically modified plants. EFSA J 8:1879–1990

Folmer J, Grant RJ, Milton CT, Beck J (2002) Utilization of Bt corn residues by grazing beef steers and Bt corn silage and grain by growing beef cattle and lactating dairy cows. J Anim Sci 80 (5):1352–1361

Gealy DR, Mitten DH, Rutger JN (2003) Gene flow between red rice (*Oryza sativa*) and herbicide-resistant rice (*O. sativa*): implications for weed management. Weed Technol 17:627–645

Giddings G (2000) Modelling the spread of pollen from *Lolium perenne*. The implications for the release of wind-pollinated transgenics. Theor Appl Genet 100:971–974

Giovannetti M, Sbrana C, Turrini A (2005) The impact of genetically modified crops on soil microbial communities. Riv Biol 98(3):393–417

Glencross BJ, Curnow W, Hawkins GW, Kissil M, Peterson D (2003) Evaluation of the feed value of a transgenic strain of the narrow-leaf lupin (*Lupinus angustifolius*) in the diet of the marine fish, *Pagrus auratus*. Aquac Nutr 9(3):197–206

Goujon A (2018) Human population growth. Encyclopedia of Ecology. March 2018

Groot AT, Dicke M (2002) Insect-resistant transgenic plants in a multi-trophic context. Plant J 31:387–406

Gustafson DI, Horak MJ, Rempel CB, Metz SG, Gigax DR et al (2005) An empirical model for pollen-mediated gene flow in wheat. Crop Sci 45:1286–1294

Halver JE, Hardy RW (2002) Fish nutrition, 3rd ed. Academic, San Diego, CA. 824 p

Herman RA, Zhuang M, Storer NP, Cnudde F, Delaney B (2019) Risk-Only assessment of genetically engineered crops is risky. Trends Plant Sci 24:158–168

Hilbeck A (2002) Transgenic host plant resistance and nontarget effects. In Letourneau DK, Burrows BE (eds) Genetically engineered organisms. Assessing environmental and human health effects. CRC, Boca Raton, FL, pp 167–185. 438p

Hyun Y, Bressner GE, Ellis M, Lewis AJ, Fischer R (2004) Performance of growing-finishing pigs fed diets containing. Roundup ready corn (event nk603), a non-transgenic genetically similar corn, or conventional corn lines. J Anim Sci 82:571–580

International Service for the Acquisition of Agri-biotech Applications (ISAAA). Brief 53-2017

Irtwange S (2006) Application of biological control agents in pre- and post-harvest operations. Agric Eng Int CIGR E J 8:1–10

Ismail K, Azhar TNT, Yong CY, Aslan AS, Omar WZ, Majid I, Ajagbe AM (2012) Problems on commercialization of genetically modified crops in Malaysia. Procedia Soc Behav Sci 40:353–357

Jialin Z, Juan H, Jianqun N, Qingwen Z, Xiaoxia L (2013) Influence of wind direction on pollen-mediated gene flow in transgenic insect-resistant cotton. Acta Ecologica Sinica 33:6803–6812

Johannessen MM, Andersen BA, Jorgensen RB (2006) Competition affects gene flow from oil seed rape (♀) to *Brassica rapa* (♂). Heredity 96:360–367

Key S, Ma JK-C, Drake PMW (2008) Genetically modified plants and human health. J R Soc Med 101(6):290–298

Kramkowska M, Grzelak T, Czyzewska K (2013) Benefits and risks associated with genetically modified food products. Ann Agric Environ Med 20:413–419

Ladics GS, Cressman RF, Herouet-Guicheney C, Herman RA, Privalle L, Song P, Ward JM, Mcclain S (2011) Bioinformatics and the allergy assessment of agricultural biotechnology products: industry practices and recommendations. Regul Toxicol Pharmacol 60:46–53

Lazebnik J, Arpaia S, Baldacchino F, Banzato P, Moliterni S, Vossen JH, Zande EM, Loon JJA (2017) Effects of a genetically modified potato on a non-target aphid are outweighed by cultivar differences. J Pest Sci 90:855–864

Lu B, Snow AA (2005) Gene Flow from genetically modified rice and its environmental consequences. BioScience 55(8):669–678

Lu B, Yang C (2009) Gene flow from genetically modified rice to its wild relatives: assessing potential ecological consequences. Biotechnol Adv 27:1083–1091

Lusser M, Davies HV (2013) Comparative regulatory approaches for groups of new plant breeding techniques. New Biotechnol 30(5):437–446

Lutz B, Wiedemann S, Einspanier R, Mayer J, Albrecht C (2005) Degradation of Cry1Ab protein from genetically modified maize in the bovine gastrointestinal tract. J Agric Food Chem 53:1453–1456

Lynch JM, Benedetti A, Insam H, Nuti MP, Smalla K, Torsvik V, Nannipieri P (2004) Microbial diversity in soil: ecological theories, the contribution of molecular techniques and the impact of transgenic plants and transgenic microorganisms. Biol Fert Soils 40:363–385

Macfadyen S, Davies AP, Zalucki MP (2015) Assessing the impact of arthropod natural enemies on crop pests at the field scale. Insect Sci 22:20–34

Martinez-Ghersa MA, Worster CA, Radosevich SR (2003) Concerns a weed scientist might have about herbicide-tolerant crops: a revisitation. Weed Technol 17:202–210

Mejia L, Jacobs CM, Utterback PL, Parsons CM, Rice D (2010) Evaluation of the nutritional equivalency of soybean meal with the genetically modified trait DP-305423-I when fed to laying hens. Poult Sci 89:2634–2639

Mohanta RK, Singhal KK, Tyagi AK, Rajput YS, Prasad S (2010) Nutritional evaluation of transgenic cottonseed in the ration of lactating dairy cows. Trop Animal Health Prod 42:431–438

Moon HS, Eda S, Saxton AM, Ow DW, Stewart CN (2011) An efficient and rapid transgenic pollen screening and detection method using flow cytometry. Biotechnol J 6:118–123

Ndakidemi B, Mtei K, Ndakidemi PA (2016) The potential of common beneficial insects and strategies for maintaining them in bean fields of Sub Saharan Africa. Am J Plant Sci 7:425–436

Neher DA (1999) Soil community composition and ecosystem processes: comparing agricultural ecosystems with natural ecosystems. Agrofor Syst 45:159–185

Nicholls CI, Altieri MA (2013) Plant biodiversity enhances bees and other insect pollinators in agroecosystems, a review. Agron Sustain Dev 33:257–274

O'Callaghan M, Glare TR, Burgess EPJ, Malone LA (2005) Effects of plants genetically modified for insect resistance on non-target organisms. Annu Rev Entomol 50:271–292

Onkin SS, Velez JC, Totten AK, Stanisiewski EP, Hartnell GF (2003) Effects of feeding silage and grain from glycophosphate-tolerant or insect-protected corn-hybrids on feed intake, ruminal digestion, and milk composition in dairy cattle. J Dairy Sci 86:1780–1788

Paul MJ, Nuccio ML, Basu SS (2018) Are GM crops for yield and resilience possible? Trends Plant Sci 23(1):10–16

Pellegrino E, Bedini S, Nuti M, Ercoli L (2018) Impact of genetically engineered maize on agronomic, environmental and toxicological traits: a meta-analysis of 21 years of field data. Sci Rep 8:3113

Phelinas P, Choumert J (2017) Is GM soybean cultivation in Argentina sustainable? World Dev 99:452–462

Pu DQ et al (2014) Flower-visiting insects and their potential impact on transgene flow in rice. J Appl Ecol 51:1357–1365

Ramesh T et al (2019) Soil organic carbon dynamics: impact of land use changes and management practices: a review. Adv Agron 156:1–107

Rieben S, Kalinina O, Schmid B, Zeller SL (2011) Gene flow in genetically modified wheat. PLoS One 6(12). https://doi.org/10.1371/journal.pone.0029730

Romeis J, Meissle M, Naranjo SE, Li Y, Bigler F (2014) The end of a myth—Bt(Cry1Ab) maize does not harm green lacewings. Front Plant Sci 5:1–10

Sanden M, Berntssen MH, Krogdahl A, Herme GI, McKellep AM (2005) An examination of the intestinal tract of Atlantic salmon. *Salmo salar* L., parr fed different varieties of soy and maize. J Fish Dis 28:317–330

Saxena D, Stotzky G (2000) Insecticidal toxin from Bacillus thuringiensis is released from roots of transgenic Bt corn in vitro and in situ. FEMS Microbiol Ecol 33:35–39

Saxena D, Stotzky G (2001) *Bacillus thuringiensis* (Bt) toxin released from root exudates and biomass of Bt corn has no apparent effect on earthworms, nematodes, protozoa, bacteria, and fungi in soil. Soil Biol Biochem 33:1225–1230

Saxena D, Flores S, Stoztky G (2002) Bt toxin is released in root exudates from 12 transgenic corn hybrids representing three transformation events. Soil Biol Biochem 34:133–137

Scheideler SE, Hileman RE, Weber T, Robeson L, Hartnell GF (2008) The in vivo digestive fate of the Cry3Bb1 protein in laying hens fed diets containing MON 863 corn. Poult Sci 87:1089–1097

Schuler TH (2000) The impact of insect resistant GM crops on populations of natural enemies. Antenna 24:59–65

Seralini GE, Clair E, Mesnage R, Gress S, Defarge N, Malatesta M, Hennequin D, de Vendomois JS (2014) Republished study: long-term toxicity of a Roundup herbicide and a Roundup-tolerant genetically modified maize. Environ Sci Eur 26(14):1–17

Shennan C (2008) Biotic interactions, ecological knowledge and agriculture. Phil Trans R Soc B 363(1492):717–739

Sidhu RS, Hammond BG, Fuchs RL, Mutz J, Holden LR (2000) Glyphosate-tolerant corn: the composition and feeding value of grain from glyphosate-tolerant corn is equivalent to that of conventional corn (*Zea mays*). J Agric Food Chem 48:2305–2312

Singh M, Tiwari DP, Kumar A (2003) Effects of feeding transgenic cotton seeds on nutrients utilization, milk production and its composition in lactating Murrah buffaloes. Buffalo J 19:117–126

Tripathi MK, Mondal D, Somvanshi R, Karim SA (2011) Haematology, blood biochemistry and tissue histopathology of lambs maintained on diets containing an insect controlling protein (Cry1Ac) in Bt cotton seed. J Anim Physiol Anim Nutr (Berl) 95:545–555

Tudisco R, Lombardi P, Bovera F, d'Angelo D, Cutrignelli MI (2006) Genetically modified soyabean in rabbit feeding: detection of DNA fragments and evaluation of metabolic effects by enzymatic analysis. Anim Sci 82:193–199

Verma SR (2013) Genetically modified plants: public and scientific perceptions. ISRN Biotechnol. Article ID 820671:11

Wang Z, Zemetra RS, Hansen J, Mallory-Smith C (2001) The fertility of wheat jointed goatgrass hybrid and its backcross progenies. Weed Sci 49:340–345

Warwick SI, Beckie HJ, Hall LM (2009) Gene flow, invasiveness, and ecological impact of genetically modified crops. Ann N Y Acad Sci 1168:72–99

Welbaum GE, Sturz AV, Dong Z, Nowak J (2004) Managing soil microorganisms to improve productivity of agro-ecosystems. Crit Rev Plant Sci 23:175–193

Yadav SK, Soni R, Rajput AS (2018) Role of microbes in organic farming for sustainable agro-ecosystem. In: Panpatte D, Jhala Y, Shelat H, Vyas R (eds) Microorganisms for green revolution. Microorganisms for sustainability, vol 7. Springer, Singapore, pp 241–252

Yan S et al (2015) Pollen-mediated gene flow from transgenic cotton under greenhouse conditions is dependent on different pollinators. Sci Rep 5:15917

Zamana QU, Lia C, Chenga H, Hua Q (2018) Genome editing opens a new era of genetic improvement in polyploid crops. Crop J 7(2):141–150

Zhang CJ, Yook MJ, Park HR, Lim SH, Kim JW, Nah G, Song HR, Jo BH, Roh KH, Park S, Kim DS (2018) Assessment of potential environmental risks of transgene flow in smallholder farming systems in Asia: *Brassica napus* as a case study in Korea. Sci Total Environ 640–641:688–695

Application of Biochar in Agriculture: A Sustainable Approach for Enhanced Plant Growth, Productivity and Soil Health

6

Shahrukh Nawaj Alam, Zaira Khalid, Sweta, Bhaskar Singh, Abhishek Guldhe, D. K. Shahi, and Kuldeep Bauddh

Contents

Abstract

Soil quality degradation is one of the major outcomes of environmental pollution that can be characterized by deficient nutrients and concurrent presence of toxic substances in the agroecosystems. To enhance crop productivity, farmers use uncontrolled synthetic fertilizers that further deteriorate the soil health. Several

Shahrukh Nawaj Alam and Zaira Khalid contributed equally to this chapter.

S. N. Alam · Z. Khalid · Sweta · B. Singh · A. Guldhe · K. Bauddh (✉)
Department of Environmental Science, Central University of Jharkhand, Ranchi, Jharkhand, India
e-mail: kuldeep.bauddh@cuj.ac.in

D. K. Shahi
Department of Soil Science and Agricultural Chemistry, Birsa Agricultural University, Ranchi, Jharkhand, India

K. Bauddh et al. (eds.), *Ecological and Practical Applications for Sustainable Agriculture*, https://doi.org/10.1007/978-981-15-3372-3_6

organic components have been developed to reduce the overload of chemical fertilizers. Application of biochar to the soil is one of the potential methods that not only enhance the crop productivity but at the same time improve soil quality parameters. The application of biochar has also been proved to reduce the greenhouse gaseous emission from the agroecosystems. In this chapter, the efforts have been made to explore the production, features, mode of application, and effects of biochar on crop productivity and soil health. Further, the impact of biochar in the mitigation of abiotic stresses has also been discussed.

Keywords

Biochar · Crop productivity · Soil health · Sustainable agriculture · Carbon sequestration

6.1 Introduction

In the past few decades, with the increment in industrialization and advancement in agricultural practices, there has been a massive problem of environmental degradation. This is primarily due to the use of various harmful compounds that are damaging the life force, whether they are in air, soil, or water. The rapidly increasing population demands food security. This has led to the extensive use of chemical fertilizers in modern agriculture, which has its benefits, but in the long run, they have proven themselves to be a global menace as it is deteriorating the soil quality and productivity. This intensive agricultural practice is directly affecting the food production as it is non-sustainable and deteriorates the fertility and productivity of the soil, resulting in the shortage of food in a time when there is an utmost requirement to sustain the growing population (Wei and Yang 2010). In the current era, there are more energy and land requirement, which results in huge waste generation that collectively contribute to global climate change. In the process of meeting these demands, we have subjected ourselves to pollution via harmful gaseous emission during energy production, cutting of trees to provide for industrial, residential and agricultural land and toxic gas release by waste dumpsite or by their incineration. All these activities contribute in the rising concentration of greenhouse gases (GHGs), resulting in the global temperature increase and climate change. Therefore, there is an immense need to combat these issues by employing sustainable methods and, at the same time meet the demand of the current population. There are various methods employed to manage wastes and nourish the soil at the same time such as composting, vermicomposting and biocharring. Several kinds of slow-release fertilizers have been developed and found to bear the significant potential to improve the agricultural productivity (Singh et al. 2010; Kumar et al. 2012, 2013a, b, 2014a, b, c, 2015; Rai et al. 2017). Methods like composting and vermicomposting have certain demerits, especially requirement of more time for their preparation (Boldrin et al. 2009).

Biocharring is a relatively better option as it not only can be used as an effective soil amendment but also helps in mitigating global warming. Biochar, essentially an end- or co-product of biomass thermochemical conversion processes known as pyrolysis, has been established as a suitable method to sequester carbon by more than 10%, produce energy and boost soil fertility (Woolf et al. 2010; Yuan et al. 2018). Due to all the proven beneficial properties of biochar such as soil fertility improvement (Joseph et al. 2010; Sohi et al. 2010; Kookana et al. 2011), its application in water treatment and its potential towards mitigating climate change (Galinato et al. 2011; Harvey et al. 2012), the policymakers of developing and developed countries are therefore showing great interest in research and development of biochar (Jeffery et al. 2015; Dai et al. 2017; Lin et al. 2017; Randolph et al. 2017; Zheng et al. 2017; Niazi et al. 2018).

6.2 Biochar

Biochar is a carbon-rich compound formed by heating the biomass in an oxygen deficit condition. It differs from charcoal and activated carbon only as per its end use, where biochar is produced to be used as an amendment in soil for soil nourishment and carbon sequestration. Terra preta is a form of soil which is found in the Amazon basin that is formed by the practice of slash-and-char agricultural technique used in the past. In this technique, a long-term slow burning is done as opposed to the slash-and-burn technique used at present that results in charring of biomass. It is then used as a conditioner in the soil (Zheng et al. 2010). The production of biochar is generally done via latest technological systems used for the generation of biofuel where biochar is the by-product. It is produced by the heating of biomass in an oxygen-free environment. By this process, biochar carbon is chemically altered which results in the formation of benzene-type ring structure and is highly resistant to microbial action unlike the carbon found in most organic matter. By virtue of this, chemical alteration biochar can remain stable for a very long period of time. The longevity of the presence of biochar in the soil can assist in the sequestration of carbon that is lost in the process of photosynthesis. Thus, it reduces the carbon emission and mitigates the greenhouse effect. It has been reported that biochar not only assists in the sequestration of carbon but also decreases the emissions of more potent GHGs such as nitrous oxide and methane (Spokas et al. 2009). There are a number of benefits that biochar provides when amended into the soil. It has been reported that biochar can boost the fertility of soil and improve its quality by increasing the pH of the soil, by increasing the water-holding capacity of soil, by providing habitat to beneficial microbes and fungi, by improving cation exchange capacity and also by retaining nutrient in the soil (Lehmann et al. 2006; Lehmann 2007; Lone et al. 2015; Oladele et al. 2019; Saha et al. 2019; Wang et al. 2019). The characteristic of biochar varies with the type of feedstock and also the technology used for its production. Biochar can act as a sorbent due to its large surface area, charge density and negative surface charge. It has been observed that biochar has greater sorption ability than the organic matter present in the soil naturally (Liang

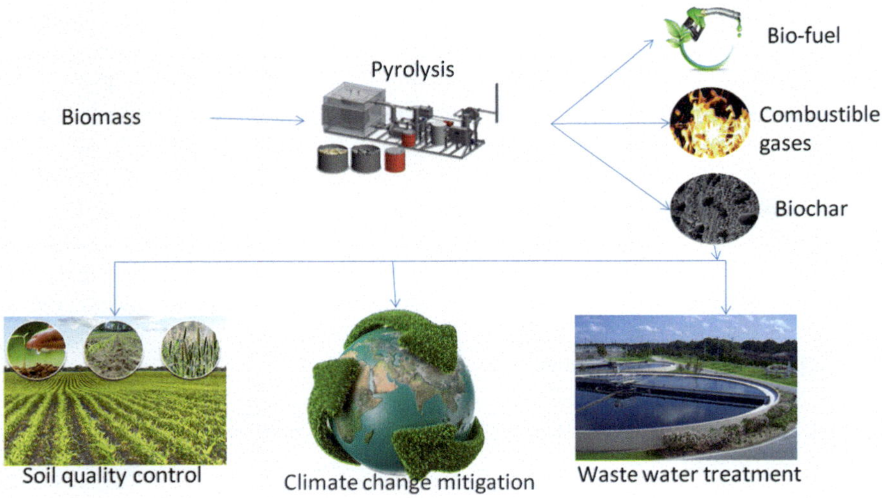

Fig. 6.1 Production and application of biochar

et al. 2010). The use of biochar is not only limited to the soil, but it is also used for the removal of contamination from water as it can be used as a sorbent (Cheng et al. 2008; Cao and Harris 2010). Biochar production is generally done in a sustainable manner with the use of sustainably obtained waste biomass (Fig. 6.1). The sustainable production and use of biochar, on the one hand, provides a means for abating climate change by the sequestration of carbon and, on the other hand, provides a way of reducing waste, improving soil quality, enhancing crop productivity and conserving natural resources (Winsley 2007; Laird 2008; Zheng et al. 2010).

6.2.1 Synthesis of Biochar

Biochar is derived from a wide range of carbon-rich biomass with different chemical and physical properties. The physical and chemical properties such as moisture content, ash content, fraction of mixed carbon, calorific value, volatile compounds, particle density and the percentage of lignin, cellulose and hemicellulose are responsible for the type of thermal conversion the biomass will undergo for the production of biochar. Biochar is a by-product in the generation of bio-oil or biofuel. Nowadays, there are several technologies that aid in the production of biochar like torrefaction, pyrolysis (slow, fast and intermediate), hydrothermal carbonization, gasification and flash carbonization. The most common technology that is used in biochar production is pyrolysis. It is a thermochemical treatment in which the biomass is exposed to elevated temperature in deficit oxygen which results in the disintegration of materials. The factors that influence the production of biochar are temperature, residence time, heating rate, cooling rate, pressure and the surrounding medium. Pyrolysis is of various types such as described by Nartey and Zhao (2014):

- Torrefaction is a pyrolytic process performed at low temperature of about 200–320 °C (Spokas et al. 2012).
- Slow pyrolysis is a continuous process in which biomass is transferred to an externally heated oxygen-free furnace. The temperature at which slow pyrolysis is carried out is always less than 450 °C with a long residence time. In this method, the production of bio-oil is about 75%, syngas 13% and biochar 12%.
- Intermediate pyrolysis is a process in which the biomass is subjected to a moderate temperature of about 450 °C and more with moderate hot vapour residence time. The production of biochar in this case is about 25%.
- Fast pyrolysis is a process in which the transfer of heat is rapid which acts on the biomass. It is done at moderate-to-high temperature of about 650 °C with a short hot vapour residence time. Here, the production of biochar is 35%.

Other technologies that are used for the production of biochar are:

- Gasification is a thermochemical process in which the biomass is heated in the presence of small amount of air at a temperature of 700 °C and more, producing bio-oil, syngas and about 10% biochar as by-products.
- Hydrothermal carbonization is a method in which the wet biomass without prior drying is transformed into biochar and energy under high atmospheric pressure and temperature.
- Flash carbonization is a method in which controlled flash fire is ignited at high pressure underneath a packed bed biomass where the flow of air is downwards at the rate of 0.8–1.5 kg of air per kg of biomass. The residence time is below 30 min, and the temperature ranges from 350 to 650 °C (Spokas et al. 2012; Nartey and Zhao 2014).

The production and the quality of biochar varies with the different technologies (different temperature, pressure and residence time) employed for its generation (Zhang et al. 2013; Luo et al. 2013) as shown in Table 6.1.

6.2.2 Characteristics of Biochar

Different types of biochar may have a distinct kind of physiochemical characteristics depending upon the nature of raw material like mineral content, elemental composition, grain or fibre size, density, calorific value, etc. and on the processing condition of the pyrolysis, e.g. final pressure, temperature, chemical or thermal pre-treatment procedure, heating rate and heating fluxes (Table 6.2). Some of these characteristics are pH, cation exchange capacity, porosity, surface area, carbon content, C:N ratio, C:H ratio, mineralogical composition and surface function group.

Table 6.1 Technologies used for the production of biochar (Spokas et al. 2012; Nartey and Zhao 2014)

Method of treatment	Temperature required (°C)	Residence time	Biochar produced (%)
Fast pyrolysis	650	Short hot vapour residence time	12
Intermediate pyrolysis	>450	Moderate hot vapour residence time	25
Slow pyrolysis	<450	Long residence time	35
Gasification	>700	Long vapour residence time	10
Hydrothermal carbonization	200–250	Not determined	Not determined
Flash carbonization	350–650	Residence time below 30 min	50

6.2.2.1 pH

pH varies depending on the processing methods used to prepare biochar. All kinds of biochar are usually alkaline in nature with some exception depending on the raw material used. Generally, pH of the biochar increases with the increasing temperature and type of raw material selected. It has been reported that increasing the temperature in pyrolysis process leads to greater ash production, which is directly correlated to the pH of the biochar (Jin et al. 2016). It has been observed that factors like increase in temperature (from 300 to 700 °C) during pyrolysis increased the cations and carbonates in biochar that led to increase in pH of the biochar residue (Yuan et al. 2011). Another reason can be vanishing of acidic functional groups at such a high temperature during pyrolysis (Al-Wabel et al. 2013).

6.2.2.2 Cation Exchange Capacity (CEC)

Generally, cation exchange capacity of biochar is high because of the negative charges and their affinity towards cationic charge; this includes cations of most heavy metals like lead, chromium, zinc, etc. It has been established through various studies that the ability of biochars to absorb heavy metal can be utilized in remediating the heavy metal contamination of the soil (Cao and Harris 2010; Jiang et al. 2012). Like other parameters, CEC also varies depending upon the raw material used for production of biochars. There is considerable difference in the values of CEC derived from terrestrial biomass and that of aquatic biomass like algae where the values have been reported to be 4.5–40 cmol/kg for terrestrial biomass (Bird et al. 2008; Uzoma et al. 2011), whereas the CEC for biomass such as of algae is 29~41 cmol/kg which is relatively higher than terrestrial biomass (Bird et al. 2008).

6.2.2.3 Porosity and Surface Area

Biochars have unique sorption capacity for pollutants such as pesticides through surface sorption. The property directly responsible for this ability is porosity and surface area. Greater porous structure and surface area means greater sorption capacity. Pore structure is usually formed during the process of pyrolysis when high temperature causes water loss during dehydration process (Bagreev et al. 2001). The pore size is highly inconsistent and according to the International Union of Pure and Applied Chemistry (IUPAC 1972) can have micropore, mesopore and macropore with the pore size of <2 mm, 2–50 mm and >50 mm, respectively. Both the porosity and surface area are directly dependent on the temperature of operation of the pyrolysis which is proven by various studies which successfully concluded that the increase in temperature produces larger pores and hence larger surface area (Cantrell et al. 2012; Liu et al. 2017). Other than temperature, the raw material of the biochar feedstock is also an important driving factor towards the degree of porosity and surface area. The surface area of biochar made of oak wood or corn stover is much greater in the range of 112–642 m^2/g than that of pig manure biochar (3.32–20.5 m^2/g) and biosolid biochar with surface area of 50.9–94.2 m^2/g (Cantrell et al. 2012; Liu et al. 2017).

6.2.2.4 Carbon Content, C:N Ratio and C:H Ratio

The carbon content of biochar is relatively very high as the residue left after pyrolysis is carbon, and the degree of residual content widely depends on the feedstock used for the process, e.g. carbon content of wood biochar (68.2%) is greater when compared to the biochar produced from grass (58.5%) in an identical laboratory controlled condition. This difference in both the values is due to the ash content in wood (about 0.1%) and leaf (7.7%) during pyrolysis (Hammes et al. 2006).

The biochar which is generally obtained after high temperature treatment is usually low in hydrogen and nitrogen content which results in relatively greater C:N and C:H ratio (Wang et al. 2018). In general, the C:N ratio of biochar may vary from 7 to 400 (Bridle and Pritchard 2004; Chan et al. 2007), whereas C:H ratio of unburned fuel composites like cellulose and lignin can be approx. 1.5. Graetz and Skjemstad (2003) have reported that the C:H ratio of 5 biochars formed after burning of biomass at temperature of above 400 °C is likely to be equal to or less than 0.5.

6.2.2.5 Mineralogical Composition

Mineral composition of biochar is directly responsible for the high CEC and is also an important parameter in determining the sorption capacity of the biochar (Dias et al. 2010). Temperature during process of pyrolysis and feedstock used are both important factors in determining direct mineral composition of any biochar. High temperature enriches mineral composition in biochars (Subedi et al. 2016). Usually, biochars made of biosolids have greater P content in the range of 1.82–3.60% in comparison to the biochar prepared from oak wood (0.03–0.06%), whereas K content in pig manure biochar has been reported higher than biochars prepared by other material (Subedi et al. 2016).

6.2.2.6 Surface Functional Groups

For the sorption capability, biochar possesses functional groups like carboxylic (−COOH), hydroxyl group, lactonic group and amine groups (Li et al. 2014; Antón-Herrero et al. 2018). In general, the final pyrolysis temperature and feedstock are the most important factors in determining the number of functional groups present in the biochar (Chen et al. 2015). It has been reported by various scientists that unlike surface area, porosity and pH which increase with increasing pyrolysis temperature, values of factors like C:H and C:N ratio decrease with increasing temperature that also results in reduction of availability of functional group on biochar surface (Liu et al. 2018).

6.2.3 Biochar as a Soil Amendment

Biochar has been found to have several qualities, and one of them is the beneficial effect on soil. It has been reported to boost soil fertility and its quality by increasing water-holding capacity, providing liming effect, providing habitat to beneficial bacteria and fungi, improving cation exchange capacity, increasing pH and retaining nutrient in soil for a long time (Lehmann et al. 2006). Biochar has the capacity to remain or persist in the soil for millennia as it is highly resistant to microbial disintegration and mineralization. Biochar as compared to other organic matter like compost and manure has greater capacity to retain nutrients and makes it available to plants. It is also a more stable source of nutrient than compost and manure (Chan et al. 2007). A study has been conducted to see the effect of biochar on crop yield, and most of them have shown positive correlation (Yamato et al. 2006). Various studies have shown a considerable increase in crop yield and improved soil quality (Chan et al. 2007; Laird et al. 2009). Biochar as a soil amendment has another major benefit as it has the ability to sequester carbon from the atmosphere and sink it into the soil by using it in the soil as soil conditioners (Winsley 2007; Laird 2008). Studies have also indicated that the use of biochar in soil is not only responsible for carbon dioxide sequestration, but it may also decrease the emissions of other greenhouse gases such as nitrous oxide and methane which are considered more potent GHGs (Spokas et al. 2009; Wang et al. 2019).

6.3 Application of Biochar in Agriculture

Biochar may also be produced by utilizing crop residues (Wang et al. 2013; Windeatt et al. 2014; Zhou et al. 2018) and may be applied in agroecosystem as an organic amendment (Fig. 6.2). Intensive agricultural practices have declined the quality of the soil and have taken away its self-cleaning property. Application of biochar is a sustainable approach to combat this problem. Biochar is gaining popularity due to its favourable effects on soil which include:

- Increase of CEC and nutrient cycling.
- Decreases agricultural GHG emission.
- Increases the carbon sequestration into the agricultural soil and also increases the organic carbon of the soil.
- Alkalinity of the biochar helps in decreasing the acidity of the soil.

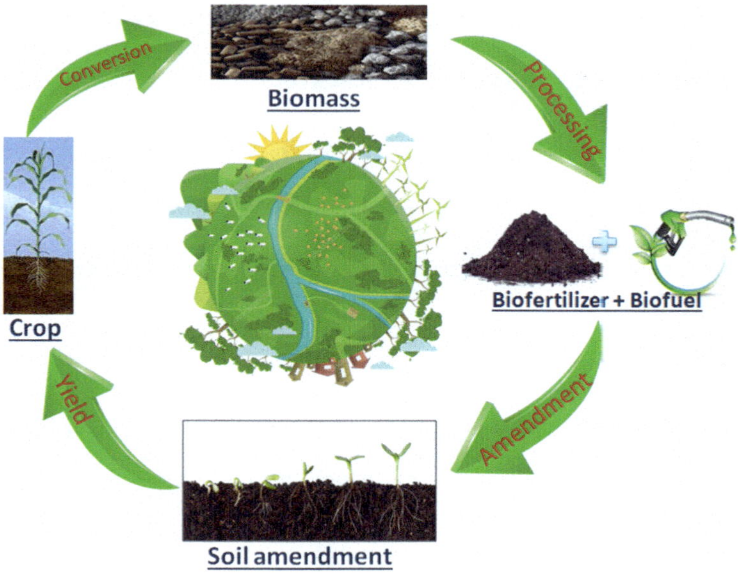

Fig. 6.2 Production of biochar by utilizing crop residues and application of it as amendment

- Helps in mitigating salinity of the soil.
- Can utilize the agricultural residue to convert again into biochar and biofuel.
- Can provide mitigating means to fight drought condition of the soil.
- Helps reduce heavy metal concentration from the soil.
- Increases water retention capacity and aeration of the soil.
- Has an application in fighting desertification and afforestation programmes.

6.3.1 Mode of Application

Biochar is applied in the soil by common incorporation through even spreading of the compound. Firstly, it is spread manually onto the soil in such a manner that it equally spreads over the whole field, and then, tilling is done to mix it with the soil either through a machine or manually. Apart from this when tilling is not possible, biochar is commonly applied just on the soil surface and then covered with the organic matter, or it can also be applied by mixing with compost or other amendment. Another method can be application in the form of slurry if the amendment area is very large by making solution of finely ground biochar. In the case of planting trees or plotted plantations, biochar can be simply mixed with backfill soil. Regardless of the mode of application, the handling of dry biochar should be done very carefully as it is very dusty in nature and can spread in nature; therefore, wetting of this powdery biochar is advised before application (Hunt et al. 2010).

6.3.2 Effects of Biochar Application on Soil Properties

6.3.2.1 Impact on Soil Physicochemical Properties

It has been reported that biochar benefits the improvement of soil and productivity of crop by three basic mechanisms:

1. By improving soil physical characteristic to help the growth of root and nutrient cycling
2. Provide chemically active surface area where several important chemical reactions take place
3. Direct alteration and modification of soil chemical properties making it fit for enhancing plant growth

The amendment of biochar in the soil alters soil characteristics like pore size, density and structure, thereby modifying the properties like water-holding capacity, nutrient cycling, soil workability and overall plant growth. Various actions of biochar may include:

- Reduction in the overall bulk density of the soil
- Net increase in the soil specific area
- Increase in porosity, thereby increasing soil aeration
- Increase in water-holding capacity along with water retention time and water permeability (Glaser et al. 2001; Chan et al. 2007; Jien and Wang 2013)

6.3.2.2 Impact on Soil Chemical Properties

Biochar has the potential to modify cation exchange capacity, sorption capacity of the surface and base saturation capability. Application of biochar may alter the chemical properties such as pH, nutrient availability, exchangeable cations, CEC, etc. pH is an important factor determining the health of the soil, and amendment and application of biochar in the soil positively increases the pH of the soil as a result of increase in CEC due to larger surface area and porosity of the biochar (Lammirato et al. 2011).

CEC of the biochar-amended soil is increased due to the high surface area of the biochar which provides greater surface for the cations to act upon. This property of biochar helps in retaining the nutrients which are then provided to the plant, thereby increasing productivity (Suliman et al. 2016, 2017).

Nutrient availability is another factor that is very important for a good quality soil, and the application of biochar successfully provides the soil with these nutrients either directly or through nutrient retention capability. This proves to be beneficial in the long run as slower release of these nutrients takes place from the organic matter (Laghari et al. 2015; Gul and Whalen 2016; Shepherd et al. 2017; Yue et al. 2017) (Table 6.2).

Table. 6.2 Physiochemical properties of different biochars

Biochar feed	Temperature of preparation (°C)	Application (t per ha/%)	Properties						References
			pH	CEC (Cmol/kg)	N	P	K	OC	
Maize stalk	350	5 t/ha	6.08	6.0	166 (kg ha^{-1})	28 (kg ha^{-1})	191 (kg ha^{-1})	4.4 (g ha^{-1})	Pandian et al. (2016)
Red gram stalk	350	5 t/ha	6.19	6.0	166 (kg ha^{-1})	28 (kg ha^{-1})	192 (kg ha^{-1})	4.4 (g ha^{-1})	Pandian et al. (2016)
Corn stover	650	156 mg/ha	6.64	19.04	–	–	–	–	Chintala et al. (2014)
Pecan shell	700	2%	6.4	5.59	–	30.5 %	35 (g/kg)	–	Novak et al. (2009)
Rice husk	–	5%	6.58	6.58	0.13%	24.35 ppm	0.42 (Cmol/kg)	0.6%	Ndor et al. (2015)
Saw dust	–	5%	6.58	6.33	0.12%	24.21 ppm	0.37 (Cmol/kg)	0.73%	Ndor et al. (2015)
Wood	–	5%	7.6	–	3.70 (mg/kg)	8.63 (mg/kg)	107.3 (mg/kg)	3.23%	Arshad et al. (2017)
Wheat straw	400	–	10.12	–	5.70 (g/kg)	0.89 (g/kg)	48.93 (g/kg)	–	Xie et al. (2013)

6.3.2.3 Impact on Soil Biodiversity

The physical and chemical characteristics of biochar and the changes made to the soil properties after the amendment of biochar can modify the soil biodiversity severely. Surface and pores of the biochar and the alteration in the physiochemical properties allow proliferation of microbes by providing them with favourable conditions and diverse habitat niche. The direct favourable changes of the amendment cause extensive growth of microorganisms which indirectly and directly results in improved plant growth. The plant growth leads to more litter production and extensive root growth which in turn provides more habitat and niches to all kinds of microorganisms.

6.3.2.3.1 Habitats for Microbial Growth in Biochar-Amended Soil

It has been well established that pores present in the biochar serve as a habitat for microorganisms (Warnock et al. 2007; Jaafar et al. 2014) and provide favourable condition for microorganisms like bacteria, fungi and protozoa which are beneficial for the growth of plant (Warnock et al. 2007; Jaafar et al. 2014). Macropores in biochar possibly represent the safest microbial habitat because they are of perfect size to harbour beneficial bacteria and fungi (Quilliam et al. 2013). The mesopores and micropores help in proliferation of these microbes with their relatively smaller pore size where water as well as dissolved substances can be stored which are essential for the metabolism of these microbes. Apart from pores, physical properties like surface area and its black colour also play important roles in microbial growth as more surface area means increased colonization of microbes and the black colour of the biochar traps more heat which is essential for microbial growth and other enzymatic activities.

6.3.2.3.2 Soil Microbial Community Structure After Biochar Amendment

Numerous researches have reported that the application of biochar has a positive effect on the microbial community growth (shift in fungal and bacteriological community structure) in the soil. On a similar note Sun et al. (2016) reported that bacterial community structure was several times more dynamic than the fungal community structure after the amendment of biochar in the brown soil in the field experiment. Gomez et al. (2014) in their study concluded that the fungi-to-bacteria ratio after 1 year of the amendment of biochar by doing fast pyrolysis biochar was considerably low. It has been reported that the fungi-to-bacteria ratio of the soil after amendment is mostly dependent on the C: N ratio of the biochar (Muhammad et al. 2014).

6.3.2.3.3 Enzymatic Activity in Biochar-Amended Soil

Enzymatic activity after the biochar amendment is directly dependent on the condition of preparation and feedstock used. The extracellular enzymatic activities that are carried out in the soil are also related to the decomposition of organic compounds and the nutrient cycle in the soil (Burns et al. 2013). It has been reported by several scientists that amendment of biochar has considerable effect on the enzymatic activities (Bailey et al. 2011; Paz-Ferreiro et al. 2012; Masto et al. 2013). The degree of impact of biochar amendment on soil's enzymatic activity is vastly dependent on:

- Surface area and porosity
- Sorption and desorption taking place on CEC and AEC sites
- Binding capacity of biochar surface for the extracellular enzymes

Generally, a high surface area and porosity reduces the extracellular enzymatic activity of the soil because all the functional groups present in the biochar have the tendency to bind up extracellular enzymes. This results in interference with the diffusion rate of the substrates (Lammirato et al. 2011).

6.3.3 Effects of Biochar Application on Plants Growth

6.3.3.1 Impacts on Plant Growth

Biochar is a novel approach to sustainable agriculture, i.e. growing of crops in such a manner which does not harm the ecosystem, and is healthy for man as well as animal. Biochar is found to be more effective than other organic matter as it has the capacity of retaining nutrients for the plants. In biochar, availability of large surface area and pore provides home to many microorganisms which assist in the plant growth. Application of biochar in soil assists in crop improvement, soil fertility leading to crop productivity. Uzoma et al. (2011) carried out an experiment in which the effect of cow dung derived manure with the addition of biochar at different rates in the sandy soil was observed on the growth of maize plant. Noticeable improvement in the growth and productivity was observed. Jeffery (2011) conducted a systematic study to check the impact of biochar on the productivity. A significant increase in productivity was seen due to the increase in water uptake capacity of soil and liming effect that the biochar provides. Liu et al. (2013) did an experiment to test the productivity of soil and reported good response. The experiment was done for potted plants as well as on the field. The responses of the potted plants were better than that on the field. Deb et al. (2016) found out that the effect of biochar on the soil that is deficient in nutrient showed better response, resulting in higher productivity of crops. Biochar when introduced in the soil increases the surface area, cation exchange capacity, soil porosity, water-holding capacity, nutrient retention and liming effect (Yamato et al. 2006; Thies and Rillig 2009; Liu et al. 2013) which ultimately increases crop productivity and plant growth (Yu et al. 2019).

6.3.3.2 Impacts on Plant Biomass

Application of biochar significantly ameliorates fertility of soil as well as the productivity of crops (Lehmann and Joseph 2015).

Wang et al. (2019) observed that dry weight of bamboo plant was increased significantly in amendment of 5% wood biochar, 5% rice straw biochar and 5% Chinese walnut shell biochar. The results also revealed that the application of 5%

straw biochar was found to be most effective and enhanced plant biomass with increase of 157, 113 and 111% in leaves, roots and stems of bamboo, respectively.

Glaser et al. (2001) observed increase in the biomass of *Oryza sativa* L. (rice) by 20% and *Vigna unguiculata* L. (cowpea) by 50%, owing to the application of biochar at 68 t/ha and at 136.75 t/ha; it was noted that the biomass of cowpea was increased by 100%. Vaccari et al. (2011) noted that amendment of biochar improved the biomass and yield in *Triticum durum L.* (durum wheat) by 30%. Similarly, Oguntunde et al. (2004) recorded an increase in biomass of grain by 91% and in maize by 44% by biochar application in the soil. Biochar considerably modified the plant biomass, shoot and root dry weight, which was increased to 52 and 36%, respectively (Abid et al. 2017). Rondon et al. (2007) recorded that by the application of 60 g/kg biochar over control treatment, there was 39% increase in dry biomass in *Phaseolus vulgaris L.* (bean). Cheng et al. (2014) hypothesized and observed a significant increase in the biomass and pod yield of peanut. Rogovska et al. (2014) recorded a substantial increase in maize biomass yield of about 11–55% after application of biochar. Wu et al. (2017) conducted an experiment to observe the effect of biochar-supported Ni/Fe nanoparticle-treated soil on *Brassica chinensis* in which the plant biomass and the root and shoot length were reported to increase considerably (Table 6.3).

6.3.3.3 Impacts on Photosynthetic Pigments

Photosynthesis is a very important process in the plants in which the photosynthetic pigments absorb the energy from light; this energy is then used to form sugar and other nutrients by using carbon dioxide and water. The process is responsible for carbon sequestration. The deficiency of nutrients in the soil reduces the production of chlorophyll, resulting in reduced crop yield and quality. Several studies have confirmed a direct correlation between leaf nitrogen content and photosynthetic rate (Field and Mooney 1986; Wright et al. 2007). Iannelli et al. (2002) observed and reported a significant increase in the length of stem of various crops by the application of biochar in metal-contaminated soils due to the improved photosynthetic condition. Sun et al. (2016) have also reported that a positive effect of biochar on photosynthetic rate of plants was observed. Cheng et al. (2014) conducted a study

Table 6.3 Biomass responses of different feedstock biochars

Biochar feedstock	Soil type	Application rate	Crops	Biomass response (%)	Reference
Eucalyptus deglupta Blume	Volcanic-ash inceptisol	25 kg/g of soil	Rice	+166	Noguera et al. (2010)
Green waste	Alfisol	50 t/ha	Radish	+91	Chan et al. (2007)
Green waste	Alfisol	100 t/ha	Radish	+130	Chan et al. (2007)
Secondary forest wood	Highly weathered ferralsol	11 t/ha	Rice	+22	Steiner et al. (2007)

in which biochar derived from peanut shell was amended in two types of soil from Queensland (Australia) to quantify the impact of biochar on the yield and the rate of photosynthesis in peanut plant. An increasing photosynthetic rate and capacity was observed that resulted in good crop yield. The improved photosynthetic ability in the reported study is direct association with leaf nitrogen. Abid et al. (2017) found that the application of biochar in soil highly influenced and enhanced all the photosynthetic and accessory pigments (chlorophyll a, b and total chlorophyll, anthocyanin, carotenoids and lycopene) in tomato (*Solanum lycopersicum* L.). The presence of contaminants in the soil affects plant growth by decreasing the surface area of the leaf, reducing the production of chlorophyll and other pigments. The application of biochar appears to alleviate the stress caused by the contaminants.

6.3.3.4 Impacts on Yield of Crops

As the impact of biochar increases plant growth, plant biomass and plant photosynthetic pigment, it is an obvious conclusion that the yield of crops will get increased by its application as well (Table 6.4). There are several experiments conducted to verify the impact on crop yield by the amendment of biochar in the soil. The amendment of biochar on nutrient-poor soil has been found to improve the crop yield (Van Zwieten et al. 2010; Gaskin et al. 2010; Pandey et al. 2015; Li et al. 2018; Wu et al. 2019; Yu et al. 2019). The crop responses have been found to be highly variable which depends on the biochar type and its application rate. The response also depends on the type of soil as well as the climatic conditions. Jeffery et al. (2011) conducted a series of experiments where it was noted that the crop yield and productivity was significantly improved by the application of biochar. Among the feedstock, the result for poultry litter was the best where 28% increase in crop yield was reported. Yamato et al. (2006) acknowledged a substantial increase in the yield of maize due to the amendment of biochar with fertilizers. Steiner et al. (2007) noticed 4–12 times increase in the yield of rice and sorghum in the harvest. It was due to the fact that the application of compost/fertilizer was done with biochar.

6.4 Role of Biochar Application in Mitigation of Gaseous Emissions

Nitrous oxide (N_2O) and methane (CH_4) are the significant greenhouse gases with the global warming potential of 298 and 28–36 times than that of carbon dioxide over a 100-year time, and together they are responsible for 6–8% of anthropogenic radiative force (Davidson 2009). Agricultural practices in total contribute to about 60% of the anthropogenic source of N_2O (Reay et al. 2012). Therefore, new agricultural practices are being developed to minimize this number, and such practice includes the use of biochar.

The results of biochar amendment on the emission of GHGs like N_2O, CH_4 and CO_2 have been widely studied by various scientists with varying results. Biochar amendment is believed to supress the agricultural emission of GHGs due to its black colour which may affect the albedo which directly affects the GHG emission

Table 6.4 Crop yield responses of different feedstock biochars

Biochar feedstock	Soil type	Application rate	Crops	Crop yield response (%)	Reference
Bark of *Acacia mangium*	Acidic soil	37 t/ha	Maize	+50	Yamato et al. (2006)
Eucalyptus deglupta Blume	Volcanic-ash inceptisol	25 g/kg soil	Rice	+294	Noguera et al. (2010)
Eucalyptus deglupta Blume	Clay-loam oxisol	90 g/kg soil	Phaseolus vulgaris L.	+46	Rodon et al. (2007)
Poultry litter	Alfisol	50 t/ha	Radish	+96	Chan et al. (2008)
Secondary forest wood	Highly weathered xanthic ferralsol	11 t/ha	Rice/ sorghum	+17	Steiner et al. (2007)
Wheat straw	Calcareous loam soil	20 t/ha	Corn	+18.2	Zhang et al. (2012)

resulting from the soil. Most of the studies regarding N_2O have shown a fairly consistent decrease in N_2O emission (Chen et al. 2017; Fungo et al. 2019; Song et al. 2019; Weldon et al. 2019). Although it has been established that amendment of biochar reduces N_2O emission, other gases like CO_2 emission have been fairly controversial with wide variation in the CO_2 emission of the biochar-amended soil. Spokas and Reicosky (2009) conducted a study in the period of 100 incubation days where three different types of soil were amended with 16 types of biochar. The study concluded that all three types had three different effects with actions like suppression, neutral results and an increase in CO_2 respiration as a result of biochar amendment. It has been reported that amendment of biochar results in suppression of CH_4 emission into the atmosphere. A study conducted by Liu et al. (2012) showed that the biochar-amended paddy soil reduced the CH_4 emission by 91.2% in comparison with the soil without biochar-amended soil. In contrast to these findings, Castaldi et al. (2011) conducted a study on biochar-amended wheat crop and the impact of the amendment on greenhouse gas emission. They reported that the amendment had no significant difference on the emission chart in comparison with other biochar concentrations or the control.

6.5 Mitigation of Acidity of Soil

Acid soil (<5.5 pH) is a worldwide problem wherein approx. a considerable share of agricultural land suffers from this problem. Acidity of the soil can prove to be a menace as it may severely affect the productivity of the crop (Goulding 2016).

Acidity is mainly caused by acid deposition (Duan et al. 2016), use of ammonium-based fertilizer, etc. (Cai et al. 2014). Nowadays, different techniques are applied to control the acidity of the soil, some of which include the use of lime, industrial by-products (alkaline), organic wastes, etc. The use of biochar for the control of acidity of the soil is a relatively new technique but has been reported to be effective because of its other benefits such as waste disposal, mitigation of global warming, reduction of heavy metal toxicity and energy production. Generally, biochars are alkaline in nature and therefore when applied help control the pH of the acidic soil (Shi et al. 2015). This mechanism also helps in increasing the crop productivity which has been proven in numerous studies conducted by scientists all over the world in various soils with different types of feedstocks (Jeffery et al. 2011).

The characteristics of biochar may vary considerably depending upon the feedstock and method of production, but all the biochar with specific property of high pH buffering capability and alkalinity helps to control the acidity of soil. It has been established that amendment of biochar increases the pH of soil as most of the biochars have pH > 7. Amendment of biochar in the alkaline soils shows no significant change and, in many cases, has been found to even decrease the pH level to suitable condition due to its alkaline nature (Dai et al. 2016; Fidel et al. 2017).

6.6 Role of Biochar in Carbon Sequestration

Carbon sequestration may be defined as long-term entrapment of environmental carbon-based compounds like CO_2 or other forms which may contribute in increasing the global temperature of the planet. Storage or sequestration of carbon by biochar amendment is an effective method and at the same time results in enhancement of the carbon content in the agroecosystem (Jung et al. 2018). The amendment of biochar incorporated with production of bioenergy crop can efficiently sequester more carbon from the atmosphere than being produced (Roberts et al. 2010). Lehmann (2007) in his study stated that amendment of biochar increases the carbon sequestration with a value of around 20%. Biochar is a very stable compound in the soil and may persist for more than 1000 years because of the resistance towards the microbial decomposition and mineralization which leads to the sequestration of CO_2, and this may be the basic criterion for selecting the biochar over other amendments for the purposes of carbon sequestration. The other reason of persistence of biochar in the environment can be resistance to chemical decomposition even in the tropics where there is high probability of weathering (Schneider et al. 2011). It has proven itself to be one of the most appropriate solutions to reduce the CO_2 emission into the atmosphere solely based on the fact that it has potential to sequester about 400 billion tons of carbon by the year 2100 and also to reduce the atmospheric CO_2 concentration by 37 ppm (Van Zwieten et al. 2009). The method by which biochars sequester carbon or reduce CO_2 level is by reducing the requirement of additional fertilizer during the time of microbial life proliferation which ultimately results in carbon entrapment in the soil.

6.7 Conclusion

The application of biochar is a cost-effective and environment-friendly tool because it can be produced from naturally derived feedstock like crop residues, wood, waste from industries, etc. Biochar has several abilities to enhance soil characteristics like nutrient level, water-holding capacity, organic matter, and microbial population along with a reduction in the bioavailability of toxic substances and also to help the plants to fight against the majority of biotic and abiotic stresses. The productivity of numerous crops like *Zea maize, Oryza sativa, Phaseolus vulgaris L., Vigna radiata, Vigna unguiculata, Triticum durum, Brassica chinensis*, etc. enhanced substantially by application of different types of biochar. Further, more field-based studies should be done to assess its long-term impacts on soil quality and to explore its use as a sustainable alternative to the existing non-friendly fertilizers.

Acknowledgement Kuldeep Bauddh is thankful to the Science and Engineering Research Board (SERB), India, for the award of Research Grant (EEQ/2017/000476). Author Abhishek Guldhe is thankful to the Department of Biotechnology, Govt. of India, for Ramalingaswami Fellowship.

References

Abid M, Danish S, Zafar-ul-Hye M, Shaaban M, Iqbal MM, Rehim A, Qayyum MF, Naqqash MN (2017) Biochar increased photosynthetic and accessory pigments in tomato (*Solanum lycopersicum* L.) plants by reducing cadmium concentration under various irrigation waters. Environ Sci Pollut Res 24(27):22111–22118

Al-Wabel M, Al-Omran A, El-Naggar AH, Nadeem M, Usman ARA (2013) Pyrolysis temperature induced changes in characteristics and chemical composition of biochar produced from conocarpus wastes. Bioresour Technol 131:374–379

Antón-Herrero R, García-Delgado C, Alonso-Izquierdo M, García-Rodríguez G, Cuevas J, Eymar E (2018) Comparative adsorption of tetracyclines on biochars and stevensite: looking for the most effective adsorbent. Appl Clay Sci 160:162–172

Arshad M, Khan AHA, Hussain I, Anees M, Iqbal M, Soja G, Yousaf S (2017) The reduction of chromium (VI) phytotoxicity and phytoavailability to wheat (*Triticum aestivum* L.) using biochar and bacteria. Appl Soil Ecol 114:90–98

Bagreev A, Bandosz TJ, Locke DC (2001) Pore structure and surface chemistry of adsorbents obtained by pyrolysis of sewage sludge-derived fertilizer. Carbon 39(13):1971–1979

Bailey VL, Fansler SJ, Smith JL, Bolton H Jr (2011) Reconciling apparent variability in effects of biochar amendment on soil enzyme activities by assay optimization. Soil Biol Biochem 43 (2):296–301

Bird MI, Ascough PL, Young IM, Wood CV, Scott AC (2008) X-ray microtomographic imaging of charcoal. J Archaeol Sci 35(10):2698–2706

Boldrin A, Andersen JK, Møller J, Christensen TH, Favoino E (2009) Composting and compost utilization: accounting of greenhouse gases and global warming contributions. Waste Manag Res 27:789–799

Bridle TR, Pritchard D (2004) Energy and nutrient recovery from sewage sludge via pyrolysis. Water Sci Technol 50(9):169–175

Burns RG, DeForest JL, Marxsen J, Sinsabaugh RL, Stromberger ME, Wallenstein MD, Weintraub MN, Zoppini A (2013) Soil enzyme in changing environment: current knowledge and future directions. Soil Biol Biochem 58:216–234

Cai Z, Wang B, Xu M, Zhang H, Zhang L, Gao S (2014) Nitrification and acidification from urea application in red soil (Ferralic Cambisol) after different long-term fertilization treatments. J Soil Sediment 14:1526–1536
Cantrell KB, Hunt PK, Uchimiya M, Novak JM, Ro KS (2012) Impact of pyrolysis temperature and manure source on physicochemical characteristics of biochar. Bioresour Technol 107:419–428
Cao X, Harris W (2010) Properties of dairy-manure-derived biochar pertinent to its potential use in remediation. Bioresour Technol 101(14):5222–5228
Castaldi S, Riondino M, Baronti S, Esposito FR, Marzaioli R, Rutigliano FA, Miglietta F (2011) Impact of biochar application to a Mediterranean wheat crop on soil microbial activity and greenhouse gas fluxes. Chemosphere 85(9):1464–1471
Chan KY, Van Zwieten L, Meszaros I, Downie A, Joseph S (2007) Agronomic values of greenwaste biochar as a soil amendment. Soil Research 45(8):629
Chan KY, Van Zwieten L, Meszaros I, Downie A, Joseph S (2008) Agronomic values of greenwaste biochar as a soil amendment. Soil Res 45(8):629–634
Chen T, Zhou Z, Xu S, Wang H, Lu W (2015) Adsorption behavior comparison of trivalent and hexavalent chromium on biochar derived from municipal sludge. Bioresour Technol 190:388–394
Chen L, Chen XL, Zhou CH, Yang HM, Ji SF, Tong DS, Reay CMQ (2017) Environmental-friendly montmorillonite-biochar composites: facile production and tunable adsorption-release of ammonium and phosphate. J Clean Prod 156:648–659
Cheng CH, Lehmann J, Thies JE, Burton SD (2008) Stability of black carbon in soils across a climatic gradient. J Geophys Res Earth Surf 113(G2):G02027. https://doi.org/10.1029/2007JG000642
Cheng CH, Lin TP, Lehmann J, Fang LJ, Yang YW, Menyailo OV, Lai JS (2014) Sorption properties for black carbon (wood char) after long term exposure in soils. Org Geochem 70:53–61
Chintala R, Mollinedo J, Schumacher TE, Malo DD, Julson JL (2014) Effect of biochar on chemical properties of acidic soil. Arch Agron Soil Sci 60(3):393–404
Dai Z, Meng J, Shi Q, Xu B, Lian Z, Brookes P, Xu JM (2016) Effects of manure-and lignocellulose-derived biochars on adsorption and desorption of zinc by acidic types of soil with different properties. Eur J Soil Sci 67:40–50
Dai Z, Zhang X, Tang C, Muhammad N, Wu J, Brookes PC, Xu J (2017) Potential role of biochars in decreasing soil acidification-a critical review. Sci Total Environ 581:601–611
Davidson EA (2009) The contribution of manure and fertilizer nitrogen to atmospheric nitrous oxide since 1860. Nat Geosci 2:659–662
Deb D, Kloft M, Lässig J, Walsh S (2016) Variable effects of biochar and P solubilizing microbes on crop productivity in different soil conditions. Agroecol Sustatin Food Syst 40(2):145–168
Dias BO, Silva CA, Higashikawa FS, Roig A, Sánchez-Monedero MA (2010) Use of biochar as bulking agent for the composting of poultry manure: effect on organic matter degradation and humification. Bioresour Technol 101(4):1239–1246
Duan L, Chen X, Ma X, Zhao B, Larssen T, Wang S, Ye Z (2016) Atmospheric S and N deposition relates to increasing riverine transport of S and N in Southwest China: implications for soil acidification. Environ Pollut 218:1191–1199
Fidel RB, Laird DA, Thompson ML, Lawrinenko M (2017) Characterization and quantification of biochar alkalinity. Chemosphere 167:367–373
Field C, Mooney HA (1986) The photosynthesis nitrogen relationship in wild plants. In: Givnish TJ (ed) On the economy of plant form and function. Cambridge University Press, Cambridge
Fungo B, Chen Z, Butterbach-Bahl K, Lehmannn J, Saiz G, Braojos V, Neufeldt H (2019) Nitrogen turnover and N2O/N2 ratio of three contrasting tropical soils amended with biochar. Geoderma 348:12–20
Galinato SP, Yoder JK, Granatstein D (2011) The economic value of biochar in crop production and carbon sequestration. Energy Policy 39(10):6344–6350

Gaskin JW, Speir RA, Harris K, Das KC, Lee RD, Morris LA, Fisher DS (2010) Effect of peanut hull and pine chip biochar on soil nutrients, corn nutrient status, and yield. Agron J 102:623–633
Glaser B, Haumaier L, Guggenberger G, Zech W (2001) The 'Terra Preta' phenomenon: a model for sustainable agriculture in the humid tropics. The Sci Nature 88:37–41
Gomez JD, Denefa K, Stewart CE, Zhengd J, Cotrufo MF (2014) Biochar addition rate influences soil microbial abundance and activity in temperate soils. Eur J Soil Sci 65:28–39
Goulding K (2016) Soil acidification and the importance of liming agricultural soils with particular reference to the United Kingdom. Soil Use Manag 32:390–399
Graetz RD, Skjemstad JO (2003) The charcoal sink of biomass burning on the Australian continent. CSIRO Atmos Res 64:1–61
Gul S, Whalen JK (2016) Biochemical cycling of nitrogen and phosphorus in biochar-amended soils. Soil Biol Biochem 103:1–15
Hammes K, Smernik RJ, Skjemstad JO, Herzog A, Vogt UF, Schmidt MWI (2006) Synthesis and characterisation of laboratory-charred grass straw (Oryza sativa) and chestnut wood (Castanea sativa) as reference materials for black carbon quantification. Org Geochem 37(11):1629–1633
Harvey OM, Herbert BE, Kuo LJ, Louchouarn P (2012) Generalized two-dimensional perturbation correlation infrared spectroscopy reveals mechanisms for the development of surface charge and recalcitrance in plant-derived biochars. Environ Sci Technol 46:10641–10650
Hunt J, DuPonte M, Sato D, Kawabata A (2010) The basics of biochar: a natural soil amendment. Soil Crop Manag 30(7):1–6
Iannelli MA, Pietrini F, Fiore L, Petrilli L, Massacci A (2002) Antioxidant response to cadmium in *Phragmites australis* plants. Plant Physiol Biochem 40(11):977–982
IUPAC (1972) Manual of symbols and terminology. Pure Appl Chem 31:578
Jaafar NM, Clode PL, Abbott LK (2014) Microscopy observations of habitable space in biochar for colonization by fungal hyphae from soil. J Integ Agri 13(3):483–490
Jeffery S, Verheijen FGA, van der Velde M, Bastos AC (2011) A quantitative review of the effects of biochar application to soils on crop productivity using meta-analysis. Agric Ecosyst Environ 144(1):175–187
Jeffery S, Meinders MB, Stoof CR, Bezemer TM, van de Voorde TF, Mommer L, van Groenigen JW (2015) Biochar application does not improve the soil hydrological function of a sandy soil. Geoderma 251:47–54
Jiang TY, Jiang J, Xu RK, Li Z (2012) Adsorption of Pb (II) on variable charge soils amended with rice-straw derived biochar. Chemosphere 89(3):249–256
Jien SH, Wang CS (2013) Effects of biochar on soil properties and erosion potential in a highly weathered soil. Catena 110:225–233
Jin J, Kang M, Sun K, Pan Z, Wu F, Xing B (2016) Properties of biochar-amended soils and their sorption of imidacloprid, isoproturon, and atrazine. Sci Total Environ 550:504–513
Joseph SD, Camps-Arbestain M, Lin Y, Munroe P, Chia CH, Hook J, van Zwieten L, Kimber S, Cowie A, Singh BP, Lehmann J, Foidl N, Smernik RJ, Amonette JE (2010) An investigation into the reactions of biochar in soil. Aust J Soil Res 48:501–515
Jung K, Lee SY, Lee YJ (2018) Facile one-pot hydrothermal synthesis of cubic spinel-type manganese ferrite/biochar composites for environmental remediation of heavy metals from aqueous solutions. Bioresour Technol 261:1–9
Kookana RS, Sarmah AK, Van Zwieten L, Krull E, Singh B (2011) Biochar application to soil: agronomic and environmental benefits and unintended consequences. In: Sparks DL (ed) Advances in agronomy, vol 112, pp 103–143
Kumar M, Bauddh K, Sainger M, Sainger PA, Singh JS, Singh RP (2012) Increase in growth, productivity and nutritional status of rice (*Oryza sativa* L. c.v. basmati) and enrichment in soil fertility applied with an organic matrix entrapped urea. J Crop Sci Biotechnol 15(2):137–144
Kumar M, Bauddh K, Kumar S, Sainger M, Sainger PA, Singh RP (2013a) Increase in growth, productivity and nutritional status of wheat (*Triticum aestivum* L. C.v. Wh-711) and enrichment in soil fertility applied with organic matrix entrapped urea. J Environ Biol 34:1–9
Kumar S, Bauddh K, Barman SC, Singh RP (2013b) Evaluation of conventional and Organic matrix entrapped urea and diammonium phosphate for growth and productivity of *Triticum aestivum*

L. and mobilization of NO_3^-, NO_2^-, NH_4^+ and PO_4^{-3} from soil to plant leaves. Int J Agron Plant Prod 4(6):1357–1368

Kumar S, Bauddh K, Barman SC, Singh RP (2014a) Amendments of microbial biofertilizers and organic substances reduces requirement of urea and DAP with enhanced nutrient availability and productivity of wheat (*Triticum aestivum* L.). Ecol Eng 71:432–437

Kumar S, Bauddh K, Barman SC, Singh RP (2014b) Organic matrix entrapped bio-fertilizers increase growth, productivity and yield of *Triticum aestivum* L. and mobilization of NO_3^-, NO_2^-, NH_4^+ and PO_4^{-3} from soil to plant leaves. J Agri Sci Technol 16:315–329

Kumar M, Bauddh K, Sainger M, Sainger PA, Singh RP (2014c) Increase in growth, productivity and nutritional status of wheat (*Triticum aestivum* L.) and enrichment in soil microbial population applied with biofertilizers entrapped with organic matrix. J Plant Nutr 38:260–276

Kumar M, Bauddh K, Kumar S, Sainger M, Sainger PA, Singh RP (2015) Enhancing efficacy of Azotobactor and Bacillus entrapping in organic matrix for rice cultivation. Agroecol Sustain Food Syst 39:907–923

Laghari M, Mirjat MS, Hu Z, Fazal S, Xiao B, Hu M, Guo D (2015) Effects of biochar application rate on sandy desert soil properties and sorghum growth. Catena 135:313–320

Laird DA (2008) The charcoal version: a win-win scenario for simultaneously producing bioenergy, permanently sequestering C, while improving soil and water quality. Agron J 100:178–181

Laird DA, Brown RC, Amonette JE, Lehmann J (2009) Review of the pyrolysis platform for coproducing bio-oil and biochar. Biofuels Bioprod Biorefin 3(5):547–562

Lammirato G, Miltner A, Kaestner M (2011) Effects of wood char and activated carbon on the hydrolysis of cellobiose by b-glucosidase from *Aspergillus niger*. Soil Biol Biochem 43:1936–1942

Lehmann J (2007) Bio-energy in the black. Front Ecol Environ 5:381–387

Lehmann J, Joseph S (eds) (2015) Biochar for environmental management: science, technology and implementation. Routledge, New York

Lehmann J, Gaunt J, Rondon M (2006) Bio-char sequestration in terrestrial ecosystems–a review. Mitig Adapt Strat Glob Chang 11(2):403–427

Li H, Zhang D, Han X, Xing B (2014) Adsorption of antibiotic ciprofloxacin on carbon nanotubes: pH dependence and thermodynamics. Chemosphere 95(1):150–155

Li R, Liang W, Huang H, Jiang S, Guo D, Li M, Zhang Z, Ali A, Wang JJ (2018) Removal of cadmium(II) cations from an aqueous solution with aminothiourea chitosan strengthened magnetic biochar. J Appl Polym Sci 135(19):46239. https://doi.org/10.1002/app.46239

Liang B, Lehmann J, Sohi SP, Thies JE, O'Neill B, Trujillo L, Gaunt J, Solomon D, Grossman J, Neves EG, Luizão FJ (2010) Black carbon affects the cycling of non-black carbon in soil. Org Geochem 41(2):206–213

Lin L, Jiang W, Xu P (2017) Comparative study on pharmaceuticals adsorption in reclaimed water desalination concentrate using biochar: impact of salts and organic matter. Sci Total Environ 601:857–864

Liu X, Zhang A, Ji C, Joseph S, Bian R, Li L, Paz-Ferreiro J (2013) Biochar's effect on crop productivity and the dependence on experimental conditions—a meta-analysis of literature data. Plant Soil 373(1–2):583–594

Liu Z, Chen X, Jing Y, Li Q, Zhang J, Huang Q (2014) Effects of biochar amendment on rapeseed and sweet potato yields and water stable aggregate in upland red soil. Catena 123:45–51

Liu W, Huo R, Xu J, Liang S, Li J, Zhao T, Wang S (2017) Effects of biochar on nitrogen transformation and heavy metals in sludge composting. Bioresour Technol 235:43–49

Liu Y, Lonappan L, Brar SK, Yang S (2018) Impact of biochar amendment in agricultural soils on the sorption, desorption, and degradation of pesticides: a review. Sci Total Environ 645:60–70

Lone AH, Najar GR, Ganie MA, Sofi JA, Ali T (2015) Biochar for sustainable soil health: a review of prospects and concerns. Pedosphere 25(5):639–653

Luo S, Qin PF, Shao JH, Peng L, Zeng QR, Gu JD (2013) Synthesis of reactive nanoscale zero valent iron using rectorite supports and its application for Orange II removal. Chem Eng J 223:1–7

Masto RE, Ansari MA, George J, Selvi VA, Ram LC (2013) Co-application of biochar and lignite fly ash on soil nutrients and biological parameters at different crop growth stages of *Zea mays*. Ecol Eng 58:314–322

Muhammad N, Dai Z, Xiao K, Meng J, Brookes PC, Liu X, Xu J (2014) Changes in microbial community structure due to biochars generated from different feedstocks and their relationships with soil chemical properties. Geoderma 226:270–278

Nartey OD, Zhao B (2014) Biochar preparation, characterization, and adsorptive capacity and its effect on bioavailability of contaminants: an overview. Adv Mat Sci Eng 2014:1–12. https://doi.org/10.1155/2014/715398

Ndor E, Dauda SN, Azagaku ED (2015) Response of maize varieties (*Zea mays*) to biochar amended soil in Lafia, Nigeria. J Experim Agri Inter 5(6):525–531. https://doi.org/10.9734/AJEA/2015/12375

Niazi NK, Bibi I, Shahid M, Ok YS, Burton ED, Wang H, Lüttge A (2018) Arsenic removal by perilla leaf biochar in aqueous solutions and groundwater: an integrated spectroscopic and microscopic examination. Environ Pollut 232:31–41

Noguera D, Rondon M, Laossi KR, Hoyos V, Lavelle P, de Carvalho MHC, Barot S (2010) Contrasted effect of biochar and earthworm on rice growth and resource allocation in different soils. Soil Biol Biochem 42:1017–1027

Novak JM, Lima I, Xing B, Gaskin JW, Steiner C, Das KC, Schomberg H (2009) Characterization of designer biochar produced at different temperatures and their effects on a loamy sand. Annals of Environ Sci 3:195–206

Oguntunde PG, Fosu M, Ajayi AE, Van de Giesen N (2004) Effects of charcoal production on maize yield, chemical properties and texture of soil. Biol Fertil Soil 39:295–299

Oladele SO, Adeyemo AJ, Awodun MA (2019) Influence of rice husk biochar and inorganic fertilizer on soil nutrients availability and rain-fed rice yield in two contrasting soils. Geoderma 336:1–11

Pandey DS, Pan I, Das S, Leahy JJ, Kwapinski W (2015) Multi-gene genetic programming based predictive models for municipal solid waste gasification in a fluidized bed gasifier. Bioresour Technol 179:524–533

Pandian K, Subramaniayan P, Gnasekaran P, Chitraputhirapillai S (2016) Effect of biochar amendment on soil physical, chemical and biological properties and groundnut yield in rainfed Alfisol of semi-arid tropics. Arch Agron Soil Sci 62(9):1293–1310

Paz-Ferreiro J, Gascó G, Gutiérrez B, Méndez A (2012) Soil biochemical activities and the geometric mean of enzyme activities after application of sewage sludge and sewage sludge biochar to soil. Biol Fertil Soils 48(5):511–517

Quilliam RS, Glanville HC, Wade SC, Jones DL (2013) Life in the 'charosphere'–does biochar in agricultural soil provide a significant habitat for microorganisms? Soil Biol Biochem 65:287–293

Rai A, Kumar S, Bauddh K, Singh N, Singh RP (2017) Improvement in growth and alkaloid content of Rauwolfia serpentina on application of organic matrix entrapped biofertilizers (*Azotobacter chroococcum*, Azospirillum brasilense and Pseudomonas putida). J Plant Nutr 40:2237–2247

Randolph P, Bansode RR, Hassan OA, Rehrah D, Ravella R, Reddy MR, Ahmedna M (2017) Effect of biochars produced from solid organic municipal waste on soil quality parameters. J Environ Manag 192:271–280

Reay DS, Davidson EA, Smith KA, Smith P, Melillo JM, Dentener F, Crutzen PJ (2012) Global agriculture and nitrous oxide emissions. Nat Clim Chang 2(6):410–416

Roberts KG, Gloy BA, Joseph S, Scott NR, Lehmann J (2010) Life cycle assessment of biochar systems: estimating the energetic, economic, and climate change potential. Environ Sci Technol 44(2):827–833

Rogovska N, Laird DA, Rathke SJ, Karlen DL (2014) Biochar impact on Midwestern Mollisols and maize nutrient availability. Geoderma 230:340–347

Rondon MA, Lehmann J, Ramírez J, Hurtado M (2007) Biological nitrogen fixation by common beans (*Phaseolus vulgaris* L.) increases with bio-char additions. Biol Fertil Soils 43:699–708

Schneider D, Escala M, Supawittayayothin K, Tippayawong N (2011) Characterization of biochar from hydrothermal carbonization of bamboo. Int J Energ Environ 2:647–652

Saha A, Basak BB, Gajbhiye NA, Kalariya KA, Manivel P (2019) Sustainable fertilization through co-application of biochar and chemical fertilizers improves yield, quality of Andrographis paniculata and soil health. Ind Crop Prod 140:111607

Shepherd JG, Joseph S, Sohi SP, Heal KV (2017) Biochar and enhanced phosphate capture: mapping mechanisms to functional properties. Chemosphere 179:57–74

Shi K, Xie Y, Qiu Y (2015) Natural oxidation of a temperature series of biochars: opposite effect on the sorption of aromatic cationic herbicides. Ecotoxicol Environ Saf 114:102–108

Singh RP, Sainger M, Bauddh K, Senger RS, Jaiwal PK (2010) Sustained nutrient supply reduced nutrient loss and high plant productivity with slow release fertilizers. In: Senger RS, Sharma AK (eds) Stable food production and sustainable agriculture. Studium Press, Lanham, pp 62–79

Sohi SP, Krull E, Lopez-Capel E, Bol R (2010) A review of biochar and its use and function in soil. In: Advances in agronomy, vol 105. Academic, Burlington, pp 47–82

Song Y, Li Y, Cai Y, Fu S, Luo Y, Wang H, Chang SX (2019) Biochar decreases soil N2O emissions in Moso bamboo plantations through decreasing labile N concentrations, N-cycling enzyme activities and nitrification/denitrification rates. Geoderma 348:135–145

Spokas KA, Reicosky DC (2009) Impacts of sixteen different biochars on soil greenhouse gas production. Annals Environ Sci 3:179–193

Spokas KA, Koskinen WC, Baker JM, Reicosky DC (2009) Impacts of woodchip biochar additions on greenhouse gas production and sorption/degradation of two herbicides in a Minnesota soil. Chemosphere 77:574–581

Spokas KA, Cantrell KB, Novak JM, Archer DW, Ippolito JA, Collins HP, Boateng AA, Lima IM, Lamb MC, McAloon AJ, Lentz RD, Nichols KA (2012) Biochar: a synthesis of its agronomic impact beyond carbon sequestration. J Environ Qual 41(4):973–989

Steiner C, Teixeira WG, Lehmann J, Nehls T, de Macedo JLV, Blum WEH, Zech W (2007) Long term effects of manure, charcoal and mineral fertilization on crop production and fertility on a highly weathered central Amazonian upland soil. Plant Soil 291:275–290

Subedi R, Taupe N, Ikoyi I, Bertora C, Zavattaro L, Schmalenberger A, Grignani C (2016) Chemically and biologically-mediated fertilizing value of manure-derived biochar. Sci Total Environ 550:924–933

Suliman W, Harsh JB, Abu-Lail NI, Fortuna AM, Dallmeyer I, Garcia-Perez M (2016) Influence of feedstock source and pyrolysis temperature on biochar bulk and surface properties. Biomass Bioenergy 84:37–48

Suliman W, Harsh JB, Abu-Lail NI, Fortuna AM, Dallmeyer I, Garcia-Pérez M (2017) The role of biochar porosity and surface functionality in augmenting hydrologic properties of a sandy soil. Sci Total Environ 574:139–147

Sun J, He F, Pan Y, Zhang Z (2016) Effects of pyrolysis temperature and residence time on physicochemical properties of different biochar types. Acta Agric Scand Sect B: Soil Plant Sci 67(1):12–22

Thies JE, Rillig MC (2009) Characteristics of biochar: biochar properties. In: Lehmann J, Joseph S (eds) Biochar for environmental management – science and technology. Earthscan, London, pp 85–105

Uzoma KC, Inoue M, Andry H, Fujimaki H, Zahoor A, Nishihara E (2011) Effect of cow manure biochar on maize productivity under sandy soil condition. Soil Use Manag 27:205–212

Vaccari FPS, Baronti E, Lugato L, Genesio S, Castaldi F, Fornasier F, Miglietta F (2011) Biochar as a strategy to sequester carbon and increase yield in durum wheat. Eur J Agron 34:231–238

Van Zwieten L, Kimber S, Downie A, Morris S, Petty S, Rust J, Chan KY (2009) A glasshouse study on the interaction of low mineral ash biochar with nitrogen in a sandy soil. Soil Res 48 (7):569–576

Van Zwieten L, Kimber S, Morris S, Chan KY, Downie A, Rust J, Joseph S, Cowie A (2010) Effects of biochar from slow pyrolysis of papermill waste on agronomic performance and soil fertility. Plant Soil 327(1–2):235–246

Wang S, Zhao X, Xing G, Yang L (2013) Large-scale biochar production from crop residue: a new idea and the biogas-energy pyrolysis system. Bio Res 8(1):8–11

Wang RZ, Huang DL, Liu YG, Zhang C, Lai C, Zeng GM, Luo H (2018) Investigating the adsorption behavior and the relative distribution of Cd^{2+} sorption mechanisms on biochars by different feedstock. Bioresour Technol 261:265–271

Wang Y, Zhong B, Shafi M, Ma J, Guo J, Wu J, Ye Z, Liu D, Jin H (2019) Effects of biochar on growth, and heavy metals accumulation of moso bamboo (Phyllostachy pubescens), soil physical properties, and heavy metals solubility in soil. Chemosphere 219:510–516

Warnock DD, Lehmann J, Kuyper TW, Rillig MC (2007) Mycorrhizal responses to biochar in soil–concepts and mechanisms. Plant Soil 300(1–2):9–20

Wei B, Yang L (2010) A review of heavy metal contaminations in urban soils, urban road dusts and agricultural soils from China. Microchem J 94(2):99–107

Weldon S, Rasse DP, Budai A, Tomic O, Dörsch P (2019) The effect of a biochar temperature series on denitrification: which biochar properties matter? Soil Biol Biochem 135:173–183

Windeatt JH, Ross AB, Williams PT, Forster PM, Nahil MA, Singh S (2014) Characteristics of biochars from crop residues: potential for carbon sequestration and soil amendment. J Environ Manag 146:189–197

Winsley P (2007) Biochar and bioenergy production for climate change mitigation. N Z Sci Rev 64 (1):5–10

Woolf D, Amonette JE, Street-Perrott FA, Lehmann J, Joseph S (2010) Sustainable biochar to mitigate global climate change. Nat Commun 1:56. https://doi.org/10.1038/ncomms1053

Wright AL, Dou F, Hons FM (2007) Soil organic C and N distribution for wheat cropping systems after 20 years of conservation tillage in Central Texas. Agric Ecosyst Environ 121:376–382

Wu S, He H, Inthapanya X, Yang C, Lu L, Zeng G, Han Z (2017) Role of biochar on composting of organic wastes and remediation of contaminated soils—a review. Environ Sci Pollut Res 24 (20):16560–16577

Wu C, Shi L, Xue S, Li W, Jiang X, Rajendran M, Qian Z (2019) Effect of sulfur-iron modified biochar on the available cadmium and bacterial community structure in contaminated soils. Sci Total Environ 647:1158–1168

Xie Z, Xu Y, Liu G, Liu Q, Zhu J, Tu C, Amonette J, Cadisch G, Yong JH, Hu S (2013) Impact of biochar application on nitrogen nutrition of rice, greenhouse gas emissions and soil organic carbon dynamics in two paddy soils of China. Plant Soil 370:527–540

Yamato M, Okimori Y, Wibowo IF, Anshori S, Ogawa M (2006) Effects of the application of charred bark of Acacia mangium on the yield of maize, cowpea and peanut, and soil chemical properties in South Sumatra, Indonesia. Soil Sci Plant Nutr 52(4):489–495

Yu H, Zou W, Chen J, Chen H, Yu Z, Huang J, Gao B (2019) Biochar amendment improves crop production in problem soils: a review. J Environ Manag 232:8–21

Yuan JH, Xu RK, Zhang H (2011) The forms of alkalis in the biochar produced from crop residues at different temperatures. Bioresour Technol 102:3488–3497

Yuan P, Wang J, Pan Y, Shen B, Wu C (2018) Review of biochar for the management of contaminated soil: preparation, application and prospect. Sci Total Environ 659:473–490

Yue Y, Cui L, Lin Q, Li G, Zhao X (2017) Efficiency of sewage sludge biochar in improving urban soil properties and promoting grass growth. Chemosphere 173:551–556

Zhang M, Gao B, Yao Y, Xue Y, Inyang M (2012) Synthesis, characterization, and environmental implications of graphene-coated biochar. Sci Total Environ 435:567–572

Zhang X, Wang H, He L, Lu K, Sarmah A, Li J, Bolan NS, Pei J, Huang H (2013) Using biochar for remediation of soils contaminated with heavy metals and organic pollutants. Environ Sci Pollut Res 20:8472–8483

Zheng H, Guo W, Li S, Chen Y, Wu Q, Feng X, Yin R, Ho S-H, Ren N, Chang J-S (2017) Adsorption of p-nitrophenols (PNP) on microalgal biochar: analysis of high adsorption capacity and mechanism. Bioresour Technol 244:1456–1464

Zheng W, Guo M, Chow T, Bennett DN, Rajagopalan N (2010) Sorption properties of green waste biochar for two triazine pesticides. J Hazard Mater 181:121–126

Zhou Q, Ba O, Tong T, Gao T, Liu G (2018) An in-situ technique for producing low-cost agricultural biochar. Pedosphere 28:690–695

Role of Starch Polymer-Coated Urea in the Mitigation of Greenhouse Gas Emissions from Rice and Wheat Ecosystems

7

Khushboo Gupta, Nirmali Bordoloi, Kushal Kumar Baruah, and Dipti Gorh

Contents

Abstract

The excessive use of synthetic nitrogen fertilizer in agriculture interferes in natural ecosystem functioning and pollutes the environment through nitrogen leaching, volatilization, nitrification, and denitrification. The consumption of

K. Gupta · N. Bordoloi (✉)
Department of Environmental Science, Central University of Jharkhand, Brambe, Ranchi, Jharkhand, India

K. K. Baruah
School of Earth and Environmental Sciences, Royal Global University, Guwahati, Assam, India

D. Gorh
Department of Environmental Science, Tezpur University, Tezpur, Assam, India

K. Bauddh et al. (eds.), *Ecological and Practical Applications for Sustainable Agriculture*, https://doi.org/10.1007/978-981-15-3372-3_7

conventional fertilizer in the agricultural field has increased nitrous oxide (N_2O) emission, an important trace gas, which is responsible for stratospheric ozone layer depletion and global warming. Alternatively, the use of slow-release fertilizer (SRF) is considered to ameliorate crop nitrogen use efficiency (NUE) and yields while minimizing problems related to environment caused by inorganic N-fertilizer. For SRF, the coating of urea particles can be done with various coating materials; thus, nutrients release slowly from coated fertilizer and plant can efficiently use it. The degree of N release to soil depends on the coating material being used. Starch is one type of encapsulating polymer which can be used to coat or encapsulate the urea granules to slow the rate of fertilizer release. Although starch is mineralizable, the addition of different cross-linking agents reduces the release rate of entrapped urea from their cross-linking structure. The presence of coating directly retards the ammonium (NH_4^+) oxidation by reducing the activity of nitrifying bacteria and indirectly reduces the substrate availability for denitrification in soil. In this chapter, we have briefly reviewed about the different types of SRF and its use in the different agroecosystems. The efficiency of SRF for the reduction of greenhouse gas from rice and wheat ecosystems, mainly N_2O emission by affecting nitrification and denitrification process in soil, has also been discussed.

Keywords

Slow-release fertilizer · Starch-coated urea · Nitrogen use efficiency · Greenhouse gas · Grain yield

7.1 Introduction

The application of synthetic fertilizer largely contributes to agricultural production, and worldwide more than one-third growth of cereal production is depending on it (FAO 2013). The utilization of N fertilizer in crop production increases day by day to fulfill the demand for food of the expanding population. Globally nitrogen fertilizer application in agriculture increased at a rate of approximately 6–7% year^{-1} during the 1990s (Mosier 2004), and fertilizer production increased rapidly from 152.9 million tons in 1990 to 157 million tons in 2000. Based on the production capacity, worldwide fertilizer demand is estimated to rise up to 208 million tons by the year 2020 (Bumb and Baanante 1996).

In intensive agricultural system, plants use fertilizer least efficiently due to various nitrogen losses through runoff, ammonia volatilization, leaching, and denitrification, resulting in reduced nitrogen use efficiency, and plants cannot use up to 50% of added fertilizer (Asghari and Cavagnaro 2011). In the world, India is the second highest purchaser of the synthetic fertilizer (FAOSTAT 2014). Annually, approximately 165 Tg fertilizer is consumed in the Indian subcontinent, and about 41–58% and 2.88% of applied fertilizer is lost due to gaseous emission and leaching to groundwater, respectively. This intensity of N losses is influenced by the environmental condition, type of fertilizer, fertilizer rate, timing, and application method, characteristics of soil, and crop status (Peng et al. 2011).

In future, the fertilizer utilization will continue to increase to fulfill the food demand due to the rapid population growth and lower nitrogen use efficiency by crop. The worldwide population is expected to rise by 2.9 billion between 2009 and 2050, and globally production of cereals is required to be increased approximately 1000 Mt by 2050 to accomplish the requirement of food of the increasing population (FAO 2009).

Excessive use of inorganic and organic fertilizers enhanced the productivity of nitrous oxide (N_2O) and in the soil through the biochemical process nitrification and denitrification under aerobic and anaerobic conditions, respectively (Yan et al. 2000). N_2O is a powerful greenhouse gas (trace gas) of growing concern and accountable for stratospheric ozone layer destruction. The global concentration of atmospheric N_2O is rising at a rate of about 0.26% year^{-1} (Forster et al. 2007). In the atmosphere, a minute rise in the concentration of N_2O is a cause for global concern, as the global warming potential (GWP) of one N_2O molecule is 298 times higher in comparison to one molecule of CO_2 over a period of 100 years, resulting in global warming and climate change. It has been investigated that crop growth and grain productivity are considerably affected by the increasing concentration of greenhouse gases via rising temperature (Chauhan et al. 2014). The use of N fertilizer in cropland has been frequently adopted for better grain productivity; however, there is a strong need to enhance the nitrogen uptake capacity by plant, since N is a key limiting factor for crop production.

Excessive use of inorganic N fertilizer decreased the NUE, causing a series of economic and environmental problems by N loss as leaching or emission of N_2O. A new kind of controlled-release or slow-release nitrogen fertilizer can be used as an effective option to decrease the adverse effect of N fertilizer and ameliorate the NUE. The use of slow-release fertilizer can delay the nitrogen availability for plant uptake or sustain for longer period than that of "readily available" N fertilizers. The urea slow-release fertilizer by covering the urea granules with such types of materials, which show the property of water resistance, controls the nutrient release, thereby synchronizing the plant nitrogen demand and increasing NUE and minimizing environmental pollution and health hazards (Yang et al. 2012). Nowadays, various types of coating materials are available for coating urea, viz., polymer, sulfur, resin, phosphogypsum, etc. Starch is the most abundant encapsulating polymer used to encapsulate the fertilizer granules (urea) to slow the rate of fertilizer release. Starch polymer contains a unit of glucose which is connected by the glycosidic bonds. It is the cheapest biopolymer and biodegradable in nature (Han et al. 2009). Native starch is not suitable in slow-release systems because of the rapid enzymatic degradation, substantial swelling, and low cold water solubility. Physical and chemical modification of starch can overcome these problems. The release intensity of the urea from urea coating can be controlled by cross-linking of starch to provide a network structure. The intensity of the cross-linking density of starch matrix depends on the various cross-linking agents. Various cross-link agents like glutaraldehyde, calcium chloride, and citric acid can be used for this purpose. The interest in the use of starch polymer-coated urea in the agricultural field for reduction of greenhouse gas emission has been increasing day by day. In this section, we have extensively discussed the application of polymer-coated urea in agriculture and subsequently their effectiveness in reduction of agricultural greenhouse gases for reduction of future climate change by rising temperature.

7.2 Global Overview of Polymer-Coated Urea in Agriculture

The use of uncoated urea or commercial urea is the principal wellspring of nitrogen filling in the environment. About 10 million Mg N is reduced from the 16 million Mg urea-N by the mechanisms of volatilization, leaching, denitrification, etc. in the paddy field of Southeast Asia (Patil et al. 2010). Therefore, inefficient use of nitrogen, which is increasing in the environment causing quick loss of uncoated fertilizers, is the chief issue in the lowland rice paddy field. Polymer membrane encapsulating fertilizer can reduce or minimize environmental pollution through decreasing the fertilizer loss. Polymer-coated urea (PCU) is intended to synchronize nitrogen release from soil and uptake with the least amount of side effect. There are various PCUs which have already been trialed for crops, like wheat (Malhi et al. 2003; El-Sirafy et al. 2006), rice (Singh 2007; Tang et al. 2007; Bordoloi et al. 2016), potato (Pack et al. 2006; Hutchinson et al. 2003), etc. Nowadays, researchers are more worried about the formation and design of the fertilizers according to the crop requirement as PCU is spreading widely in crops so as to attain the maximum revival of N efficiency.

Polymer coating materials include bio-based coating materials and conventional synthetic polymers such as polyolefins, acrylic resins, and glycerol ester (Wei et al. 2017; Sabadini et al. 2008). In modern times, improved polymers which are natural, namely, modified corn stover, cellulose, starch, castor, and wheat straw, are used as controlled-release materials in fertilizer (Kumbar et al. 2001; Chen et al. 2008; Han et al. 2009; Qiao et al. 2016; Zhang et al. 2017). Biodegradable cellulose is an abundant and inexhaustible material. However, the application of cellulose is restricted because it is insoluble in most solvents. Ethyl cellulose (EC) is a cellulose derivative that is thermoplastic, nontoxic, and stable and is soluble in organic solvents. EC has been used in pill coatings and fertilizer coatings due to its excellent film-forming abilities and hydrophobicity (Desai et al. 2006). The different materials used for manufacturing coatings of polymers are polyacrylonitrile (PAN), cellulose acetate (CA), polysulfone (PSF), etc. (Jarosiewicz and Tomaszewska 2003).

The controlled-release fertilizers (CRF) used to create a physical barrier around the granules of urea at its outer circumference, which retarded the rate of dissolution and proposed a gentle release of nitrogen as per the crop demand (Guo et al. 2005). Split fertilization program is replaced by the controlled-released fertilizer as it is a saver of time and money (Naz and Sulaiman 2016). The controlled-release formulations depend upon the coating of inorganic polymer (e.g., sulfur) or organic polymer (e.g., resins or thermoplastic materials) or upon the hydrophobic matrix (Naz and Sulaiman 2016). But such coating of mineral/polymer is found to be very expensive, which shows an inconsistent discharge pattern and criticism for the environmental concern (Shaviv 2001; Timilsena et al. 2015). Several biopolymers are being studied in a controlled-release fertilizer such as starch, sodium alginate (SA), cellulose, and polyvinyl alcohol (PVA) (Wang et al. 2012; Jin et al. 2013). These can often be mixed into a polysaccharide matrix and are eco-friendly in nature, water retentive, and mechanically stable (Han et al. 2009; Wang et al. 2012). Some prior studies were suggested that in order to develop water absorption speed or mechanical rigidity and ion exchange capacity in CRF, minerals were added for the preparation of nanocomposite hydrogels (Bortolin et al. 2013).

In the current years, the growth and application of controlled- and slow-release fertilizer have been speedily increased because such kind of fertilizer assists in the reduction of environmental impact as well as surface water contamination by preventing or slowing the nutrient release behavior (Davidson and Gu 2012; Azeem et al. 2014). The nutrient release rate from CRF mostly depends upon the thickness and coating material. Shaviv (2005) and Trenkel (2010) have reported that in the case of slow-release fertilizer, the way the nutrient is liberated is nearly unpredictable and depends upon the alteration in the climatic conditions and soil type, whereas for controlled-released fertilizer, it is easy to predict the nutrient released pattern, time of nutrient release, and quantity. Resin-coated fertilizer (RCU) and sulfur-coated fertilizer (SCU) are the most commonly sold in the market; some previous studies have shown that sulfur-coated urea has considerably boosted the rice yield and NUE as compared to the conventional urea with the same nitrogen rate (Zhang et al. 2007; Kiran et al. 2010). However, previous studies have reported that some soil parameters like soil temperature and soil moisture affect the nutrient release rate from sulfur-coated urea. The increase in these parameters can enhance the release of nutrient behavior rate (Gu et al. 2011). One of the major drawbacks of SCU is high N liberation were takes place at the prior stage, which cause inadequate supply of N at the end and leads to low accumulation of dry matter (Miao et al. 2016). The nutrient-release behaviour of polyolefin-coated fertilizer (one type of polymer coated urea) is controlled by temperature (Rajani and Sen 2017). The efficiency of N used in rice field has been improved by resin or sulfur-coated urea (Inubushi et al. 2002; Arguissain and Livore 2003). In the rice field of Valencia (Spain), the usage of urea coated with polymer (40% nitrogen) enhanced the efficiency of N recovery and grain yield in comparison with urea or ammonium sulfate during the three consecutive rice growing seasons (Carreres et al. 2003).

Several authors have investigated the environmental aspects of slow-release and polymer-coated fertilizer and reported that lowering impact on environment and increasing NUE can be affected by N release characteristics of coated fertilizer as per the crop demand (Shaviv and Mikkelsen 1993; Shaviv 1995). Dou and Alva (1998) reported from their studies that the outcome of various controlled-release fertilizers (CRF) in comparison with the urea is found in the rootstock of citrus seedlings mainly in sandy soil. They indicated that CRF has a greater N uptake by seedling than for urea, and therefore, N losses are less in CRF as it acts as a source of N to a soluble N fertilizer. The development of agro-environment takes place due to modern farming systems along with the application of CRF (Shoji 2005). For instance, rice culture where no tillage is performed can efficiently improve the water and biological and atmospheric conditions of rice fields. Masuda et al.

(2003) reported that the application of polymer-coated N fertilizers could decrease the nitrate leaching when used in the field of sugarcane and demonstrated that the use of N-based fertilizer could be reduced up to 40% without compromising on the yield of sugar.

Zhang (2007) reported that the use of excessive conventional fertilizers, in the regions where there is a severe rise in the nonpoint source pollution, could be diminished noticeably. Several research centers of agriculture in Japan have reported that without affecting the yield of several crops like rice, tea, potatoes, sugarcane, maize, and various vegetables, nutrient use rates could considerably be reduced (by 20–60%) by applying slow- or controlled-release fertilizers. Shoji et al. (2001) have made a comparison between polyolefin-coated urea with nitrification inhibitor and unamended urea on flood irrigated barley, center-pivot irrigated maize, and potato grown in lysimeter on a large scale. Emission of nitrous oxide from the CRF applied plots is one-third achieved as urea, and N recovery was almost twice that attained with urea. From the above study it was found that the application of CRF and nitrification inhibitors may protect the environment through larger NUE and reduction in N rate. In a country like Japan, a campaign started by authorities aimed at reducing the large utilization rates of N fertilizers. Therefore, use of mineral fertilizer has reduced about 30% over the last 30 years. Thus, a program has been launched by the Japanese Ministry for Agriculture, Forestry, and Fishery (MAFF) in order to replace the use of traditional fertilizers of N by SRF and CRF; primarily, it is a coated fertilizer, and utilization of these particular types of fertilizers has developed slowly. SRF and CRF are mostly applied to the staple crops like rice and high-value vegetables.

7.3 Preparation of Cross-Linked Starch-Coated Urea

The main method to prepare the slow-release fertilizer is associated with covering a soluble conventional fertilizer with a defensive coating, (encapsulation) which is water-insoluble, and coating material with semipermeable or impervious pores. The nature of the coating material determines the dissolution rate and synchronizes the release of nutrient with plant demand. The important methods involve the (1) release of nutrient from materials via a physical barrier (sulfur-coated urea, starch-coated urea), (2) release of nutrient from materials via microbial decomposition of compounds with low solubility (urea-formaldehyde), and (3) release of nutrient from materials assimilated in the medium and may be covered itself (gel-based matrices) (Sempeho et al. 2014). Preparation of cross-linked starch-coated urea is by the following steps.

7.3.1 Starch-Urea Preparation

Starch (5 g) was taken and dissolved in distilled water (25 mL), and solution was heated at 70–75 °C until it became clear. Then, 95 g of urea was taken and dissolved in 30 mL of distilled water, and the mixture was heated in order to obtain a clear solution. Then, it is followed by the addition of starch mixture to the urea solution under constant stirring to form starch-urea polymer.

7.3.2 Cross-Linked Starch-Coated Urea Preparation

Cross-linked starch-coated urea was prepared by following the procedure of Chowdary and Chaithanya (2010) with little modification. A cross-linking agent $CaCl_2$ (1 g dissolved in a minimum amount of water) was added in the starch-urea solution under constant stirring. This mixture was heated further under constant stirring for 30 min to obtain a cross-linking starch-urea polymer. The whole matrix was then cooled and then dried polymer was powered for use.

7.4 Classification of Slow-Release Fertilizer (SRF)

Nutrients are slowly released from SRF throughout the season as compared to normal or common fertilizer. The nutrient release rate, pattern, and duration are strongly affected by the handling conditions, soil conditions, and biological activity in the soil (Shaviv 2001). On the basis of the mode of action, SLFs are divided into three main groups. These are physically, chemically, and biochemically altered type.

7.4.1 Physically Altered

This includes water-soluble fertilizers with some physical barriers. The physical properties of N fertilizer are changed by coating and encapsulating urea in a membrane to control the nutrient release rate. The slow-release fertilizers are categorized into coated and uncoated products. Uncoated products depend on the inherent physical characteristics of fertilizer like low solubility in slow release. Coated fertilizer consists of a variety of coating agents, which are applied to fertilizer particles to control their release rate and solubility in soil. Controlling the nutrient release rate from the coated fertilizer can be beneficial for environmental and economic point. The coated fertilizer exits either as granules/cores coated with hydrophobic polymer or as fertilizer matrix. Further, on the one hand, the coated granular fertilizers can be subdivided into fertilizers coated with organic polymer mainly consisting of natural high molecular materials (e.g., starch, fibrin, thermoplastic, or resins) and fertilizers coated with inorganic materials (sulfur, gypsum, silicate, and phosphoric acid). On the other hand, the matrix materials are subcategorized into hydrophobic materials (polyolefins and rubber) and gel-forming polymers, which are hydrophilic and can reduce the dissolution rate of soluble fertilizer because of their high ability to retain water (Table 7.1).

Table 7.1 Different types of coated fertilizers used in the agricultural field

Coated fertilizer	Characteristics
Osmocote	Osmocote particles are small, round, and known as prills. Osmocote is composed of a semipermeable membrane surrounded by water-soluble nitrogen and other nutrients. Nutrient release rate of osmocote is dependent on the moisture, temperature, and thickness of the coating. It is recommended for nursery stock, floriculture, and row crops
Sulfur-coated urea	Sulfur-coated urea is granules of urea, which is manufactured by molten sulfur coating to hot urea and sealed with a microcrystalline wax or polyethylene oil. Nitrogen is released after the breakdown of sealant or by diffusion, and the rate of discharge to soil is dependent upon the sealant weight or the thickness of the coating. Sulfur-coated urea is mainly applicable in the area of high rainfall or irrigation and used for grass forages, cultivation of strawberries, and ornamental plants
Nutricote	The nitrate compound fertilizers are coated by a special resin to manufacture the nutricote. Most of the nutricote granules are 3–4 mm in diameter. After application in soil, the nutricote granules absorb soil moisture via the coating membrane and break down the nutrient, and finally the nutrient can steadily release through the coating

Source: Varadachari and Goertz (2010)

7.4.2 Chemically Altered

Chemical alteration of slow-release fertilizer depicts that urea is chemically altered to a water-insoluble form up to some extent. These can be categorized into compounds of biologically decomposed, like urea-formaldehyde (UF), crotonylidene diurea (CDU), and urea-triazone (UT), and compounds of chemically decomposed, like urea-acetaldehyde/cyclo-diurea (CDU) or isobutylidene-diurea (IBDU) (Trenkel 2010; Azeem et al. 2014).

7.4.3 Biochemical Type

Biochemical alteration like addition of nitrification inhibitors and urease inhibitors inhibit or slow down the soil microbial nitrification process, resulting in decrease in the possibility of nitrate loss and thereby reduce the denitrification process and production of N_2O in soil (Scheer et al. 2017). These inhibitors inhibit or depress the soil microorganisms' activities and enzymes responsible for urea decomposition and control the N release process (Yan et al. 2008).

7.5 Nutrient Release Behavior of SRF

Nutrient release may be evident with the modification in the applied fertilizer or the chemical substance into plants' accessible form. When low-cost and easily available urea is incorporated to the soil, it undergoes changes via physical, biological, and chemical modification. During the early stage of growth, plants required only a small quantity of nutrient; therefore, excess nutrients are lost through leaching, gaseous emission such as N_2O, and runoff (Azeem et al. 2014). The pictorial representation of nutrient transformations is depicted in Fig. 7.1.

In case of slow-release fertilizer, the nutrient discharge rate is slower than that of the synthetic fertilizer, and a plant can uptake readily available nutrient before it is leached. In the "coated fertilizers," the core of soluble fertilizer is enclosed with insoluble water coating. The water penetration rate into the soluble fertilizer core and the rate of release solubilized fertilizer to soil are managed by the coating materials. The mechanism involved in nitrogen release for coated urea is by penetration of water via microspores and defection of the coating; thus, quick release of urea that dissolves from the fertilizer core occurrs. Figures 7.2 and 7.3 represent the behavior

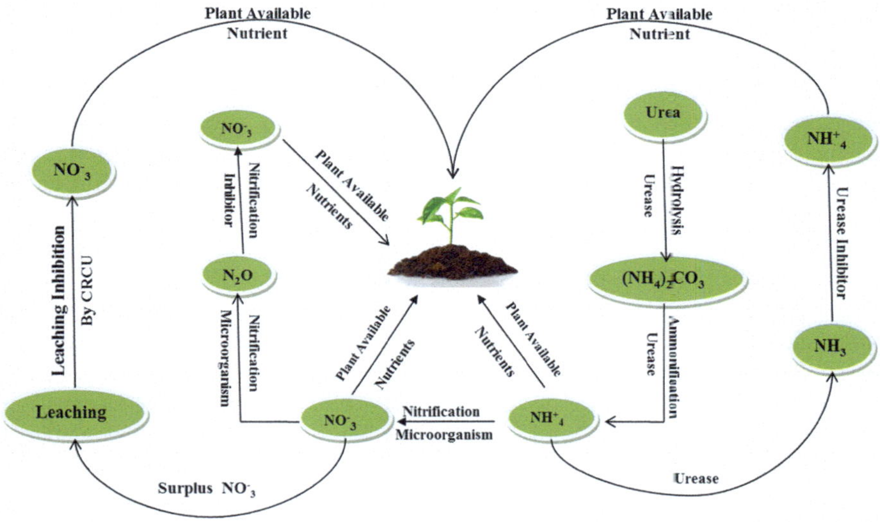

Fig. 7.1 Modification of urea after applying into the soil

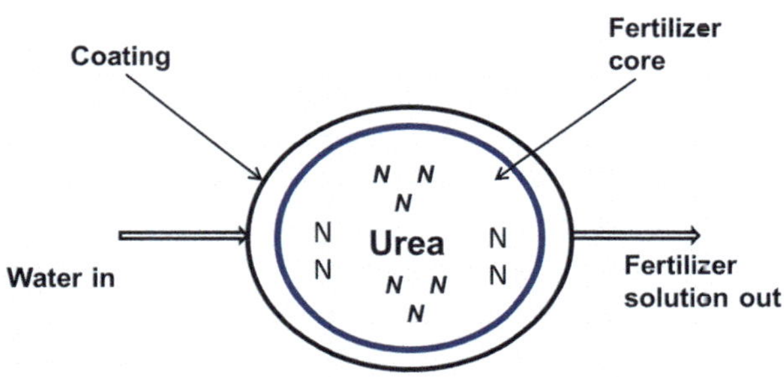

Fig. 7.2 Presentation of a slow or control-release fertilizer prill absorbing water and releasing the fertilizer as solution

Fig. 7.3 Cross-linked structure of starch-urea matrix

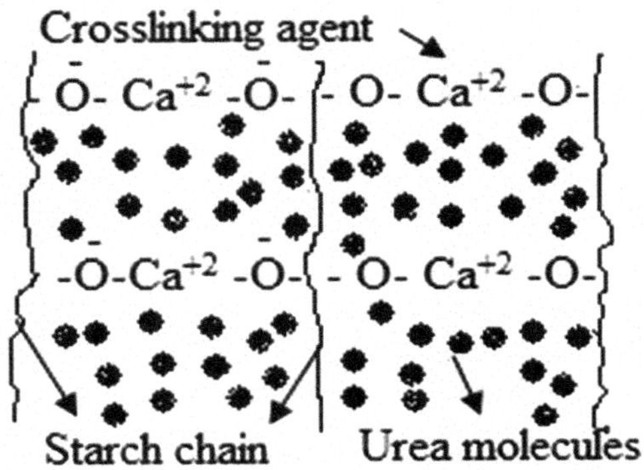

of nutrient released from the slow-release fertilizer and cross-linked structure of polymer-coated urea.

The most technically advanced technique in the control-release system is the use of polymer-coated fertilizers, which can control the product longevity and increase the nutrient use efficiency of crop. Coatings of polymer can be of impervious or semipermeable membranes with many small pores (Fujita and Shoji 1999). The fertilizer products coated with polymer discharge their nutrient through diffusion by means of semi-impermeable membrane, and the rate of release can be changed by the coating thickness and constitution of the coating materials. The release of nutrient via polymer membrane is significantly influenced by the permeability of the coating of polymer and temperature rather than soil properties, viz., texture, pH, redox potential, ionic strength, and microbial activity (Shaviv 2005). The starch-coated urea reduces the oxidation of NH_4^+ because of the presence of the starch coating and can slow down the nitrifying bacteria's activity, thus reducing the N_2O emission (Akiyama et al. 2010). The slow or control release of NH_4^+ from starch-coated urea is because of the presence of hydroxyl group (OH^-) on the starch chains, and the addition of the cross-linking agent like glutaraldehyde, citric acid, and calcium chloride helps in the formation of cross-linking bonds between the neighboring chains of the starch molecules. Then, the urea molecule entrapped by the cross-linking agent gets released slowly from the network-like structure. Although starch is mineralizable, the formation of such ionic cross-links tightens the starch matrix, thus reducing the urea release rate (Bordoloi et al. 2016; Bordoloi and Baruah 2017).

The release of nitrogen from urea-formaldehyde (UF) is based on the following three steps (Guo et al. 2006). In the first step, the process of absorption of water takes place through soil which causes swelling of the coating materials and thus gets modified into hydrogels which increases the network of the coating materials. In the step two, cross linked polymers structure and soluble parts of UF which get dissolves and slow diffusion of water takes place with exchange of the water in the hydrogel and in the soil. The nutrient release occurs in the last step, where soil microbes attack the swollen coverings and gather around the granules of UF; consequently, the part of nitrogen which is insoluble in UF granule gets degraded into ammonia and urea and is slowly released to the soil for plant use. An alkyd-resin-coated fertilizer is named osmocote, in which water enters through the minute pores in the coating (Hauck 1985).This leads to an increase in the osmotic pressure inside the pore that becomes enlarged and through which nutrients get discharged.

7.6 Efficiency for Reduction of Greenhouse Gas

The FAO reported that, by 2020, the overall requirement for nitrogen-based fertilizer will be greater than 200 million tons with an annual increasing rate of 1.9%, which connotes how crucial is urea for the incoming global agriculture scenario (FAO 2018). But, the wholesale application of urea has been claimed as an important factor of producing pollution in the environment. Due to less nitrogen use efficiency (barely go beyond 50%), urea is released into the environment either in the form of NO_3^- by means of leaching or as N_2O to the atmosphere via microbial nitrification and denitrification, causing health hazard and global warming, respectively (Richardson et al. 2009). To enhance the urea use efficiency and reduction of N_2O emission, numerous technologies were suggested from which the most promising one is polymer-coated urea (Choudhury and Kennedy 2005).

The slow-release urea and organic fertilizers may fluctuate greatly in their constituents, and the method of application, as a result, causes variation in the mitigation potential of greenhouse gas (N_2O and CH_4) emission from agriculture (Meijide et al. 2007). Several studies have broadly described about SRU treatments which lower the N_2O emission, due to the urease inhibitor which suppresses the alteration of amide nitrogen into ammonium hydroxide and ammonium (Turner et al. 2008; Trenkel 2010). Polymer-coated urea is highly dependent on soil temperature and moisture in order to release N into the soil (Gao et al. 2015). First, with adequate soil moisture, water hydrates the granule and dissolves N within the granule. The granule does not disperse due to the polymer coating. As soil temperature increases, the N will diffuse out of the granule for crop uptake. Akiyama et al. (2010) found a reduction in N_2O emissions from nitrification inhibitor and polymer-urea fertilizers by 35 and 38%, respectively, as compared to the traditional fertilizers. A study conducted in the wheat field of China has observed that the use of urea-formaldehyde reduced 42% of N_2O emissions as compared to the commercial fertilizer (Jiang et al. 2010). Soon et al. (2011) reported that coated polymer can occasionally increase the N availability and decrease N_2O emission and has improved crop production as compared to the conventional urea in wheat agriculture. A four-year experiment on the use of thermoplastic resin-coated urea (240 kg N ha^{-1}) in paddy field found approximately 13% reduction in greenhouse gas emission over urea from rice fields located in Jurong City, China (Ji et al. 2013). Amendments of green urea (urea covered with Neb26) and orange urea (urea covered by Agrotain) in rice paddy significantly ($p < 0.05$) lowered the total N_2O flux as compared to white urea (conventional fertilizer), but no major disparities were found related to CH_4 fluxes (Trinh et al. 2017). The application of starch-coated urea minimizes the N_2O emissions by 23 and 22% as compared to NPK in wheat and summer rice ecosystems of Assam (state of north east India) without affecting the productivity (Bordoloi et al. 2016; Bordoloi and Baruah 2017).

7.7 Advantage of Slow-Release Fertilizer

The slow-release fertilizer liberates nutrient to the plants at a very slow rate over a long duration on the basis of microbial activity which is completely based on the soil temperature. They are generally dry mixed which is simple to spread and can cover a wide area. The organic SRF is completely dependent on the activity of microorganism and stays for a longer duration into the soil. Increasing NUE in plant and reducing problems related to the environment mainly depend upon the supply of nutrients according to plant requirement and continuing nutrient accessibility during the cropping period.

Some benefits of SRF are as follows:

- SRF has long-term persistence in soil; therefore, nutrient availability for the plant remains at the later stages of plant growth.
- It detracts the loss of fertilizers from seepage or runoff so that plants can uptake nutrients continuously according to their need.
- These also promote in increasing the crop production in a zone where moisture deficit or drought condition exists. As they meet with water, they become reactive and break down into soluble products.
- This technology assists in measuring out nutrients over a period of time under controlled moisture conditions.
- A pellet near the root under most favorable conditions can supply high dose of fertilizer at the place where the plant needs it.
- SRF is cost-effective and also easy to apply as compared to fast- or controlled-release fertilizers.
- They do not pollute the soil as they remain inactive for longer periods and do not have deleterious impact on the soil and the environment.
- Slow-release nitrogen fertilizer assists in combating the N_2O emission from the agricultural field. The use of urea with biochemical inhibitor or urea-formaldehyde could diminish N_2O emissions.
- In terms of the impact on the environment, the best results were obtained at fertilization with SRF in the row technique and at conventional fertilization (Marcin and Monika 2018).
- SRF is applied to decrease the overall amount of nutrients introduced in the environment and to limit the energy utilization for fertilization.

7.7.1 Impact on Farming System

SRF is accessible and applicable on all types of plants, turf grasses, annuals, perennials, shrubs, and trees. After each irrigation, the slow-release fertilizers release their nutrients which are beneficial for the farmers because of the slow release of nutrients being used to fulfill the plant needs instead of the large release which causes toxicity to the plants. The release of nutrient rate depends upon the solubility of the ingredients of the fertilizer, soil constituents, and its weather conditions.

Coated urea and nitrification inhibitor (DCD) with urea considerably decrease the N_2O emission as compared to conventional urea (Cheng et al. 2006). Some studies established that urea with urease inhibitor, urease combined with urea and nitrification inhibitor (DCD), and covered urea had a narrow impact on the emission of N_2O from extensive grassland (Dobbie and Smith 2003). According to Delgado and Mosier (1996), urea coated with polyolefin named as polyolefin-coated urea (POCU) could lower the N_2O emission in comparison with the traditional urea in the initial 21 days after incorporation, but after that emissions have remained higher within the next 60 days. The total emission of N_2O over 90 days demonstrated that there is no major variation between POCU and traditional urea treatments.

By using SRF, the production efficiency can be increased and dispersion of biogens in the environment can be diminished, in certain soil and climatic conditions (Niemiec and Komorowska 2018). Applying SRF as a method of fertilization optimization has slowly and steadily increased in recent years (Chen et al. 2018). High-cash crops grow abundantly having maximum yields and utilize bulk amounts of nutrient effectively by using SRFs (Landels 1994). These crops include strawberries, melons, etc. Some crops are effectively used for high income and nutrient demand like citrus, oil palm, banana, and kiwi. Countries like the USA, Florida and California, are the major consumers of SRFs for citrus.

7.7.2 Economic Aspect

The economic aspect is the most important parameter for the assessment of agricultural system that depicts the quantity of yield and its technological parameter that is a unit mass of the product (Niemiec and Komorowska 2018). Fertilization with SRF, using both direct and row methods, has caused a statistically considerable increase in the commercial yield. There are certain economic advantages associated with controlling the supply of nutrients.

7.7.2.1 Potential for Reducing Nutrient Losses

Practically, several processes like runoff and leaching lead to the nutrient losses that are assumed as irreversible at least for a short duration. These methods are the main reason for less retrieval of nitrogen in paddy rice (Ghosh and Bhat 1998). Many studies demonstrated that there is a possibility of reducing such losses by applying polymer-coated fertilizer or bio-amended ammonium fertilizers (nitrification inhibitors) (Shaviv 2005; Das and Adhya 2014). Wang et al. (2015) observed that the application of particular thermoplastic polymer-coated urea can reduce the nitrogen deficiency and improve NUE in rice agroecosystem.

7.7.2.2 Cost of Fertilizer Application

SRF can fulfill the demand of crop nutrients for the whole season through one-time application, which involves saving of money. Moreover, different bio-amendments, namely, nitrification inhibitors, can save the cost of additional application (Trenkel 1997). The slower release of nutrient from polymer-coated urea can raise the economy by avoiding the additional in-season N fertilizer application (Hyatt et al. 2010). Optimizing the fertilizer rate, source, application timing, and placement are the important factors to maintain the balance between economic profitability and nitrogen use efficiency of crop.

7.8 Conclusion

The rice and wheat cultivation is the primary source of greenhouse gases, mainly CH_4 and N_2O. The maximum applied conventional fertilizer in the agricultural fields may lose because of leaching, surface runoff, vaporization, and low crop productivity due to less nitrogen utilization efficiency of crop. Many researchers indicate that the polymer-coated fertilizer has the opportunity to raise the nitrogen use efficiency for plant by diminishing N loss in soil. These eco-friendly coated fertilizers have great potential in agriculture to reduce the greenhouse gas emission with an optimum grain productivity, which contributes to mitigate the climate change through increasing temperature. Further, from the economic point of view and better environmental safeguard, long-term studies on the type and application rate of urea covered with polymer in wheat and rice agriculture should be carried out in different regions.

References

Akiyama H, Yan X, Yagi K (2010) Evaluation of effectiveness of enhanced-efficiency fertilizers as mitigation options for N_2O and NO emissions from agricultural soils: meta-analysis. Glob Change Biol 16:1837–1846

Arguissain GG, Livore AB (2003) Eficiencia de fertilizante de liberaci_n controlada para arroz irrigado. In: Proceedings of the 3rd international temperate rice conference, Punta del Este, Uruguay.

Asghari HR, Cavagnaro TR (2011) Arbuscular mycorrhizas enhance plant interception of leached nutrients. Funct Plant Biol 38:219–226

Azeem B, KuShaari K, Man ZB, Basit A, Thanh TH (2014) Review on materials and methods to produce controlled release coated urea fertilizer. J Control Release 181:11–21

Bordoloi N, Baruah KK (2017) A two-year field assessment on the effect of slow release of nitrogenous fertiliser on N_2O emissions from a wheat cropping system. Soil Res 55:191–200

Bordoloi N, Baruah KK, Maji TK (2016) Nitrous oxide emission from transplanted rice field in alluvial soil as influenced by management of nitrogen fertilizer. Soil Use Manage 32:573–582

Bortolin A, Aouada FA, Mattoso LHC, Ribeiro C (2013) Nanocomposite PAAm/methyl cellulose/montmorillonite hydrogel: evidence of synergistic effects for the slow release of fertilizers. J Agric Food Chem 61:7431–7439

Bumb BL, Baanante CA (1996) World trends in fertilizer use and projections to 2020. International Food Policy Research Institute

Carreres T, Sendra J, Ballesteros R, Valiente FE, Quesada A, Carrasco D, Legans F, Cuadra JG (2003) Assessment of slow release fertilizers and nitrification inhibitors in flooded rice. Biol Fertil Soils 39:80–87

Chauhan BS, Kaur P, Mahajan G, Randhawa RK, Singh H, Kang MS (2014) Global warming and its possible impact on agriculture in India. Adv Agron 123:65–123

Chen L, Xie Z, Zhuang X, Chen X, Jing X (2008) Controlled release of urea encapsulated by starch-g-poly (L-lactide). Carbohy Polym 72:342–348

Chen Q, Qin J, Sun P, Cheng Z, Shen G (2018) Cow dung-derived engineered biochar for reclaiming phosphate from aqueous solution and its validation as slow-release fertilizer in soil-crop system. J Clean Prod 172:2009–2018

Cheng W, Sudo S, Tsuruta H, Yagi K, Hartley A (2006) Temporal and spatial variations in N_2O emissions from a Chinese cabbage field as a function of type of fertilizer and application. Nutr Cycling Agroecosyst 74(2):147–155

Choudhury ATMA, Kennedy IR (2005) Nitrogen fertilizer losses from rice soils and control of environmental pollution problems. Commun Soil Sci Plant Anal 36:1625–1639

Chowdary KPR, Chaithanya K (2010) Preparation and evaluation of crosslinked starch urea—a new polymer for control release of aceclofenac. Asian J Chem 22:4265–4270

Das S, Adhya TK (2014) Effect of combine application of organic manure and inorganic fertilizer on methane and nitrous oxide emissions from a tropical flooded soil planted to rice. Geoderma 213:185–192

Davidson D, Gu FX (2012) Materials for sustained and controlled release of nutrients and molecules to support plant growth. J Agric Food Chem 60:870–876

Desai J, Alexander K, Riga A (2006) Characterization of polymeric dispersions of dimenhydrinate in ethyl cellulose for controlled release. Int J Pharm 308:115–123

Delgado JA, Mosier AR (1996) Mitigation alternatives to decrease nitrous oxides emissions and urea-nitrogen loss and their effect on methane flux. J Environ Qual 25(5):1105–1111

Dobbie KE, Smith KA (2003) Impact of different forms of N fertilizer on N_2O emissions from intensive grassland. Nutr Cycl Agroecosyst 67(1):37–46

Dou H, Alva AK (1998) Nitrogen uptake and growth of two citrus rootstock seedlings in a sandy soil receiving different controlled-release fertilizer sources. Biol Fertil Soils 26(3):169–172

El-Sirafy ZM, Woodard HJ, El-Norjar EM (2006) Contribution of biofertilizers and fertilizer nitrogen to nutrient uptake and yield of Egyptian winter wheat. J Plant Nutr 29(4):587–599

FAO (Food and Agricultural Organization of the United Nations) (2009) Global agriculture towards 2050. How to feed the World 2050

FAOSTAT (2014) http://faostat3.fao.org/home/E. Accessed April 2015

Food and Agriculture Organization of the United Nations (2013) FAO statistical yearbook 2013: world food and agriculture. Rome

Food and Agriculture Organization of the United Nations (2018) World fertilizer trends and outlook 2018

Forster P, Ramaswamy V, Artaxo P, Berntsen T, Betts R, Fahey DW et al (2007) Changes in atmospheric constituents and in radiative forcing. In: Solomon S et al (eds) Climate change 2007: the physical science basis. Cambridge University Press, Cambridge, UK, pp 131–234

Fujita T, Shoji S (1999) Kinds and properties of Meister fertilizers. In: Shoji S (ed) Meister controlled release fertilizer—properties and utilization. Konno, Sendai, pp 13–34

Gao X, Asgedom H, Tenuta M, Flaten DN (2015) Enhanced efficiency urea sources and placement effects on nitrous oxide emissions. Agron J 107(1):265–277

Ghosh BC, Bhat R (1998) Environmental hazards of nitrogen loading in the wetland fields. Environ Pollut 102:123–126

Gu J, Xu K, Fu T, Zhang D, Tong G, Luo J, Tong E, Yi W, Xu Q (2011) Nitrogen release characteristics of different hermetic material sulfur coated urea and their effects on summer maize (In Chinese with English abstract). Zhiwu Yingyang Yu Feiliao Xuebao 17:630–637

Guo M, Liu M, Zhan F, Wu L (2005) Preparation and properties of a slow-release membrane-encapsulated urea fertilizer with superabsorbent and moisture preservation. Ind Eng Chem Res 44:4206–4211

Guo M, Liu M, Liang R, Niu A (2006) Granular urea-formaldehyde slow-release fertilizer with superabsorbent and moisture preservation. J Appl Polym Sci 99(6):3230–3235

Han X, Chen S, Hu X (2009) Controlled-release fertilizer encapsulated by starch/polyvinyl alcohol coating. Desalination 240:21–26

Hauck RD (1985) Slow release and bio-inhibitor-amended nitrogen fertilisers. In: Engelstad OP (ed) Fertiliser technology and use, 3rd edn. SSSA, Madison, WI, pp 293–322

Hutchinson C, Simonne E, Solano P, Meldrum J, Livingston-Way P (2003) Testing of controlled release fertilizer programs for seep irrigated Irish potato production. J Plant Nutr 26 (9):1709–1723

Hyatt CR, Venterea RT, Rosen CJ, McNearney M, Wilson ML, Dolan MS (2010) Polymer-coated urea maintains potato yields and reduces nitrous oxide emissions in a Minnesota loamy sand. Soil Sci Soc Am J 74:419–428

Inubushi K, Acquaye S, Tsukagoshi S, Shibahara F, Komatsu S (2002) Effects of controlled-release coated urea (CRCU) on soil microbial biomass N in paddy fields examined by the 15N tracer technique. Nutr Cycl Agroecosyst 63:291–300

Jarosiewicz A, Tomaszewska M (2003) Controlled-release NPK fertilizer encapsulated by polymeric membranes. J Agric Food Chem 51:413–417

Ji Y, Liu G, Ma J, Zhang G, Xu H, Yagi K (2013) Effect of controlled-release fertilizer on mitigation of N2O emission from paddy field in South China: a multi-year field observation. Plant Soil 371:473–486

Jiang J, Hu Z, Sun W, Huang Y (2010) Nitrous oxide emissions from Chinese cropland fertilized with a range of slow-release nitrogen compounds. Agric Ecosyst Environ 135:216–225

Jin S, Wang Y, He J, Yang Y, Yu X, Yue G (2013) Preparation and properties of a degradable interpenetrating polymer networks based on starch with water retention, amelioration of soil, and slow release of nitrogen and phosphorus fertilizer. J Appl Polym Sci 128:407–415

Kiran JK, Khanif YM, Amminuddin H, Anuar AR (2010) Effects of controlled release urea on the yield and nitrogen nutrition of flooded rice. Commun Soil Sci Plant Anal 41:811–819

Kumbar SG, Kulkarni AR, Dave AM, Aminabha TM (2001) Encapsulation efficiency and release kinetics of solid and liquid pesticides through urea formaldehyde crosslinked starch, guar gum, and starch guar gum matrices. Appl Polym Sci 82:2863–2866

Landels SP (1994) Controlled release fertilizer: supply and demand trends in US nonfarm markets. Proceedings of ACS National Meeting, Division of Fertilizers and Soil Chemistry

Malhi S, Oliver E, Mayerle G, Kruger G, Gill KS (2003) Improving effectiveness of seedrow-placed urea with urease inhibitor and polymer coating for durum wheat and canola. Commun Soil Sci Plant Anal 34(11–12):1709–1727

Marcin N, Monika K (2018) The use of slow-release fertilizers as a part of optimization of celeriac production technology. Agric Eng 22(2):59–68

Masuda T, Katsuta Y, Sugahara K, Banzai K, Shibano K (2003) Improved sugarcane cultivation in the subtropical islands of Japan using controlled-release N-fertilizers. JIRCAS Research Highlights

Meijide A, Diez JA, Sanchez-Martin L, Lopez-Fernandez S, Vallejo A (2007) Nitrogen oxide emissions from an irrigated maize crop amended with treated pig slurries and composts in a Mediterranean climate. Agric Ecosyst Environ 121:383–394

Miao X, Xing X, Ding Y, Ke J, Liu Z, Tang S, Li G (2016) Yield and nitrogen uptake of bowl-seedling machine-transplanted rice with slow-release nitrogen fertilizer. Agron J 108:313–320

Mosier AR (2004) Agriculture and the nitrogen cycle. Assessing the impacts of fertilizer use on food production and the environment. SCOPE series no. 65. Island Press, Washington, DC

Naz MY, Sulaiman SA (2016) Slow release coating remedy for nitrogen loss from conventional urea: a review. J Control Release 225:109–120

Niemiec M, Komorowska M (2018) The use of slow-release fertilizers as a part of optimization of celeriac production technology. Agric Eng 22(2):59–68

Pack JE, Hutchinson CM, Simonne EH (2006) Evaluation of controlled-release fertilizers for northeast Florida chip potato production. J Plant Nutri 29(7):1301–1313

Patil MD, Das BS, Barak E, Bhadoria PB, Polak A (2010) Performance of polymer-coated urea in transplanted rice: effect of mixing ratio and water input on nitrogen use efficiency. Paddy Water Environ 8(2):189–198

Peng SZ, Yang SH, Xu JZ, Luo YF, Hou HJ (2011) Nitrogen and phosphorus leaching losses from paddy fields with different water and nitrogen managements. Paddy Water Environ 9:333–342

Qiao D, Liu H, Yu L, Bao X, Simon GP, Petinakis E, Chen L (2016) Preparation and characterization of slow-release fertilizer encapsulated by starch-based superabsorbent polymer. Carbohydr Polym 147:146–154

Rajani, Sen A (2017) Nitrogen release dynamic of different coated urea under puddle lowland rice cultivation. Int J Curr Microbiol App Sci 6(10):2638–2641

Richardson D, Felgate H, Watmough N, Thomson A, Baggs E (2009) Mitigating release of the potent greenhouse gas N_2O from the nitrogen cycle—could enzymic regulation hold the key? Trends Biotechnol 27:388–397

Sabadini RC, Silva MM, Pawlicka A, Kanicki J (2008) Gellan gum-O, O′-bis (2-aminopropyl) polyethylene glycol hydrogel for controlled fertilizer release. J Appl Polym Sci 135 (2):45636–45642

Scheer C, Rowlings D, Firrell M, Deuter P, Morris S, Riches D, Porter I, Grace P (2017) Nitrification inhibitors can increase post-harvest nitrous oxide emissions in an intensive vegetable production system. Sci Rep 7:43677

Sempeho SI, Kim HT, Mubofu E, Hilonga A (2014) Meticulous overview on the controlled release fertilizers. Adv Chem. https://doi.org/10.1155/2014/363071

Shaviv A (1995) Plant response and environmental aspects as affected by rate and pattern of nitrogen release from controlled release N fertilizers. Prog Nitrogen Cycling Stud 68:285–291

Shaviv A (2001) Advances in controlled-release fertilizers. Adv Agron 71 1–49

Shaviv A (2005) Controlled release fertilizers, In: Shaviv A (ed) Proceedings of the international workshop on enhanced-efficiency fertilizers, Frankfurt, Germany. International Fertilizer Association, Paris, pp 28–30

Shaviv A, Mikkelsen RL (1993) Slow release fertilisers for a safer environment maintaining high agronomic use efficiency. Fert Res 35:1–12

Shoji S (2005) Innovative use of controlled availability fertilizers with high performance for intensive agriculture and environmental conservation. Sci China Ser C Life Sci 48:912–920

Shoji S, Delgado J, Mosier A, Miura Y (2001) Use of controlled release fertilizers and nitrification inhibitors to increase nitrogen use efficiency and to conserve air and water quality. Commun Soil Sci Plant Anal 32:1051–1070

Singh Y (2007) Efficient management of fertilizer nitrogen in wet direct-seeded rice (Oryza sativa) in northwest India. Indian J Agric Sci 77:561–564

Soon YK, Malhi SS, Lemke RL, Lupwayi NZ, Grant CA (2011) Effect of polymer-coated urea and tillage on the dynamics of available N and nitrous oxide emission from Gray Luvisols. Nutr Cycl Agroecosyst 90:267–279

Tang SH, Yang SH, Chen JS, Xu PZ, Zhang FB, Shao-ying AI, Huang X (2007) Studies on the mechanism of single basal application of controlled-release fertilizers for increasing yield of rice (*Oryza safiva L.*). Agri Sci China 6(5):586–596

Timilsena YP, Adhikari R, Casey P, Muster T, Gill H, Adhikari B (2015) Enhanced efficiency fertilisers: a review of formulation and nutrient release patterns. J Sci Food Agric 95:1131–1142

Trenkel ME (1997) Controlled release and stabilized fertilizers in agriculture. IFA, Paris

Trenkel ME (2010) Slow-and controlled-release and stabilized fertilizers: an option for enhancing nutrient use efficiency in agriculture, 2nd edn. International Fertilizer Industry Association, Paris

Trinh MV, Tesfai M, Borrell A, Nagothu US, Bui TPL, Quynh VD, Thanh LQ (2017) Effect of organic, inorganic and slow-release urea fertilisers on CH_4 and N_2O emissions from rice paddy fields. Paddy Water Environ 15(2):317–330

Turner D, Edis R, Chen D, White B, Christie R (2008) Nitrogen loss from top-dressing fertilisers to wheat. BGC 2000 season research results. http://www.farmtrials.com.au/trial/13971

Varadachari C, Goertz HM (2010) Slow-release and controlled-release nitrogen fertilizers. In: Singh B (ed) ING bulletins on regional assessment of reactive nitrogen, Bulletin no. 11. SCON-ING, New Delhi, pp 1–42

Wang Y, Liu M, Ni B, Xie L (2012) κ-Carrageenan–sodium alginate beads and superabsorbent coated nitrogen fertilizer with slow-release, water-retention, and anticompaction properties. Ind Eng Chem Res 51:1413–1422

Wang S, Zhao X, Xing G, Yang Y, Zhang M, Chen H (2015) Improving grain yield and reducing N loss using polymer-coated urea in southeast China. Agron Sustain Dev 35:1103–1115

Wei Y, Li J, Li Y, Zhao B, Zhang L, Yang X, Chang J (2017) Research on permeability coefficient of a polyethylene controlled-release fi lm coating for urea and relevant nutrient release pathways. Polym Test 59:90–98

Yan XY, Du L, Shi S, Xing GX (2000) Nitrous oxide emission from wetland rice soil as affected by the application of controlled-availability fertilizers and mid-season aeration. Biol Fertil Soils 32:60–66

Yan X, Jin JY, He P, Liang MZ (2008) Recent Advances on the technologies to increase fertilizer use efficiency. Agric Sci 7(4):469–479

Yang J, Liu G, Ma J, Xu H, Yagi K (2012) Effect of controlled-release fertilizer on nitrous oxide emission from a winter wheat field. Nutr Cycl Agroecosyst 94:111–122

Zhang Y, Wang J, Liang Y, Xu M, Sun L, Jiang Y, Hu Y (2007) Effect of sulfur-coated urea on growth and nitrogen use efficiency of rice (In Chinese with English abstract). J Soil Water Conserv 21:108–111

Zhang SG, Yang YC, Gao B, Li YC, Li Z (2017) Superhydrophobic controlled-release fertilizers coated with bio-based polymers with organosilicon and nano-silica modifications. J Mater Chem A 5:19943–19953

Suitability of Coupling Application of Organic and Inorganic Fertilizers for Crop Cultivation

8

Susmita Karmakar, Arijita Bhattacharyya, Bappa Ghosh, Rajeshwari Roy, Sanjeev Kumar, Biswajit Kar, and Gautam Saha

Contents

S. Karmakar (✉) · S. Kumar
Department of Environmental Science, Central University of Jharkhand, Ranchi, India

A. Bhattacharyya · B. Ghosh
Department of Agricultural Chemicals, Bidhan Chandra Krishi Viswavidyalaya, Mohanpur, West Bengal, India

R. Roy
Department of Environmental Studies, Rabindra Bharati University, Kolkata, West Bengal, India

B. Kar
Department of Environmental Sciences, Kanchrapara College, Kanchrapara, West Bengal, India

G. Saha
Department of Agricultural Meteorology and Physics, Bidhan Chandra Krishi Viswavidyalaya, Mohanpur, West Bengal, India

K. Bauddh et al. (eds.), *Ecological and Practical Applications for Sustainable Agriculture*, https://doi.org/10.1007/978-981-15-3372-3_8

Abstract

In recent times, achieving sustainability as a development goal is a global concern. In agricultural system, a conflict regarding accomplishing higher production with diminishing environmental risks is always being remained. Rapid growth in world population put higher pressure to increase the agricultural production which triggered the use of inorganic fertilizers to attain higher yield. Inorganic fertilization intensified the crop productivity by too many folds. However, this leads to gradual worsening in soil fertility. Unplanned land use pattern and improper inorganic fertilization practices degrade the soil quality. Moreover, misuses of inorganic fertilizers are resulting in health hazards and environmental problems. In consequences of these, concept of organic fertilization has been introduced as it can improve the soil health. Organic agriculture has the potentiality to sustain the ecosystem by ameliorating the detrimental effects of inorganic fertilizers. Yet organic fertilizers have some limitations regarding nutrient availability and sufficient food production issues which are essential to feed the increasing population. So advancement in agricultural management practices is required for food security. A judicious approach combining the inorganic and organic nutrient sources can be an effective alternative to maintain the soil characteristics which in turn increase the yield efficiency and curtail the environmental risks. Various researches on suitability of coupling application of inorganic and organic fertilizers to optimize the productivity and to achieve sound agricultural practices are ongoing worldwide. Such agricultural management practices have already been recognized by means of ensuring food security and safeguarding the environment.

Keywords

Coupling fertilization · GHG emission · Inorganic fertilizer · Organic fertilizer · Productivity · Soil health

8.1 Introduction

Global goals of sustainable development aimed at achieving zero hunger and stated "end hunger, achieve food security and improved nutrition and promote sustainable agriculture." It is quite frustrating to perceive that more than 820 million people are undernourished throughout the globe (FAO 2019). Agricultural production is necessarily needed to obtain a 50% increase in the current production to meet the hunger of the exponentially increasing population. The hunger scenario is even worse in this country. India is currently reported to have almost 15% as underfed among the entire population (FAO 2019). Agriculture along with its allied sectors has been the largest contributor of Indian GDP for the last six to seven decades. A fall in agriculture-based economy in India has been observed for a few decades. This might be due to the intensive resource exploitations, resource in the sense of the unplanned use of freshwater, unrestricted deforestation, massive land degradation due to improper uses of lands, and so on. Besides, a few societal hindrances have been adverse for the agricultural developments. Precisely, an efficiently managed agricultural practice with limited resources, to feed the hungry and populated nation, is the prime need of this hour.

An efficient agricultural system must have two basic aims: first is the amount or productivity and the next one is obviously the quality or food value. A well-fertilized soil is the major requirement for the both. The green revolution in the first phase of the 1960s has commenced several researches, technologies, high-yielding crop varieties, pesticides, and chemical fertilizers. This had resolved the previous agriculture-related adversities. The artificially applied chemical fertilizers worked remarkably as the nutrient sources for the crop. Although the apparent problems of the farmers were gone, the externality in form of soil and water pollution became the major problem which directly started to damage the ecosystem. In addition, emissions of greenhouse gases from unplanned chemical fertilization became a serious concern. Besides, inorganic or chemical fertilizers were mostly imported, and thus the costs were higher.

Practically, continual application of chemical fertilizer deteriorates the soil quality in terms of affecting the soil physicochemical and microbial population (Mahajan et al. 2008). Synthetic fertilizer amendments also influence the amount and qualitative values of crop yield. Excessive inorganic fertilizer use leads to reduce the soil pH which in turn decreases the vegetation's nutrient availability (Liu et al. 2014). Reduced decomposition rate of organic materials within crop soil due to overuse of inorganic fertilizers often resulted in diminishing the essential soil micronutrients affecting the nutritional quality of the crop (Rashid et al. 2016). Inorganic fertilization has already been proven to slash down the plants' nitrogen use efficiency, increment soil salinity, and pollute the ground water (Chang et al. 2013; Badr et al. 2016). In case of intensive agri-practices, inorganic fertilization could not be suited perfectly because of its cost and its influence on production quality, soil quality, and natural environment (Obi and Ebo 1995; Natsheh and Mousa 2014).

Simultaneously, some conventional practices and indigenous knowledge have come into action again. These are the organic alternatives of chemical fertilizers that may provide the essential nutrients to crops and can maintain the soil health. Moreover, organic manuring is least harmful to the environment. Incorporation of organic matter can sustain the crop productivity through better nutrient cycling. Both the soil microorganisms and vegetative cover are benefitted from organic matters within soil as it comprises various nutrients and energy (Majumdar et al. 2014). Soil health can be improved by intensifying the use of organic manures (Ojeniyi 2000; Maritus and Vleic 2001). It is documented that soil buffering abilities can be enhanced after use of organic manures (Cooper 2017). Soil organic carbon stock can be increased due to application of farm-based manure, compost, or any soil amendment originated from biological origin which improves disease resistance in plants and helps them grow healthy (Liu et al. 2007; D'Hose et al. 2016). However, a number of studies indicated that sustainable productivity is often not depended on sole inorganic or sole organic amendment (Satyanarayana et al. 2002; Jobe 2003).

In recent times, organic fertilizers, coupled with inorganic or chemical fertilizers in precalculated ratio, are widely being experimented to acquire an expected amount and quality of crops and by disturbing the nature the least. The requirements are needed to be preassessed by soil testing or the other researches and a particular mixture of fertilizer for a particular type of crop might be the best solution that pushes toward attaining the sustainable development goals in the near future. An immediate approach toward the application of synthetic fertilizers in combination with organic fertilizers is of urgent need to achieve higher crop yield and environmental security (Yamoah et al. 2002; Kumar et al. 2013a, b). Ipimoroti et al. (2002) address the sustainability in prolonged cropping pattern in tropical regions where farm management practices must include the integrated application of synthetic and organic fertilizers. Optimal availability of necessary plant nutrients at proper time from coupling inorganic and organic amendments enriches the soil biota and enhances the productivity (Zhao et al. 2014). Besides, a lot of field-based studies established the fact that the combined use of inorganic and organic fertilizers has a strong influence to ameliorate the emission of greenhouse gases from agroecosystems (Meng et al. 2005; Zhang et al. 2012; Skinner et al. 2014).

8.2 Fertilizers in Agricultural System

World population is growing fast. With a rapid increase in population, there is a need to ensure the food security. For this, it is imperative that under a complete infrastructure of sustainable development, agricultural systems must be developed. We all mainly depend on plant for our food. Their growth depends on mineral nutrient which is supplied from soil environment. Soil is heterogeneous in nature. Hence, there is a lack of consistency in soil fertility around the globe. Moreover, existing fertile soil loses their quality and tends to decrease gradually due to different environmental condition and human interferences. Degradation of soil fertility is often associated with threats to food security and poverty. To attain a sustainable productivity, mineral fertilizers in forms of inorganic or organic must be incorporated to the soil. So, maintenance and enrichment of soil fertility is a key parameter in this respect by proper application of plant nutrients. Since the earlier decades, our food security mainly depends on mineral fertilization along with improved farm management technologies. Here, fertilizers play a crucial role to conserve our natural resources. Thus, present and future need to meet the challenge of food security will also be resolved.

8.2.1 Why Fertilizers?

- Food is the major need to live; global nutrition statistics indicate each nation (especially with agricultural economy) must produce more quality foods using optimum resources. Fertilizers are in high demand as the demand of food increases.
- The potentiality of quality production from the least fertile land is only possible by using fertilizers.
- Lesser time of crop production is an important utility of fertilizer application.
- Due to the changing climatic conditions, the rainfall-dependent crops are under massive stress. Application of certain fertilizers enhances the water-retaining capacity of soil, thus solving the crisis to a certain extent.

8.2.2 The Drawbacks

- The chemicals often washed away and ultimately fall into adjacent water bodies and contribute largely to the water pollution. Moreover, high nutrient in water body causes algal blooms, resulting in eutrophication, sharp increase in biochemical oxygen demand.
- Uncontrolled application of chemicals hampers soil biodiversity and soil ecosystem as well.
- Soil's natural fertility gets entirely lost, and this leads to land degradation.
- Chemical or inorganic fertilizers are economically expensive.

8.2.3 The Alternatives

- The only alternative can be environment-friendly fertilizers having the potential to increase the productivity. The conventional uses of fertilizers from plant parts, animal parts or excretes, minerals, and sludges do not have much effect on the environment. These are good nutrients too.
- Ash, bones, excretes, compost, farm yard manure, household organic waste, poultry manure, biochar, etc. are largely used as organic fertilizers.
- Slow release, controlled release, and organic matrix-entrapped fertilizers can also use as an alternative of conventional fertilizers.
- Integrated nutrient management is an emerging practice to ensure the food production.

8.2.4 But Not the Appropriate One

- Uncertainty is the major problem of using organic fertilizers, since the amount of nutrient contents is unknown.
- The production is much delayed while using organic fertilizers.

8.2.5 Still, There Is a Ray of Hope

The agricultural scientists from different corners of the earth are performing several researches, using the best technologies on this issue. Experiments are going on, targeting some particular crops and applying a proportional mixture of chemical and organic fertilizers to achieve the best possible result. The following table represents some experiments that infer the applicability of combined fertilizers for the best yields of a few major crops as well as adjust the emission of greenhouse gases (Table 8.1).

8.3 Inorganic Fertilizers

Inorganic fertilizers are one of the major sources of plant nutrients. They are also called high analysis chemical fertilizers with few impurities. Inorganic fertilizers are defined as the chemical fertilizers, prepared artificially with different synthetic ingredients of the inorganic materials, and enriched with essential plant nutrients in appreciable amount (Jaja and Barber 2017). Prior to the application of this, soil must be tested to know the actual need of soil fertility with exact amount of fertilizer. Addition of unwanted nutrients can disturb the soil environment by disrupting the pH levels which facilitate accumulations of salts and other harmful elements (Guo et al. 2017).

There are different sources of chemical fertilizers of which N–P–K, i.e., nitrogen (e.g., sodium nitrate, ammonium sulphate, ammonium nitrate, urea, ammonium phosphate, dibasic, ammonium phosphate, monobasic), phosphorus (e.g. di-calcium phosphate, anhydrous, bone meal, rock phosphate (fluoroapatite), single superphosphate, triple superphosphate), and potassium (e.g. potassium chloride, potassium nitrate, potassium sulphate), are three pillars (Tisdale et al. 1993; Iwami 1995). Inorganic fertilizers also include commercially available products of calcium (e.g., lime, dolomite, calcium sulphate), magnesium (e.g. Epsom salt), and sulphur (e.g., elemental sulphur). Also, there are some macronutrients such as iron sulphate, zinc sulphate, and copper sulphate salts in hydrated form (Silva 2000).

Inorganic fertilizers were introduced in the last century followed by initial increase in agricultural production. However, with time, it started to show their side effects which include water basin pollution, nutrient-leaching problems, destruction of different microorganisms, and friendly insects. Also, it reduced the soil fertility (Savci 2012). Therefore, there are some disadvantages of inorganic fertilizers with its wide range of advantages. It is well known to us that too much are not good for anything. Hence, chemical fertilizers must be used properly.

Quantitative analysis of inorganic fertilizers is discussed here using SWOT analysis (Strengths—Weaknesses—Opportunities—Threats).

Table 8.1 Experiments on coupled fertilizers

Experimental crop	Fertilizers used	Inferences	References
Maize (*Zea mays* L.)	Manure + chemical fertilizer	Higher seed yield	Rahman et al. (2012)
Rice (*Oryza sativa* L.)	Farm yard manure + inorganic fertilizer	Enhanced productivity and yield, shortened days to maturity	Tadesse et al. (2013)
Maize (*Zea mays* L.)	NPK + organic manure	Improve the fertility of black soil and enhance the productivity	Zhao and Zhou (2011)
Soybean (*Glycine max* L.)	Vermicompost + inorganic fertilizers	Improve the crop biometric parameters and increase yield in terms of pod numbers and seed numbers in each plant and that of the value of seed index	Devi et al. (2013)
Chickpea (*Cicer arietinum* L.)	Mineral nitrogen fertilizer + *Rhizobium* inoculation	Enhanced physiological traits (chlorophyll content, leaf area index, and protein content in grain) and agronomical traits	Namvar et al. (2013)
Wheat (*Triticum aestivum* L.) cultivar IPA 99	Biofertilizers (*Azotobacter* sp., *Azospirillum* sp. and *Pseudomonas* sp.) + mineral fertilizers (NP)	Significant positive effects on agronomic traits and yield components of wheat	Utoor and Ayyad (2014)
Maize (*Zea mays* L.)	Farmyard manure + NPK	Increase the crop performance and balance the soil chemical properties	Usman et al. (2015)
Walnut (*Juglans regia* L.)	NPK + vermicompost; NPK + poultry manure	Improved yield of walnut	Wani et al. (2017)
Mustard (*Brassica campestris* cv. B_9)	N:P:K (100:50:50) + biofertilizer N:P:K (100:50:50) + vermicompost	Enhancement in crop production by means of adjusting morphophysiological characteristics and biochemical attributes	Mondal et al. (2017)
Peanut (*Arachis hypogeal* L.)	Compost + manure + *Bradyrhizobium* inoculation + NP application	Improved nodulation and yielded higher biomass	Argaw (2017)

8.3.1 Strengths (Advantages of Commercial Chemical Fertilizers)

- Inorganic fertilizers are convenient to use and affordable.
- They are predictable and reliable.
- As the nutrients required for plant are readily available in chemical fertilizers, hence it works immediately.
- Commercial products of inorganic fertilizer consist a balanced distribution of three main nutrient ratio, i.e., N (nitrogen), P (phosphorus), and K (potassium). Some other minerals are also present as stated in the composition.
- These are obtained as solid or liquid form and easy to store.
- Accurately blended formulations for different crops are readily available.
- Growers can prepare their own mixture by mixing different formulations separately.
- It takes a while. With an inorganic fertilizer, you save a lot of time and effort.
- It is easily available in market.
- All the necessary nutrients are available in ready-to-use condition.
- It is easier to apply with uniform distribution.

8.3.2 Weaknesses (Disadvantages of Commercial Chemical Fertilizers)

- Commercially available, some inorganic fertilizers having special formulations are sometimes not affordable due to its higher cost.
- Wastage of essential mineral nutrients often occurs due to leaching.
- Applied inorganic fertilizers breakdown to their basic units in maximum cases and can easily be uptaken by plants through absorption. However, these elements may be washed away by the process of overwatering plants.
- Application of inorganic fertilizers needs proper measuring with careful mixing of different formulations in accurate proportion on the basis of a specific plant requirement.
- Overdose may cause different harmful effects on plants. Direct application of it in excess amount can kill the plants by disrupting the plant delicacy which is known as fertilizers burn.
- Another point of concern is that its composition not only is with the basic element but also is associated with different salts and other compounds. These might be toxic to human health. Soil is a complex matrix. Human health as well as natural environmental balance mostly depends on soil environment. If these toxic substances or different salts are accumulated in soil environment, then this will disturb the soil chemistry affecting the plant growth (Acosta-Motos et al. 2017).
- Improper application of inorganic fertilizers can potentially disturb the soil pH balance and fertility equilibrium in a long-term basis.
- Heavy metal accumulation from different salts is also a serious matter of concern.
- Inorganic fertilizers are unable to replace the depleted trace minerals in soil.

- Chemical fertilizer preparation is based on different synthetic ways which face resource limitations.
- Fertilizer run-off, mainly from nitrogen-enriched compounds, promotes reduction in oxygen level in aquatic region creating lack of dissolved oxygen affecting the aquatic ecosystem which is commonly known as eutrophication (Bhateria and Jain 2016).
- Inappropriate and excess utilization of synthetic fertilizers to achieve higher crop yield indirectly affect the global climate change through the agricultural emissions of greenhouse gases especially methane and nitrous oxide (Rahman et al. 2019).
- Chemical fertilization affects the birth rate of some insect pests which safeguard the crops.

8.3.3 Opportunities

8.3.3.1 Demand of Nitrogenous Fertilizers

Many developing countries are relying on conventional nitrogen fertilizer for crop production, and it is consuming more than half of the world's nitrogenous fertilizer. Global fertilizer demand for nitrogen-rich fertilizers, viz., ammonium fertilizers, nitrate fertilizers, combined ammonium and nitrate fertilizers, and amide fertilizers, has been directly related to the demand for food and fibre. China and India are at centre of the world's attraction due to their growing population as well as the increasing per capita income. Both these countries are the major contributors to the increase of nitrogenous fertilizer market.

8.3.3.2 Demand of NPK and Urea-Based Fertilizers

According to the International Fertilizer Association (IFA), an increasing growth rate is predicted for urea. Hence, most of the N-capacity additions occurred as urea. Nitrogen fertilizer demand growth is at 1.1% per year through 2021. Annual growth rate of 1.6% is assessed for phosphate and 2.2% for potassium It is reported that demand of nitrogen fertilizer will rich 1.1% per year by 2021 (Fertilizer outlook 2018–2022, IFA 2018). Also annual growth rate for phosphate and potassium is estimated as 1.6% and 2.2%, respectively.

8.3.3.3 Geographical Variances in Fertilizer Application

The need of fertilizers is monitored by different factors of which land area for plantation, crop production potentiality, mixing of crops, fertilizer-to-crop ratio, nutrient scheduling, and recycling practices are the most important.

8.3.3.4 Growth in Demand for Food Grains

It is important to understand the whole scenario of fertilizer market which eventually depends on grain market. Generally, grains are known as end market for fertilizer applications.

8.3.3.5 Huge Export Potential

Low-cost producer's status especially in India is the primary opportunity to increase fertilizer exports.

8.3.4 Threats

8.3.4.1 Threat from Genetically Modified Seeds

There might be negative impacts on the trades on agrochemicals due to self-immunity of genetically modified (GM) seeds towards natural adversaries.

8.3.4.2 Integrated Pest Management and Rising Demand for Organic Farming

The demand for organic food seeks much attention. Different states also reduce the practice of chemical-based fertilization and introduce organic farming. For instance, in Kerala, India, this is more applicable. Thus, several approaches towards sustainable agriculture like integrated pest management (IPM), zero budget farming, carbon neutral farming, organic farming, and popularization of bio-pesticides by different governments and NGOs are gaining momentum.

8.3.4.3 Limited Farmland Availability and Growing Exports

The available area for farming is reducing day by day globally. Though India and other South Asian countries have higher gross cultivated land, extension of this scope is very limited.

In a nutshell, chemical fertilizers have both advantages and disadvantages. Hence, a balanced and proper method should be followed during their use.

8.4 Organic Fertilizers

Organic fertilizers consist of different components that originated from fresh as well as dried plant materials to animal matter, animal excreta, and human excreta and dissipate to related agrisubstances (Das and Jana 2003; Kumar et al. 2004). It is applied to soils to enhance essential plant nutrients involving major and minor nutrients required for various physiological activities through increasing the primary nutrient supply to the host plant. They also enrich the soil's nutritive status and preserve the soil fertility. The nutrient content varies significantly according to their sources and materials which are easily biodegradable and can create better nutrient sources. Organic fertilizer enhances the microbial growth of the soil and tends to be decomposed into plant's essential mineral nutrients as the microorganisms grow.

The presence of nitrogen and phosphorus content is less in organic amendments. Chemical fertilizers are synthetic which are hazardous, whereas organic fertilizers are eco-friendly in nature. Chemical fertilization is causing unbelievable harm to human health and environment. Reliability on chemical fertilizers for increasing the yield will promote further decrease of soil quality and contamination of soil as well as groundwater subsequent in the loss of ecological balance. Hence, one of the major concerns is pollution and contamination of soil. Selecting organic fertilizers instead of chemical ones could be an effective alternative as they safeguard the natural environment and a good example for sustaining the agricultural practices.

Different types of fertilizers, such as chemical fertilizers, organic amendments, and natural fertilizers, are available on the market. However, farmers around the globe depend on chemical fertilizers, but many are now shifting to organic fertilizers due to the apparent long-term benefits.

8.4.1 Benefits of Organic Fertilizers

Organic fertilizers are based on carbon components which strengthen the production quality. Different advantages of organic amendments over chemical fertilizers are as follows.

8.4.1.1 Plant Growth

Organic fertilizers provide nutrients necessary for plant growth (Jo 1990) and release their nutrients more slowly compared to chemical fertilizer in soil. Organic fertilizers cannot be immediately taken up by plants. Soil microbes break them up and make it available for plant. Unlike chemical fertilizers, the nutrients of organic fertilizer are contained in complex molecules that are not easily run-off away in a heavy rainstorm or irrigation session. Hence, the plants gradually adsorb the essential nutrients over an extended time period.

8.4.1.2 Slow Release of Nutrients

Plant community uses their roots to absorb the nutrients from soil which were incorporated through fertilization. In chemical fertilizer, these nutrients are readily available which are taken by the plant roots. Hence, absorbed nutrients are greater than their requirements, causing the roots as well as plant to burn up. In this case, organic fertilizers do not have the nutrient in freely usable form. Soil microbial biota breaks down the added amendments and releases the nutrients (Ni et al. 2010). Organic fertilizers strengthen the soil aggregates, hence managing the soil fertility (Ball et al. 2004). In this slow release process, there is no risk of higher availability of nutrients to the plant. Application of organic fertilizers creates fewer chances for 'plant damage'. Several formulations have been produced to control the excess release of nutrients and are commonly popularized as slow release fertilizers produced by using either microbial consortium or chemical fertilizers (Singh et al. 2010; Kumar et al. 2012a, b, 2013a, b, 2014a, 2015).

8.4.1.3 Improvement of Soil Health

Organic matters present in organic fertilizers improve the soil texture, hence allowing it to hold water longer, and increase the bacterial and fungal activity within the soil. Organic fertilizers enhance soil health by nurturing the soil biota that helps to make soil nutrients available to plants. It ensures that the complex structure of soil and nutrient content of farms remain fertile over a long time. Some fertile ancient land areas are present in India and China where agriculture is being practiced for more than thousand years. Fertility is sustained as it was amended with organic fertilizers. However, rise in chemical fertilization practices resulted in infertility which forced the existing farmers to have that entire place or become more dependent on chemical fertilizers. Chemical fertilizers basically add water-soluble chemicals which are either absorbed by the plant roots or leached away, potentially contaminating water resources, while addition of organic matters through organic fertilizers helps the soil to retain moisture and nutrients. In particular, sandy soils get the benefit of organic applications.

8.4.1.4 Microbes Thrive

Chemical fertilizer consists of chemical molecules without carbon which are not accessible to soil microbes. On the other hand, organic fertilizer having higher amount of organic matter sustains the microbes thrive. Carbon is already included in their chemical makeup; and the soil microbes feed on carbon, along with nitrogen, phosphorus, and potassium in a natural bioprocess (Liu et al. 2013; Bargaz et al. 2018).

8.4.1.5 Long-Term Benefits to the Soil

Synthetic fertilizers are formed with the purpose of serving to enhance the plant growth. They contain a balanced proportion of essential plant nutrients along with some hazardous components causing the soil acidity. This can be harmful for the useful soil microbes. Numerous evidence have been seen regarding the harmful effects of long-term chemical fertilization on soil, while organic amendments stimulate the microbial growth ensuring better soil fertility (Diacono and Montemurro 2011).

8.4.1.6 Environmental Impact

The organic amendments in agriculture have various positive as well as negative impacts on environment, i.e. enhance soil health, plant and wildlife biodiversity, nutrient stocking, and climate change as well as greenhouse gas emission (Lynch 2009). Hazardous components associated with chemical fertilizers can contaminate the soil and water resource. Overdose of inorganic chemical fertilizers deteriorates the soil (Hazra 2014; Singh et al. 2017). However, organic fertilizers always contain biodegradable elements. Organic fertilizers do not harm the ecological balance in the soil because they do not leave behind any artificial compounds.

8.4.1.7 Cost-Effective On-Farm Production and Easy to Apply

Organic fertilizers are generally produced at home or on farms by using a mix of cow, sheep, chicken, and horse manure along with wastes like vegetable leaves and dead plants (Westerman and Bicudo 2005; Jenkins 2009). This is an option to get rid of domestic and farm-originated waste. Hence, the cost of these fertilizers is much lower than the cost of chemical fertilizers. The application of organic fertilizers is easier. They can act as soil conditioners also and are applied prior to planting or seeding. Still others are sprayed on the leaves of the plant as a foliar fertilizer (Fageria et al. 2009).

8.4.1.8 Sustainability and Environmental Safety

Organic fertilizers are biodegradable and insoluble in water and safeguard the environment (Dunsin and Odeghe 2015; Hazra 2016). However, inorganic fertilization resulted in environmental pollution, affecting the human health and the extinction of some species. However, local wildlife may often be affected by the use of inorganic fertilizers. Organic fertilization creates a safe environment for our family and pets also.

8.4.1.9 Health Security

Organic fertilizers ensure that the food products produced are free of toxicologically significant chemicals (Zhu and Jin 2013). As a result, people who consume organically produced food products are less susceptible to diseases than those who ate food items produced using chemical fertilizers.

8.4.1.10 Low-Capital Investment for On-Site Production

Organic fertilizers are made from mined rock minerals and natural plant and animal materials. Organic material includes ingredients like manure, guano, dried and powdered blood, ground bone, crushed shells, finely pulverized fish, phosphate rock and wood, etc. Moreover, organic amendments maintain the soil structure and help to increase its nutrient-holding ability (Mumpton 1999; Mar and Okazaki 2012). Therefore, long-term organic farmers require less fertilizer inputs as the soil is already enriched with essential nutrients.

8.4.1.11 Reduce the Use of chemical Fertilizers and Pesticides

Organic fertilizer can reduce the prerequisite for pesticides and the overall major nutrient requirements. Because of the reductions, organic fertilizer can be cost neutral and sometimes a cost saving (Chen 2006).

8.4.1.12 Employment

Organic fertilizer production facilities require only smaller land area, and it can be prepared locally which leads to create employment for local people, especially in rural and marginal areas where earning opportunities are sometimes be bleak.

8.4.2 Limitations of Organic Fertilizers

The major limitation of using an organic fertilizer is that it may not contain sufficient amount of major nutrients like nitrogen, phosphorous, and potassium, also known as NPK. They are just soil amendments. NPK directly affects plant growth by feeding the plant. However, if a fertilizer does not include any NPK, it affects plant growth indirectly.

8.4.2.1 Product Divergent

All the organic products produced are dissimilar, and many products create inconsistent results. Farmers are not sure that the selected organic fertilizer is reviewed or certified by any third-party organization.

8.4.2.2 Slow and Sustained Release of Nutrients

Some plants are malnourished, even on the approach of dying. At this stage, plants need an immediate involvement of high doses of nutrients fast. This is possible with chemical fertilization but not with an organic fertilizer. In order to meet the needs of a dying plant, the slow and sustained release of nutrients by the organic fertilizer cannot be speed up (Adegbidi et al. 2003). Hence, the plant may die. Organic fertilizers are degraded into basic elements like nitrogen, phosphorus, and potassium which cannot meet the appropriate proportions, i.e. nitrogen is more abundant than the remaining (Savci 2012). The plants may absorb a lot of one specific nutrient over the others and may have an unbalanced diet.

8.4.2.3 Low Nutrient Content

The level of nutrients present in organic fertilizer is often low (Davis 2015). In organic amendments, the nutrients are present as complex structure which indicates that organic fertilization is a procedure rather than an event.

8.4.2.4 Not Enough Crop Yield

Crop production under organic farming is low as compared to conventional farming practices mainly due to less nutrient content, limited chances to enhance the soil productivity as earliest, and slow release of nutrients. Numerous field studies as well as meta-analysis of global datasets already discussed about the insufficient production potentiality of organic agricultural system to feed the society (Badgley et al. 2007; Kirchmann et al. 2008; de Ponti et al. 2012; Seufert et al. 2012).

8.4.2.5 High Demand, Low Supply

The use of organic fertilizer in a big farm or field is not so much costly, but availability is limited. A huge aomunt of raw material is required to prepare organic fertilizers through composting. The household wastes or farm based wastes are not enough to make adequate organic fertilizer.

8.4.2.6 Simple but Untidy and Inconvenient

It is very easy to prepare compost. However, preparation of organic fertilizers from the rotten organic ingredients results in unpleasant odour. Sometimes, it becomes a difficult task.

8.5 Coupling Application of Inorganic and Organic Fertilizers

Fertilizer amendments enhance the crop growth through immediate supplying of adequate major nutrients, viz. nitrogen, phosphorus, and potassium to crop soil. Basically, the chemical fertilizers have the capability to increase the agricultural production potentiality. However, indiscriminate and wide use of chemical fertilizers in agroecosystems instigates the worsening of soil quality (Liu et al. 2010). Soil fertility can be declined due to unscrupulous application of inorganic fertilizers (Savci 2012; Kumar et al. 2013a, b; Singh et al. 2017). Emissions of greenhouse gases—another important area of concern—are also associated with non-judicious application of inorganic fertilizers. On other hand, amendment of organic fertilizers to crop soil can potentially improve nutrient balance and physicochemical characteristics of soil. Organic supplements were not only recognized as source of major nutrients and organic matter but also can intensify the microbial activities within crop soil (Albiach et al. 2000). Yet, solely organic fertilization is not adequate to accomplish the sustainability in crop yield (Prasad 1996). To achieve environmentally sound and sustainable agricultural systems, judicious fertilization practices are of urgent need. Amendment of inorganic fertilizers in combination with organic manure is an effective strategy to sustain the grain yield with least effect on environmental health.

8.5.1 Soil Health and Crop Productivity in Response to Inorganic Fertilization Integrated with Organic Manures

Fertilizer amendments provide essential nutrients to crop soil (Masarirambi et al. 2012). Integrated application of inorganic fertilizers with organic nutrient sources, viz. farm yard manure, vermicompost, dairy manure, and/or poultry manure, may increase the crop yield with maintaining the soil health. Sustainable agriculture depends upon balanced soil health. Combined use of inorganic and organic nutrient sources greatly influences the soil pH, organic carbon content, and availability of nitrogen, phosphorus, and potassium. Electrical conductivity—an indicator of soil microbial activities—is also affected by such coupling fertilization. Hati et al. (2007) reported significant influence of continuous inorganic and organic fertilization on soil EC. Incorporation of chemical fertilizers along with green manure, farm yard manure, and rice straw has the potentiality to diminish the EC level in crop soil under rice–wheat rotation (Kumar et al. 2012a, b). Soil pH indicates the soil chemical characteristics leading to control the crop growth performance through influencing the soil microbial community and availability of nutrients (Dalal and Moloney 2000).

Fertilization practices, standing crop, cropping sequence, and soil type determine the amplitude of effect. Numerous studies argued that soil pH can be maintained through long duration application of integrated inorganic and organic source of nutrients (Sharma et al. 1998; Tirol-Padre et al. 2007; Kumar et al. 2012a, b). It is a proven fact that soil organic carbon content is a governing parameter for agricultural productivity and climate change (Li et al. 2007). Instead of traditional nutritional practices, coupling application of inorganic and organic nutrient sources can significantly enhance the organic carbon stock within crop soil (Lal 2004; Rai et al. 2015). Literatures recognized the importance of combined fertilizer amendments including inorganic and organic sources to improve status of soil organic carbon (Moharana et al. 2012; Benbi and Senapati 2010). Long-term crop field management with balanced and integrated nutrient amendments can potentially modify the soil microbial dynamics and physicochemical characteristics (Masto et al. 2006). Conjoint application of conventional fertilizers and organic manuring have positive effect on nutrient cycling in crop soil and availability of essential crop nutrients, i.e. nitrogen, phosphorus, and potassium. Significant enhancement in available nitrogen in crop soil after integrated nutrient application has been documented by Nath et al. (2011). Increment in available phosphorus and potassium level in various crop soil was found to be improved after long-term use of chemical fertilizers in combination with manure and/or enriched compost (Reddy et al. 2000; Bednarek et al. 2012; Hemalatha and Chellamuthu 2013).

Soil microbial functions are another aspect of soil health. Soil biological population and enzyme activity were observed to be improved under integrated application of mineral fertilizers, straw, and swine manure (Juan et al. 2008). Positive impact of combined addition of chemical and organic fertilizers on soil microbes was recognized in previous study (Dutta et al. 2003). In another field-based research on pearl-millet and wheat cropping sequence, an improved soil biological health was observed after conjoint application of farm yard manure, sugarcane filter cake, poultry manure, and chemical fertilizer (Kaur et al. 2005). Long-term agricultural practices including coupling application of inorganic and organic fertilizers can help maintain the soil pH while increasing the availability of soil nitrogen, phosphorus, exchangeable potassium, and soil organic carbon content (Yadav et al. 2019).

It is reported that sole application of synthetic as well as organic fertilizers is not sufficient to improve the soil fertility and crop productivity (Godara et al. 2012; Getachew et al. 2014). MoA of Ethiopia (2012) already declared that soil fertility could be managed sustainably through integrated use of chemical fertilizers and farmyard manure. The importance of coupling application of inorganic and organic fertilizers to enhance the fertility of degraded soil with subsequent increase in crop productivity has already been documented (Shata et al. 2007; Gete et al. 2010; Getachew and Tilahun 2017).

Judicious application of mineral fertilizers in combination with organic source of plant nutrients is an attainable practice to tackle the problems regarding soil health and sustainable food production systems. Recent experiments on identifying the role of coupling application of inorganic and organic fertilizers to improve soil conditions and agricultural productivity are enumerated in Table 8.2.

Table 8.2 Field experiments on the effect of coupling application of inorganic and organic fertilizers on crop productivity and soil health

Crop	Fertilizer application in combination form	Country	Inferences	References
Sugarcane (*Saccharum officinarum*)	Organic fertilizers (press mud), farmyard manure (FYM) and green manure (*Crotalaria juncea*) along with inorganic fertilizers (urea, TSP, MOP, gypsum and zinc sulphate)	Sandy loam soil of Bangladesh	Maximize the yield	Bokhtiar and Sakurai (2005)
Wheat (*Triticum aestivum*)	Biofertilizer (*Azotobacter* + *Azospirillum* in 1:1 ratio) + nitrogen	Alluvial soil, India	Increase in crop growth, yield-attributes and yield of wheat	Kachroo and Razdan (2006)
Wheat (*Triticum aestivum*)	10 tonnes FYM + 120 kg N/ha	India	Higher productivity	Ram and Mir (2006)
Rice (*Oryza sativa* L.) and sorghum (*Sorghum bicolor* L.)	NPK + Charcoal amendment	Oxisol of Brazil	Soil enriched with essential nutrients and higher crop yield	Steiner et al. (2007)
Aman rice (*Oryza sativa* L.)	Organic manures (rice straw, dhaincha, mungbean residue, cow dung, poultry manure) + recommended dose of chemical fertilizers	Bangladesh (non-calcareous dark Grey floodplain)	Maximizing the yield; soil fertility enhancement	Ali et al. (2009)
Durum wheat (*Triticum turgidum*)	50% NPK + *Azotobacter* spp. (AS), 50% NPK + *Azospirillum brasilense* (AB), 50% NPK + *Proteus vulgaris* (PV), 50% NPK + *Kurthia* spp. (KS), 50% NPK + *Klebsiella planticola* (KP), 50% NPK + *Bacillus subtilis* (PSB)	Central India (black soil)	Improvement in yield and quality of grain	Behera and Rautaray (2010)
Winter wheat (*Triticum aestivum* cv. Shiraz)	Urea + municipal waste compost	Silty loam soil of Iran	Maximizing the yield and improve the protein composition of wheat grain	Abedi et al. (2010)

(continued)

Table 8.2 (continued)

Crop	Fertilizer application in combination form	Country	Inferences	References
Sugarcane plant-ratoon system	Organic manures (farm yard manure, vermicompost, press mud) + recommended dose of inorganic fertilizer	India (clay loam soil)	Increase in cane production and sugar yield	Rama Lakshmi et al. (2011)
Oat (*Avena sativa* L.)	Biochar + recommended dose of mineral fertilizer	Infertile sandy soil in the greenhouse in Germany	Increase in soil cation exchange capacity and pH and improve overall plant growth	Schulz and Glaser (2012)
Maize (*Zea mays*)	NPK + cow dung	Sandy loam soil of Japan	Higher number of grains per cob and increased yield	Amoah et al. (2012)
Wheat (*Triticum aestivum*)-soybean (*Glycine max*) cropping system	Recommended NPK + PM @ 5 t/ha recommended NPK + FYM @10 t/ha	India (Vertisol)	Higher productivity improved soil health	Behera and Rautaray (2013)
Wheat (*Triticum durum* L. cv. Vitron)	Biochar (agricultural waste and olive tree pruning) + mineral fertilization	Acidic loamy sandy soil of southern Spain	Enhanced soil fertility and improved wheat yield	Alburquerque et al. (2013)
Rice (*Oryza sativa* L.)	Different organic fertilizers (farmyard manure, poultry manure, rice straw, sesbania, compost, and mungbean residues) in combinations with 50% of recommended dose of fertilizer (RDF)	Silty clay loam soil of Pakistan	Better crop growth and higher productivity	Arif et al. (2014)
Rice (*Oryza sativa*)	Organic manures integrated with recommended dose of mineral fertilizers	Silt loam soil of Bangladesh	Higher grain yield and improved grain quality	Islam et al. (2014)
Wheat (*Triticum aestivum* L.)	Organic matrix + urea and DAP + microbial fertilizers	India	Higher productivity and improved soil quality	Kumar et al. (2014b)
Wheat (*Triticum aestivum* L.)	Organic matrix + microbial fertilizers (*Azotobacter* + *Bacillus*)	India	Higher productivity and enhanced soil quality	Kumar et al. (2014c)
Maize–wheat rotation	100% NPK + farm yard manure (150 kg N, 32.7 kg P, 31.2 kg K ha^{-1} and 10 mg ha^{-1} FYM)	India	Enriched SOC stock and soil physical condition along with higher yield	Brar et al. (2015)

Rice (*Oryza sativa* L.)	Different levels of vermicompost and NPKS fertilizer	Silty clay loam soil of Bangladesh	Higher growth and yield	Mahmud et al. (2016)
Tomato	Integrated plant nutrient system (IPNS)/ mixed fertilizers (vermicompost + N + P + K + S + Zn + B)	Bangladesh	Best response on plant growth, fruit yield	Islam et al. (2017)
Peanut (*Arachis hypogea* L.)	Combined application of compost, manure, *Bradyrhizobium* inoculation, and NP application	Sandy soil of Ethiopia	Increase in yield	Argaw (2017)
Rice (*Oryza sativa*)	Farmyard manure in combination with three levels of chemical fertilizers [80:40:30, 120:60:45 and 160:80:60 kg N, P_2O_5 and K_2O ha^{-1}, respectively]	India	Increase in yield and yield component	Joshi et al. (2017)
Boro and Aman rice (*Oryza sativa*)	Organic manure (cow dung, cow dung slurry, trichocompost, vermicompost, poultry manure and PM slurry) + RDF	Old Brahmaputra floodplain of Bangladesh	Higher crop yield and soil health improvement	Bilkis et al. (2017)
Barley (*Hordeum vulgare* L.)	Integrated use of lime, compost and mineral fertilizer	Acid soils of Ethiopia	Increase in yield and yield components	Demissie et al. (2017)
Barley (*Hordeum vulgare* L.)	50% vermicompost + 50% recommended NP, 50% conventional compost + 50% recommended NP, 50% FYM + 50% recommended NP	Western Ethiopia (brown clay loam ultisols)	Sustenance in barley production	Abera et al. (2018)
Wheat (*Triticum aestivum* L.)	RDF + vermicompost	Gangetic alluvial soil of India	Improved productivity	Singh et al. (2018)
Turmeric (*Curcuma longa* L.)	Mg fertilizer in combination with poultry manure (PM) and NPK 15-15-15 fertilizer (NPK)	Nigeria Alfisol	Improved performance of growth, yield, and quality of crops	Adekiya et al. (2019)
Wheat (*Triticum aestivum* L.)	Systemic combination of RDF, neem-coated urea and vermicompost	India	Better vegetative growth	Dey et al. (2019)

8.5.2 Coupling Application of Inorganic and Organic Fertilizers: An Approach to Mitigate Greenhouse Gas Emissions

Sustainable agricultural system using proper dose of inorganic and organic fertilizers can satisfactorily improve the soil conditions and ameliorate the emissions of greenhouse gases. Actually, improper fertilization often leads to nutrient imbalances in agroecosystems (Vitousek et al. 2009). It is a difficult task for recent high-yielding agricultural system to slash down the emission of greenhouse gases without affecting the productivity. The concept of coupling application of inorganic and organic fertilizers is aimed to decrease the nutrients loss from crop field and increase the crops' nitrogen use efficiency through adjusting the crop's requirement (Janssen 1993).

Nitrification and denitrification within crop soil is enhanced due to the presence of excess mineral nitrogen leading to increase the production and emission of N_2O (Baggs and Blum 2004). As per IPCC, 2013, at 100 year timescale, global warming potential of N_2O is higher than CO_2 (298 times). It is a proven fact that maximum N_2O tends to be released following the nitrogenous fertilization (Leahy et al. 2004; Skiba et al. 2012; Saha et al. 2015). Decomposition of organic matter under anaerobic soil promotes the production and emission of methane, and flooded rice paddy is recognized as higher CH_4 emitter (Regina et al. 2007; Saha et al. 2017; Kar et al. 2019). There is always a chance of CO_2 emissions from crop soil mainly due to decomposed organic components and disturbances in land use patterns (Janzen 2004; Bernhardt et al. 2006).

Balanced fertilization practices comprising both the organic and inorganic amendments can lessen the emission of greenhouse gases (Saha et al. 2015; Brenzinger et al. 2018). However, some experiments reported negative responses of combined fertilization on emission of greenhouse gases from agricultural field (Bhatia et al. 2005). Worldwide researches provide strong evidences regarding this, some of which are given further (Table 8.3).

Basically, the efficiency and applicability of coupled fertilizers to reduce the emissions of greenhouse gases is tested in several experiments which need some improvisation. The fertilizer applications must be followed by concrete estimation of the needs of the nutrient requirements. The proportion of organic and inorganic fertilizer requirement would always differ from crop to crop, soil to soil, and climate to climate. However, different experiments involving conjoint application of organic and inorganic fertilizers on different cropping systems and various soil types are still ongoing to find out crop- and soil-specific proper dose of inorganic and organic amendments.

8.6 Conclusions

One who has the in-depth cognition of the soil nutrient content, which may highly vary with time and knowledge about the nutrient requirement for each particular crop to be cultivated, is able to produce a good yield in both the qualitative and

Table 8.3 Experiments on greenhouse gas emission trends from coupling application of inorganic and organic fertilizers

Crops	Fertilizer used	Inferences	References
Maize (*Zea mays* L.)–groundnut (*Arachis hypogaea* L.) rotation	Inorganic N + crop residues (NC), inorganic N only (RN), and half of inorganic N + crop residues + chicken manure (NCM)	Higher N_2O emissions from NCM as compared to other practices	Khalil et al. (2002)
Maize (*Zea mays* L.)	Organic (Rye residues), inorganic (NH_4NO_3), rye residues + NH_4NO wheat residues, wheat residues + NH_4NO_3	Emission trend of N_2O-Org, Inorg < INM Rye < winter wheat	Sarkodie-Addo et al. (2003)
Maize (*Zea mays*)-winter wheat (*Triticum aestivum*)	Manure compost, urea, integrated application of manure and urea	No significant differences in N_2O emission pattern	Meng et al. (2005)
Rice (*Oryza sativa*)	Different mixture of chemical fertilizer, pig manure compost, Chinese medicine residue compost and rapeseed cake.	Increase in global warming potential and greenhouse gas (GHG) intensity	Wang et al. (2013)
Maize (*Zea mays*), winter wheat (*Triticum aestivum*)	Cattle manure, urea, cattle manure + urea	No significant differences in greenhouse gas emission pattern	Cai et al. (2013)
Rice (*Oryza sativa*)	Compost + urea-N	Reduction in greenhouse gas intensity (GHGI), higher carbon efficiency ratio (CER) and maximize the yield	Das and Adhya (2014)
Double-rice cropping	Pig manure + chemical fertilizers	Increase the GWP of CH_4 and N_2O emissions during rice growing seasons	Wang et al. (2014)
Rape (*Brassica napus*)	Cattle manure, ammonium nitrate, combined organic and inorganic	Reduced emission of N_2O in combined fertilizer application	Nyamadzawo et al. (2014)
Winter wheat (*Triticum aestivum*)	Inorganic (NPK), inorganic (50% NPK) + organic (50% vermicompost)	Higher N_2O emission in combined fertilizer management	Karmakar et al. (2014)
Rice (*Oryza sativa*)	Inorganic fertilizers in combination with various organics (cow dung, green manure (*Sesbania aculeate*) Azolla compost, rice husk)	Combined application of inorganic (NPK) with Sesbania aculeate increase the global warming potential as compare to others; NPK and Azolla compost was effective in improving	Bharali et al. (2018)

(continued)

Table 8.3 (continued)

Crops	Fertilizer used	Inferences	References
		soil carbon and soil carbon storage capacity with high carbon efficiency ratio	
Jasmin (*Jasminum sambac* L.)	Straw + standard fertilizer treatment	Improvement in soil water retention capacity, soil acid, and mitigate the emission of greenhouse gases	Wang et al. (2019)

quantitative facets. This is the time to undergo an intensive application of researches and subsequent communications of the outcomes, to the farmers to make their job easier. Besides, governments and other organizations have to work to promote awareness as well as educate and motivate the farmers and the scientists. The question of sustainability still remains unanswered, though the most effective production, causing minimal effect on the natural environment, is the most favourable condition of sustainability. Such coupling application of fertilizers, if properly implemented, would definitely leave a remarkable contribution to mitigate the world's food problem and thus meet the Sustainable Development Goal—2030.

References

Abedi T, Alemzadeh A, Kazemeini SA (2010) Effect of organic and inorganic fertilizers on grain yield and protein banding pattern of wheat. Aust J Crop Sci 4:384–389

Abera T, Tufa T, Midega T, Kumbi H, Tola B (2018) Effect of integrated inorganic and organic fertilizers on yield and yield components of Barley in Liben Jawi district. Int J Agron 2018:1–7. https://doi.org/10.1155/2018/2973286

Acosta-Motos JR, Ortuño MF, Bernal-Vicente A, Diaz-Vivancos P, Sanchez-Blanco MJ, Hernandez JA (2017) Plant responses to salt stress: adaptive mechanisms. Agronomy 7(1):18. https://doi.org/10.3390/agronomy7010018

Adegbidi HG, Briggs RD, Volk TA, White EH, Abrahamson LP (2003) Effect of organic amendments and slow-release nitrogen fertilizer on willow biomass production and soil chemical characteristics. Biomass Bioenergy 25(4):389–398

Adekiya AO, Alori ET, Aboyeji CM, Dunsin O, Adegbite KA, Aremu CO, Bamiro O, Ejue WS, Okunlola FO, Adesola OO (2019) MgO fertilizer sole and combined with organic and inorganic fertilizers: effect on soil chemical properties, turmeric performance, and quality in a tropical Alfisol. Sci World J. https://doi.org/10.1155/2019/8140276

Albiach R, Canet R, Pomares F, Ingelmo F (2000) Microbial biomass content and enzymatic activities after the application of organic amendments to a horticultural soil. Bioresour Technol 75:43–48

Alburquerque JA, Salazar P, Barrón V, Torrent J, Campillo MDCD, Gallardo A, Villar R (2013) Enhanced wheat yield by biochar addition under different mineral fertilization levels. Agron Sustain Dev 33:475–484

Ali ME, Islam MR, Jahiruddin M (2009) Effect of integrated use of organic manures with chemical fertilizers in the rice-rice cropping system and its impact on soil health. BJAR 34(1):81–90

Amoah AA, Senge M, Miyagawa S, Itou K (2012) Effects of soil fertility management on growth, yield, and water-use efficiency of maize (*Zea mays L.*) and selected soil properties. Commun Soil Sci Plan 43:924–935

Argaw A (2017) Organic and inorganic fertilizer application enhances the effect of Bradyrhizobium on nodulation and yield of peanut (*Arachis hypogea* L.) in nutrient depleted and sandy soils of Ethiopia. Int J Recycl Org Waste Agric 6:219–231

Arif M, Tasneem M, Bashir F, Yaseen G, Iqbal RM (2014) Effect of integrated use of organic manures and inorganic fertilizers on yield and yield components of rice. J Agric Res 52 (2):197–206

Badgley C, Moghtader J, Quintero E, Zakem E, Chappell MJ, Avilés-Vázquez K, Samulon A, Perfecto I (2007) Organic agriculture and the global food supply. Renew Agric Food Syst 22 (2):86–108

Badr MA, Abou-Hussein SD, El-Tohamy WA (2016) Tomato yield, nitrogen uptake and water use efficiency as affected by planting geometry and level of nitrogen in an arid region. Agric Water Manag 169:90–97

Baggs EM, Blum H (2004) CH_4 oxidation and emissions of CH_4 and N_2O from *Lolium perenne* swards under elevated atmospheric CO_2. Soil Biol Biochem 36:713–723

Ball BC, McTaggart IP, Scott A (2004) Mitigation of greenhouse gas emissions from soil under silage production by use of organic manures or slow-release fertilizer. Soil Use Manag 20:287–295

Bargaz A, Lyamlouli K, Chtouki M, Zeroual Y, Dhiba D (2018) Soil microbial resources for improving fertilizers efficiency in an integrated plant nutrient management system. Front Microbiol 9:1606. https://doi.org/10.3389/fmicb.2018.01606

Bednarek W, Slawomir D, Traczyk P, Hanaka A (2012) Available forms of nutrients in soil fertilized with liquid manure and NPK. J Elem Sci 38:169–180

Behera UK, Rautaray SK (2010) Effect of biofertilizers and chemical fertilizers on productivity and quality parameters of durum wheat (*Triticum turgidum*) on a Vertisol of Central India. Arch Agron Soil Sci 56(1):65–72

Behera A, Rautaray SK (2013) Integrated use of organic manures and inorganic fertilizer on the productivity of wheat-soybean cropping system in the vertisols of central India. J Soil Water Conserv 12(1):59–65

Benbi DK, Senapati N (2010) Soil aggregation and carbon and nitrogen stabilization in relation to residue and manure application in rice-wheat systems in northwest India. Nutr Cycl Agroecosyst 87:233–247

Bernhardt ES, Barber VJJ, Pippen JS, Taneva L, Andrews JA, Schlesinger WH (2006) Long-term effects of free air CO_2 enrichment (FACE) on soil respiration. Biogeochemistry 77:91–116

Bharali A, Baruah KK, Baruah SG et al (2018) Impacts of integrated nutrient management on methane emission, global warming potential and carbon storage capacity in rice grown in a northeast India soil. Environ Sci Pollut Res 25(6):5889–5901

Bhateria R, Jain D (2016) Water quality assessment of lake water: a review. Sustain Water Resour Manag 2:161–173

Bhatia A, Pathak H, Jain N, Singh PK, Singh AK (2005) Global warming potential of manure amended soils under rice-wheat system in the Indo-Gangetic Plains. Atmos Environ 39:6976–6984

Bilkis S, Islam M, Jahiruddin M, Rahaman M (2017) Integrated use of manure and fertilizers increases rice yield, nutrient uptake and soil fertility in the boro-fallow-t. aman rice cropping pattern. SAARC J Agric 15:147–161

Bokhtiar S, Sakurai K (2005) Effects of organic manure and chemical fertilizer on soil fertility and productivity of plant and ratoon crops of sugarcane. Arch Agron Soil Sci 51:325–334

Brar BS, Singh J, Singh G, Kaur G (2015) Effects of long term application of inorganic and organic fertilizers on soil organic carbon and physical properties in maize–wheat rotation. Agronomy 5:220–238

Brenzinger K, Drost SM, Korthals G, Bodelier PLE (2018) Organic residue amendments to modulate greenhouse gas emissions from agricultural soils. Front Microbiol 9:3035. https://doi.org/10.3389/fmicb.2018.03035

Cai Y, Ding W, Luo J (2013) Nitrous oxide emissions from Chinese maize–wheat rotation systems: a 3-year field measurement. Atmos Environ 65:112–122

Chang TT, Shao XH, Zhang J, Mao JG, Wei YG, Yin C, Wang WN (2013) Effects of bio-organic fertilizer application combined with subsurface drainage in secondary salinized greenhouse soil. J Food Agric Environ 11:457–460

Chen JH (2006) The combined use of chemical and organic fertilizers and/or biofertilizer for crop growth and soil fertility. International workshop on sustained management of the soil-rhizosphere system for efficient crop production and fertilizer use, Bangkok, 16–20 October 2006, pp 1–11

Cooper GD (2017) Long-term effect of tillage and crop rotation practices on soil C and N in the Swartland, Western By. University of Stellenbosch, March 2017

D'Hose T, Ruysschaert G, Viaene N, Debode J, Vanden Nest T, Van Vaerenbergh J, Cornelis W, Willekens K, Vandecasteele B (2016) Farm compost amendment and non-inversion tillage improve soil quality without increasing the risk for N and P leaching. Agric Ecosyst Environ 225:126–139

Dalal RC, Moloney D (2000) Sustainability indicators of soil health and biodiversity. In: Hale P, Petrie A, Moloney D, Sattler P (eds) Management for sustainable ecosystems. Centre for Conservation Biology, Brisbane, p 10108

Das S, Adhya T (2014) Effect of combine application of organic manure and inorganic fertilizer on methane and nitrous oxide emissions from a tropical flooded soil planted to rice. Geoderma 213:185–192

Das SK, Jana BB (2003) Pond fertilization regimen: state-of-the-art. J Appl Aquac 13:35–66

Davis DA (ed) (2015) Feed and feeding practices in aquaculture. Woodhead, Oxford

de Ponti T, Rijk B, van Ittersum MK (2012) The crop yield gap between organic an conventional agriculture. Agric Syst 108:1–9

Demissie W, Kidanu S, Raghavaiah CV (2017) Effect of integrated use of lime, blended fertilizer and compost on productivity, nutrient removal and economics of barley (*Hordeum vulgare* L.) on acid soils of high lands in West Showa Zone of Ethiopia. IJLSCI 5(3):311–322

Devi KN, Singh TB, Athokpam HS, Singh NB, Shamurailatpam D (2013) Influence of inorganic, biological and organic manures on nodulation and yield of soybean (*Glycine max Merril* L.) and soil properties. AJCS 7(9):1407–1415

Dey SR, Barman R, Kandpal G (2019) Effect of combined application of organic and inorganic fertilizers on growth attributes of wheat (*Triticum aestivum* L.). J Pharmacogn Phytochem 8 (3):576–578

Diacono M, Montemurro F (2011) Long-term effects of organic amendments on soil fertility. Sustain Agric 2:761–786

Dunsin O, Odeghe TO (2015) Response of sweet bell pepper to moringa leaf extract and organo-bio degradable fertilizer. Asian J Agric Biol 3:117–123

Dutta S, Pal R, Chakeraborty A, Chakrabarti K (2003) Influence of integrated plant nutrient phosphorus and sugarcane and sugar yields. Field Crop Res 77:43–49

Fageria NK, Filho MB, Moreira A, Guimarães CM (2009) Foliar fertilization of crop plants. J Plant Nutr 32(6):1044–1064

FAO, IFAD, UNICEF, WFP, WHO (2019) The state of food security and nutrition in the world 2019. Safeguarding against economic slowdowns and downturns. Rome, FAO. ISBN: 978-92-5-131570-5

Getachew A, Tilahun A (2017) Integrated soil fertility and plant nutrient management in tropical agro-ecosystems: a review. Pedosphere 27(4):662–680

Getachew A, Lakew B, Paul NN (2014) Cropping sequence and nitrogen fertilizer effects on the productivity and quality of malting barley and soil fertility in the Ethiopian highlands. Arch Agron Soil Sci 60(9):1261–1275

Gete Z, Agegnehu G, Abera D, Shahidur R (2010) Fertilizer and soil fertility potential in Ethiopia: constraints and opportunities for enhancing the system. IFPRI, Addis Ababa, Ethiopia

Godara S, Gupta US, Singh R (2012) Effect of integrated nutrient management on herbage, dry fodder yield and quality of oat (*Avena sativa* L). Forage Res 38(1):59–61

Guo R, Shi L, Yan C, Zhong X, Gu F, Liu Q, Xia X, Li H (2017) Ionomic and metabolic responses to neutral salt or alkaline salt stresses in maize (*Zea mays* L.) seedlings. BMC Plant Biol 17:41. https://doi.org/10.1186/s12870-017-0994-6

Hati KM, Swarup A, Dwivedi A, Misra A, Bandyopadhyay K (2007) Changes in soil physical properties and organic carbon status at the topsoil horizon of a vertisol of Central India after 28 years of continuous cropping, fertilization and manuring. Agric Ecosyst Environ 119:127–134

Hazra G (2014) Slow or controlled release fertilizers for the holistic approach to economical and environmental issues: a review. IJMER 3:190–208

Hazra G (2016) Different types of eco-friendly fertilizers: an overview. Sustain Environ 1(1):54. https://doi.org/10.22158/se.v1n1p54

Hemalatha S, Chellamuthu S (2013) Impacts of long-term fertilization on soil nutritional quality under finger millet-maize cropping sequence. J Environ Res Dev 7:1571–1576

IFA (2018) Fertilizer Outlook 2018–2022. PIT and Agriculture Services, IFA Annual Conference, Berlin, 18–20 June 2018

Ipimoroti RR, Daniel MA, Obatolu CR (2002) Effect of organic mineral fertilizer on tea growth at Kusuku Mabila plateau Nigeria. Moor J Agric Res 3:180–183

Islam MR, Rashid MB, Siddique AB, Afroz H (2014) Integrated effects of manures and fertilizers on the yield and nutrient uptake by BRRI dhan49. J Bangladesh Agric Univ 12(1):67–72

Islam MA, Islam S, Akter A, Rahman Md H, Nandwani D (2017) Effect of organic and inorganic fertilizers on soil properties and the growth, yield and quality of tomato in Mymensingh, Bangladesh. Agriculture 7:18. https://doi.org/10.3390/agriculture7030018

Iwami I (1995) Brewer environmental industries, Inc. Personal communication

Jaja ET, Barber LI (2017) Organic and inorganic fertilizers in food production system in Nigeria. J Biol Agric Healthc 7(18):51–55

Janssen BH (1993) Integrated nutrient management: the use of organic and mineral fertilizers. In: van Reuler H, Prins WH (eds) The role of plant nutrients for sustainable food crop production in sub-Saharan Africa. Leidschendam, VKP, pp 89–105

Janzen HH (2004) Carbon cycling in earth systems – a soil science perspective. Agric Ecosyst Environ 104:399–417

Jenkins WK (2009) Evaluation of four organic fertilizers for vegetable and herb transplant production (Doctoral dissertation, University of Georgia)

Jo IS (1990) Effect of organic fertilizer on soil physical properties and plant growth. Tech Bull. ASPAC/FFTC, p 119

Jobe (2003) Integrated nutrient management for increased rice production in the Inland Valleys of the Gambia. In: Sanyang S, Ajayi A, Sy AA (eds) Proceedings of the 2nd biennial regional rice research review. WARDA Proceedings Series No 2(1), pp 35–41

Joshi HC, Joshi B, Guru SK, Shankdhar SC, Kumar A, Mahapatra BS, Nautiyal MK, Singh P (2017) Consequences of integrated use of organic and inorganic fertilizers on yield and yield elements of rice. IJASR 7(5):163–166

Juan L, Bing-Qiang Z, Xiu-ying L, Rui-bo J, Bing SH (2008) Effects of long-term combined application of organic and mineral fertilizers on microbial biomass, soil enzyme activities and soil fertility. Agric Sci China 7(3):336–343

Kachroo D, Razdan R (2006) Growth, nutrient uptake and yield of wheat (*Triticum aestivum*) as influenced by biofertilizers and nitrogen. Indian J Agron 51(1):37–39

Kar B, Karmakar S, Saha G, Bhattacharya R (2019) Estimation of methane flux from rice field as influenced by plant-climate under varied management practices. J Agrometeorol 21(Special Issue-'NASA 2014' part-III):56–5\

Karmakar S, Saha G, Bhattacharya R, Kar B (2014) Green house gas emission potentiality of wheat as influenced by microclimate and ambient sunshine under varied climatic conditions. J Crop Weed 10(2):231–239

Kaur K, Kapoor KK, Gupta AP (2005) Impact of organic manures with and without mineral fertilizers on soil chemical and biological properties under tropical conditions. J Plant Nutr Soil Sci 168:117–122

Khalil MI, Rosenani AB, Cleemput O, Van Fauziah CI, Shamshuddin J (2002) Nitrous oxide emissions from an ultisol of the humid tropics under maize–groundnut rotation. J Environ Qual 31(4):1071–1078. https://doi.org/10.2134/jeq2002.1071

Kirchmann H, Bergström L, Kätterer T, Andrén O, Andersson R (2008) Can organic crop production feed the world? In: Kirchmann H, Bergström L (eds) Organic crop production—ambitions and limitations. Springer, Dordrecht, pp 39–74

Kumar MS, Burgess SN, Luu LT (2004) Review of nutrient management in freshwater polyculture. J Appl Aquac 16:17–44

Kumar M, Bauddh K, Sainger M, Sainger PA, Singh JS, Singh RP (2012a) Increase in growth, productivity and nutritional status of rice (*Oryza sativa* L. c.v. Basmati) and enrichment in soil fertility applied with an organic matrix entrapped urea. J Crop Sci Biotechnol 15(2):137–144

Kumar V, Kumar T, Gupta PK, Singh RV, Singh RA, Singh SB (2012b) Effect of puddling and organic sources of nutrients on productivity of rice-wheat cropping system in lowlands. Ann Plant Soil Res 14:95–100

Kumar M, Bauddh K, Kumar S, Sainger M, Sainger P, Singh RP (2013a) Increase in growth, productivity and nutritional status of wheat (*Triticum aestivum* L. cv. Wh-711) and enrichment in soil fertility applied with organic matrix entrapped urea. J Environ Biol 34:1–9

Kumar S, Bauddh K, Barman SC, Singh RP (2013b) Evaluation of conventional and organic matrix entrapped urea and diammonium phosphate for growth and productivity of *Triticum aestivum* L. and mobilization of NO3-, NO2-, NH4+ and PO4-3 from soil to plant leaves. Int J Agron Plant Prod 4(6):1357–1368

Kumar S, Bauddh K, Barman SC, Singh RP (2014a) Amendments of microbial biofertilizers and organic substances reduces requirement of urea and DAP with enhanced nutrient availability and productivity of wheat (*Triticum aestivum* L.). Ecol Eng 71:432–437

Kumar S, Bauddh K, Barman SC, Singh RP (2014b) Organic matrix entrapped bio-fertilizers increase growth, productivity and yield of *Triticum aestivum* L. and mobilization of NO_3^-, NO_2^-, NH_4^+ and PO_4^{-3} from soil to plant leaves. J Agric Sci Technol 16(2):315–329

Kumar M, Bauddh K, Sainger M, Sainger PA, Singh RP (2014c) Increase in growth, productivity and nutritional status of wheat (*Triticum aestivum* L.) and enrichment in soil microbial population applied with biofertilizers entrapped with organic matrix. J Plant Nutr 38:260–276

Kumar M, Bauddh K, Kumar S, Sainger M, Sainger PA, Singh RP (2015) Enhancing efficacy of Azotobacter and Bacillus entrapping in organic matrix for rice cultivation. Agroecol Sustain Food Syst 39:907–923

Lal R (2004) Soil carbon sequestration impact on global climate change and food security. Science 304:1623–1627

Leahy P, Kiely G, Scanlon TM (2004) Managed grasslands: a greenhouse gas sink or source? Geophys Res Lett 31:L20507. https://doi.org/10.1029/2004GL021161

Li XX, Han CS, Delgado JA, Zhang YM, Ouyang ZY (2007) Increased nitrogen use efficiencies as a key mitigation alternative to reduce nitrate leaching in north China plain. Agric Water Manag 89:137–147

Liu B, Gumpertz ML, Hu S, Ristaino JB (2007) Long-term effects of organic and synthetic soil fertility amendments on soil microbial communities and the development of southern blight. Soil Biol Biochem 39:2302–2316

Liu E, Yan C, Mei X, He W, Bing SH, Ding L, Liu Q, Liu S, Fan T (2010) Long-term effect of chemical fertilizer, straw, and manure on soil chemical and biological properties in northwest China. Geoderma 158:173–180

Liu L, Zhang T, Gilliam FS, Gundersen P, Zhang W, Chen H, Mo J (2013) Interactive effects of nitrogen and phosphorus on soil microbial communities in a tropical forest. PLoS One 8(4): e61188. https://doi.org/10.1371/journal.pone.0061188

Liu CW, Yu S, Bo-Ching C, Hung L (2014) Effects of nitrogen fertilizers on the growth and nitrate content of lettuce (*Lactuca sativa* L.). Int J Environ Res Public Health 11:4427–4440

Lynch D (2009) Environmental impacts of organic agriculture: a Canadian perspective. Can J Plant Sci 89:621–628

Mahajan A, Bhagat RM, Gupta RD (2008) Integrated nutrient management in sustainable rice wheat cropping system for food security in India. SAARC J Agric 6(2):149–163

Mahmud AJ, Shamsuddoha ATM, Haque Md N (2016) Effect of organic and inorganic fertilizer on the growth and yield of rice (*Oryza sativa* L.). Nat Sci 14(2):45–54

Majumdar K, Dey P, Tewatai RK (2014) Current nutrient management approaches: issues and strategies. Indian J Fert 10(5):14–27

Mar SS, Okazaki M (2012) Investigation of Cd contents in several phosphate rocks used for the production of fertilizer. Microchem J 104:17–21

Maritus CHT, Vleic PLG (2001) The management of organic matter in tropical soils. What are the priorities? Nutr Cycl Agroecosyst 61:1–6

Masarirambi MT, Mandisodza FC, Mashingaidze AB, Bhebhe E (2012) Influence of plant population and seed tuber size on growth and yield components of potato (*Solanum tuberosum*). Int J Agric Biol 14(4):545–549

Masto RE, Chhonkar PK, Singh D, Patra AK (2006) Changes in soil biological and biochemical characteristics in a long-term field trial on a sub-tropical Inceptisol. Soil Biol Biochem 38:1577–1582

Meng L, Ding W, Cai Z (2005) Long-term application of organic manure and nitrogen fertilizer on N_2O emissions, soil quality and crop production in a sandy loam soil. Soil Biol Biochem 37:2037–2045

Moharana PC, Sharma BM, Biswas DR, Dwivedi BS, Singh RV (2012) Long-term effect of nutrient management on soil fertility and soil organic carbon pools under a 6-year old pearl millet-wheat cropping system in an Inceptisol of subtropical India. Fields Crop Res 136:32–41

Mondal T, Datta JK, Mondal NK (2017) Chemical fertilizer in conjunction with biofertilizer and vermicompost induced changes in morphophysiological and bio-chemical traits of mustard crop. J Saudi Soc Agric Sci 16:135–144

Mumpton FA (1999) La roca magica: uses of natural zeolites in agriculture and industry. Proc Natl Acad Sci USA 96(7):3463–3470

Namvar A, Seyed Sharifi R, Khandan T, Jafari Moghadam M (2013) Seed inoculation and inorganic nitrogen fertilization effects on some physiological and agronomical traits of chickpea (*Cicer arietinum* L.) in irrigated condition. J Central Eur Agric 14(3):28–40

Nath DJ, Ozah B, Baruah R, Borah DK (2011) Effect of integrated nutrient management on soil enzymes, microbial biomass carbon and bacterial populations under rice- wheat sequence. Ind J Agric Sci 81:1143–1148

Natsheh B, Mousa S (2014) Effect of organic and inorganic fertilizers application on soil and cucumber (*Cucumis sativus* L.) plant productivity. Int J Agric Forest 4(3):166–170

Ni B, Liu M, Lü S, Xie L, Wang Y (2010) Multifunctional slow-release organic -inorganic compound fertilizer. J Agric Food Chem 58:12373–12378

Nyamadzawo G, Wuta M, Nyamangara J, Smith JL, Rees RM (2014) Nitrous oxide and methane emissions from cultivated seasonal wetland (dambo) soils with inorganic, organic and integrated nutrient management. Nutr Cycl Agroecosyst 100:161–175

Obi ME, Ebo PO (1995) The effect of different management practices on the soil physical properties and maize production in severely degraded soil in /Southern Nigeria. Biol Res Technol 51:117–123

Ojeniyi SO (2000) Effect of goat manure on soil nutrients and okra yield in a rain forest area of Nigeria. Appl Trop Agric 5:20–23

Prasad R (1996) Cropping systems and sustainability of agriculture. Ind Farming 46:39–45

Rahman M, Islam M, Jahiruddin M, Haque M (2012) Management of organic manure and inorganic fertilizer in the maize-mungbean/dhaincha-t. aman rice cropping pattern for increased crop production. Bangladesh J Agric Res 37:225–234

Rahman N, Bruun TB, Giller KE, Magid J, Ven GWJ, Neergaard A (2019) Soil greenhouse gas emissions from inorganic fertilizers and recycled oil palm waste products from Indonesian oil palm plantations. GCB Bioenergy. https://doi.org/10.1111/gcbb.12618

Rai A, Kumar S, Bauddh K, Singh N, Singh RP (2015) Improvement in growth and alkaloid content of *Rauwolfia serpentina* on application of organic matrix entrapped biofertilizers (*Azotobacter chroococcum*, *Azospirillum brasilense* and *Pseudomonas nutida*). J Plant Nutr 40 (16):2237–2247

Ram T, Mir MS (2006) Effect of integrated nutrient management on yield and attributing characters of wheat (*Triticum aestivum*). Indian J Agron 51(3):189–192

Rama Lakshmi CS, Sreelatha T, Usha Rani T, Rao SRK, Naidu NV (2011) Effect of organic manures on soil fertility and productivity of sugarcane in north coastal zone of Andhra Pradesh. Indian J Agric Res 45(4):307–313

Rashid MI, Mujawar LH, Shahzade T, Almeelbi T, Ismail IMI, Oves M (2016) Bacteria and fungi can contribute to nutrients bioavailability and aggregate formation in degraded soils. Microbiol Res 183:26–41

Reddy D, Subbrao A, Rupa TR (2000) Effect of continuous use of cattle manure and fertilizer phosphorus on crop yields and soil organic phosphorus in a Vertisol. Bioresour Technol 75:113–118

Regina K, Pihlatie M, Esala M, Alakukku L (2007) Methane fluxes on boreal arable soils. Agric Ecosyst Environ 119:346–352

Saha G, Karmakar S, Kar B, Bhattacharya R, Singh G (2015) Studies on emission potentiality of nitrous oxide from wheat field under changed climate. Curr Sci 109(4):768–774

Saha G, Kar B, Karmakar S (2017) Methane and nitrous oxide emission from Kharif rice field as influenced by nutrients and moisture regimes in new alluvial agroclimatic region of West Bengal. Curr Sci 112(5):989–995

Sarkodie-Addo J, Lee HC, Baggs EM (2003) Nitrous oxide emissions after application of inorganic fertilizer and incorporation of green manure residues. Soil Use Manag 19:331–339

Satyanarayana V, Prasad PV, Murthy VRK, Boote KJ (2002) Influence of integrated use of farmyard manure and inorganic fertilizers on yield and yield components of irrigated lowland rice. J Plant Nutr 25(10):2081–2090

Savci S (2012) An agricultural pollutant: chemical fertilizer. Int J Environ Sci Dev 3(1):73–80

Schulz H, Glaser B (2012) Effects of biochar compared to organic and inorganic fertilizers on soil quality and plant growth in a greenhouse experiment. J Plant Nutr Soil Sci 175:410–422

Seufert V, Ramankutty N, Foley JA (2012) Comparing the yields of organic and conventional agriculture. Nature 485(7397):229–232

Sharma N, Srivastava B, Mishra B (1998) Changes in some physicochemical properties of an acid soil as affected by long term use of fertilizer and amendments. J Ind Soc Soil Sci 36:688–692

Shata SM, Mahmoud A, Siam S (2007) Improving calcareous soil productivity by integrated effect of intercropping and fertilizer. Res J Agric Biol Sci 3(6):733–739

Silva JA (2000) Inorganic fertilizer materials. In: Silva JA, Uchida R (eds) Plant nutrient management in Hawaii's soils, approaches for tropical and subtropical agriculture. College of Tropical Agriculture and Human Resources, University of Hawaii at Manoa, pp 117–120

Singh RP, Sainger M, Bauddh K, Senger RS, Jaiwal PK (2010) Sustained nutrient supply reduced nutrient loss and high plant productivity with slow release fertilizers. In: Senger RS, Sharma AK (eds) Stable food production and sustainable agriculture. Studium, Delhi, pp 62–79

Singh RP, Kumar S, Sainger M, Sainger PA, Barnawal D (2017) Eco-friendly nitrogen fertilizers for sustainable agriculture. In: Rakshit A, Abhilash PC, Singh HB, Ghosh S (eds) Adaptive soil management : from theory to practices. Springer, pp 227–246

Singh G, Kumar S, Sidhu GS, Kaur R (2018) Effect of integrated nutrient management on yield of wheat (*Triticum aestivum* L.) under irrigated conditions. Int J Chem Stud 6(2):904–907

Skiba U, Jones SK, Dragosits U, Drewer J, Fowler D, Rees RM, Pappa VA, Cardenas L, Chadwick D, Yamulki S, Manning AJ (2012) UK emissions of the greenhouse gas nitrous oxide. Phil Trans R Soc B 367:1175–1185

Skinner C, Gattinger A, Muller A, Mader P, FlieBbach A, Stolze M, Ruser R, Niggli U (2014) Greenhouse gas fluxes from agricultural soils under organic and non organic management—a global meta-analysis. Sci Total Environ 468–469:553–563

Steiner C, Teixeira WG, Lehmann J, Nehls T, Macêdo JLVD, Blum WEH, Zech W (2007) Long term effects of manure, charcoal and mineral fertilization on crop production and fertility on a highly weathered central Amazonian upland soil. Plant Soil 291:275–290

Tadesse T, Dechassa N, Bayu W, Gebeyehu S (2013) Effects of farmyard manure and inorganic fertilizer application on soil Physico-chemical properties and nutrient balance in rain-fed lowland rice ecosystem. Am J Plant Sci 04:309–316

Tirol-Padre A, Ladha JK, Regmi AP, Inubushi K (2007) Organic amendments affect soil properties in two long-term rice-wheat experiments. Soil Sci Soc Am J 71:442–452

Tisdale SL, Nelson WL, Beaton JD, Havlin JL (1993) Soil fertility and fertilizers, 5th edn. Macmillan, New York, p 634

Usman M, Madu VU, Alkali G (2015) The combined use of organic and inorganic fertilizers for improving maize crop productivity in Nigeria. IJSRP 5(10):1–7

Utoor HA, Ayyad WA (2014) Effect of mineral and bio-fertilizer application on growth and yield of wheat (*Triticum aestivum* L.). Iraqi J Sci 55(4A):1484–1495

Vitousek PM, Naylor R, Crews T, David MB, Drinkwater LE, Holland E, Johnes PJ, Katzenberger J, Martinelli LA, Matson PA, Nziguheba G, Ojima D, Palm CA, Robertson GP, Sanchez PA, Townsend AR, Zhang FS (2009) Nutrient imbalances in agricultural development. Science 324(5934):1519–1520

Wang J, Chen Z, Ma Y, Sun L, Xiong Z, Huang Q, Sheng Q (2013) Methane and nitrous oxide emissions as affected by organic–inorganic mixed fertilizer from a rice paddy in southeast China. J Soil Sediment 13. https://doi.org/10.1007/s11368-013-0731-1

Wang C, Shen JL, Zheng L, Liu JY, Qin HL, Li Y, Wu JS (2014) Effects of combined applications of pig manure and chemical fertilizers on CH_4 and N_2O emissions and their global warming potentials in paddy fields with double-rice cropping. J Environ Sci (China) 35(8):3120–3127

Wang C, Min Q, Abid AA, Sardans J, Wu H, Lai DYF, Penuelas J, Wang W, Honghui W (2019) Optimal coupling of straw and synthetic fertilizers incorporation on soil properties, active Fe dynamics, and greenhouse gas emission in *Jasminum sambac* (L.) field in Southeastern China. 2019. Sustainability 11:1092. https://doi.org/10.3390/su11041092

Wani IA, Mehraj S, Ali MT, Hassan A, Wani SA, Hussain S, Bisati IA (2017) Effect of inorganic and organic fertilisers on yield and soil nutrient status of walnut orchard Int J Plant Soil Sci 16 (2):1–13

Westerman PW, Bicudo JR (2005) Management considerations for organic waste use in agriculture. Bioresour Technol 96:215–221

Yadav SK, Benbi DK, Prasad R (2019) Effect of continuous application of organic and inorganic sources of nutrients on chemical properties of soil. Int J Curr Microbiol App Sci 8(4):2455–2463

Yamoah CF, Bationo A, Shapiro B, Koala S (2002) Trend and stability analyses of millet yields treated with fertilizer and crop residues in the Sahel. Field Crops Res 75:53–62

Zhang QC, Shamsi IH, Xu DT, Wang GH, Lin XY, Jilani G, Chaudhry AN (2012) Chemical fertilizer and organic manure inputs in soil exhibit a vice versa pattern of microbial community structure. Appl Soil Eco l57:1–8

Zhao J, Zhou L (2011) Combined application of organic and inorganic fertilizers on black soil fertility and maize yield. J Northeast Agric Univ (English Edition) 18:24–29

Zhao Z, Yan S, Fen L, Ji P, Wang X, Tong Y (2014) Effects of chemical fertilizer combined with organic manure on Fuji apple quality, yield and soil fertility in apple orchard on the loess plateau of China. Int J Agric Biol Eng 7(2):45–55

Zhu ZL, Jin JY (2013) Fertilizer use and food security in China. Plant Nutr Fert Sci 19(2):259–273

Composting: An Eco-friendly Technology for Sustainable Agriculture 9

Chukwudi O. Onwosi, Johnson K. Ndukwe, Godwin O. Aliyu, Kenechi O. Chukwu, Flora N. Ezugworie, and Victor C. Igbokwe

Contents

C. O. Onwosi (✉) · J. K. Ndukwe · G. O. Aliyu · K. O. Chukwu · F. N. Ezugworie · V. C. Igbokwe
Department of Microbiology, Faculty of Biological Sciences, University of Nigeria, Nsukka, Enugu State, Nigeria

Bioconversion and Renewable Energy Research Unit, University of Nigeria, Nsukka, Enugu State, Nigeria
e-mail: chukwudi.onwosi@unn.edu.ng

K. Bauddh et al. (eds.), *Ecological and Practical Applications for Sustainable Agriculture*, https://doi.org/10.1007/978-981-15-3372-3_9

Abstract

The effective recovery of organic waste through composting has provided mankind with a multipurpose product (compost) which, apart from being used as biofertilizer and biopesticide, can find application in erosion management and restoration of hydrocarbon-polluted soil. Interestingly, composting technology could be indispensable in the actualization of sustainable agriculture due to its environmental friendliness, cost-effectiveness, and sustainability. However, for the essential concepts of sustainable agriculture—care, health, fairness, and ecology—to be realized, the composting process should be effectively managed. Therefore, in this chapter, we addressed the various methods of composting management, challenges, and application in agriculture.

Keywords

Composting · Biofertilizer · Biopesticide · Sustainable agriculture

9.1 Introduction

Management of waste generated from various human activities has become a huge global concern (Lim et al. 2019). Efforts made by different sectors to handle these wastes have been hampered by ever increasing world population along with economic growth. While the waste containment has been adequately tackled in some industrialized economy, the developing countries have not been successful. To overcome these issues, composting technology has become a reliable option for handling organic portion of wastes (Lim et al. 2019). Effective waste management (e.g. composting) requires an organized handling of wastes: collection, stabilization, and disposal (Johari et al. 2014). The composting process is a relatively easy technology, environmental friendly, and economical compared with other methods of waste management. In this process, there is a bioconversion of organic wastes to value-added products such as nutrient-rich compost (biofertilizer). For instance, compost is a good source of nitrogen (N), phosphorus (P), and potassium (K) which enhance the restoration of fertility of the soil and contribute to plant growth (Wang et al. 2016a, b; Pergola et al. 2018; Nakasaki et al. 2019). Owing to the presence of large pool of humic substances and volatile fatty acids (VFA), compost can also be effectively used as biopesticides to control plant diseases when applied to the agricultural soil (Hestmark et al. 2019; Lim et al. 2019).

Sustainable agriculture describes the food production from animal or plants through methods that ensure environmental care, public health, and human and animal welfare (Singh 2018). Thus, the essential concepts (care, health, fairness, and ecology) of sustainable agriculture can be realized through composts or organic manure (Nielsen 2019). With a view to ensuring sustainability, Hestmark et al. (2019) reported that inoculation of VFA-rich compost (produced by solarization) into the soil enhanced pest control. Introduction of compost into the soil improves the rate of germination of different plants (Sanasam and Talukdar 2017; Milinković et al. 2019). Milinković et al. (2019) demonstrated the inhibition of fungal phytopathogens (*Pythium debaryanum*, *Rhizoctonia* sp., *Fusarium oxysporum*) by compost and its associated products.

9.2 What Is Composting?

While composting methods may vary, however, common in all is the degradation of organic feedstock by the synergistic interaction between microorganisms and some invertebrates. For instance, this solid-state composting process (Kaiser 1996) is normally induced by aerobic mesophiles and followed up by a succession of aerobic thermophiles, then mesophiles again, and finally curing mesophilic anaerobes (Pepe et al. 2013; Bhatia et al. 2013). Despite the thought that composting is an oxygen-ridden process that involves the production of humic substances from organic matter (Pepe et al. 2013; Bialobrzewski et al. 2015), anaerobic organisms (*Clostridium* sp.) have been associated with the process (Bhatia et al. 2013). The rich and complex biodiversity during composting process has been described as food web. The reason for this theory is that succession of microorganisms (fungi, bacteria) and invertebrates (millipedes, nematodes, slugs, earthworms, etc.) ensures higher degradation efficiency of the organic residues (Michel et al. 1996).

9.2.1 Microbiology of Composting

During composting, the number of different species of microorganisms changes momentarily, with each group thriving when environmental conditions are favourable, and then extinct competitively during an unfavourable condition, thus allowing the succession of new set of microorganisms (Michel et al. 1996).

9.2.1.1 Bacteria

Feedstock decomposition and heat generation in compost are usually influenced by bacteria. This is due to the nutritional diversity of this group of microorganism and the amount of myriads of enzymes produced by this group which can transform complex organic materials into simple forms. The physiological roles of diverse microbial communities resulting in enriched compost are instrumental in enhanced agricultural output and sustainability. For instance, the nitrogen-rich compost has been attributed to the presence of different nitrogen-fixing bacteria (e.g. *Caulobacter*, *Pseudomonas*, *Stenotrophomonas*, etc.) (Pepe et al. 2013). Mesophilic bacteria, such as hydrogen oxidizers, sulphur oxidizers, nitrifiers, and nitrogen fixers, often predominate at the early period of the composting process. As a result of the availability of simple sugars and polysaccharides at the initial phases of composting, heat generation is inevitable because of proliferation of microbial populations and activities. However, at ⍰40 °C, the population of mesophilic bacteria greatly reduces, allowing the thermophilic species (e.g. *Bacillus*) predominate during this phase. However, at >60 °C, there is a decline in the population of thermophiles (Michel et al. 1996). According to Onwosi et al. (2017), the transition in temperature regimes ensures effective decomposition of organic matter by the thermophiles. Bhatia et al. (2013) and Bialobrzewski et al. (2015) also noted that the prevailing environmental conditions at different composting stages regulate the biodiversity of the participating microbes. The compost temperature then begins to decline after a while as the nutrient available begins to deplete, thus allowing the recolonization of the compost field by mesophiles as the thermophilic bacteria begin to give way. This phase involves cooling and maturation which is called CURING, and stabilizes the compost for agricultural uses (Onwosi et al. 2017).

9.2.1.2 Actinomycetes

Actinomycetes are special group of aerobic chains and filaments forming bacteria that actively participate in the decomposition of complex organic residues (e.g. lignin, cellulose, etc.). However, these organisms are outcompeted by other bacterial group that utilizes the simple carbohydrates that exist in abundance at the initial stages of composting. They possess enzymes (cellulases) that can chemically break down complex materials that are inaccessible to other group of organisms (Michel et al. 1996). While some groups of actinomycetes are present in the active (thermophilic) stage of composting, others (looking like gry spider webs stretching through the compost bin) manifest when only complex or recalcitrant organic molecules are available (e.g. curing or maturation phase) (Michel et al. 1996).

9.2.1.3 Fungi

Fungi possess the ability to transform complex carbon-rich plant polymers that their bacterial counterparts are unable to degrade. In general, they hydrolyse complex materials by the secretion of extracellular hydrolytic enzymes. During composting, the majority of fungal activities are limited to the outer part of the compost as they hardly thrive under high temperature. The appearance of white fuzzy colonies on compost surface is as result of moulds, which are strict aerobes (Michel et al. 1996).

9.2.1.4 Protozoa

Although found as small percentage of the total microbial population during composting process, protozoa presence in composting is dependent on the type of feedstocks as they are predators of most fungi and bacteria especially during the curing stage.

9.2.1.5 Stage and Curing Indicator Microbes

Pepe et al. (2013) and Bhatia et al. (2013) pointed out that the presence of certain microbes could be essential in monitoring the progress of composting process (e.g. composting stage or maturity). For instance, a good composting process can be evaluated by the presence of the Firmicutes (e.g. certain phyla Bacillales or Actinobacteria) (Sundberg et al. 2013). Mayende et al. (2006) isolated thermophilic *Bacillus* strains at optimal temperatures of 70 °C and 40 °C, respectively, from compost materials capable of producing cellulases and polyphenol oxidases. Also, Habbeche et al. (2014) characterized an actinomycete thermostable keratin-degrading enzyme from *Actinomadura keratinilytica*, isolated from poultry compost.

9.2.2 The Invertebrates

Invertebrates, though not a regular organism during composting, are mostly seen in bin and outdoor composting where the temperature remains almost within the mesophilic zone. They aid microorganisms during the breakdown of organic matter by shredding the organic matter, thus making it more readily available form for the microorganisms. A number of these invertebrates can be found in composts at different points during composting process. These may include the following.

9.2.2.1 Annelida

These include Oligochaeta like pot worms which feed on fungi mycelia, decaying vegetation, and earthworms (vital in vermicomposting). Also, the detritivores are highly active in composts, feeding majorly on undecomposed plant material. While some species of worms inhabit the litter layer of soil, others populate the deep soil burrows, reaching to the surface only when they need to feed. Therefore, in order to access nutrients in the compost, the worms rely majorly on the beneficial interaction with the microbes (Edwards and Bohlen 1996).

9.2.2.2 Arthropods

Arthropods include arachnida (e.g. mites). They are common in compost as they participate in all levels of the compost food web. Whereas most mites feed commonly on the organic debris (i.e. leaves and rotten wood), others feed on fungi and bacteria that decompose organic matter. Other common arachnids (usually predatory in nature) found in compost piles are pseudoscorpions and spiders. Insecta is another common class of arthropods found in compost piles which synthesize cellulase enzyme used to breakdown complex materials. They include springtails which feed majorly on fungi and nematodes and flies. Adult flies are often attracted to rotting food, making them cause a nuisance around compost piles. Ants, beetles, and earwigs are also common insects found in compost piles. Millipedes, centipedes, and crustaceans are other common arthropod found in compost piles (Michel et al. 1996).

Other invertebrates that might be found include Nematoda, Mollusca, etc.

9.3 Types and Composting Methods

9.3.1 Composting Technologies

Composting technologies and systems vary widely due to different factors affecting the composting process. The key difference among the different composting technologies is that biomass decomposition may be carried out in closed system, bioreactor, or an open pile. Each of the options has merits or demerits, and these should be considered during the design of the methods. For instance, contrary to the uncontrolled emission of gases during open composting, in the enclosed technologies as well as the reactor technology, gases are collected and treated. Another major difference between composting technologies is that in static methods, feedstocks are heaped in piles, while their dynamic counterparts require movement of feedstocks. Often times, these technologies are combined to ensure that wastes are effectively stabilized.

9.3.1.1 Open Composting Technologies

9.3.1.1.1 Static Pile

Static pile method has been described as important composting option (Leton and Stentiford 1990; Luo et al. 2008; Nasini et al. 2016). Under this method, there is a limitation in the emission of NH_3 due to N conservation. There is the presence of outer insulating layer as well as absence of turning, thus, keeping labour requirement relatively low (Fournel et al. 2019). The absence of agitation in static pile composting means that adequate porosity has to be maintained for an extended duration of the process to encourage aeration. If feedstock of the adequate porosity cannot be found, bulking agents are normally added to the feedstocks (Arrigoni et al. 2018). The fleece, which contains air-permeable membranes, helps deal with odour and run-off. One of the demerits of this technology is that it is a slow process due to lack of agitation; hence, it is hardly used in isolation for waste treatment. It is often utilized at the curing stage to control odorous emission.

9.3.1.1.2 Windrow Composting

Windrow composting is a simple technique whereby compost piles are mechanically aerated by the windrow turners. However, in order to realize the optimization of the process, an important factor that should be considered is the height of the compost pile. This is usually 3–4 m high for a maximum output. Unlike the static pile composting, windrow can be used during the entire composting process. One of the problems associated with this method is odour emission which is an immediate outcome of frequent agitation of the compost pile. However, turning of the pile must be controlled because its absence or low level could result in anaerobiosis. This condition may favour the emission of methane which is an obnoxious greenhouse gas.

The popularity of windrow is because of its simplicity and relative cost-effectiveness and could be a reliable composting option for handling large volume of waste materials (Makan 2015; Zhu-Barker et al. 2017; De Silva and Yatawara 2017; Kong et al. 2018). In this approach, composting is achieved by placing the compostable materials (organic wastes) in long and narrow rows (about 15–115 m length, 2 m high, and 5 m wide) with regular turning (Andersen et al. 2010; Gould et al. 2013; Chen et al. 2015a; Zhu-Barker et al. 2017). To ensure evenness in the composting process as well as inflow of O_2, windrow turners are used for the mixing of the piles. Another advantage of turning the compost piles is that inflow of O_2 reduces the emission of GHG which is favoured by anaerobic conditions (Beck-Friis et al. 2000). Agitation of the pile also aids the temperature redistribution and regulation of moisture content (Gould et al. 2013). However, the turning frequency and the general operation of windrow composting vary among facilities (Zhu-Barker et al. 2017; Arriaga et al. 2017). One of the drawbacks of this procedure, as pointed out by De Silva and Yatawara (2017), is that its operation requires large space.

Turning has to be done constantly for there to be a significant effect on the oxygen supply to the windrow. The front-end loader and the straddle turners are some of the commonly used turners; however, some equipment keep the feedstock agitated by using a rotating drum with welded scrolls. Turning frequency is usually much higher at the initial decomposition phase and tends to decrease towards the curing phase; only a few composting facilities do not use turners. Although higher turning frequencies result in a decreased retention time, it also increases the operating costs.

Food wastes are usually excluded from feedstock during windrow composting; this is due in part to the foul odour released during their decomposition.

9.3.1.1.3 Enclosed Aerated Windrow

This composting method fundamentally shares the attributes of two or more composting techniques. It is particularly a blend of aerated static pile and windrow method. Thus, there is an acceleration of the composting process by a good control of oxygen and temperature levels, using both forced air and pile agitation (Boldrin et al. 2009). Aerated windrows are more commonly used after initially processing the feedstock in rotary drums or with in-vessel containers.

9.3.1.1.4 Vermicomposting

The vermicomposting (VC) technology is a composting option which relies on earthworms and microbial activities to decompose organic residues with a view to producing value-added, nutrient-rich compost (biofertilizer or soil conditioner) (Ali et al. 2015; Komakech et al. 2016; Mupambwa and Mnkeni 2018; Mupambwa et al. 2016). Thus, VC is eco-friendly and ultimately recycles nutrients (enhances the micro- and macronutrients content) in the course of organic waste stabilization (Ali et al. 2015; Sharma and Garg 2018; Ramnarain et al. 2019; Zhi-Wei et al. 2019). Different conditions (e.g. pH, moisture, feedstocks, C/N ratio, temperature, etc.) affect the overall VC process (Ali et al. 2015). VC is always a faster procedure than the traditional composting. Abbasi et al. (2018) described a new, inexpensive VC technology called the Flippable Units Vermireactor Train System (FLUVTS) which rapidly decomposes waste within a short period of time. Earthworms used in VC are affected by the moisture content, temperature, and aeration (O_2) of the compostable materials. They should have interesting attributes such as faster rate of degradation of organic residue, tolerance to harsh composting conditions, and enhanced rate of reproduction. Different earthworm types have been reportedly used in VC. These comprise anecic (deep burrowers, e.g. *Lampito mauritii*) (Tripathi and Bhardwaj 2004), endogeic (*Dichogaster bolaui*) (Jouquet et al. 2010), and epigeic (surface dwellers, e.g. *Eisenia foetida*) (Zhi-Wei et al. 2019) earthworms. The latter have been described as the most appropriate for VC because their occupation of the organic soil zone allows them to access the OM and effectively decompose them (Garg et al. 2006; Tripathi and Bhardwaj 2004; Ramnarain et al. 2019; Zhi-Wei et al. 2019). However, one of the shortcomings is possible survival of pathogenic organism because of the absence of thermophilic stage during waste stabilization. Besides, vermicompost may not be effectively used as an organic fertilizer because the high concentration of soluble salt may impair the growth of plants. It has the tendency to emit high amount of GHG into the surrounding.

9.3.1.2 Enclosed Composting Technologies

In these methods, composting is carried out in enclosed facilities—mainly inside buildings. They require short retention time and are only used for preliminary or temporary treatment of organic wastes. Afterwards, the partially stabilized organic materials are then fully cured or stabilized using the open static piles or windrows (Boldrin et al. 2009). This approach most times deploys biofilters to control the exhaust gases; hence, it minimizes the atmospheric gas emission.

9.3.1.2.1 Cell or Channel Composting

Cell or channel composting is a form of enclosed windrow technique where organic materials placed in several long and narrow piles are separately managed (Boldrin et al. 2009). In strict terms, the longer piles are referred as channels, while the shorter ones can be described as cells. In uncovered channels, the starting materials could be piled up to 2.5 m high (Haug 1993). The compost piles are frequently turned by moving organic materials to opposite end or across channels. Since, this method handles large volumes of wastes and requires aeration; the channels are always partitioned by walls to ease the movement of machines used for aeration. Although two channels are enough to start up a facility, there is a possibility of additional channels as the capacity increases. Furthermore, aeration rate and water circulation in the feedstock are controlled independently and managed in each set-up. Some technologies are put in place to make-up reduction in the size of the compost during turning within the same channel or during restacking in another channel. Most of compost's retention time in these channels ranges from 1 to 2 months (Schmitz and Meier-Stolle 1995).

9.3.1.2.2 Automated: Turning Aerated Pile

Whereas in channel composting feedstock is divided by walls, in automated-turning-aerated pile composting feedstock is placed in one large pile (up to 3.3 m high) channel which can be controlled individually (Kugler et al. 1995). Here, the turning machine frame spans over the whole composting hall. The structural orientation is that the aeration system is placed under the compost bed. The air circulation and moisture distribution are independently operated and depend largely on the facility. Feedstock can be preprocessed at one cell and then transferred to the finishing processing unit in another cell using a turning machine and then to the exit point. In considerable sized feedstocks, compost retention times range between 1 and 3 months. In cases where the organic waste stabilization was not accomplished by this process, it is usually accompanied by windrows.

9.3.1.3 Composting Reactors

This, also known as in-vessel composting, is an enclosed technology. However, this approach is smaller than enclosed systems, and the reduced compost head space (most times installed with biofilters) enhances the management of the exhaust (emitted) gases.

9.3.1.3.1 Tunnel Composting

There are different forms of this composting technique (e.g. fixed or agitated reactors) adapted for the stabilization of organic wastes such as livestock manure, sewage sludge, and organic portion of MSW (Boldrin et al. 2009). This technology offers adequate control of the composting process by ensuring that there is an adequate recirculation of exhaust gases to minimize their emission. Due to recirculation of exhaust gases, one important feature of this system is its ability to maintain a relative unique moisture content and temperature. Each tunnel reactor is independently controlled, and the configuration depends largely on the facility. Thus, the retention time is influenced by the system's agitation rate (Zachaus 1995). Shorter retention time favours least agitation although with additional curing in windrows.

9.3.1.3.2 Box and Container Composting

The box and container composting shares some similarities with windrow techniques as well as tunnel reactors. About 50–250 m^3 compost boxes exist with the entire front side having a hatch. Open-top containers are filled. The structural orientation of the containers allows the control of leachate and aeration to be done independently. By adding fresh air and recirculation, temperature profile and moisture content are enhanced. The retention time usually varies with the waste volume compostable organic matter. Second and third backfilling after the first run reduces the total retention times to 6–7 weeks. Agitation can be carried out in some systems about four times per week with a retention time reducing to just between 2 and 7 weeks.

9.3.1.3.3 Rotary Drum Composting

This is an active composting technique widely used in the management of MSW, livestock manure, sewage sludge, and food wastes (Kalamdhad and Kazmi 2009a; Li et al. 2019). It is a biofilm-driven composting procedure with some promising attributes such as simplified structural configuration, cost-effective, and environmental compatibility (Li et al. 2019). The drum which acts like a ball mill is constantly rotated to ensure optimal mixture and fluffing, thus ensuring proper aeration and odour reduction (Kalamdhad and Kazmi 2008, 2009b). One good thing here is that the removal of unused materials is easier as the compost will be drier and properly homogenized. To ensure that homogeneity is realized, it is needful to fill the reactor completely. The removal of the undegraded portion can be done by slanting the drum (Haug 1993; Kasberger 1995). The rotating drum can be compartmentalized to deal with short circuiting of compost stages which can lead to incomplete processing of material. In this type of reactor, the nature and rate of agitation can be effectively controlled (Kasberger 1995). High rotational speed can lead to compaction in wet feedstocks such as biowastes. In most cases, aeration is done by fanning and venting through the perforations in the drum to reduce odour. In general, the composting time varies with the amount of organic materials and the size of the composter (Rodrı́guez et al. 2012). Sometimes, a reduced retention time could result in partial high-rate degradation, while high retention time often leads to high-rate degradation and more stable product. Other composting forms (e.g. windrow technique) could be used further to stabilize (cure) organic waste after rotary drum treatment.

9.4 Choosing Compost (Biofertilizer) Over Conventional Chemical Fertilizers

9.4.1 Nutrient Restoration

Although chemical fertilizers provide easily available nutrients (e.g. nitrogen) essential to plant's growth, they do not stay long in the soil as they are easily leached into the underground waters (Claassen and Carey 2007). It affects topsoil fertility, which is vital to agricultural success. Introduction of compost (humus-rich) helps improve fertility of disturbed soil for sustainable agricultural activities (Al-Bataina et al. 2016; Solaiman et al. 2019). The presence of compost ultimately sustains the biological and physicochemical status of soil by enhancing the level of soil organic carbon (Solaiman et al. 2019). This could be explained by the fact that soil microbes require C for energy, and increasing SOC increases the microbial population. This could be further mineralized by these organisms making the nutrients readily available for plant's uptake. However, in compost these nutrients take longer time to release than chemical fertilizers and remain longer in the soil, therefore being advantageous for long-term land maintenance. Humus production during composting improves the soil nutrient content, leading to improved growth and plant health. Al-Bataina et al. (2016) have demonstrated that nutrient availability from compost is a slow process thereby reducing nutrient loss. Fertile soil is important in our biosphere: it ensures healthy plant growth, increases fruit and vegetable yields, improves root systems, and increases biodiversity.

It is essential to note that the recovery of nutrient-starved soil is a complex process. However, by composting, soil's lost nutrients are replenished, creating a healthier, more nutrient-rich produce (Al-Bataina et al. 2016).

9.4.2 Land Conditioning

Although inorganic fertilizers improve the nutrient content of soil, it has not been profitable as soil conditioner (Mtaita 2003). Organic fertilizer (compost) on the other hand supplies vital nutrients to the soil and improves the soil's physical characteristics (i.e. soil texture and structure) (Huang et al. 2016). Compost, as a soil amendment improves the soil's aggregate stability and organic C and lowers the bulk density and porosity (i.e. enhances air and water flow) (Arthur et al. 2011; Paradelo et al. 2019). Mature compost (biofertilizer) has been described by Rawoteea et al. (2017) and Chen et al. (2019) as effective soil conditioner.

9.4.3 Moisture Management

Compost improves both land drainage by surface percolation and moisture retention and distribution just like a spongy soil cover. These composting water retention benefits lead to reduced water usage during plant irrigation, as well as reduced detrimental water run-off (Eden et al. 2019).

9.4.4 Filtration

Through its desorptive capacity, compost could be used in filtration of pollutants from contaminated water. This ultimately leads to reduction of pollutants from surface and underground water (Faucette et al. 2013). This process, thus, helps to revamp both the water body and the soil surrounding it.

9.4.5 Public Health

Exposure to chemical fertilizers could trigger certain forms of allergy in susceptible individuals which leads to illness. Unlike compost which is eco-friendly, the chemical fertilizer can also be injurious to health by the leachates which run into water bodies.

9.4.6 Solid Waste Reduction and Other Environmental Benefits

Adequately controlled composting processes could help stabilize organic wastes as well as produce nutrient-rich compost (biofertilizer) that enhances soil ecological properties. Composting reduces the amount of GHGs released into the atmosphere especially in cases where waste management was done in landfills. Most times, nitrogen and hydrogen are blended to produce chemical fertilizers (Pollan 2006). The overall process demands great deal of energy which is generated from the fossil fuels. Unlike fertilizer, composting needs little or no energy based on the scale.

9.5 Composting: As Relating to Conventional and Organic Farming

The application of compost to enhance habitat restoration is an interesting technique. Revitalization of soil, though often ignored, is a vital foundation upon which ecological restoration of soil pH, macro-, and micro-nutrient levels can be realized. It also encourages soil microbial population, improves soil nutrient holding and distribution capacity, and reduces the frequency and intensity of water usage. The similarity in the organic matter as well as microbial population of compost with wetland soils makes composting potential in wetland restoration.

9.6 Factors Influencing Composting Process (Considerations on Optimization Strategies)

9.6.1 Temperature

This is an important parameter that drives the composting process and could also be relevant in monitoring (Raut et al. 2008; Chen et al. 2015a; Zhao et al. 2016). In general, it helps to evaluate the rate of organic matter stabilization and provides optimal condition for the composting process (Turan 2008; Awasthi et al. 2014). Different enzymatic activities that drive microbial metabolism show excellent performance at their various optimal temperatures, so there is always a need for temperature adjustment to suit the particular process. In composting piles, significant temperature gradient occurs due to nonlinear mass and energy balances (Zhang et al. 2012; Kulikowska 2016). It determines the relative advantages of a particular group of microbial population over others, which ultimately selects the succession in a compost pile. The thermophilic and thermotolerant organisms usually succeed the mesophilic ones which generate the initial rise in temperature that initiates composting processes. This succession of mesophilic microbes by the thermophiles enables the degradation of complex organic materials like lignin, celluloses, and hemicelluloses.

Temperature in composting can be grouped into four phases (mesophilic, thermophilic, cooling, and maturation stages), and these determine the type of microorganisms to be found in the compost (Chen et al. 2015a). Two stages (thermophilic or active phase and maturing phase) were proposed by Lazcano et al. (2008), and the active phase is dominated by bacteria (Sundberg et al. 2004). Most composting processes occur at both mesophilic and thermophilic temperatures between 105 and 160 °F which helps to destroy pathogens, kill weed seeds, and aid faster degradation of the wastes (Zhang and Sun 2014).The optimal composting temperature is believed to be between 130 and 150 °F, and temperatures above 160 °F is believed to be harmful to composting microbes, and this decreases the efficiency of the process.

During the mesophilic stage, the microbes present in the compost materials metabolize the readily available nutrients, and this process helps to generate more heat that will be used to propagate the composting process to the next stage. The thermophilic phase is important because it helps to destroy pathogenic organisms (faecal coliforms, intestinal parasites) reducing the health risk posed by organic matter to humans and animals and ensure maximal sanitary conditions (Sundberg et al. 2004; Zhang and Sun 2014). It also indicates the decomposition and stability of the initial's organic matter content of feedstock (Paradelo et al. 2013). Different times and temperature regimes are required for the removal of coliforms and other pathogenic bacteria (Cempirkova and Soch 2007), but 1 week period for complete pathogen elimination has been suggested (Chan et al. 2016).

The optimum composting temperature is between 40 and 65 °C (Rich and Bharti 2015), and temperatures between 52 and 60 °C are considered to maintain the greatest thermophilic activity in composting systems (Vuorinen and Saharinen 1997), while Miyatake and Iwabuchi (2005) observed the highest thermophilic activity level at 54 °C. Composting temperature should be monitored so that it will not exceed the optimal level and destroy the thermophilic microbial populations during composting (Imbeah 1998; Huang et al. 2004). Again, high temperature and excessive ammonia also inhibit the growth and activity of thermophilic nitrifying bacteria during composting process (Huang et al. 2004). Sudharsan and Kalamdhad (2015) also stated that high temperature (greater than 65 °C) inactivates fungi, actinomycetes, and other composting bacteria in thermophilic stage allowing spore former to predominate leading to system failure. It is important to monitor and adjust this parameter to effectively harness the potentials of the microorganisms and achieve good results.

9.6.2 pH

The variation in pH values of composting process affects microbial activities in different capacities (Chen et al. 2015a, 2016). It usually declines at the initial stages of composting and rises at later stages (Turan 2008), and this increases NH_3/NH_4 ratio resulting to increase in rates of volatilization indicative of biological activity. The decrease in pH can be caused by microbial decomposition of organic matter into organic and inorganic acids, volatilization of NH_4/NH_3 by nitrifying bacteria, evolution of carbon dioxide (CO_2), and the process of nitrogen and phosphate mineralization (Huang et al. 2004; Lazcano et al. 2008; Wang et al. 2016a, b). It also occurs slightly higher in aerobic condition than in anaerobic conditions due to release of potassium and under optimal pH conditions and changes from acidic to neutral pH due to complete breakdown of the organic acids in the substrates (Kalemelawa et al. 2012; Chan et al. 2016).

Composting at a very low pH level negatively affects the transition from mesophilic to thermophilic phase, and there is a decline in microbial activities (Paradelo et al. 2013; Sundberg et al. 2004). The ammonia build-up due to proteolysis increases the pH and may affect the survival of pH-sensitive organisms necessary for the composting process (Hachicha et al. 2009). Studies by Chan et al. (2016) suggested pH 7.8 as the optimal composting pH, but other studies showed different figures. Zhang and Sun (2016a, b) opined the optimal composting pH to be 7.5–8.5, and microbial activities can be hindered outside the optimal pH range, while Rich and Bharti (2015) noted the support of good microbial activity at 6.7–9 range. Chen et al. (2015b) also reported an optimal pH range of 5.5–8. However it is important to know the optimal pH of a particular composting microorganism in order to effectively harness its potentials.

Although the compost pH has been noted by Rich and Bharti (2015) not to be the key deciding factor in compost stability, Hachicha et al. (2009) suggested matured compost to have a pH of 6.0–8.5 contradicting the range of 8.0–8.5 as noted by Fernandez-Delgado et al. (2015). Composting process can be accelerated when cooled to 40 °C, and the pH increased after complete utilization of the organic component of the substrates (Paradelo et al. 2013; Chan et al. 2016). The factor pH which influences composting is dependent on temperature due to the different combinations of optimal temperature and pH necessary for various microbial group survivals (Sundberg et al. 2004). This parameter could be adjusted to accommodate the microbes involved in the process by well-known alkaline method (e.g. addition of $CaCO_3$) or wood ash (Fernandez-Delgado et al. 2015), but the choice of alkaline amendments can cause the problem of increased ammonia emission (Sundberg et al. 2004). Alternatively, zeolite amendment can be employed, but this method depends on the available soluble ions like ammonium and zeolites that can reduce the ammonium-based buffering capacity of the composting mass (Chan et al. 2016).

9.6.3 Moisture Content

This factor is important in microbial life because most of the chemical reactions do better in liquid phases, and it is critical in composting process. The optimal moisture content (MC) in composting influences oxygen uptake, microbial activities, free air space, and the process temperature (Petric et al. 2012). It is determined by the nature of the substrates, and Luangwilai et al. (2011) suggested the optimal MC of compost materials to be within the range of 50–60%. However, Onwosi et al. (2017) opined that the optimal MC for biological activity during composting is 40–70% of the compost material's weight. Composting of poultry manure and wheat straw by Petric and Selimbasic (2008) was reported at 70% optimum water content, while Ros et al. (2006) in their work observed optimal water holding capacity of 60–70% throughout pig slurry composting. The exceeding 60% water content limit according to

Abdullah and Chin (2010) makes the process tend to anaerobiosis due to restricted oxygen movement.

Increase in the moisture content reduces gas diffusion rates, thus decreasing the rate of oxygen uptake in the process and restricting microbial activity (Mohammad et al. 2012). There is an inverse relationship between temperature and moisture content as observed by Sudharsan and Kalamdhad (2015), because increase in temperature will definitely lead to increased vapourization of water resulting in water content decrease. Turan (2008) also observed high water content at low temperature, and it is a useful tool in determining the water availability which can slow down microbial activity at lower levels or cause water logging at higher levels of moisture content leading to fouling due to anaerobiosis (Makan et al. 2013; Abdullah and Chin 2010).

The optimum moisture content of composting process depends on biological and physicochemical nature of the compost material, and it is vital for soluble nutrient distribution required for optimal microbial metabolic activity (Guo et al. 2012). Moisture content loss during composting suggests strong decomposition rate (Sudharsan and Kalamdhad 2015), but at very low level, dehydration may set in and microbial activities hindered (Makan et al. 2013).

9.6.4 Type of Feedstocks

The feedstock can range from homogenous wastes (such as sawdust) to mixed heterogeneous waste such as municipal solid waste (MSW). Feedstock from plants can include the following: vegetable scraps, yard waste, grass chippings, leaves, brush, woodchips, straw, apple pomace, etc. Those from animal origin include fish waste, meat waste, horse manure, poultry manure, pig/swine manure, etc. Those from home can include simple home waste, food waste, sewage sludge, etc.

9.6.5 Electrical Conductivity

The electrical conductivity (EC) is a numerical expression of conduction of electrical current by an aqueous solution, and it reflects the salinity of an organic amendment (Lazcano et al. 2008). During composting, salt concentration unavoidably increases as a result of complex organic matter decomposition (Chan et al. 2016). It shows the total salt content of the compost suggesting its quality and application as a fertilizer (Jiang et al. 2015; Awasthi et al. 2014). The formation of mineral salts (phosphates and ammonium ions) from organic matter transformation can increase EC, and its precipitation or ammonium volatilization could decrease EC (Huang et al. 2004), and low EC value of matured compost best supports plant growth (Yang et al. 2015). High EC value compost negatively affects plant growth (Awasthi et al. 2014; Yang et al. 2015) and will only be used when mixed well with low EC value soil or other materials (Gao et al. 2010). The salinity of the compost affects seed germination or plant growth negatively especially at seedling stage (Niccolo and Eugenio 2013).

Different authors reported several optimal values for EC limits ideal for compost application. Awasthi et al. (2014), Chowdhury et al. (2014), and Zhang and Sun (2016a, b) adopted 4 ms cm^{-1}. However, Mohee et al. (2015) disagreed with this and suggested EC range of <3.5 ms cm^{-1} to be safe for compost agricultural application. Furthermore, Fernandez-Delgado et al. (2015) reported 3 ms cm^{-1} EC value as criterion for soil amendment with compost. Mulec et al. (2016) opined EC value lower than 2.5 ms cm^{-1}, and Turan (2008) reported values less than 2 ms cm^{-1} as the ideal value for composts. The EC value of compost can be monitored for effective composting using zeolite to reduce the composting salinity Chan et al. (2016) because of its ability to accommodate and allow free exchange of ions, adsorbing ions to its surface leading to decrease in the EC values of the compost.

9.6.6 Nutrients

Microbes degrade organic components of feedstock during composting to acquire nutrients (N, P, K) and metabolic energy (Chen et al. 2011), and C, N, P, and K are the major nutrients required by composting microbes for better efficiency. Some carbon compounds such as lignin with cellulose and aromatics embedded in them resist degradation, but nitrogenous compounds are considered easily degradable with the exception of keratin and other few similarly resistant components. However, C and N are most crucial in microbial transformation of organic matter for energy and building of cell structures (Chen et al. 2011; Iqbal et al. 2015). A reduction in the amount of nitrogen in feedstock decreases microbial activity, and slow decomposition of available carbon results (Igoni et al. 2008). However, if the nitrogen available is beyond the microbial requirement, the excess nitrogen will be volatilized as ammonia gas. Carbon on the other hand evolved as carbon dioxide (CO_2) during its bio-oxidation (Lazcano et al. 2008). The C/N ratio normally decreases during organic matter decomposition (Yang et al., 2015) and is used as decomposition degree indicator. According to Huang et al. (2004), feedstock with 25–30:1 C/N ratio is ideal for active composting, but Pace (1995) stated that 20–40:1 C/N ratio has regularly given good compost yield. Petric et al. (2015) adopted initial C/N values between 25 and 40 even up to 50, but these ratios suggest that microbes use 30–35 time available carbon more than they convert nitrogen (Igoni et al. 2008). Low C/N ratio results to nitrogen accumulation released as ammonia with malodorous characteristic, and huge amount of basic soluble salts will be liberated making the soil non-conducive for growing plants (Awasthi et al. 2014; Mohee et al. 2015). High C/N ratio, however, suggests inadequate availability of nitrogen for optimal microbial growth causing advancement in composting at a slow rate (Chen et al. 2011). The initial C/N ratio of composting feedstock is important because it determines the mineralization rate of its constituent organic matter and also nitrification processes in the system (Ros et al. 2006).

The C/N ratio of compost feedstock can be adjusted and enhanced with different bulking agents (rice husk, wood chip, sawdust, and peanut shells) to improve its porosity (Wang et al. 2016a, b; Zhang and Sun 2016a, b), and they are also degraded during composting. Reports by Imbeah (1998) suggest that co-composting of feedstock with these bulking agents also serves as odour control measure that resulted from excess moisture absorption. As Zhang and Sun (2016a, b) also reported, urea can also be used for the purpose of C/N adjustment during composting. A high degradation level of the compost feedstock suggests lower C/N ratio (Lazcano et al. 2008).

9.6.7 Aeration/Oxygen Content/Feed Particle Size

Composting is fundamentally an aerobic process and also important factor in composting (Chen et al. 2015b) which releases gaseous water (H_2O) and carbon dioxide (CO_2) through molecular oxygen (O_2) consumption (Awasthi et al. 2014). The efficiency of composting process is evaluated by aeration rate which is directly linked with microbial dynamics (Nakasaki et al. 2009). Aeration helps maintain optimal process temperature for thermophilic organic waste bioconversion in composting (Raut et al. 2008). It also provides sufficient O_2 needed for the biodegradation of organic matters, and helps to evaporate excess moisture from the substrate (Petric and Selimbasic 2008). It is greatly influenced by the physicochemical attributes (MC, organic matter content, temperature, and pH) and a major stabilizing factor in finished compost (Rich and Bharti 2015; Guo et al. 2012). Low level of aeration results in fouling due to anaerobiosis, and excessive level overcools the compost pile causing a decrease in decomposition rate (Gao et al. 2010). Sundberg and Jonsson (2008) were of the opinion that increased aeration rate also increases evaporation rate and cools and dries larger compost faster.

In order to attain adequate aeration rate during composting, composters employ different turning regimes, and the airflow rate effect on rate of organic matter decomposition, temperature distribution, and other composting parameters is dependent on ambient condition and location inside the composting mass (Bari and Koeing 2012). Li et al. (2015) suggested once per day turning for adequate aeration of the piles although successful use of natural aeration in composting pig manure. This method is cost-effective and less energy intensive in static pile composting of animal excreta than in forced aeration method (Oudart et al. 2015).

9.7 Effects on Soil's Physical, Biological, and Chemical Properties

9.7.1 Effects on Soil's Microbial Interactions and Activities

Composting process of organic waste encourages the production of beneficial microorganisms (mainly bacteria and fungi) which in turn break down organic matter to create humus. Thus, soil microbiota interactions and activities are enhanced due to the high concentration of organic materials.

9.7.2 Effects on Soil Structure

Compost addition to soils enhances the water and nutrient-holding capacity of the soil; it also improves the microbial activity and ensures a gradual release of essential nutrient and the moisture content of the soil (Eden et al. 2019). In sandy soils, compost greatly improves the moisture holding capacity to enhance plant's transpiration ability even under drought conditions. It also improves aeration and water conduit in heavy clay soil type, thus reducing the effect of excessive moisture on plants. Plant growth-promoting microbes and other desirable microorganisms are enhanced for further conversion of organic wastes into humus. The entire composting process and its product improve the soil structure, moisture content, and nutrient content of amended soils.

Soil amendment with compost could be used to remediate metal-contaminated sites, because of its tendency to bind metals and reduce metal uptake by plants (Shuman et al. 2001). Also, The US National Park Service (NPS) reported that composting process possesses the ability to remediate volatile and semi-volatile organic compounds like poly-aromatic hydrocarbons, heating fuels, and explosives. Also complete degradation of pesticides, preservatives, and chlorinated and non-chlorinated hydrocarbons by processes has been reported in soils contaminated by these environmental pollutants (USNPS 2009). The continuous application of urban waste-derived compost contaminated by metal traces and other recalcitrant materials can accumulate these pollutants in the amended soil (Powers and McSorley 2000). However, low nutrient compost materials will yield end product with lower nutrient value, thus requiring the need for further nutrient addition to the finished product. Amendment of soil with compost improves porosity, water retention capacity, and the overall soil texture. However, there is need for caution during the process to avoid loss of volatile nutrients such as ammonia and nitrogen through volatilization or by leaching of non-volatile nutrient into ground water.

9.7.3 Effects on Pollutants in the Soil

Considerable amount of organic pollutants (e.g. petroleum, polycyclic aromatic hydrocarbon, and pesticides) are discharged into the soil through different anthropogenic activities (Ren et al. 2018). Thus, they constitute environmental concern due their negative effect. Efforts to restrain these pollutants by conventional methods have been challenging, because they were very expensive and often result in secondary pollution. However, composting technology holds great promise in bioremediation of polluted soil since it is eco-friendly and cost-effective. For instance, Xiang et al. (2019) demonstrated that the combination of compost and biochar effectively reduced plant uptake of polybrominated diphenyl ethers (flame retardants) in a polluted soil. Soobhany (2018) showed that extracts from MSW composting or vermicomposting mitigated the ecotoxicity of heavy metal-polluted soil. The major complaint about solid-phase bioremediation is that it is a slow process and could be costly. However, addition of matured compost to the bio-pile during operation or the use of composting process reduces the clean-up time.

9.7.4 Effects on Soil pH and CEC

The buffering capacity of compost-amended soils helps to maintain the soil pH to provide optimum pH for the growth of beneficial microorganisms for plant's growth and metabolism. It also aids in easy exchange of relevant cations between the plants, soils, and microorganisms.

9.7.5 Effects on Essential Micronutrients

Micronutrients essential for proper growth and development of plants and beneficial for soil microbes are easily made available in the soil by the added compost. The compost traps essential micronutrients and macronutrients which can easily be leached out of the soil.

9.7.6 Effects on Climate Change

Food wastes which are major constituents of household garbage and important component of municipal waste contain a lot of organic matter and will not be ideally disposed in landfills (USDA 1997). Food delivery and handling into landfills is costly, creates a lot of health hazards, and pollutes the environment when allowed to accumulate in the environment (Means et al. 2005). The wastes in landfills compact, liquefy, and are mixed with other toxic compounds and seep down the ground to pollute ground waters, rivers, lakes, and coastal waters. To alleviate these problems, composting of food wastes removes them from getting to landfills and causing these aforementioned environmental and health hazards. This will also reduce methane production in landfills and leachate pollution of ground waters (University of Colorado Recycling Services 2002).

9.7.7 Effects on Soil-Borne Pathogens and Plant Diseases

Composting process increases the humus content of the soil thereby increasing its nutrient content. This nutrient causes rapid growth in the soil microbes, but plant growth-promoting bacteria outgrow the pathogenic ones due to their high competitive ability. According to Traversa et al. (2010), the humic-like content of composts improves plant health and growth and is also exerting biocontrol on soil phytopathogens such as fungi. Also, composting renders *Parascaris equorum* eggs non-viable making them nonpathogenic to horses when ingested (Gould et al. 2013). Variation in temperatures during composting is an important component of the process, and it eliminates pathogenic microbes from the finished compost (Pandey et al. 2016).

9.7.8 Effects on the Qualities of Agricultural Yield

Due to the ability of composts to make nutrients and other conditions necessary for plants and soil microbes to be available in the soil, it aids in the improvement of crop yield. The proper use of biological amendments and compost improves soil quality, increases tuber yields, reduces diseases in plants, and helps in soil sustainability and disease management programs (Bernard et al. 2012). In metal-contaminated soils, compost binds metal pollutant and makes them unavailable for plant uptake by edible plants, thus remediating the metal pollutant (Shuman et al. 2001).

9.7.9 Socioeconomic Effects

For a composting network to be able to become a reality, it has to be economically viable (e.g., job creation, economic incentives, and the use of local knowledge to boost the economy). It is important to make the compost market available and affordable to local consumers. Miller and Angiel (2009) showed in their findings that viable composting market can be achieved in Amherst town, New York, where the equivalent of \$22.8 billion (more than twice the community-invested financial resources) was generated from composting facility. The Californian State Integrated Waste Management Board saves money through water conservation, reduction in chemical fertilizers, and herbicide use and also creates market for local producers of compost. Food waste diversion for environmental benefits of composting creates beneficial cooperation for both haulers and composters (Connolly 2006), and dumping fees at composting site are cheaper compared to landfills.

9.7.10 Challenges of Using Compost (Organic Fertilizer)

One of the major challenges of uncontrolled composting process is the nitrogen loss through NH_3 and N_2O emission. This results in production of compost with low nutrient (nitrogen) quality and may not effectively support sustainable agriculture. Also, there could be issue of heavy metal in compost especially when MSWs are used for compost production (Powers and McSorley 2000). This could jeopardize the productivity of soil receiving this compost due to phytotoxicity potential of heavy metal-laden soil. Additionally, low-quality compost could be produced when low-nutrient organic wastes are stabilized. This is a direct result of unmanaged composting process.

Another major drawback of composting is odour generation. Aside from the rate of decomposition, odour is imparted on the compost based on the type of feedstock used. Biosolids which decompose faster generate more odours than clean yard trimmings which take time to decompose. The composting system or the operational facility contributes to odour generation during composting. Indoor composting in an airtight building allows the use of biofilters to absorb the odorous gases. Temperatures and moisture are another odour effect in pile composting systems. Odour generation tends to increase with a decrease in temperature and an increase in moisture level. Rise in temperature favours most enzyme activity as well as enhances the growth of thermotolerant microbes which are known to produce least anaerobic by-product. High moisture content tends to clog the pore spaces as oxygen is less soluble in water, thus creating an anaerobic condition which favours 'anaerobiosis' where most of the odours are generated.

9.8 Conclusion

Composting technology is indispensable in the management of wastes. Its significance in sustainable agriculture is underlined by its nutrient-rich by-product (compost) which ensures proper distribution of nutrients when applied to soil as organic manure (biofertilizer). Composting has other benefits in the society such as its environmental friendliness, cost-effectiveness, and pathogen reduction. Different composting methods have been used in the stabilization of wastes. The applicability of composting in sustainable agricultural technology requires effective management of organic wastes. However, to outcompete inorganic fertilizer, the nutrient content of compost and its availability to plants should be reconsidered. To make composting more competitive, there is need to adopt the mechanized techniques to make compost available to farmers. In the rural or developing countries, there is need to intensify awareness programmes to make the techniques needed for effective composting available to local farmers so that compost with improved quality could be produced. Governments especially in the developing countries should map out policies that would enhance the profitability of composting technology.

References

Abbasi SA, Hussain N, Tauseef SM, Abbasi T (2018) A novel FLippable units vermireactor train system—FLUVTS—for rapidly vermicomposting paper waste to an organic fertilizer. J Clean Prod 198:917–930

Abdullah N, Chin NL (2010) Simplex-centroid mixture formulation for optimized composting of kitchen waste. Bioresour Technol 101:8205–8210

Al-Bataina BB, Young TM, Ranieri E (2016) Effects of compost age on the release of nutrients. Int Soil Water Conserv Res 4(3):230–236

Ali U, Sajid N, Khalid A, Riaz L, Rabbani MM, Syed JH, Malik RN (2015) A review on vermicomposting of organic wastes. Environ Prog Sustain 34(4):1050–1062

Andersen JK, Boldrin A, Samuelsson J, Christensen TH, Scheutz C (2010) Quantification of greenhouse gas emissions from windrow composting of garden waste. J Environ Qual 39:713–724

Arriaga H, Viguria M, López DM, Merino P (2017) Ammonia and greenhouse gases losses from mechanically turned cattle manure windrows: a regional composting network. J Environ Manag 203:557–563

Arrigoni JB, Paladino G, Garibaldi LA, Laos F (2018) Inside the small-scale composting of kitchen and garden wastes: thermal performance and stratification effect in vertical compost bins. Waste Manag 76:284–293

Arthur E, Cornelis WM, Vermang J, De Rocker E (2011) Amending a loamy sand with three compost types: impact on soil quality. Soil Use Manag 27:116–123

Awasthi MK, Pandey AK, Khan J, Bundela PS, Wong JWC, Selvam A (2014) Evaluation of thermophilic fungal consortium for organic municipal solid waste composting. Bioresour Technol 168:214–221

Bari Q, Koeing A (2012) Application of a simplified mathematical model to estimate the effect of forced aeration on composting in a closed system. Waste Manag 32(11):2037–2045

Beck-Friis B, Pell M, Sonesson U, Jonsson H, Kirchmann H (2000) Formation and emission of N_2O and CH_4 from compost heaps of organic household waste. Environ Monit Assess 62:317–331

Bernard E, Larkin RP, Tavantzis SM, Erich MS, Alyokhin A, Sewell G, Lannan A, Gross S (2012) Compost, rapeseed rotation, and biocontrol agents significantly impact soil microbial communities in organic and conventional potato production systems. Appl Soil Ecol 52:29–41

Bhatia A, Madan S, Sahoo J, Ali M, Pathania R, Kazmi AA (2013) Diversity of bacterial isolates during full scale rotary drum composting. Waste Manag 33:1595–1601

Bialobrzewski I, Miks-Krajnik M, Dach J, Markowski M, Czekała W, Głuchowska K (2015) Model of the sewage sludge-straw composting process integrating different heat generation capacities of mesophilic and thermophilic microorganisms. Waste Manag 43:72–83

Boldrin A, Andersen JK, Moller J, Christensen TH, Favoino E (2009) Composting and compost utilization: accounting of greenhouse gases and global warming contributions. Waste Manag Res 27(8):800–812

Cempirkova R, Soch M (2007) The analysis of real microbiological risks for dissociated slurry. Agric Trop Subtrop 40:164–171

Chan M, Selvam A, Wong JWC (2016) Reducing nitrogen loss and salinity during 'struvite' food waste composting by zeolite amendment. Bioresour Technol 200:838–844

Chen L, De Haro MM, Moore A, Falen C (2011). The composting process: dairy compost production and use in Idaho CIS 1179. University of Idaho

Chen R, Wang Y, Wang W, Wei S, Jing Z, Lin X (2015a) N_2O emissions and nitrogen transformation during windrow composting of diary manure. J Environ Manag 160:121–127

Chen Z, Zhang S, Wen Q, Zheng J (2015b) Effect of aeration rate on composting of penicillin mycelial dreg. J Environ Sci 37:172–178

Chen M, Huang Y, Liu H, Xie S, Abbas F (2019) Impact of different nitrogen source on the compost quality and greenhouse gas emissions during composting of garden waste. Process Saf Environ 124:326–335

Chowdhury AKMMB, Michailides MK, Akratos CS, Tekerlekopoulou AG, Pavlou S, Vayenas DV (2014) Composting of three phase olive mill solid waste using different bulking agents. Int Biodeterior Biodegr 91:66–73

Claassen VP, Carey JL (2007) Comparison of slow-release nitrogen yield from organic soil amendments and chemical fertilizers and implications for regeneration of disturbed sites. Land Degrad Dev 132:119–132

Connolly J (2006) Economics of supermarket organics diversion. BioCycle 47(3):30–36

De Silva S, Yatawara M (2017) Assessment of aeration procedures on windrow composting process efficiency: a case on municipal solid waste in Sri Lanka. Environ Nanotechnol Monit 8:169–174

Eden M, Martínez I, Keller T, Houot S (2019) Soil physical properties of a Luvisol developed on loess after 15 years of amendment with compost. Soil Tillage Res 191:207–215

Edwards CA, Bohlen PJ (1996) Biology and ecology of earthworms. Chapman and Hall, London

Faucette B, Cardoso F, Mulbry W, Millner P (2013) Performance of compost filtration practice for green infrastructure stormwater applications. Water Environ Res 85(9):806–814

Fernandez-Delgado JM, Prahauser B, Walter A, Insam H, Franke-Whittle IH (2015) Co-composting of biowaste and wood ash, influence on a microbially driven-process. Waste Manag 46:155–164

Fournel S, Godbout S, Ruel P, Fortin A, Duquette-Lozeau K, Létourneau V, Généreux M, Lemieux J, Potvin D, Côté C, Duchaine C, Pellerin D (2019) Production of recycled manure solids for use as bedding in Canadian dairy farms: II. Composting methods. J Dairy Sci 102 (2):1847–1865

Gao MC, Li B, Yu A, Liang FY, Yang LJ, Sun YX (2010) The effect of aeration rate on forced-aeration composting of chicken manure and sawdust. Bioresour Technol 101(6):1899–1903

Garg P, Gupta A, Satya S (2006) Vermicomposting of different types of waste using *Eisenia foetida*: a comparative study. Bioresour Technol 97(3):391–395

Gould JC, Rossano MG, Lawrence LM, Burk SV, Ennis RB, Lyons ET (2013) The effects of windrow composting on the viability of *Parascaris equorium* eggs. Vet Parasitol 191 (1–2):73–80

Guo R, Li G, Jiang T, Schuchardt F, Chen T, Zhao Y, Shen Y (2012) Effect of aeration rate, C/N ratio and moisture content on the stability and maturity of compost. Bioresour Technol 112:171–178

Habbeche A, Saoudi B, Jaouadi B, Haberra S, Kerouaz B, Boudelaa M, Badis A, Ladjama A (2014) Purification and biochemical characterization of a detergent stable keratinase from a newly thermophilic actinomycete *Actinomadura keratinilytica* strain Cpt29 isolated from poultry compost. J Biosci Bioeng 117(4):413–421

Hachicha S, Sellami F, Cegarra J, Hachicha R, Drira N, Medhioub K, Ammar E (2009) Biological activity during co-composting of sludge issued from the OMW evaporation ponds with poultry manure-physico-chemical characterization of the processed organic matter. J Hazard Mater 162:402–409

Haug RT (1993) The practical handbook of compost engineering. CRC, Boca Raton

Hestmark KV, Fernández-Bayo JD, Harrold DR, Randall TE, Achmon Y, Stapleton JJ, Simmons CW, VanderGheynst JS (2019) Compost induces the accumulation of biopesticidal organic acids during soil biosolarization. Resour Conserv Recyl 143:27–35

Huang GF, Wong JWC, Wu QT, Nagar BB (2004) Effect of C/N on composting of pig manure with sawdust. Waste Manag 24:805–813

Huang M, Zhu Y, Li Z, Huang B, Luo N, Chun Liu C, Zeng G (2016) Compost as a soil amendment to remediate heavy metal-contaminated agricultural soil: mechanisms, efficacy, problems, and strategies. Water Air Soil Pollut 227:359

Igoni HA, Ayotamuno MJ, Eze CL, Ogaji SOT, Probert SD (2008) Design of anaerobic digesters for producing biogas from municipal solid-waste. Appl Energy 85(6):430–438

Imbeah M (1998) Composting piggery waste, a review. Bioresour Technol 63:197–203

Iqbal MK, Nadeem A, Sherazi F, Khan RA (2015) Optimization of process parameters for kitchen waste composting by response surface methodology. Int J Environ Sci Technol 12 (5):1759–1768

Jiang J, Liu X, Huang Y, Huang H (2015) Inoculation with nitrogen turnover bacterial agent appropriately increasing nitrogen and promoting maturity in pig manure composting. Waste Manag 39:78–85

Johari A, Alkali H, Hashim H, Ahmed SI, Mat R (2014) Municipal solid waste management and potential revenue from recycling in Malaysia. Mod Appl Sci 8:37–49

Jouquet P, Plumere T, Thu TD, Rumpel C, Duc TT, Orange D (2010) The rehabilitation of tropical soils using compost and vermicompost is affected by the presence of endogeic earthworms. Appl Soil Ecol 46(1):125–133

Kaiser J (1996) Modelling composting as a microbial ecosystem, a simulation approach. Ecol Model 91:25–37

Kalamdhad AS, Kazmi AA (2008) Mixed organic waste composting using rotary drum composter. Int J Environ Waste Manag 2:24–36

Kalamdhad AS, Kazmi AA (2009a) Effects of turning frequency on compost stability and some chemical characteristics in a rotary drum composter. Chemosphere 74(10):1327–1334

Kalamdhad AS, Kazmi AA (2009b) Rotary drum composting of different organic waste mixtures. Waste Manag Res 27(2):129–137

Kalemelawa F, Nishihara E, Endo T, Ahmad Z, Yeasmin R, Tenywa MM, Yamamoto S (2012) An evaluation of aerobic and anaerobic composting of banana peels treated with different inoculums for soil nutrient replenishment. Bioresour Technol 126:375–382

Kasberger P (1995) Das dynamisch gesteuerte LESCHA-Verfahren (The dynamically controlled LESCHA-process; in German). In: Wiemer K, Kern M (eds) Abfallwirtschaft Neues aus Forschung und Praxis – HerstellerforumBioabfall – Verfahren der Kompostierung und anaeroben Abfallbehandlung im Vergleich. M.I.C. Baeza, Witzenhausen, pp 120–131

Komakech AJ, Zurbrügg C, Miito GJ, Wanyama J, Vinneras B (2016) Environmental impact from vermicomposting of organic waste in Kampala, Uganda. J Environ Manag 181:395–402

Kong Z, Wang X, Liu Q, Li T, Chen X, Chai L, Liu D, Shen Q (2018) Evolution of various fractions during the windrow composting of chicken manure with rice chaff. J Environ Manag 207:366–377

Kugler R, Hofer H, Leisner R (1995) Das Wendelin-Tafelmieten-Kompostierungsverfahren (The Wendelin pile composting process; in German). In: Wiemer K, Kern M (eds) Abfallwirtschaft Neues aus Forschung undPraxis – Herstellerforum Bioabfall – Verfahren der Kompostierung und anaeroben Abfallbehandlung im Vergleich. M.I.C. Baeza, Witzenhausen, pp 13–23

Kulikowska D (2016) Kinetics of organic matter removal and humification progress during sewage sludge composting. Waste Manag 49:196–203

Lazcano C, Gomez-Brandon M, Domínguez J (2008) Comparison of the effectiveness of composting and vermicomposting for the biological stabilization of cattle manure. Chemosphere 72:1013–1019

Leton TG, Stentiford EI (1990) Control of aeration in static pile composting. Waste Manag Res 8 (4):299–306

Li Y, Liu B, Zhang X, Gao M, Wang J (2015) Effects of Cu exposure on enzyme activities and selection for microbial tolerances during swine-manure composting. J Hazard Mater 283:512–518

Li N, Zeng W, Yang Y, Wang B, Li Z, Peng Y (2019) Oxygen mass transfer and post-denitrification in a modified rotating drum biological contactor. Biochem Eng J 144:48–56

Lim LY, Lee CT, Bong CPC, Lim JS, Klemes JJ (2019) Environmental and economic feasibility of an integrated community composting plant and organic farm in Malaysia. J Environ Manag 244:431–439

Luangwilai T, Sidhu HS, Nelson MI, Chen X (2011) Modelling the effects of moisture content in compost piles. In: CHEMECA 2011, Australian Chemical Engineering Conference Australia, Engineers Australia

Luo W, Chen TB, Zheng GD, Gao D, Zhang YA, Gao W (2008) Effect of moisture adjustments on vertical temperature distribution during forced-aeration static-pile composting of sewage sludge. Resour Conserv Recycl 52(4):635–642

Makan A (2015) Windrow co-composting of natural casings waste with sheep manure and dead leaves. Waste Manag 42:17–22

Makan AH, Assobhei O, Mountadar M (2013) Effects of initial moisture content on in-vessel composting under air pressure of organic fraction of municipal solid wastes in Morocco. Iran J Environ Health Sci Eng 10(1):3

Mayende L, Wilhelmi BS, Pletschke BI (2006) Cellulases (CMCases) and polyphenoloxidases from thermophilic *Bacillus* spp. isolated from compost. Soil Biol Biochem 38:2963–2966

Means NE, Starbuck CJ, Kremer RJ, Jett LW (2005) Effects of a food waste-based soil conditioner on soil properties and plant growth. Compost Sci Util 13(2):116–121

Michel FC Jr, Forney LJ, Huang AJ-F, Drew S, Czuprenski M, Lindeberg JD, Reddy CA (1996) Effects of turning frequency, leaves to grass mix ratio and windrow vs. pile configuration on the composting of yard trimmings. Compost Sci Util 4:26–43

Milinković M, Lalević B, Jovičić-Petrović J, Golubović-Ćurguz V, Kljujev I, Raičević V (2019) Biopotential of compost and compost products derived from horticultural waste – effect on plant growth and plant pathogens' suppression. Process Saf Environ 121:299–306

Miller L, Angiel J (2009) Municipal yard trimmings compo sting benefit cost analysis. BioCycle 50 (7):21–24

Miyatake F, Iwabuchi K (2005) Effect of high compost temperature on enzymatic activity and species diversity of culturable bacteria in cattle manure compost. Bioresour Technol 96 (16):1821–1825

Mohammad M, Alam M, Kabbashi NA, Ahsan A (2012) Effective composting of oil palm industrial waste by filamentous fungi, a review. Resour Conserv Recycl 58:69–78

Mohee R, Boojhawon A, Sewhoo B, Rungasamy S, Somaroo GD, Mudhoo A (2015) Assessing the potential of coal ash and bagasse ash as inorganic amendments during composting of municipal solid wastes. J Environ Manag 159:209–217

Mtaita TA (2003) Food. In: Hazeltine B, Bull C (eds) Field guide to appropriate technology. Academic, San Diego, pp 277–480

Mulec AO, Mihelic R, Walochnik J, Bulc TG (2016) Composting of the solid fraction of blackwater from a separation system with vacuum toilets – effects on the process and quality. J Clean Prod 112:4683–4690

Mupambwa HA, Mnkeni PNS (2018) Optimizing the vermicomposting of organic wastes amended with inorganic materials for production of nutrient-rich organic fertilizers: a review. Environ Sci Pollut Res Int 25(11):10577–10595

Mupambwa HA, Ravindran B, Mnkeni PNS (2016) Potential of effective micro-organisms and *Eisenia fetida* in enhancing vermi-degradation and nutrient release of fly ash incorporated into cow dung-paper waste mixture. Waste Manag 48:165–173

Nakasaki K, Tran LTH, Idemoto Y, Abe M, Rollon AP (2009) Comparison of organic matter degradation and microbial community during thermophilic composting of two different types of anaerobic sludge. Bioresour Technol 100:676–682

Nakasaki K, Hirai H, Mimoto H, Quyen TNM, Koyama M, Takeda K (2019) Succession of microbial community during vigorous organic matter degradation in the primary fermentation stage of food waste composting. Sci Total Environ 671:1237–1244

Nasini L, De Luca G, Ricci A, Ortolani F, Caselli A, Massaccesi L, Regni L, Gigliotti G, Proietti P (2016) Gas emissions during olive mill waste composting under static pile conditions. Int Biodeterior Biodegrad 107:70–76

Niccolo P, Eugenio C (2013) The effects of heavy metals, ammonia and electrical conductivity in compost derived from swine solid fraction on seed germination and root elongation of *Lepidium sativum* L. Poster presented at the "15th international conference of the network of recycling of agricultural, municipal and industrial residues in agriculture (RAMIRAN)", June 2013

Nielsen KM (2019) Organic farming. Encyclopedia of ecology, 2nd edn. Reference module in earth systems and environmental sciences, vol 4, pp 550–558

Onwosi CO, Igbokwe VC, Odimba JN, Eke IE, Nwankwoala MO, Iroh IN, Ezeogu LI (2017) Composting technology in waste stabilization: on the methods, challenges and future prospects. J Environ Manag 190:140–157

Oudart D, Robin P, Paillat JM, Paul E (2015) Modelling nitrogen and carbon interactions in composting of animal manure in naturally aerated piles. Waste Manag 46:588–598

Pace MG (1995) The composting process. Utah State University Cooperative Extension, Utah, pp 1–2

Pandey PK, Cao W, Biswas S, Vaddella V (2016) A new closed loop heating system for composting of green and food wastes. J Clean Prod 133:1252–1259

Paradelo R, Moldes AB, Barral MT (2013) Evolution of organic matter during the mesophilic composting of lignocellulosic winery wastes. J Environ Manag 116:18–26

Paradelo R, Eden M, Martinez I, Keller T, Houot S (2019) Soil physical properties of a Luvisol developed on loess after 15 years of amendment with compost. Soil Tillage Res 191:207–215

Pepe O, Ventorino V, Blaiotta G (2013) Dynamic of functional microbial groups during mesophilic composting of agro-industrial wastes and free-living (N_2)-fixing bacteria application. Waste Manag 33:1616–1625

Pergola M, Persiani A, Palese AM, Di Meo V, Pastore V, D'Adamo C, Celano G (2018) Composting: the way for a sustainable agriculture. Appl Soil Ecol 123:744–750

Petric I, Selimbasic V (2008) Development and validation of mathematical model for aerobic composting process. Chem Eng J 139(2):304–317

Petric I, Helic A, Avdic EA (2012) Evolution of process parameters and determination of kinetics for co-composting of organic fraction of municipal solid waste with poultry manure. Bioresour Technol 117:107–116

Petric I, Avdihodzic E, Ibric N (2015) Numerical simulation of composting process for mixture of organic fraction of municipal solid waste and poultry manure. Ecol Eng 75:242–249

Pollan M (2006) The omnivore's dilemma: a natural history off our meals. Penguin Group, New York

Powers LE, McSorley R (2000) Ecological principles of agriculture. Thompson Delmar Learning, Albany

Ramnarain YI, Adil Ansari AA, Ori L (2019) Vermicomposting of different organic materials using the epigeic earthworm *Eisenia foetida*. Int J Recyl Org Waste Agric 8(1):23–33

Raut MP, William SMPP, Bhattacharyya JK, Chakrabarti T, Devotta S (2008) Microbial dynamics and enzyme activities during rapid composting of municipal solid waste – a compost maturity analysis perspective. Bioresour Technol 99:6512–6519

Rawoteea SA, Mudhoo A, Kumar S (2017) Co-composting of vegetable wastes and carton: effect of carton composition and parameter variations. Bioresour Technol 227:171–178

Ren X, Zeng G, Tang L, Wang J, Wan J, Wang J, Deng Y, Liu Y, Peng B (2018) The potential impact on the biodegradation of organic pollutants from composting technology for soil remediation. Waste Manag 72:138–149

Rich N, Bharti A (2015) Assessment of different types of in-vessel composters and its effect on stabilization of MSW compost. Int Res J Eng Technol 2(3):1–6

Rodríguez L, Cerrillo MI, García-Albiach V, Villaseñor J (2012) Domestic sewage sludge composting in a rotary drum reactor: optimizing the thermophilic stage. J Environ Manag 112:284–291

Ros M, Klammer S, Knapp B, Alchberger K, Insam H (2006) Long term effects of compost amendment of soil on functional and structural diversity and microbial activity. Soil Use Manag 22:209–218

Sanasam SD, Talukdar NC (2017) Quality compost production from municipality biowaste in mix with rice straw, cow dung, and earthworm *Eisenia Fetida*. Compost Sci Util 25:141–151

Schmitz T, Meier-Stolle G (1995) Das Biotin-UndeKompoflex-verfahren (The Biofix and Kompoflex process; in German). In: Wiemer K, Kern M (eds) Abfallwirtschaft Neuesaus Forschung and Praxis-Herstellerforum Bioafall-Verfahren der kompostierung und anaeroben Abfallbehandlung in vergleich. M.I.C. Baeza, Witzenhausen, pp 183–192

Sharma K, Garg VK (2018) Comparative analysis of vermicompost quality produced from rice straw and paper waste employing earthworm *Eisenia fetida* (Sav.). Bioresour Technol 250:708–715

Shuman L, Dudka S, Das K (2001) Zinc forms and plant availability in a compost amended soil. Water Air Soil Pollut 128(1/2):1–11

Singh J (2018) Role of earthworm in sustainable agriculture. In: Sustainable food systems from agriculture to industry. Improving production and processing. Elsevier, Amsterdam, pp 83–122

Solaiman ZM, Yang H, Archdeacon D, Tippet O, Tibi M, Whiteley AS (2019) Humus-rich compost increases lettuce growth, nutrient uptake, mycorrhizal colonisation, and soil fertility. Pedosphere 29(2):170–179

Soobhany N (2018) Remediation potential of metalliferous soil by using extracts of composts and vermicomposts from municipal solid waste. Process Saf Environ 118:285–295

Sudharsan VV, Kalamdhad AS (2015) Evolution of chemical and biological characterization during thermophilic composting of vegetable waste using rotary drum composter. Int J Environ Sci Technol 12:2015–2024

Sundberg C, Jonsson H (2008) Higher pH and faster decomposition in biowaste composting by increased aeration. Waste Manag 28:518–526

Sundberg C, Smars S, Jonsson H (2004) Low pH as an inhibiting factor in the transition from mesophilic to thermophilic phase in composting. Bioresour Technol 95:145–150

Sundberg C, Yu D, Franke-Whittle I, Kauppe S, Smars S, Insam H, Romantschuk M, Jonsson H (2013) Effect of pH and microbial composition on odour in food waste composting. Waste Manag 33:204–211

Traversa A, Loffredo E, Gattullo CE, Senesi N (2010) Water-extractable organic matter of different composts, a comparative study of properties and allelochemical effects on horticultural plants. Geoderma 156:287–292

Tripathi G, Bhardwaj P (2004) Decomposition of kitchen waste amended with cow manure using an epigeic species (*Eisenia fetida*) and an anecic species (*Lampito mauritii*) Bioresour Technol 92 (2):215–218

Turan NG (2008) The effects of natural zeolite on salinity level of poultry litter compost. Bioresour Technol 99:2097–2101

University of Colorado Recycling Services (2002) Colorado institutional food waste composting guide. Retrieved from http://recycling.colorado.edu/files/1423d363613bc4311dcaS076ce69t71edec22b1b.pdf

USDA (1997) Estimating and addressing America's food losses. Retrieved from http://www.ers.usda.gov/Publications/FoodReview/JanI997/Jan97a.pdf

USNPS (2009) Composting. Retrieved from http://www.nps.gov/climate friendly parks/Mitigation/Composting.html

Vuorinen AH, Saharinen MH (1997) Evolution of microbiological and chemical parameters during manure and straw co-composting in a drum composting system. Agric Ecosyst Environ 66:19–29

Wang X, Selvam A, Wong JWC (2016a) Influence of lime on struvite formation and nitrogen conservation during food waste composting. Bioresour Technol 217:227–232

Wang Y, Pang L, Liu X, Wang Y, Zhou K, Luo F (2016b) Using thermal balance model to determine optimal reactor volume and insulation material needed in a laboratory-scale composting reactor. Bioresour Technol 206:164–172

Xiang L, Sheng H, Gu C, Marc R-G, Wang Y, Bian Y, Jiang X, Wang F (2019) Biochar combined with compost to reduce the mobility, bioavailability and plant uptake of 2,2′,4,4-′-tetrabrominated diphenyl ether in soil. J Hazard Mater 374:341–348

Yang F, Li G, Shi H, Wang Y (2015) Effects of phosphor-gypsum and super phosphate on compost maturity and gaseous emissions during kitchen waste composting. Waste Manag 36:70–76

Zachaus D (1995) Kompostierung (composting; in German). In: Thome-Kozmiensky KJ (ed) Biologische Abfallbehandlung. EF-Verlag fur Energie- und Umwelttechnik GmbH, Berlin, pp 215–353

Zhang L, Sun X (2014) Changes in physical, chemical, and microbiological properties during the two-stage co-composting of green waste with spent mushroom compost and biochar. Bioresour Technol 171:274–284

Zhang L, Sun X (2016a) Improving green waste composting by addition of sugarcane bagasse and exhausted grape marc. Bioresour Technol 218:335–343

Zhang L, Sun X (2016b) Influence of bulking agents on physical, chemical, and microbiological properties during the two-stage composting of green waste. Waste Manag 48:115–126

Zhang H, Li C, Li G, Zang B, Yang Q (2012) Effect of spent air reusing (SAR) on maturity and greenhouse gas emissions during municipal solid waste MSW composting with different pile height. Procedia Environ Sci 16:59–69

Zhao X, Li B, Ni J, Xie D (2016) Effect of four crop straws on transformation of organic matter during sewage sludge composting. J Integr Agric 15:232–240
Zhi-Wei S, Tao S, Wen-Jing D, Jing W (2019) Investigation of rice straw and kitchen waste degradation through vermicomposting. J Environ Manag 243:269–272
Zhu-Barker X, Bailey SK, Paw UKT, Burger M, Horwath WR (2017) Greenhouse gas emissions from green waste composting windrow. Waste Manag 59:70–79

Nanoagroparticles: An Emerging Trend in Modern Agriculture System 10

Ritu Singh, Avimanu Sharma, Nisha Kumari, Monalisha Behera, Sanjeev Kumar, and Arif Jamal Siddiqui

Contents

Abstract

The application of nanotechnology in agricultural domain has drastically transformed the traditional methods of agriculture. The development of nanoagroparticles is not only enhancing the crop production but also improving the food quality, keeping in check the environmental pollution. Several nano-

R. Singh (✉) · A. Sharma · N. Kumari · M. Behera
Department of Environmental Sciences, School of Earth Science, Central University of Rajasthan, Ajmer, India
e-mail: ritu_ens@curaj.ac.in

S. Kumar
Department of Environmental Science, Central University of Jharkhand, Ranchi, India

A. J. Siddiqui
Department of Biology, College of Science, University of Hail, Hail, Saudi Arabia

K. Bauddh et al. (eds.), *Ecological and Practical Applications for Sustainable Agriculture*, https://doi.org/10.1007/978-981-15-3372-3_10

based agroparticles that have led to the smart agricultural systems include nanofertilizers, nanopesticides, nanoweedicides, nanosensors, etc. Apart from increasing the food productivity, these nano-based agroparticles also perform other functions such as controlled release of nutrients, delivery at target sites, sensing of disease in crops, and killing them, sense the stress conditions in the environment, minimize the release of excess chemicals into environmental matrices, etc. It is also observed that different modes of application of these nanoagroparticles influence the growth and crop productivity enormously, thus giving the alternatives to choose best course of action for crop growth. Moreover, nanotechnology in seed science has improved the quality of seeds producing healthy plants and increased crop yield. However, there are various constraints that are still needed to be addressed to bring the nanoagroparticles to commercial platforms. Mainly, knowledge about its environmental pathways and its long-term impacts on human and animal communities are the major area of concern and investigation. In the present-day scenario, the discovery of nano-integrated agroparticles is a miracle for increasing food demand, and its further enhancement could be a step toward the goal of sustainable agriculture.

Keywords

Nanotechnology · Nanoparticles · Nutrients · Nanoagroparticles · Sustainable agriculture

10.1 Introduction

The soaring population of the present-day world demands a secure and flourishing agricultural sector to eliminate the rising tensions of food shortages and lift a nation's economy in the process. It is essential to increase the crop productivity substantially to meet the growing food demand, making the food items available to each and every individual of the country. In the context of India, an economically developing country with an uncontrolled population upsurge, agriculture forms a major part of the country's economy, making up about 14.2% of the Indian GDP and generating employment and food security for the nation. Hence, the rising pressure on the agricultural sector to increase the crop productivity has led to the extensive use of chemical fertilizers and pesticides in the farm lands to grow the crops within a short time span and maximize the crop yield. Though this over-the-counter use of synthetic agroparticles has increased the crop yield, they have left the environment polluted and the farmland barren resulting in more severe consequences like poisoning the water bodies and soil, destroying the beneficial microbial communities, and affecting the human and other animals exposed to the chemical pesticides and fertilizers. Since the fertilizers are toxic and lethal in nature, they affect human health as well as damage the environment to a great extent. However, fertilizers can be termed as both assets and liabilities at the same time depending on the nature of their use. The synthetic chemicals can be used to increase productivity as well as the fertility of the soil, while on the other hand, it can too disrupt the nutrient balance of the soil and impose substantial threat on the environment by causing water pollution through chemical leach outs. Despite their contribution in increasing crop yield, chemical leach outs have turned out as one of the major aftereffects of the extensive application of toxic agroparticles. The reason for these chemical leach outs can be attributed to larger grain size and smaller surface area of pesticides and fertilizers, which make them escape easily into the environment through rainfalls. The leaching of these commercially available chemicals pollutes the nearby lakes and rivers causing the water nutrient rich otherwise known as eutrophication. Eutrophication then led to the invasive algal blooms increasing the biological oxygen demand of the water thereby threatening the lives of aquatic organisms. A report by the United States Department of Agriculture 2012 suggests that USA alone experienced acceleration in pesticides consumption from 196 to 516 million pounds. Moreover, the commercial agroparticles have downgraded the quality of foods produced and striped off the soil of its fertility (Alfadul et al. 2017). Therefore, there is a dire need to improve the former conventional methods of agriculture with new and advanced ones which can not only enhances the crop production but also would pose no threats on the environment and its ecosystems as well as preserve the soil fertility at the same time.

10.2 Sustainable Agriculture

The concept of sustainable agriculture expanded prominently after publication of the Brundtland Report in 1987. Although the meaning of sustainable agriculture is a little bit ambiguous, United States Department of Agriculture 2012 defined it as 'agriculture which is strongly emphasized on producing long-term crops and livestock, having marginal consequences on the environment'. The basic idea behind sustainable approach is to make an upright balance between the constant need to produce food and safeguarding the ecological systems from adverse impacts. Besides, there are several other objectives allied with sustainable agriculture like conservation of water, lowering the usage of agroparticles (fertilizers and pesticides), promoting healthy biodiversity, reducing waste production, minimizing environmental impact, self-sustaining, resource conservation, etc. (Mahmoud and Taha 2018; Velten et al. 2015) (Fig. 10.1). The application of environmental fertilizers like compost, vermicompost, slow release fertilizers, etc. has been also proven to enhance crop productivity along with enhanced soil health (Singh et al. 2010; Kumar et al. 2012, 2014a, b, c, 2015; Rai et al. 2017). Sustainable agriculture may also focus on retaining financial stability of farms, helping farmers to adopt superior farming practices and improving their quality of life (Campbell et al. 2014). Numerous farming techniques can be used to make agriculture more sustainable. Nanotechnology is one of these advanced methods that has been successfully incorporated in developing fertilizers with high efficiency and no prominent risks on the biodiversity. Nanotechnology is defined as the branch of Science that deals with extremely small particles or systems, i.e. 1–100 nm that has revolutionized various sectors since decades. Agriculture is among one of these sectors that is completely

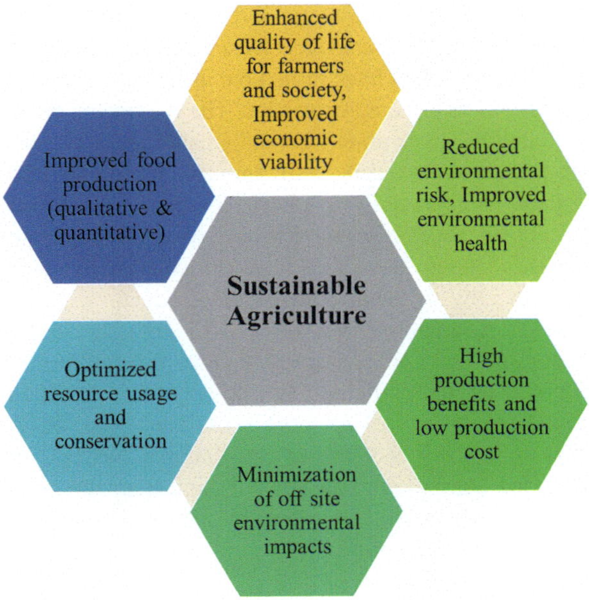

Fig. 10.1 Various components of sustainable agriculture

transformed with the onset of nanotechnology in the conventional farming practices. Nanotechnology has replaced the synthetic agroparticles with nanoagroparticles and developed new nanodevices that have not only augmented the crop production but also encouraged sustainable agriculture to a great extent.

10.3 Nanoagriculture: A Way Towards Sustainable Agriculture

The need of present-day agriculture is the production of crops that can be sustainable and can show a magnificent yield at the same time without using synthetic pesticides, weedicides, and fertilizers. Nevertheless, the agroparticles available in the market like fertilizer, herbicides/weedicides, and pesticides have been reported to consist of deleterious concerns associated with it. They are found to contaminate the water bodies through leaching, risking the life of living organisms and leading to various undesirable outputs. Hence, the precise control and management of input chemicals should be encouraged to significantly reduce the associated risks (Kah 2015). The development of advanced agricultural system with nano based tools could be an effective approach to mark a revolution in the domain of agriculture which can not only reduce the destructive consequences of modern agriculture system on the environment but also boost the crop yield in both quantity and quality (Liu and Lal 2015). Another contribution of nanotechnology in the field of agriculture is the development of nanobiosensors. Nanobiosensors have distinct disease-sensing capacities in crop plants that help reduce the crop diseases thereby increasing the productivity (Fraceto et al. 2016). Besides, features like signal transduction are being

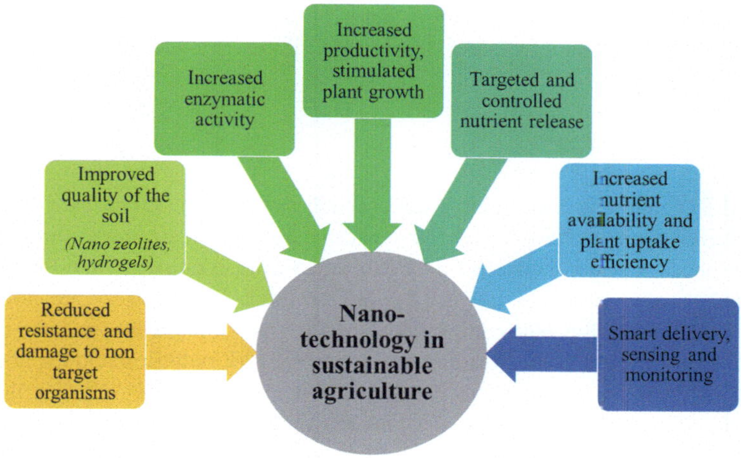

Fig. 10.2 Role of nanotechnology in sustainable agriculture

introduced into nanosensors for quick sensing of microbial toxins in crop plants. Furthermore, miniature size of nanosensors makes them more smart devices for its easy use and placement in agricultural fields and farm lands (Viswanathan and Radecki 2008; Fraceto et al. 2016). These nanobiosensors are also found to be helpful in detecting fungal toxins in several foods (Sertova 2015). With controlled release of nutrients and all other advanced properties, nanoagroparticles can be stated as one of the cutting-edge discoveries for enhancing crop production (Gruere 2012; Mukhopadhyay 2014). Hence, the incorporation of nanoagroparticles in the traditional agriculture has given rise to a new type of agricultural system, termed as nanoagriculture. This agriculture system has shown significant contribution in different aspects of sustainable agriculture which has been illustrated briefly in Fig. 10.2.

Nanoagriculture is basically the use of nanoparticles in agriculture to provide controlled release of fertilizers to crops. These nanoparticles minimize the nutrient loss from fertilizers during field applications thereby increasing the yield through streamline management of nutrients (Sekhon 2014; Wilson et al. 2008). Despite the development in transforming the fertilizers, the present state of the agricultural sector is quite negligible and is losing sight to make its way to the bigger markets. The integration of nanotechnology in developing tools and efficient fertilizers, therefore, is encouraged to strengthen the agricultural sector considerably (Kandasamy and Prema 2015). Nanodevices and tools, such as nanoparticles, nanocapsules, viral capsids, and nanosensors, are few instances that are being used in crop fields for sensing diseases, efficient release of nutrients, easy absorption of nutrients, smooth delivery of specific ingredient to target sites, etc. Target-specific nanofertilizers can minimize the damage to nontargeted plant tissues and prevent the massive release of chemicals into soil matrices (Chhipa 2017a). Sustainable agricultural practices with nanoagroparticles are gaining focus presently for its decreased harm to the soil and more eco-friendly nature. Here is a brief description of different nanoagroparticles and nanotools used in nanoagriculture for enhanced crop production.

10.3.1 Nanofertilizers

Any fertilizer that is in nanometer scale and delivers nutrients to crops is known as nanofertilizer. This technology is the newest invention in the agricultural sector which replaces commercial fertilizers for its smart release nutrients into the soil in a controlled manner (Liu and Lal 2015). The commercial fertilizers are not target specific and release the nutrients quickly, and excess amount leaches out in high concentration to water bodies leading to the problems of pollution. On the contrary, when nanofertilizers are added to the soil, it releases the nutrients in a controlled manner keeping the plant healthy and preventing damage to the soil by preventing excessive nutrient concentration in the soil and supporting microbial activities. Moreover, nanoforms have better solubility and availability to plants. Also, nanofertilizers not only help in streamline management of nutrients but also decrease the pollution of soil as well as water bodies (Nagula and Usha 2016). Figure 10.3 (a) displays the effects of nanofertilizers on soil and plant and (b) illustrates the benefits of using nanofertilizers over conventional fertilizers. Through site-targeted delivery, these nanofertilizers precisely release the agroparticles which in turn decrease the toxicity and improve nutrient utilization by the plants. High absorption of nutrients by plants enhances its performance in terms of increased productivity, high photosynthesis, and substantial expansion in leaf surface area. Additionally, the controlled discharge of nutrients in soil matrices also contributes to reduced eutrophication and pollution of water and soil substrates (Liu and Lal 2015).

Several nanoparticles have been used for the preparation of nano-based fertilizers through biological and chemical methods which have displayed remarkable benefits as fertilizers and promoted plant growth (Table 10.1). Nanofertilizers are synthesized by encapsulation of nutrients in nanoparticles which are then coated with a thin protective film made up of different natural polymers. Nanofertilizers can also be delivered in the form of emulsion in targeted sites. Therefore, nanofertilizers can play a unique role in the production of food crops, increasing the productivity by 35–40% (Chen and Yada 2011). Without the fertilizer input, it is impossible to sustain the productivity of agricultural sector, and nanofertilizer can act as medium to improve crop yields promoting sustainable agriculture. Available reports and patented researches strongly propose that there is a huge scope for the formulation of nanofertilizers and coating them with suitable organic and inorganic polymers for slow release of nutrients (Subramanian and Tarafdar 2011). The mesoporous nature of polymer nanoparticles makes them efficient carrier systems for desired nutrients increasing the overall efficiency of nanofertilizers. Corradini et al. (2010) observed that chitosan nanoparticles stabilized itself in the presence of NPK nanoparticles and found it a suitable polymer for entrapment of nutrients. Likewise, Kottegoda et al. (2011) prepared urea nanoparticles modified with hydroxyapatite nanoparticles for slow release of nitrogen to the soil, consequently increasing the crops growth.

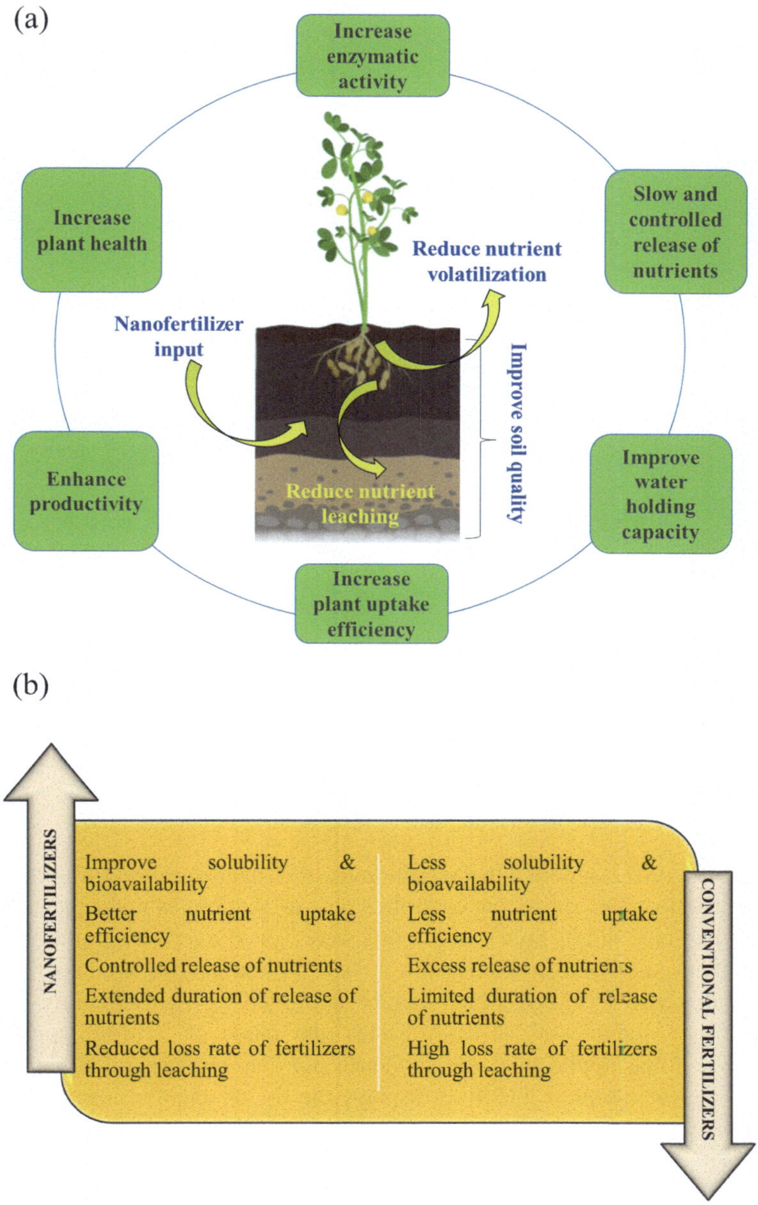

Fig. 10.3 Nanofertilizers. (**a**) Effects on soil and plant, (**b**) comparison with conventional fertilizers

Table 10.1 List of nanoparticles used as nanofertilizers

Nanoparticle(s)	Method of synthesis	Plant	Effect(s)	Reference(s)
Zinc oxide (ZnO)	Chemical synthesis	*Cicer arietinum*	Increased root-shoot and total biomass	Burman et al. (2013)
Zinc oxide (ZnO)	Biological	*Brassica nigra*	Root elongation	Zafar et al. (2016)
Zinc (Zn)	Chemical	*Oryza sativa*	Increased wet weight, dry weight, and total biomass	Upadhyaya et al. (2015)
Copper oxide (CuO)	Biological	*Zea mays*	Increased growth rate	Adhikari et al. (2016)
Copper (Cu)	Chemical	*Lactuca sativa*	Increased root-shoot length	Shah and Belozerova (2009)
Sulfur (S)	Chemical	*Arachis hypogea*	Increased oil content in the kernel and increased overall length of the plant	Thirunavukkarasu et al. (2018)
Iron and Magnesium (Mg)	Chemical	*Vigna unguiculata*	Increased chlorophyll content	Delfani et al. (2014)
Molybdenum (Mo)	Chemical	*Cicer arietinum*	Increased biomass	Taran et al. (2014)
Hydroxyapatite	Chemical	*Glycine max*	Increased growth rate and biomass	Liu and Lal (2015)
Manganese (Mn)	Chemical	*Vigna radiata*	Increased shoot length and chlorophyll rate	Pradhan et al. (2013)

Wanyika et al. (2012) reported that mesoporous silica nanoparticles can entrap about 15.5% urea in its pores enhancing controlled release of nutrients to the plants. When ZnO nanoparticles and its bulk counterparts were coated over a macronutrient, it is observed that coated nanofertilizers show a faster dissolution rate as compared to the commercial macro fertilizer, which can be attributed to high solubility of ZnO nanoparticles (Milani et al. 2012). Besides, the mode of fertilizer application, i.e. direct or foliar application, can influence the delivery system of nutrients to plants to a great extent. Table 10.2 displays the list of nanofertilizers which are commercially available in the market for use.

10.3.2 Nanopesticides

Pesticides in nanosized range are known as nanopesticides. These nanosized particles carry the same properties as the commercial pesticides but are considered

Table 10.2 Commercially available nanofertilizers

Nanofertilizer	Content	Manufacturer company
Nano Green	Extracts obtained from corn, grains, soybean, potatoes, coconut, etc.	Nano Green Sciences, Incorporation of India
The Nano-Ag Answer®	Sea kelp, microorganism, and mineral electrolyte	Urth Agriculture, CA, US
Biozar Nano-Fertilizer	Combination of organic materials, macromolecules, and micronutrients	Fanavar Nano-Pazhoohesh Markazi Company, Iran
Nano Max NPK Fertilizer	Numerous organic acids that are chelated with macronutrients, amino acids, organic carbon, organic micronutrients, vitamins, etc.	JU Agri Sciences Pvt. Ltd., Janakpuri, New Delhi, India
Master Nano Chitosan Organic Fertilizer	Water soluble chitosan, organic acids, phenolic compounds, salicylic acids	Pannaraj intertrade, Thailand
TAG NANO (NPK, PhoS, Zinc, Cal, etc.) fertilizers	Proteino-lacto-gluconate chelated with micronutrients, vitamins, extracts of seaweed, humic acid, probiotics	Tropical Agrosystem India(P) Ltd., India
Nano-Gro™	Plant growth promoter and immunity enhancer	Agro nanotechnology corporation, US

Adapted from Prasad et al. 2017 (Prasad R, Bhattacharyya A, Nguyen QD (2017) Nanotechnology in sustainable agriculture: recent developments, challenges, and perspectives. Frontiers in Microbiology 8:1014)

as more efficient as they have several advantages over conventional pesticides such as (1) controlled release of active ingredients, (2) low dosage requirement, (3) protection from degradation and leaching, (4) reduced phytotoxicity, (5) enhanced bioavailability, etc. (Chhipa 2017a, b). Nanopesticides slowly release the active ingredients to kill the target pests without harming the nontargeted species and the soil matrices in the process. Moreover, the chemicals present in nanopesticides are environmentally safe and less toxic as compared to commercial pesticides. Till date, several types of nanopesticides are developed like nanogel, nanocapsules, nanoemulsion, and nanospheres. Pesticides in the form of nanoparticles are easily absorbed by the plants and delivered to the target area of infestation. Delivery that is site-specific and ingredients that are environment friendly, unified in a nanoranged particle, are a boon for plant protection (Nair et al. 2010). There are several reports that tested the effectiveness of nanopesticides against the plant pathogens. Zn nanoparticles showed antifungal activity against *Penicillium expansum*, *Aspergillus niger*, and different other microbes (He et al. 2011). Copper nanoparticles have been studied against gram-positive and gram-negative bacteria by Cioffi et al. (2005). Silver nanoparticles were also reported to be effective against plant pathogens like *Bipolaris sorokiniana*, *Colletotrichum species*, etc. (Mishra et al. 2014; Lamsal et al. 2011; Park et al. 2006). Several other examples of nanopesticides are listed in Table 10.3.

Table 10.3 Nanopesticides and their applications

Nanoparticles	Disease(s) and pathogen(s)	Observations	References
Silver nanoparticle	Powdery mildews of pumpkin by fungi	Complete inhibition of fungal disease at 3 ppm concentration	Park et al. (2006)
	Anthracnose in pepper by *Colletotrichum species*	Highest rate of inhibition of the growth of fungal hyphae and conidial germination was observed at 100 ppm of nanoparticles	Lamsal et al. (2011)
	R. solani, *S. sclerotiorum*, and *S. minor*	Inhibition of the hyphal growth resulting in plasmolysis and collapse of hyphae	Min et al. (2009)
	Fusarium oxysporum	Hyphae fragmentation and spores disruption	Gopinath and Velusamy (2013)
Titanium dioxide nanoparticles	Bacterial leaf spot of rose by *Xanthomonas* sp.	Complete inhibition of bacterial growth	Paret et al. (2013)
Zinc oxide nanoparticles	Conidiophores and conidia of *P. expansum* in plants	Prevents the development of conidiophores and conidia of *P. expansum*	He et al. (2011)
Copper nanoparticles	Bacterial blight of *Punica granatum*, caused by *X. axonopodis pv. punicae (Xap)*	Suppress Xap growth at 0.2 ppm	Mondal and Mani (2012)
Sulphur nanoparticles	*Erysiphe cichoracearum* (Powdery mildew fungi)	Inhibited conidial germination of *Erysiphe cichoracearum* (Powdery mildew fungi) at 1000 ppm	Gopal et al. (2012)
Chitosan Nanoparticles (CNPs)	*A. alternata*, *M. phaseolina*, and *R. solani*	Cu-CNPs inhibit pathogens growth at 0.1% followed by increase in seed germination percentage, seedling length, fresh weight, and dry weight	Saharan et al. (2013, 2015)

10.3.3 Nanoherbicides

High productivity in agriculture needs crop fields that remain free from weeds, as they take up major share of the provided fertilizers from the soil and deprive the crops from the essential nutrients resulting in low productivity. Apart from this, weeds also have other disadvantages too like damaging the health of soil, killing the insects and microbes that are beneficial for plant growth, etc. To minimize these adverse impacts, a new type of herbicide is introduced into the market, i.e. nanoherbicides. This new generation of herbicides are target specific in action and eco-friendly and do not affect the crop plants or soil microbial species. Their high penetration properties helps in eliminating the weeds before resistance could develop. They kill the target weeds without harming other species including the crop plants. Moreover, the nanosize of the herbicides or weedicides makes them penetrate through the soil easily and attack the weed seeds, killing them at the initial stage (Abigail et al. 2016; Abigail and Chidambaram 2017). Surfactants, organic polymers, and mineral nanoparticles are some of the basic components in the formulation of nanoweedicides (Schnoor et al. 2018). Although, nanoformulations of herbicides are quite effective, further research is still required for improvement in their release behaviour and characteristics.

10.3.4 Nanosensors

Nanosensors are the advanced tools which are of much significance in agriculture system and food production. These sensors are basically analytical devices which have at least one sensing dimension within 100 nm. They can monitor physico-chemical properties of the places which are otherwise difficult to reach (Fraceto et al. 2016). The applications of nanosensors in modern agriculture turned the conventional agricultural practices into smart-agriculture, which is an environmental-friendly approach for meeting the goals of sustainable agriculture. They are found to be more efficient as compared to the conventional biosensors in terms of speed, sensitivity, selectivity, precision, accuracy, detection limits, cost, etc. Nanosensors can be used for various functions such as (1) measurement of soil parameters such as pH, temperature, and moisture content, (2) nutritional status of soil, (3) crop health, crop monitoring, and time of crop harvesting, (4) diagnosis of diseases caused by pests and pathogens (virus, bacteria, and fungi), (5) analysis of nutrients and pesticides in soil and crop, (6) prediction of nitrogen uptake, (7) estimation of soil water tension, (8) sensing food freshness, etc. (Joyner and Kumar 2015; Fraceto et al. 2016). Several metal nanomaterials have been utilized for nanosensor fabrication. These include gold nanoparticles, carbon nanotubes, quantum dots, nanocomposites with polymers etc. (Cesarino et al. 2012; Liu et al. 2012; Talarico et al. 2016; Zheng et al. 2011). Although nanosensors represent one of the emerging technologies devoted to the control of quality, bio/security, and safety in agriculture field as well as along the food supply chain, however, most of the analysis and detections on field are conducted using conventional methods as nanosensors are presently in the developing stage only (Tothill 2011). Thus, there is a huge scope of development in this field that could address the issue of sustainability.

10.4 Nanoparticles and Plant Disease Control

Nanoparticles of carbon, silver, silica, and alumino-silicates are mostly used in farm lands to control crop diseases. Silver nanoparticles are the most common antimicrobial agents that are frequently used to control the plant diseases due to their long history of strong inhibitory and bactericidal effects (Gajbhiye et al. 2009; Gopinath and Velusamy 2013; Lee et al. 2013; Swamy et al. 2015). Nanosilver has high surface area in comparison to the bulk silver, showing excellent antimicrobial activities against various pathogens. Since silver displays enormous modes of action in nanorange inhibiting the growth of microorganisms, it is found be a suitable agent in agricultural applications (Lee et al. 2013). Moreover, silver nanoparticles are reported to destroy the plant pathogens producing no extra threat to the environment as compared to commercial fungicides and other germicides available in the market (Park et al. 2006). Nanosilver affects the biochemical processes and metabolic activities of microorganisms and disrupts their plasma membrane leading to the decrease of infection in diseased plants (Pal et al. 2007). The silver nanoparticles also prevent the expression of ATP in associated proteins of microorganisms and destroy them completely (Yamanaka et al. 2005). Though the exact mechanism of biomolecule inhibition, by silver nanoparticles, is yet to be fully understood, their application in diseased plants is so far an effective and eco-friendly approach to control crop diseases (Swamy et al. 2015). Likewise, zinc oxide (ZnO) and magnesium oxide (MgO) nanoparticles can also be used as antibacterial and anti-odour agents. Their properties like dispensability, optical transparency, and smoothness make them an attractive antibacterial ingredient in varied products. They have also been recognised as an antibacterial preservative for food and wood products (Aruoja et al. 2009). Polymer-coated nanoparticles penetrate the cuticles of target weeds slowly and steadily in a controlled manner and kill the unwanted weeds effectively. The application of nanoparticles for prevention and control of plant diseases is more acceptable than the use of synthetic pesticides and herbicides as they do not harm the host crop plant and minimizes water and soil pollution (Barik et al. 2008).

10.5 Smart Delivery Systems Using Nanotechnology for Advanced Farming

Quick delivery of fertilizers and pesticides in a controlled and precise manner is defined as 'smart delivery systems' in advanced farming. It majorly includes controlled and timely release of nutrients, spatially targeted region, self-regulatory action, and other multifunctional properties. It can regulate the delivery of different nutrients, bioactive molecules, and pesticides, allowing judicious use of agroparticles in farming practices. The following discussion gives a brief idea on smart delivery actions of nanoagroparticles.

10.5.1 Delivery of Fertilizers

Nanofertilizers should be prepared in a way that it could possess all the desired traits within it, like high solubility, stability, efficiency, timely release of nutrients, target-specific action, and less ecotoxicity, easy mode of delivery, and safe disposal (Green and Beestman 2007). In target-specific nanofertilizers, nutrients are incorporated with nanoparticles to deliver them at specific sites. There are various modes of nanofertilizer applications. Some of the nutrient-loading methods are absorption of

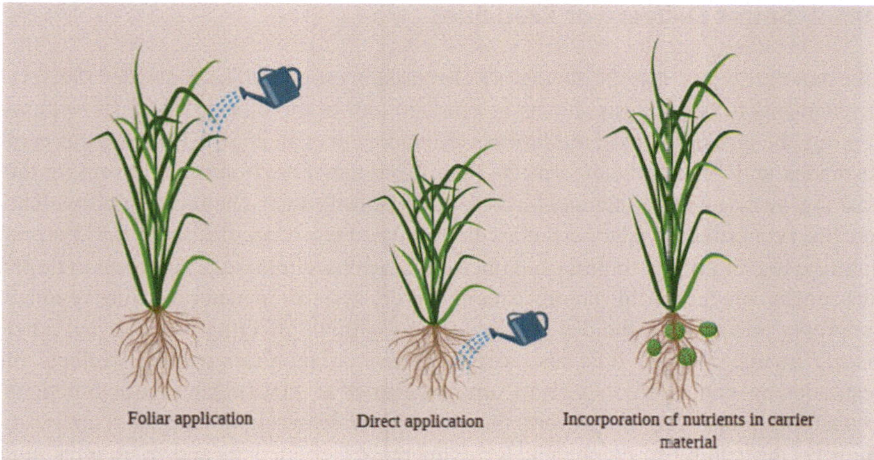

Fig. 10.4 Different modes of nanofertilizer application

required nutrients on surface of nanoparticles, attachment of nutrients on ligand-mediated nanoparticles, encapsulation of nutrients in nanopolymeric shell, entrapment of polymer nanoparticle in nutrients, and spraying topically on plants (Fig. 10.4). A brief description about the applications of nanofertilizer is as follows:

(a) *Direct application*: In direct mode of application, nanofertilizers are directly applied to the soil near the roots of the plants. The nanosized fertilizers are either applied in single nutrient form or in a composite form (Sekhon 2014). The composite fertilizer is the combination of two or more nutrients having nano-scale dimension. The nanofertilizers are applied in definite quantity to the soil according to the requirement of soil and the plants grown.
(b) *Incorporation of nanoparticles in carrier material*: Nanonutrients can be encapsulated in carrier material to provide nutrients to the plants in a requisite amount. The carrier material is a thin film of organic or inorganic polymer that controls the release of nutrients (De Oliveira et al. 2014). This is the best way of delivering nutrients to the plants without the loss of fertilizer. It reduces leaching of nutrients and prevents water pollution.
(c) *Foliar spray*: Nanofertilizers in the form of liquid can be sprayed topically over the plants and not to the soil. When certain plants suffer from nutrient deficiencies, the foliar spray helps in providing nutrients directly to the leaves for recovering of plants quickly. The leaves absorb the nutrients promptly through their stomata and epidermis cells. As compared to soil applications, foliar spray helps the plants to take up the nutrients in higher percentages and at a faster rate. Studies show that foliar spray is as much as 95% effective as an application technique (Prasad et al. 2012). Secondary nutrients like calcium, magnesium, zinc, iron, manganese, etc. are provided through foliar spray technique to plants, to recover them faster from these nutrient deficiencies (Subramanian and Tarafdar 2011).

10.5.2 Smart Delivery of Pesticides

Like nanofertilizers, nanopesticides are also used in crop field for controlled delivery of chemicals to target pests. It acts as efficient tool in monitoring the pests without harming the environment in the process (Khandelwal et al. 2016). The integration of pesticides with nanoparticles protects them from harsh environment preventing the washing away of toxic chemicals, thereby increasing their chemical stability. The specific properties of nanopesticides that make them more desirable are thermal stability, higher affinity to pests and microbes, increased release of chemical to target sites, and its biodegradable nature. Smart delivery systems in nanopesticides regulate the proper release of pesticides into target sites without affecting the crop and other insects in the vicinity. It is also very effective in monitoring the ill effects of pesticides on nontargeted species (Simon-Delso et al. 2017). The major functions of nanopesticides are to prevent the pesticide release in nontargeted areas, to improve the solubility of chemicals for increased efficiency, and to modify and control the functions of active ingredients incorporated in nanopesticides according to the surrounding environment. The insects secretes varied cuticular lipids to create barrier against the pesticides and protect them from death, however, there are some smart nanopesticides, like nanosilica incorporated pesticides, that can adsorb the cuticular lipids through physisorption, causing the death of the insects (Sahayaraj 2014).

10.6 Nanotechnology in Seed Science and in Advanced Formulation of Agroparticles

A seed is the most important entity that determines the productivity of any crop. Conventionally, seeds were first tested in the laboratories for germination and then distributed to farmers for sowing in their respective fields. Since seed testing was done in well-equipped laboratories or agricultural departments, it could be hardly reproduced in the fields due to their moisture deficiencies under rainfed conditions. In the present-day agriculture, nanoscience has evolved to revolutionize this field and play an important role in enhancing crop production. As stated before, many agroparticles incorporated engineered nanoparticles to enhance their efficiency in farm lands such as nanofertilizers, nanopesticides, nanosensors, etc. thereby making agriculture sustainable (Mahakham et al. 2017). In a similar fashion, researchers combined seed science with nanotechnology to increase the efficiency of seeds and elevate the productivity in the long run.

Zheng et al. conducted a study in 2005 to compare the effect of nano-TiO_2 and non nano-TiO_2 on the germination and growth of spinach seed. They concluded that the nano-TiO_2 treatment in the concentration of 0.25–4% enhanced the germination rate of the aged seed and chlorophyll formation and improved ribulose bisphosphate carboxylase/oxygenase activity, boosting its vigour. On the other hand, the effect of non-nano-TiO_2 was not effective for seedling. Another study conducted on nano-priming of seed improved the crop yield as well as its tolerance capacity against environmental stress. Nano-priming is a useful technique employed to increase the quality of low vigour seeds and make them resistant against environmental stress (Ibrahim 2016; Chen and Arora 2013). Raja et al. (2019) conducted an in vitro study using *Vigna mungo* (black gram) as a model plant, treated with several concentrations of biogenic ZnO and Cu nanoparticles. The report suggests that phyto (*C. sativum* and *N. oleander*)-mediated ZnO and Cu nanoparticles significantly improved the seed quality of black gram. Jasmine rice (*Oryza sativa* L. cv. KDML 105), a nutritious rice species, is found to be very sensitive to different environmental stress conditions, resulting in reduction of crop yield. To enhance the germinating capacity and starch metabolism of these seeds, silver NPs synthesized by extracts of kaffir lime leaves were applied as nano-priming agent over the seeds. Ag NPs increase the expression of aquaporin genes, facilitate the diffusion of water and H_2O_2, and stimulate the process of seed germination (Mahakham et al. 2017). Similarly, Anandaraj and Natarajan (2017) modified the onion seeds with zinc oxide (ZnO), silver (Ag), copper oxide (CuO), and titanium oxide (TiO_2) nanoparticles with different concentrations, ranging from 750, 1000, and 1250 to 1500 mg kg^{-1} via dry dressing techniques. The dose of 1000 mg kg^{-1} of NPs showed a better performance than the control in accelerating the germination rate of seeds by 72% and thereby increasing the shoot length by 7.5 cm, root length by 6.4 cm, and vigour index by 998. Some other nanoparticles used in seed coatings are displayed in Table 10.4.

10.7 Future Perspectives

Due to unique physicochemical properties of nanoagroparticles, their application for plant growth and protection is consistently being explored. Several collaborative and funded projects are focused on developing safer nanofertilizers and nanopesticides for effective crop yields with an environment-friendly approach. Integration of nanotechnology in agricultural domain is still at a primary stage but is evolving quite rapidly. However, it is crucial to understand several modes of action of different nanoagroparticles before making their use a general practice on farmlands. Moreover, the use of nanoagroparticles should be carried out according to the regulatory frameworks to monitor their release into the environment.

10.8 Conclusion

Since the demand of food is growing at a faster rate, it is crucial to increase the productivity of crops. For the enhanced crop production, it is important to find an advanced solution that can accelerate the productivity and does not pollute the environment. The incorporation of nanotechnology in agriculture has provided the ultimate solution required for the development of agricultural sector. The nanoagroparticles have replaced the conventional agroparticles as they are target specific, control the release of chemicals, do not harm the nontargeted species

Table 10.4 List of nanoparticles used in seed coatings

Plant species	Nanoparticles	Observations	References
Thymus kotschyanus	Silica and silver (Ag) NPs	Improved all early growth characteristics of plant seedlings AgNPs (20% concentration) show increased germination rate than silica and control treatments	Khalaki et al. (2016)
Agropyron elongatum L. (wheatgrass)	SiO_2 NPs	Application of SiO_2 NPs significantly enhanced the germination of seed from 58% to 86.3 and 85.7% in 40 and 60 mg L^{-1} respectively	Azimi et al. (2014)
Lycopersicum esculentum L. (Tomato), *Allium cepa* L. (Onion), and *Raphanus sativus* L. (Radish)	TiO_2 NPs	In tomato and onion, highest germination % (100% and 30%, respectively) was observed at 100 mgL^{-1} of NPs and in radish, 100% germination was observed at 400 mgL^{-1}. In the greenhouse, tall seedlings were observed after exposure to 400 and 200 mg L^{-1} for tomato and onion, respectively, and 400 and 100 mg L^{-1} for radish	Haghighi and da Silva (2014)
Solanum lycopersicum (tomato) and *Vigna radiata* (mungbean)	Activated carbon-based TiO_2 (AC-TiO_2) nano-composite	Enhanced the seed germination in both plant species	Singh et al. (2016)
Trigonella foenum-graecum L. (Fenugreek seed)	AgNPs	Improved percent seed germination, mean germination time, seed germination index, seed vigour index, seedling fresh weight and dry weight	Hojjat and Hojjat (2015)
Solanum lycopersicum (Tomato)	Carbon-based nanomaterials (CBNMs) $C_{60}(OH)_{20}$ (fullerol) and multiwalled nanotubes (MWNTs)	The germination % and seedling length and weight were enhanced in the presence of MWNTs. CBNMs do not alter the germination percentage	Ratnikova et al. (2015)
Cajanus cajan (pigeon pea)	Zn and Fe NPs	Among both the treatments, seed coated with 750 ppm of Zn NPs showed significant	Korishettar et al. (2016)

(continued)

Table 10.4 (continued)

Plant species	Nanoparticles	Observations	References
		elevation in seed germination, seedling length and dry weight, field emergence, seedling vigour index. The dehydrogenase activity and amylase activity also levels up. (25.67 mm)	

including soil microbes and crop plants, and last but not the least are eco-friendly in nature. Since the nanofertilizers release the nutrients in a slow and steady manner, they also reduce high cost of continuous application of fertilizers. Moreover, the nanofertilizers are highly efficient to control the active ingredients according to the environmental stress and requirement of the soil. Another nanotechnology application is the development of nanosensors. Nanosensors are increasingly helping to sense the plant diseases and kill them keeping the plants healthy. These smart nanosensors are programmed to sense the metabolism of plants, detection of diseases, environmental stress, etc. and stimulate actions to protect the plants accordingly. Though these unique properties of nano-based agroparticles have made them a potential candidate, nanotechnology is still at its nascent stage in agricultural sector. The detailed study of the environmental pathways of these advanced species is still unknown and is getting explored. The full potential of nanoagroparticles can only be exploited, when their adverse impacts on ecosystems are completely understood and their long-term effects are well investigated.

References

Abigail MEA, Chidambaram R (2017) Nanotechnology in herbicide resistance. In: Seehra MS (ed) Nanostructured materials: fabrication to applications. IntechOpen, London, p 207. https://doi.org/10.5772/intechopen.68355

Abigail MEA, Samuel M, Chidambaram R (2016) Application of rice husk nanosorbents containing 2,4-dichlorophenoxyacetic acid herbicide to control weeds and reduce leaching from soil. J Taiwan Inst Chem Eng 63:318–326

Adhikari T, Sarkar D, Mashayekhi H, Xing B (2016) Growth and enzymatic activity of maize (*Zea mays* L.) plant: solution culture test for copper dioxide nano particles. J Plant Nutr 39(1):99–115

Alfadul MS, Altahir OA, Khan AM (2017) Application of nanotechnology in the field of food production. Acad J Sci Res 5(7):143–154

Anandaraj K, Natarajan N (2017) Effect of nanoparticles for seed quality enhancement in onion [*Allium cepa* (Linn) cv. CO (On)]5. Int J Curr Microbiol Appl Sci 6(11):3714–3724

Aruoja V, Dubourguier HC, Kasemets K, Kahru A (2009) Toxicity of nanoparticles of CuO, ZnO and TiO_2 to microalgae *Pseudokirchneriella subcapitata*. Sci Total Environ 407:1461

Azimi R, Borzelabad MJ, Feizi H, Azimi A (2014) Interaction of SiO_2 nanoparticles with seed prechilling on germination and early seedling growth of tall wheatgrass (*Agropyron elongatum* L.). Polish J Chem Technol 16(3):25–29

Aziz N, Faraz M, Pandey R, Sakir M, Fatma T, Varma A (2015) Facile algae-derived route to biogenic silver nanoparticles: synthesis, antibacterial and photocatalytic properties. Langmuir 31:11605–11612

Barik TK, Sahu B, Swain V (2008) Nanosilica-from medicine to pest control. Parasit Res 103:253–258

Burman U, Saini M, Kumar P (2013) Effect of zinc oxide nanoparticles on growth and antioxidant system of chickpea seedlings. Toxicol Environ Chem 95(4):605–612

Campbell BM, Thornton P, Zougmoré R, Van AP, Lipper L (2014) Sustainable intensification: what is its role in climate smart agriculture? Curr Opin Environ Sustain 8:39–43

Cesarino I, Moraes FC, Lanza MR, Machado SA (2012) Electrochemical detection of carbamate pesticides in fruit and vegetables with a biosensor based on acetylcholinesterase immobilised on a composite of polyaniline–carbon nanotubes. Food Chem 135(3):873–879

Chen K, Arora R (2013) Priming memory invokes seed stress-tolerance. Environ Exp Bot 94:33–45

Chen H, Yada R (2011) Nanotechnologies in agriculture: new tools for sustainable development. Trend Food Sci Technol 22:585–594

Chhipa H (2017a) Nanofertilizers and nanopesticides for agriculture. Environ Chem Lett 15 (1):15–22

Chhipa H (2017b) Nanopesticide: current status and future possibilities. Agric Res Technol 5 (1):1–4

Cioffi N, Torsi L, Ditaranto N, Tantillo G, Ghibelli L, Sabbatini L (2005) Copper nanoparticle/ polymer composites with antifungal and bacteriostatic properties. Chem Mater 17:5255–5262

Corradini E, Moura MR, Mattoso LHC (2010) A preliminary study of the incorporation of NPK fertilizer into chitosan nanoparticles. Exp Polym Lett 4(8):509–515

De Oliveira JL, Campos EVR, Bakshi M, Abhilash PC, Fraceto LF (2014) Application of nanotechnology for the encapsulation of botanical insecticides for sustainable agriculture: prospects and promises. Biotechnol Adv 32(8):1550–1561

Delfani M, Firouzabadi MB, Farrokhi N, Makarian H (2014) Some physiological responses of black-eyed pea to iron and magnesium nanofertilizers. Commun Soil Sci Plant Anal 45:11

Fraceto LF, Grillo R, deMedeiros GA, Scognamiglio V, Rea G, Bartolucci C (2016) Nanotechnology in agriculture: which innovation potential does it have? Front Environ Sci 4:20

Gajbhiye M, Kesharwani J, Ingle A, Gade A, Rai M (2009) Fungus mediated synthesis of silver nanoparticles and their activity against pathogenic fungi in combination with fluconazole. Nanomedicine 5:382–386

Gopal M, Kumar R, Goswami A (2012) Nano-pesticides—a recent approach for pest control. J Plant Prot Sci 4(2):1–7

Gopinath V, Velusamy P (2013) Extracellular biosynthesis of silver nanoparticles using Bacillus sp. GP-23 and evaluation of their antifungal activity towards *Fusarium oxysporum*. Spectrochim Acta Part A Mol Biomol Spect 106:170–174

Green JM, Beestman GB (2007) Recently patented and commercialized formulation and adjuvant technology. Crop Prot 26(3):320–327

Gruere GP (2012) Implications of nanotechnology growth in food and agriculture in OECD countries. Food Policy 37(2):191–198

Haghighi M, da Silva JAT (2014) The effect of N-TiO_2 on tomato, onion, and radish seed germination. J Crop Sci Biotechnol 17(4):221–227

He L, Liu Y, Mustapha A, Lin M (2011) Antifungal activity of zinc oxide nanoparticles against *Botrytis cinerea* and *Penicillium expansum*. Microbiol Res 166:207–215

Hojjat SS, Hojjat H (2015) Effect of nano silver on seed germination and seedling growth in fenugreek seed. Int J Food Eng 1(2):106–110

Huang J, Li Q, Sun D, Lu Y, Su Y, Yang X, Wang H, Wang Y, Shao W, He N, Hong J, Chen C (2007) Biosynthesis of silver and gold nanoparticles by novel sundried *Cinnamomum camphora* leaf. Nanotechnol 18:105104

Ibrahim EA (2016) Seed priming to alleviate salinity stress in germinating seeds. J Plant Physiol 192:38–46

Joyner JJ, Kumar DV (2015) Nanosensors and their applications in food analysis: a review. Inter J Sci Technol 3(4):80

Kah M (2015) Nanopesticides and nanofertilizers: emerging contaminants or opportunities for risk mitigation? Front Chem 3:64

Kandasamy S, Prema RS (2015) Methods of synthesis of nano particles and its applications. J Chem Pharm Res 7:278–285

Khalaki AM, Ghorbani A, Moameri M (2016) Effects of silica and silver nanoparticles on seed germination traits of *Thymus kotschyanus* in laboratory conditions. J Rangeland Sci 6 (3):221–231

Khandelwal N, Barbole RS, Banerjee SS, Chate GP, Biradar AV, Khandare JJ, Giri AP (2016) Budding trends in integrated pest management using advanced micro-and nano-materials: challenges and perspectives. J Environ Manag 184:157–169

Korishettar P, Vasudevan SN, Shakuntala NM, Doddagoudar SR, Hiregoudar S, Kisan B (2016) Seed polymer coating with Zn and Fe nanoparticles: an innovative seed quality enhancement technique in pigeonpea. J Appl Nat Sci 8(1):445–450

Kottegoda N, Munaweera I, Madusanka N, Karunaratne V (2011) A green slow-release fertilizer composition based on urea-modified hydroxyapatite nanoparticles encapsulated wood. Curr Sci 101:73–78

Kumar M, Bauddh K, Sainger M, Sainger PA, Singh JS, Singh RP (2012) Increase in growth, productivity and nutritional status of rice (*Oryza sativa* L. c.v. Basmati) and enrichment in soil fertility applied with an organic matrix entrapped urea. J Crop Sci Biotechnol 15(2):137–144

Kumar S, Bauddh K, Barman SC, Singh RP (2014a) Amendments of microbial biofertilizers and organic substances reduces requirement of urea and DAP with enhanced nutrient availability and productivity of wheat (*Triticum aestivum* L.). Ecol Eng 71:432–437

Kumar S, Bauddh K, Barman SC, Singh RP (2014b) Organic matrix entrapped bio-fertilizers increase growth, productivity and yield of *Triticum aestivum* L. and mobilization of NO3-, NO2-, NH4+ and PO4-3 from soil to plant leaves. J Agric Sci Technol 16:315–329

Kumar M, Bauddh K, Sainger M, Sainger PA, Singh RP (2014c) Increase in growth, productivity and nutritional status of wheat (*Triticum aestivum* L.) and enrichment in soil microbial population applied with biofertilizers entrapped with organic matrix. J Plant Nutr 38:260–276

Kumar M, Bauddh K, Sainger M, Sainger PA, Singh RP (2015) Enhancing efficacy of Azotobactor and Bacillus entrapping in organic matrix for rice cultivation. Agroecol Sustain Food Syst 39:907–923

Lamsal K, Kim SW, Jung JH, Kim YS, Kim KS, Lee YS (2011) Application of silver nanoparticles for the control of colletotrichum species in vitro and pepper anthracnose disease in field. Mycobiology 39(3):194–199

Lee KJ, Park SH, Govarthanan M, Hwang PH, Seo YS, Cho M, Oh BT (2013) Synthesis of silver nanoparticles using cow milk and their antifungal activity against phytopathogens. Mater Lett 105:128–131

Liu R, Lal R (2015) Potentials of engineered nanoparticles as fertilizers for increasing agronomic productions. Sci Total Environ 514:131–139

Liu D, Chen W, Wei J, Li X, Wang Z, Jiang X (2012) A highly sensitive, dual readout assay based on gold nanoparticles for organophosphorus and carbamate pesticides. Anal Chem 84 (9):4185–4191

Mahakham W, Sarmah AK, Maensiri S, Theerakulpisut P (2017) Nanopriming technology for enhancing germination and starch metabolism of aged rice seeds using phytosynthesized silver nanoparticles. Sci Rep 7(1):8263

Mahmoud AWM, Taha SS (2018) Main sulphur content in essential oil of *Eruca sativa* as affected by nano iron and nano zinc mixed with organic manure. Agriculture 64(2):65–79

Milani N, McLaughlin MJ, Stacey SP, Kirby JK, Hettiarachchi GM, Beak DG, Cornelis G (2012) Dissolution kinetics of macronutrient fertilizers coated with manufactured zinc oxide nanoparticles. J Agric Food Chem 60(16):3991–3998

Min JS, Kim KS, Kim SW, Jung JH, Lamsal K, Kim SB, Jung M, Lee YS (2009) Effects of colloidal silver nanoparticles on sclerotium-forming phytopathogenic fungi. Plant Pathol J 25 (4):376–380

Mishra S, Singh BR, Singh A, Keswani C, Naqvi AH, Singh HB (2014) Biofabricated silver nanoparticles act as a strong fungicide against Bipolaris sorokiniana causing spot blotch disease in wheat. PLoS One 9(5):e97881

Mondal KK, Mani C (2012) Investigation of the antibacterial properties of nanocopper against *Xanthomonas axonopodis* pv. punicae, the incitant of pomegranate bacterial blight. Ann Microbiol 62:889–893

Mukhopadhyay SS (2014) Nanotechnology in agriculture: prospects and constraints. Nanotechnol Sci Appl 7:63

Nagula S, Usha PB (2016) Application of nanotechnology in soil and plant system with special reference to nanofertilizers. Adv Life Sci 5(14):5544–5548

Nair R, Varghese SH, Nair BG, Maekawa T, Yoshida Y, Kumar DS (2010) Nanoparticulate material delivery to plants. Plant Sci 179:154–163

Pal S, Tak YK, Song JM (2007) Does the antibacterial activity of silver nanoparticles depend on the shape of the nanoparticle? A study of the gram-negative bacterium *Escherichia coli*. App Environ Microbiol 73:1712–1720

Paret ML, Vallad GE, Averett DR, Jones JB, Olson SM (2013) Photocatalysis: effect of light-activated nanoscale formulations of TiO_2 on *Xanthomonas perforans* and control of bacterial spot of tomato. Phytopathology 103:228–236

Park HJ, Kim SH, Kim HJ, Choi SH (2006) A new composition of nanosized silica-silver for control of various plant diseases. Plant Pathol J 22(3):295–302

Pradhan S, Patra P, Das S, Chandra S, Mitra S, Dey KK, Akbar S, Palit P, Goswami A (2013) Photochemical modulation of biosafe manganese nanoparticles on *Vigna radiata*: a detailed molecular, biochemical, and biophysical study. Environ Sci Technol 47:9

Prasad TNVKV, Sudhakar P, Sreenivasulu Y, Latha P, Munaswamy V, Reddy KR, Sreeprasad TS, Sajanlal PR, Pradeep T (2012) Effect of nanoscale zinc oxide particles on the germination, growth and yield of peanut. J Plant Nutr 35(6):905–927

Rai A, Kumar S, Bauddh K, Singh N, Singh RP (2017) Improvement in growth and alkaloid content of *Rauwolfia serpentina* on application of organic matrix entrapped biofertilizers (*Azotobacter chroococcum, Azospirillum brasilense and Pseudomonas putida*). J Plant Nutr 40 (16):2237–2247

Raja K, Sowmya R, Sudhagar R, Moorthy PS, Govindaraju K, Subramanian KS (2019) Biogenic ZnO and Cu nanoparticles to improve seed germination quality in blackgram (*Vigna mungo*). Mater Lett 235:164–167

Ratnikova TA, Podila R, Rao A M, Taylor AG (2015) Tomato seed coat permeability to selected carbon nanomaterials and enhancement of germination and seedling growth. Sci World J 2015:1–9

Saharan V, Mehrotra A, Khatik R, Rawal P, Sharma S, Pal A (2013) Synthesis of chitosan based nanoparticles and their in vitro evaluation against phytopathogenic fungi. Int J biol Macromole 62:677–683

Saharan V, Sharma G, Yadav M, Choudhary MK, Sharma SS, Pal A, Raliya R, Biswas P (2015) Synthesis and in vitro antifungal efficacy of Cu–chitosan nanoparticles against pathogenic fungi of tomato. Int J Biol Macromol 75:346–353

Sahayaraj K (2014) Nanotechnology and plant biopesticides: an overview. In Advances in Plant Biopesticides (pp. 279-293). Springer, New Delhi

Schnoor B, Elhendawy A, Joseph S, Putman M, Chacón-Cerdas R, Flores-Mora D, Bravo-Moraga F, Gonzalez-Nilo F, Salvador-Morales C (2018) Engineering atrazine loaded poly (lactic-co-glycolic acid) nanoparticles to ameliorate environmental challenges. J Agri Food Chem 66 (30):7889–7898

Sekhon BS (2014) Nanotechnology in agri-food production: an overview. Nanotechnol Sci Appl 7:31–53

Sertova MN (2015) Application of nanotechnology in detection of mycotoxins and in agricultural sector. J Cent Eur Agric 16(2):117–130

Shah V, Belozerova I (2009) Influence of metal nanoparticles on the soil microbial community and germination of lettuce seeds. Water Air Soil Pollut 197:143–148

Simon-Delso N, San Martin G, Bruneau E, Delcourt C, Hautier L (2017) The challenges of predicting pesticide exposure of honey bees at landscape level. Sci Rep 7:3801

Singh RP, Sainger M, Bauddh K, Senger RS, Jaiwal PK (2010) Sustained nutrient supply reduced nutrient loss and high plant productivity with slow release fertilizers. In: Senger RS, Sharma AK (eds) Stable food production and sustainable agriculture. Studium Press (India), pp 62–79

Singh P, Singh R, Borthakur A, Srivastava P, Srivastava N, Tiwary D, Mishra PK (2016) Effect of nanoscale TiO_2-activated carbon composite on *Solanum lycopersicum* (L.) and *Vigna radiata* (L.) seeds germination. Energy Ecol Environ 1(3):131–140

Subramanian KS, Tarafdar JC (2011) Prospects of nanotechnology in Indian farming. J Agric Sci 81(10):887–893

Swamy MK, Akhtar MS, Mohanty SK, Sinniah UR (2015) Synthesis and characterization of silver nanoparticles using fruit extract of Momordica cymbalaria and assessment of their in vitro antimicrobial, antioxidant and cytotoxicity activities. Spectrochim Acta Part A: Mol Biomol Spectro 151:939–944

Talarico D, Arduini F, Amine A, Cacciotti I, Moscone D, Palleschi G (2016) Screen-printed electrode modified with carbon black and chitosan: a novel platform for acetylcholinesterase biosensor development. Anal Bioanal Chem 408(26):7299–7309

Taran NY, Gonchar OM, Lopatko KG, Batsmanova LM, Patyka MV, Volkogon MV (2014) The effect of colloidal solution of molybdenum nanoparticles on the microbial composition in rhizosphere of *Cicer arietinum* L. Nanoscale Res Lett 9:289

Thirunavukkarasu M, Subramanian KS, Kannan P, Balaji T (2018) Response of nano-sulphur to the groundnut. IJCS 6(3):2067–2072

Tothill EI (2011) Biosensors and Nanomaterials and their application for Mycotoxin determination. World Mycotoxin J 4:351–374

Upadhyaya H, Shome S, Tewari S, Bhattacharya MK, Panda SK (2015) Effect of Zn nano-particles on growth responses of rice, Nano technology – novel perspectives and prospects. McGraw Hill education, New Delhi, pp 508–512

United States Department of Agriculture (2012) Nanoscale science and engineering for agriculture and United States Department of Agriculture. Census of agriculture 2012. https://www.agcensus.usda.gov/Publications/2012/Full_Report/Volume_1,_Chapter_1_US/usv1.pdf. Accessed 20 Nov 2016

Velten S, Leventon J, Jager N, Newig J (2015) What is sustainable agriculture? A systematic review. Sustainability 7:7833–7865

Viswanathan S, Radecki J (2008) Nanomaterials in electrochemical biosensors for food analysis – a review. J Food Nutr Sci 58(2)

Wang L, Hu C, Shao L (2017) The antimicrobial activity of nanoparticles: present situation and prospects for the future. Int J Nanomed 12:1227–1249

Wanyika H, Gatebe E, Kioni P, Tang Z, Gao Y (2012) Mesoporous silica nanoparticles carrier for urea: potential applications in agrochemical delivery systems. J Nanosci Nanotechnol 12 (3):2221–2228

Wilson MA, Tran NH, Milev AS, Kannangara GK, Volk H, Lu GM (2008) Nanomaterials in soils. Geoderma 146(1–2):291–302

Yamanaka M, Hara K, Kudo J (2005) Bactericidal actions of a silver ion solution on *Escherichia coli*, studied by energy-filtering transmission electron microscopy and proteomic analysis. Appl Environ Microbiol 71:7589–7593

Zafar H, Ali A, Ali JS, Haq IU, Zia M (2016) Effect of ZnO nanoparticles on *Brassica nigra* seedlings and stem explants: growth dynamics and antioxidative response. Front Plant Sci 7:535

Zamir D (2001) Improving plant breeding with exotic genetic libraries. Nat Rev Genet 2(12):983

Zheng L, Hong F, Lu S, Liu C (2005) Effect of nano-TiO_2 on strength of naturally aged seeds and growth of spinach. Biol Trace Elem Res 104:83–91

Zheng Z, Li X, Dai Z, Liu S, Tang Z (2011) Detection of mixed organophosphorus pesticides in real samples using quantum dots/bi-enzyme assembly multilayers. J Mater Chem 21 (42):16955–16962

Agri-Nanotechnology for Sustainable Agriculture

11

Garima Pandey

Contents

Abstract

An interdisciplinary field like nanotechnology is exhaustively being explored for its benefits for the welfare of society. The encouraging outcomes in various fields paved a new scope for utilization of nanotechnology in the agri-sector also. The green revolution led to the uncontrolled use of agrochemicals, which on the one hand resulted in an increase in the productivity, but on the other hand has severe adverse effects on soil diversity and aquatic ecosystems and has negative impacts on the health of people growing and consuming these chemical-laden agriproducts. Nanotechnology emerges as a resourceful front in current agri-

G. Pandey (✉)
Department of Chemistry, SRM Institute of Science and Technology, Modinagar, India

K. Bauddh et al. (eds.), *Ecological and Practical Applications for Sustainable Agriculture*, https://doi.org/10.1007/978-981-15-3372-3_11

practices and is expected to become a key strength through new inventions. As a result, the amalgamation of nanotechnology with agriculture results in enhanced productivity, better pest and weed management, and reduced health implications to soil, water, and people, though a lot of safety, toxicity, risk assessment, and risk management issues still remain unsettled lagging behind the broad application of nanotechnology in agriculture. This chapter will talk about the green synthesis, applications of nanotechnology in agriculture and agro-economy, uncertainties, risks, and ethical concerns related to agri-nanotechnology.

Keywords

Green synthesis · Agro-economy · Risks and ethical concerns · Agri-nanotechnology

11.1 Introduction

The development of the agri-sector is an indispensable factor of economic growth in developing countries. Currently, the ever-increasing global population is leading to a decline in the ratio of demand and supply of agriproducts. To meet this issue, it is necessary to employ new tools in the agriculture sector (Abbas et al. 2016). The inclusion of biotechnology and nanotechnology in agricultural practices can play a vital role in increasing the production rate with improved food processing and packaging. Nanotechnology is extensively influencing the society through its vast applications in our day-to-day life, and from the year 2003, it has embarked into the agriculture and food industry, and over the span of the last few years, the researches in the area of agri-nanotechnology has skyrocketed (Agrawal and Rathore 2014). It influences almost all the areas of the agri-sector, including soil and plant health, irrigation management, water filtration, food processing, food packaging, and vector and pest management. Nanotechnology lets us construct nanoscale structures through modifications at the atomic level (Aouada and de Moura 2015). The objective of using nanomaterials and nanotechnology in the agri-sector is to increase the yield, manage the nutrient loss, and minimize the usage of chemicals, through pest and nutrient management. It is being approximated that by the year 2020, the worldwide economic impact of nanotechnology will be more than 3 trillion US dollars, employing around six million workers (Aziz et al. 2016). This involved people engaged in improving the production and promotion of nano-based goods, quality assessment, efficiency enhancement, and safety measurements of foodstuffs. Due to poor assessment and management of associated risks, most of the nano-based products used in food industry are not directly being induced in human food. Majority of them are the surface materials only contacting the food except the nano-oxide of iron and titanium which are being used as food color and food pigment (Baker et al. 2017).

11.2 Synthesis of Nanoparticles

The exceptional physical, chemical, mechanical, electronic, and optical characteristics of nanoparticles are continuously drawing the attention of scholars towards sustainable methods of their synthesis. To facilitate this, researchers are focusing on the green synthesis methods to lessen the environmental hazards (Baligar and Fageria 2015). The cost-efficiency, environmental affability, decreased toxicity, and rapid reaction rate are the characteristic features of green synthesis (Berekaa 2015; Bhagat et al. 2015). The use of noxious solvents, intense temperature and pressure, and endothermic nature of reactions are all non-eco-friendly and can cause grave impacts to the environmental equilibrium. In the green synthesis of nanoparticles, plant extracts and appropriate metal ions are being mixed in a fixed ratio under the required physical parameters (Camilli et al. 2014; Chhipa 2016; Cox et al. 2017). There are various chemical, physical, biological, and cross-synthesis routes to construct nanoparticles with a chosen character. The various procedures for synthesizing NPs are listed in Fig. 11.1. The extensively used physical techniques are extremely pricey, whereas the chemical methods are disadvantageous to the society and the environmental health. In addition to the environmental impacts, there are several other issues like insufficient growth rate and imprecise structures of synthesized nanoparticles and toxicity of synthesized substance (Das et al. 2015; De Matteis 2017; de Oliveira et al. 2014).

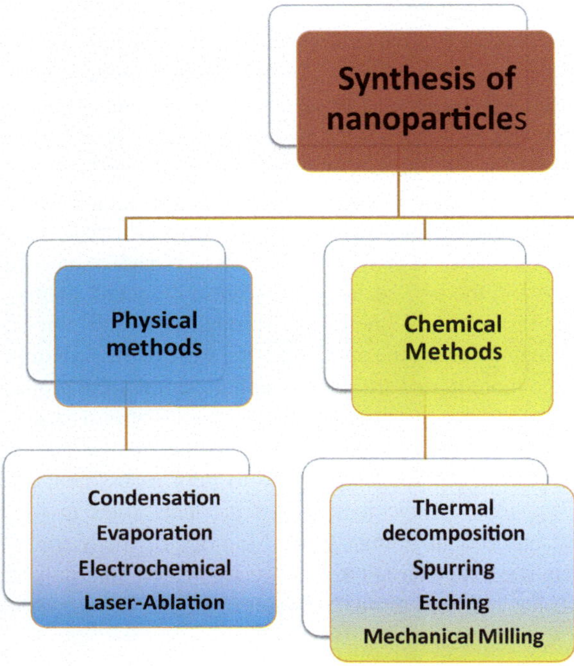

Fig. 11.1 Synthesis of nanoparticles

11.2.1 Green Synthesis of Nanoparticles

The green synthesis of nanomaterials avoids the creation of unnecessary or detrimental by-products by using sustainable, consistent, and environment-friendly synthesis routes (Ditta and Arshad 2016). The green synthesis of metal nanoparticles has been implemented to use diverse organic resources such as plant extracts, microorganisms, yeast, etc.

11.2.1.1 Nanoparticles from Plants

For the purpose of green synthesis of metallic nanoparticles, a broad biodiversity of plants and availability of various phytochemicals like terpenoids, ascorbic acids, aldehydes, flavones, ketones, amides, etc. in different plant extracts are being explored to a great extent (Ezz El-Din and Manjaiah 2017; Fraceto et al. 2016; Glenn and Florescu 2016). These phytochemicals have the capability to reduce metallic salts into their corresponding metallic nanoparticle. The green synthesis of nanoparticles is a single-step environment-friendly and economically efficient biological reduction process which involves comparatively little amount of energy to kick off in contrast to the chemical and physical methods (Hajirostamlo et al. 2015) (Table 11.1).

Table 11.1 Nanoparticles from plants (Pandey 2018c)

Name of plant	Metal nanoparticle
Macrotyloma uniflorum, Aloe vera, Sinapis arvensis, Artemisia nilagirica, Semen cassia, Zea mays, Nerium oleander, Magnolia kobus, Oryza sativa Helianthus annuus, Saccharum officinarum, Sorghum bicolour, Basella alba, Capsicum annuum var. aviculare, Medicago sativa, Callicarpa maingayi, Ficus benghalensis, Pinus eldarica, Sesbania drummondii, Pithophora oedogonia, Allium sativum, Hovenia dulcis	Silver nanoparticles
Zingiber officinale, Abelmoschus esculentus, Hamamelis, Mentha, Eucalyptus, Terminalia chebula, Morinda citrifolia, Anacardium occidentale, Jatropha waste, Stevia rebaudiana, Diospyros kaki	Gold nanoparticles
Carica papaya, Aloe vera, Green tea, Sorghum bran, Dodonaea viscosa, Azadirachta indica, Eucalyptus tereticornis, Sargassum muticum	Iron nanoparticles
Ricinus communis, Calotropis gigantean, Ocimum tenuiflorum, Ocimum sanctum, Nerium oleander, Tabernaemontana divaricata, Ficus religiosa, Punica granatum, Carica papaya	Copper nanoparticles
Ixora coccinea, Limonia acidissima, Parthenium hysterophorus, Pongamia pinnata, Trifolium pratense	Zinc nanoparticles
Solanum trilobatum, Moringa oleifera, Trigonella foenum-graecum, Nyctanthes arbor-tristis	Titanium nanoparticles

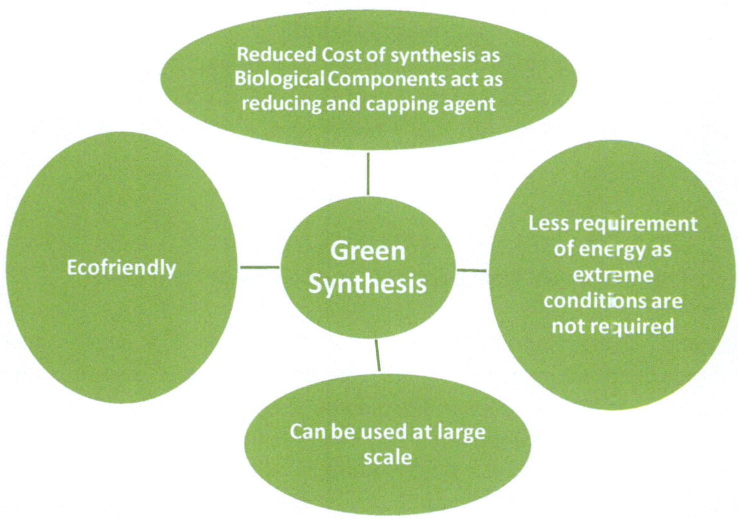

11.2.1.2 Nanoparticles from Microbes

Microbes do have the prospective of reducing metal ions into pure nanoparticles through the extracellular enzymes. Their unique metal affinity to bacteria and fungi make them useful for the green synthesis of metal nanoparticles. The microbial scheme of creating metal nanoparticles is relatively time-consuming in comparison with the plant extracts; still their easy handling, growth and culture methods, and the

Table 11.2 Nanoparticles from microorganism (Pandey 2018c)

Name of microorganism	Metal nanoparticle
Staphylococcus aureus, *Streptomyces naganishii*, *Trichoderma reesei*, *Brevibacterium casei*	Silver nanoparticles
Klebsiella pneumoniae, *Streptomyces viridogens*, *Nocardia farcinica*, *Rhodopseudomonas capsulata*, *Penicillium brevicompactum*, *Neurospora crassa*, *Thermomonospora* sp., *Aspergillus oryzae*, *Aspergillus clavatus*	Gold nanoparticles
E. coli, *Shewanella oneidensis*, *Pleurotus* sp., *Klebsiella oxytoca*	Iron nanoparticles
Stereum hirsutum, *Shewanella oneidensis*, *Streptomyces* sp., *Penicillium aurantiogriseum*, *Hypocrea lixii*	Copper nanoparticles
Candida albicans, *Lactobacillus*, *Streptomyces* sp.	Zinc nanoparticles
Aspergillus tubingensis, *Bacillus amyloliquefaciens*, *Fusarium oxysporum*	Titanium nanoparticles

low investment and lesser environmental risks make them beneficial for biosynthesis (Handford et al. 2015; Jayaseelan et al. 2011).

Fungi: Fungi are comparatively new in their utilization for synthesizing nanoparticles but are advantageous than bacteria because of their easy management, huge enzyme secretion, and hassle-free downstream process, though eukaryotic genetic alterations in fungi are not so easy (Karpachev et al. 2016).

Bacteria: Bacteria being prokaryotes easily go through genetic alteration for synthesizing nanoparticles. Nanoparticles of Ag, Au, Zn, Pb, Fe, and iron sulfide and Cd quantum dots are generally being synthesized by bacteria (Khan and Fulekar 2016) (Table 11.2).

11.3 Application of Nanotechnology in Agriculture

In developing countries agriculture is the spine of economic development. The primary test which the agri-sector faces is the uncertain climatic conditions, bioaccumulation, and rapid rate of industrialization along with the mounting demand for food to feed a projected population of six to nine billion people by the year 2050 (Kouhkan et al. 2019). The world of agriculture is facing a large array of challenge of stagnant harvest yield, nutrient deficit, declining soil organic matters, weather alterations, diminishing cultivable area, water scarcity, GMO resistance, and dearth of manual labor. The agriculture of India is yet not out of the effects of the green revolution reflecting an exponential rise in the consumption of fertilizers from 0.5 tons in the year 1960 to 23 million tons in 2008 (Kumar et al. 2015; Liu and Lal 2015; Mabe et al. 2017). Even though the crop productivity rose up to four times, still it led to decrease in the organic content of soil resulting into the stagnant yield of some crops. Due to the dependency on various natural factors like soil, water, climate, etc., the agricultural sector is very unpredictable (Mani and Mondal 2016;

Milewska-Hendel et al. 2016). Therefore, in order to meet the challenges of environmental sustainability with food managing the demands of food quality and quantity, it is very necessary to sense, record, manipulate, and store the consistent and accurate statistics of all abiotic and biotic components of the environment (Monreal et al. 2016). To fulfill this, it is very necessary to combine agriculture with technological advances and infusing the principles of nanotechnology in agricultural practices has shown to achieve enhanced outcomes. Nanotechnology has enormous potential and benefits in the agri-sector. Nanoparticles when infused in plants in precise amounts make various structural and physiological changes in them reflecting effective results in plant growth and seed germinations and rate of production (Morales-Díaz et al. 2017; Mufamadi and Sekhejane 2017). The benefits of nanotechnology comprise of pest management through nanoformulations of pesticides, amplifying the productivity through nanoencapsulation for sustained release of nutrients, developing pest resistance through nano-mediated gene transfer, creating nanobiosensors for precision farming, etc. Nanoparticles capture medicine, nutrient, and genes of concern and facilitate their sustained release for better solubility, assimilation, permanence, and diffusion of herbicides through plant cells (Nima et al. 2014).

11.3.1 Precision Farming

Precision agriculture or precision farming is the method that makes agricultural practices more precise and managed in terms of crop growth and raising of cattle. The chief constituent of precision farming is the inclusion of a broad range of objects such as GPS-based soil sampling and control systems, robotic systems, sensors, drones, and self-governing means of transportation, automatic software, and hardware tools. Precision farming is the ultimate aim to make the most of crop yield with minimum efforts (Nuruzzaman et al. 2016; Ozdemir and Kemerli 2016). For precision farming, the efficient crop yield is ensured by precise identification and location of agriculture issues with the help of technical aids like sensing devices, computers, and satellite positioning systems. With the help of technical interventions, the physiographic factors like soil health, nutrient status, and water level can be accurately assessed and managed to improve agricultural productivity (Pandey 2018a). Precision farming techniques are enabled with advanced monitoring tools and minuscule sensors help in monitoring soil conditions, crop growth, and agricultural waste management. The use of wireless nanosensors in various countries is being reported by the *Forbes Magazine*, e.g., small nanosensors are being used by Honeywell, a Tech R and D company using small nanosensors for monitoring requisition and expiry status of food articles in grocery stores in Minnesota. By using nanosensors, the requirement for herbicides, pesticides, and nutrients can precisely be estimated for every single corner of farm resulting into maximum efficiency with optimal inputs (Pandey 2018b). Smart delivery systems and sensors based on nanotechnology help in the proficient exploitation of natural agro-resources like soil, water, and the environment along with chemical nutrients and medicines through precision farming as well as help farmers detect weeds, pests, or environmental stress like drought. Once the issue is being sensed, nanosensors will make automatic adjustments for irrigation, or pesticide application. Nanosensors dispersed in the field can also sense the existence of a microorganism in soil and plants to manage them. Precision farming can as well facilitate management of agricultural wastes and thus help keep the environmental pollution to a bare minimum (Parisi et al. 2015).

11.3.1.1 Delivery of Fertilizers

A huge amount of fertilizers used to increase crop production has led to a lot of detrimental effects on the valuable microflora of the soil. Apart from this, a large amount of fertilizer gets wasted due to runoffs and cause pollution. Nanoencapsulated fertilizers can resolve this issue as they have the property of getting rapidly absorbed by plants completely. Nanoencapsulation facilitates the strong hold of nutrients and their sustained release and surface protection (Patra and Baek 2017). Inorganic fertilizers like diammonium phosphates, urea, etc. used to supplement the requirements of nitrogen, phosphorus, and potassium in soil get wasted into the environment resulting in economical loss and environmental pollution as well (Prasad et al. 2016, 2017). The approach of using more stable and efficiently absorbed nano-coated chemical nutrients helps in reducing the dissolution rate of the fertilizer and leading to its sustained release for efficient absorption by plant roots.

This nano coating of nutrients facilitate sustainability by trimming down the wastage and minimizing environmental pollution. These slow-releasing encapsulated fertilizers are exceptional substitutes to the conventionally used soluble fertilizers. Researchers performed various studies to administer the controlled-release model of 22 essential nutrients by using nanoclays and nanocomposites (Raliya and Tarafdar 2014). Nano-coated sulfur fertilizers are being used for sulfur-scarce soils. Eco-friendly kaolin and chitosan nanoparticles have shown excellent results for sustained release of N-P-K fertilizers. Nanoencapsulation of fertilizers helps in the improved absorption of nutrients from the soil. Nanosilica and SiO_2 films help protect plants from infections and unfavorable environmental conditions and enhance the growth of roots and seedlings. Nontoxic titanium oxide nanoparticles are being used as an additive to increase crop productivity. To overcome the issues of high rate of water solubility, leaching, and denitrification of N fertilizers, various slow- and controlled-release fertilizers are being designed by using nanoclay-like montmorillonites, bentonites, halloysites, and zeolites (Rossi et al. 2014; Saharan et al. 2013; Schmid and Stoeger 2016).

11.3.1.2 Nanobiosensors

Nanosensors when immobilized with a range of bio-receptors are called nanobiosensors and are designed for detecting and analyzing data on atomic scales for detection and sensing of various chemicals, pathogens, enzymes, pollutants, infections in crops, and water level of soil. The pathogenic bacteria *E. coli* can be sensed by using a nanobiosensor made up of antibodies coated on fiber-optic nanosensors (Sekhon 2014). Nanobiosensors made up of antibodies conjugated with fluorescent Si nanoparticles can help in detecting the gram(−) bacteria *Xanthomonas axonopodis*, responsible for bacterial infections in Solanaceae plants. The peculiar optical properties of gold nanoparticles make them a potential biosensor for sensing pathogens; for example, the karnal bunt disease in wheat crop can be detected by using gold nanobiosensors. Though the applications of nanobiosensors for detecting plant pathogens is still in its primitive stage, still various nanobiosensors made up of CNTs, Si nanoparticles, and various nanowires are being reported to detect and report plant pathogens very precisely (Sertova 2015).

11.3.1.3 Nanopesticides and Nanoherbicides

In a broader aspect, pesticides and herbicides enhance crop yield and plant growth either by killing the unwanted weeds, grasses, insects, or microorganisms or by making the plants resistant to them (Shweta et al. 2016). The increased usage of pesticides may lead to decline in N_2 fixation, resistance in pest and pathogens, bioaccumulation of pesticides, and reduced biodiversity of soil. The use of nanopesticides may help resolve these issues to some extent. But a major fraction of pesticides applied to plants and soils gets wasted through leaching and runoff (Siddiqui et al. 2015). Therefore, it is necessary to enclose these pesticides in some coating to have their controlled and precise release and increased solubility. Various nanoparticles made up of silver, zinc, and titanium oxides have shown promising results in controlling infections and pests in rice and silkworms (Subramanyam and Siva 2016). Citric acid and multiwalled carbon nanotube encapsulated pesticides, Mancozeb and Zineb, have shown promising results in controlling the fungi *Alternaria alternata*. The exploitation of nanoformulations opens new ways to boost the strength and constancy of natural substances. This is achieved with the help of the anti-pathogenic behavior of nanoparticles and because of specific defense mechanisms inside the plants (Tarafdar et al. 2013; Wang et al. 2017) (Table 11.3).

One more area of concern in agri-practices is when weeds are grown in between and along with the standing crops. Herbicides are used to get rid of these weeds, but usually herbicides when sprayed might have an effect on the standing crop, sourcing a considerable loss of crop production. The nanoscale dimensions and target-based

Table 11.3 Effects of nanopesticides (Wang et al. 2017)

Nanopesticides	Effect
Essential oil-filled glycol-coated polyethylene nanoparticles	Against red flour beetle (*Tribolium castaneum*)
Silver nanoparticles	Protects oak trees from Raffaelea Cucurbit family against powdery mildew
Ag NPs by *Tinospora cordifolia*	Against *Pediculus humanus*, *Anopheles subpictus*, *Culex quinquefasciatus*
Hydrophobic aluminum-silicate nanoparticles as phenolic suspension	Protects *Bombyx mori* from grasserie disease
Hydrophobic nanosilica	Controls the spread of highly resistant species *Spodoptera littoralis*

precise delivery of nanoherbicides make them easy to blend with soil particles to wipe out the weeds without distressing the major crops. Nanoencapsulation is also useful in achieving sustained delivery and controlled solubility of herbicides. For example, herbicides containing atrazine, ametryn, and triazine are being nanoencapsulated using carbon nanotubes and silver, zinc oxide, and titanium oxide nanoparticles to attain more than 80% efficiency in precise and sustained release to plants (Yang et al. 2017; Zhang et al. 2016).

11.3.1.4 Nanofiltration in Agriculture

The dearth of water has become a serious issue for agricultural practices in various regions of the world. To deal with this, it is very much essential to opt for some economically competent and sustainable methods for irrigation and to check water wastage by making apt amendments in irrigation techniques, though these modifications might be time-consuming and inapt in areas with continual lack of water. Nanotechnology can be helpful to resolve the issues related to water availability. The applications of nanofilters have shown to be a very effective tool for managing irrigation water by water treatment methods (Amin 2018). The pore dimensions of 0.5 nm to 1 nm make nanofilters highly useful for water softening and wastewater treatment methods. It is recommended in agriculture that the irrigation water should be free from particles greater than 50 μm, toxic salts, and heavy metals and should have low salinity (Yan et al. 2019). Therefore, it is necessary to treat water to remove every unwanted substance which may lead to decrease in productivity, quality, and diversity of crops. In a number of dry and hot places in some countries, solar-powered nanofilters are being used for managing desalinated water for irrigation as it is being shown by the requirement shows 25% less demand for irrigation and fertilizers, with a substantial increase in crop yield (Quist-Jensen et al. 2015).

11.3.1.5 Micronutrient Supply

Though the appropriate requirement of micronutrients is less than 100 ppm, still they have a key role in plant metabolism as activators to various enzymes. Chitosan nanoparticles have been found to be helpful in the slow release of some plant growth hormones like 1-naphthylacetic acid (Dayarathne et al. 2019). Iron oxide nanoparticles have positive effects on the growth of plants in soils rich in calcium and high in pH. It is been observed that iron nanoparticles have enhanced effects on the yield, protein content, grain weight, and spike weight when directly being applied on the leaves of wheat plant (Hans and Jana 2018). Symptoms of iron deficiency in soya bean plants can be overcome by applying nanoemulsion of iron nanoparticles. Micronutrients like Mn, Fe, Cu, B, Zn, Mo, etc. are very vital for the proper growth of plants (Feregrino-Perez et al. 2018). The mammoth increase in the crop production during the green revolution has led to a drastic change in the micronutrient balance of soil. To enhance the micronutrient availability to plants, nanoformulations of zinc, iron, molybdenum, etc. can either be infused through soil or sprayed on the plants. With the help of nanotechnology, smart seeds are being developed by making seeds absorb the nanoemulsions, which can be programmed in such a way that they only germinate when the conditions are adequate to them (Singh et al. 2018; Sun-Waterhouse and Waterhouse 2016). Nano-coated smart seeds have the ability to sense water and appropriate conditions for germination, to detect adequateness of moisture during storage. It was reported that application of nanosilicon dioxide ($nSiO_2$: size 12 nm) significantly enhanced the characteristics of tomato seed germination. Germination in tomato seeds was reported to enhance when applied with SiO_2 nanoparticles (Cushen et al. 2012).

11.3.1.6 Nanogenetic Manipulation of Agricultural Crops

Nanotechnology proffers innovative tools for manipulating plant genes using nanofibers, particles, and capsules. Appropriately designed nanomaterials work as carriers and might hold plant genes and substances controlling the movement of genetic materials. Nanofibers find their use in crop engineering, drug delivery, and environment monitoring via quick and efficient delivery of genetic material to cells. A mesoporous Si nanoparticle is shown to successfully transport foreign DNA into cells (Wakeil et al. 2017; Liu and Lal 2015). Starch nanoparticles are being reported to bind and transport genetic material through plant cells through the instant pore channels in the cell wall. Nanobiosensors can help protect crop field by sensing and releasing alerts for pollen grain contamination originating from genetically modified crops. The amalgamation of nanotechnology with biotechnology led to the breakthrough designing of three-dimensional molecular structures through constructing self-assembling synthetic DNA sequences as crystals. This method can well be used for enhancing vital crops by connecting and categorizing desired essential organic compounds like nucleic acids, protein, lipids, and carbohydrate molecules to these crystals (Liu and Lal 2015). Nanoparticles loaded with agrochemicals or genetic materials are capable of functioning as a magic bullet or gene gun, resulting in target-based precise delivery of these products. This method has given effective result in using mesoporous nanosilica or gold-capped nanoparticles for introducing specific DNA strands to corn and tobacco plants (Wang et al. 2019) (Table 11.4).

Table 11.4 Advantages of nanoformulations over conventional formulations (Nasrollahzadeh et al. 2019)

Desirable characteristics	Examples of nanofertilizer-enabled technologies
Formulation with controlled-release characteristics	The nano-designed formulations may allow the fertilizers to cleverly manage the discharge rate of the nutrients as per requirement patterns of the crops
Dispersion and solubility management of micronutrients	Nanoformulations of micronutrients might enhance the solubility and help in diffusion of non-dissolvable micronutrients in soil
New methods for controlled release	The rates and patterns of release of water-soluble nutrients can be accurately controlled by encapsulating the fertilizers in resins or polymer coatings
Affectivity of nutrient release	Nanoformulations might help in widening the period of effectivity of fertilizers in the soil
Leaching of nutrients	Nano-designed formulations help in reducing the loss of nutrients from soil through leaching

11.4 Nanotechnology in Agribusiness

11.4.1 Sustainable Water Use

Nanotechnology can be a boon in desiccated and drought-affected regions, as water scarcity leads to a great loss in crop production and agri-economy as well. Nanohydrogels can optimize the water consumption and increase the sustainability of agricultural practices by periodic absorption and discharge of water and nutrients (Kundu et al. 2019). It has been observed that soils laden with nanosilver-coated hydrogel has 7.5% more capacity to hold water. Hydrogels have the capacity to store water more than 150 times their weight (Rai et al. 2018).

11.4.2 Treatment of Seeds

Nanotreatment of seeds contributes towards the increase in the numbers and weight and weather resistance. Around 75% increase in dry weight, more than 15% increase in shelf-life, more than 85% increase in drought resistance, and three times increase in vitamin content are being observed in seeds when treated with nanosolutions which results in improved productivity and revenue generations (Baker et al. 2017; Kumar et al. 2019).

11.4.3 Pest and Disease Detection

The spread of disease, contaminants, pests, and microorganisms results into relentless harm to the agribusiness. The accurate and selective sensing of such serious threats by nanobiosensors helps in managing agricultural practices in a more healthy way through preventing outbreaks of diseases, pests, infections, and monitoring soil health resulting into the increased productivity and enhanced characteristics of food grains (Sozer and Kokini 2009; Rienzie et al. 2019).

11.4.4 Enhanced Delivery of Nutrients and Plant Protection Products

The smart delivery systems based on nanotechnology help in enhancing the reach of nutrients and precise delivery of protection products, resulting into the improvements in the quality, quantity, and life span of agriproducts (Corsi et al. 2018; Martinho 2018).

11.4.5 Decreased Pollution and Reduced Runoff

Nano-applications in agriculture help in reducing the pollution caused by chemical fertilizers and medicines and help in remediation of heavy metal polluted soils. This makes the discarded soils to be used again. Nanosolutions help in controlling the loss of agrochemicals caused by leaching and runoff which saves revenue loss as well (Dahabieh et al. 2018; Dudo et al. 2011).

11.5 Risks, Toxicity, Co-Contaminant Effects, and Impacts of Nanomaterials

The general properties and risks associated with nanoparticles are analyzed with the help of various articles being published. The characteristic properties of nanoparticles are directly or indirectly associated with their synthesis. The characteristics of nanoparticles and the challenges and risks associated are summed up in Table 11.5.

The environment and human beings are exposed to nanomaterials when these are released in the environment during their manufacture, usage, clearance, and management of harvests containing nanoproducts. The tiny dimensions of nanoparticles

Table 11.5 Properties of nanomaterials and associated risks (He et al. 2019)

Properties of nanomaterial	Risk associated
Aggregation	The agglomeration and high solubility and fusion of nanoparticles pose substantial risks by decreasing the resistance to corrosion, leading to phase change and weakening of infrastructure
Reactivity	The specific properties of agrochemicals based on the function groups might get altered due to the unprompted degradation of nanoparticles
Impurity	Owing to their highly reactive nature, nanoparticles are very prone to react with external impurities which may alter their properties and outcomes of their applications. Therefore, it becomes necessary to encapsulate them with some nonreactive substance
Contaminants	During their synthesis, most of the nanoparticles get contaminated by the precursors used for their synthesis which alters the actual properties of the nanomaterial. For example, CNTs get contaminated by metals like Rb, Yt, or Ni, and iron nanoparticles by sulfur
Size	The agglomeration and aggregation of nanoparticles hinder in retaining the size of nanoparticles. Therefore, it becomes necessary to encapsulate the synthesized nanoparticles
Clearance and recycling	There are no very clear clearance policies for the management of nanomaterials because not much data related to their exposure and applications issues is available for studies
Shape effects	Nanoparticles show specific toxic behavior at a particular aspect ratio, i.e., their toxicity is shape dependent. For example, 10-μm fiber of asbestos can cause cancer, and fibers having a length ranging 5–10 μm may cause mesothelioma, whereas fibers with length of 2 μm can cause asbestosis

make possible their translocation within the body causing organ damage, cancer, asthmatic attacks, irreversible oxidative stress, organ enlargements, organ dysfunctions, denaturation of protein, etc. (Di Sia 2017). The shape and the chemical composition of nanoparticles are the key reason for nanoparticle toxicity, and because of this, there are various nontoxic or less toxic nanoparticles, with some having positive effects.

11.6 Nanomaterial Regulations

The exceptional characteristics of nanomaterials like high biochemical activity, tissue penetration capacity, and better bioavailability make them a better resource for biomedical purposes. These virtues of nanomaterials might also bring possibilities for toxic effects on the environment and living being. Therefore, various laws, regulations, rules, and legislations are being formulated and implemented by government and nongovernment organizations to reduce or curb the risks associated (Justo-Hanani and Dayan 2015). However, there are no definite globally accepted protocols, norms, and regulations for the manufacture, management, testing, categorization, and evaluation of environmental influences of nanoparticles. Presently, the European Union and the United States have designed strong regulatory guidelines and legislations to manage the probable hazards of nanomaterials (Amenta et al. 2015). The current developments in the areas of nano-based formulations for agriculture bring along biosafety issues with them. The scientists at IFDC-USA (International Fertilizer Development Center) have highlighted that the pros and cons of broad manufacturing of nanoformulations are yet to be recognized (Coles and Frewer 2013). The most intricate and under-researched areas of nano-based agriproducts are risk assessment and management. In fact, the lacuna in the substantiation of permissible doses of nanoproducts for agriculture has deferred their broad market acceptance as compared to various other common technologies. Furthermore, there are not much statistical data available to understand the exploitation of nanonutrients from plants and the management of the metal residues. Risk assessment and management of a nano-based product corresponds to the pertinent risks and hazards that arise from that product all through its voyage from creation to consumption (Steinhäuser and Sayre 2017).

11.6.1 Existing Regulations of Agri-Nanoproducts at the World Level

As per the records maintained by NAAS (National Academy of Agricultural Sciences), more than 80% of the products, publications, and patents come from countries like United States, Japan, Germany, Switzerland, South Korea, France, and among Asian countries like China, whereas the rate of advancements and the investments being made in India are not satisfactory enough (Sayre et al. 2017) (Table 11.6).

Table 11.6 Existing regulations for nanoproducts (Radad et al. 2012)

Country	Regulating body/regulations
USA	FDA (Food and Drug Administration) US EPA (US Environmental Protection Agency)
Canada	CFIA (Canadian Food Inspection Agency) PHAC (Public Health Agency of Canada)
European Union	Regulation number1169/20119 (Provision of Food Information to Consumers) Regulation number 450/2009 (Active and Intelligent Materials and Articles) Regulation number 528/2012 (The Biocidal Products Regulation) Regulation number 1907/2006 (Registration, Evaluation, Authorisation and Restriction of Chemicals) Regulation number 1107/2009 (Plant Protection Products) Regulation number 1223/2009 (The Cosmetic Products Regulation)
Non-European countries	FOPH (Federal Office of Public Health of Switzerland) The Ministry of Environment and Urban Planning Russian Corporation of Nanotechnologies Food Standards Australia New Zealand (FSANZ)
Asian countries	Presently, there are no precise rules and legislations for regulating the applications of nanoproducts in agriculture The National Centre for Nanoscience and Technology (NCNST) In China, the application of nano-based products in food and agriculture is not permitted by regulatory bodies

11.7 Future Research

Nanotechnology has vast potential in the agriculture sector. Enhancing the crop production by precision farming governed by nanoprinciples is highly desirable to get maximum output with reduced inputs through superior sensing and by precise actions. Nanotechnology gives control to crops to make use of soil, water, fertilizers, herbicides, and pesticides in a more effective way. The future applications of nanotechnology comprise of exploiting nanoporous zeolites for sustained release of fertilizers and water in efficient amounts for plants, use of improved nano-capsules for delivering agrochemicals, production of biofuel, and better genetic manipulations in plants together with keeping the sustainability of the environment and health of living beings at first priority. This will comprise of defining the effects of nanomaterial behavior on the physiochemical and biological characteristics of the environment throughout their life cycle. Every part of all these experimental statistics is required to be shared among industries and government as well as nongovernment regulating bodies sequentially to evidently characterize the actual profile of concerned nanomaterial under changeable exposure circumstances and to design and execute an appropriate risk management strategy for sustainable development. In general, synchronized execution of all these facets may show the way for the development of an extensive regulatory consensus integrating various aspects of ethical studies and public engagement in decision making. These are essential for flourishing the applications of nanotechnology for achieving a long-lasting sustainable role of this emerging technology in the agri-sector, and we can convincingly be confident and hopeful for a brighter future of agriculture in amalgamation with nanotechnology.

11.8 Conclusion

In this chapter, a few fresh ideas related to the potential offerings of nanotechnology in agriculture are being examined. Out of those, some are very concrete, and for some, the primary investigational statistics are already existing, whereas quite a few have a strong futuristic approach. However, there are a range of apprehensions associated with the applications of nanomaterials in agri-sector that need to be dealt: how to handle nanoformulations in actual field conditions; what precautions should be taken; what are the best nano structures, instruments, and equipment to be used; and what are the safety precautions to be taken for the people working in the environment of nanostructures, etc. These and various other issues need to be regulated.

It can be concluded that the exploitation of nanomaterials in the agri-sector is still an underdeveloped area with the hope of being taken over by successful results and broad acceptance worldwide. The constant utilization of chemicals for enhancing agricultural efficiency has led to the contagion of soil crest, groundwater, and food resources, and in this reverence, nanotechnology is emerging as a broadly accepted technique for the development of agriculture in a sustainable manner. Promising results are already being recognized in the area of nutrients, medicines, herbicides, pesticides, and genetic materials by using the principles of nanotechnology. Consequently, with the exploitation of nanoprinciples, a controlled, precise, and target-oriented delivery of agrochemicals can be achieved. The employment of nanotools can tackle up the critical question of maintaining sustainable plant growth and plant protection. Apart from this, nanotechnology might potentially aid to offer more efficient methods for sensing, detecting, and remediation of environmental problems.

References

Abbas SS, Haneef M, Lohani M, Tabassum H, Khan AF (2016) Nanomaterials used as a plants growth enhancer: an update. Int J Pharm Sci Rev Res 5:17–23

Agrawal S, Rathore P (2014) Nanotechnology pros and cons to agriculture: a review. Int J Curr Microbiol Appl Sci 3:43–55

Amenta V, Aschberger K, Arena M, Bouwmeester H, Moniz FB, Brandhoff P, Gottardo S, Marvin HJP, Mech A, Pesudo LQ, Rauscher H, Schoonjans R, Vettori MV, Weigel S, Peters RJ (2015) Regulatory aspects of nanotechnology in the agri/feed/food sector in EU and non-EU countries. Regul Toxicol Pharmacol 73(1):463–476. https://doi.org/10.1016/j.yrtph.2015.06.016

Amin M (2018) Nanofiltration systems and applications in wastewater treatment: review article. Ain Shams Eng J 9:3077. https://doi.org/10.1016/j.asej.2018.08.001

Aouada FA, de Moura MR (2015) Nanotechnology applied in agriculture: controlled release of agrochemicals. In: Rai M, Ribeiro C, Mattoso L, Duran N (eds) Nanotechnologies in food and agriculture. Springer, Cham, pp 103–118. https://doi.org/10.1007/978-3-319-14024-7_5

Aziz N, Pandey R, Barman I, Prasad R (2016) Leveraging the attributes of *Mucor hiemalis*-derived silver nanoparticles for a synergistic broad-spectrum antimicrobial platform. Front Microbiol 7:1984. https://doi.org/10.3389/fmicb.2016.01984

Baker S, Volova T, Prudnikova SV, Satish S, Prasad MNN (2017) Nanoagroparticles emerging trends and future prospect in modern agriculture system. Environ Toxicol Pharmacol 53:10–17. https://doi.org/10.1016/j.etap.2017.04.012

Baligar VC, Fageria NK (2015) Nutrient use efficiency in plants: an overview. In: Rakshit A, Singh HB, Sen A (eds) Nutrient use efficiency: from basics to advances. Springer, New Delhi, pp 1–14. https://doi.org/10.1007/978-81-322-2169-2_1

Berekaa MM (2015) Nanotechnology in food industry; advances in food processing, packaging and food safety. Int J Curr Microbiol App Sci 4:345–357

Bhagat Y, Gangadhara K, Rabinal C, Chaudhari G, Ugale P (2015) Nanotechnology in agriculture: a review. J Pure App Microbiol 9:737–747

Camilli L, Pisani C, Gautron E, Scarselli M, Castrucci P, D'Orazio F et al (2014) A three-dimensional carbon nanotube network for water treatment. Nanotechnology 25:065701. https://doi.org/10.1088/0957-4484/25/6/065701

Chhipa H (2016) Nanofertilizers and nanopesticides for agriculture. Environ Chem Lett 15:15–22. https://doi.org/10.1007/s10311-016-0600-4

Coles D, Frewer LJ (2013) Nanotechnology applied to European food production – a review of ethical and regulatory issues. Trends Food Sci Technol 34(1):32–43. https://doi.org/10.1016/j.tifs.2013.08.006

Corsi I, Winther-Nielsen M, Sethi R, Punta C, Della Torre C, Libralato G, Lofrano G, Sabatini L, Aiello M, Fiordi L, Cinuzzi F, Caneschi A, Pellegrini D, Buttino I (2018) Ecofriendly nanotechnologies and nanomaterials for environmental applications: key issue and consensus recommendations for sustainable and ecosafe nanoremediation. Ecotoxicol Environ Saf 154:237–244. https://doi.org/10.1016/j.ecoenv.2018.02.037

Cox A, Venkatachalam P, Sahi S, Sharma N (2017) Reprint of: Silver and titanium dioxide nanoparticle toxicity in plants: a review of current research. Plant Physiol Biochem 110:33–49. https://doi.org/10.1016/j.plaphy.2016.08.007

Cushen M, Kerry J, Morris M, Cruz-Romero M, Cummins E (2012) Nanotechnologies in the food industry – recent developments, risks and regulation. Trends Food Sci Technol 24(1):30–46

Dahabieh MS, Bröring S, Maine E (2018) Overcoming barriers to innovation in food and agricultural biotechnology. Trends Food Sci Technol 79:204–213. https://doi.org/10.1016/j.tifs.2018.07.004

Das S, Wolfson BP, Tetard L, Tharkur J, Bazata J, Santra S (2015) Effect of N-acetyl cysteine coated CdS:Mn/ZnS quantum dots on seed germination and seedling growth of snow pea (*Pisum sativum* L.):imaging and spectroscopic studies. Environ Sci 2:203–212. https://doi.org/10.1039/c4en00198b

Dayarathne HNP, Jeong S, Jang A (2019) Chemical-free scale inhibition method for seawater reverse osmosis membrane process: air micro-nano bubbles. Desalination 461:1, 1–9

De Matteis V (2017) Exposure to inorganic nanoparticles: routes of entry, immune response, biodistribution and in vitro/in vivo toxicity evaluation. Toxics 5:29. https://doi.org/10.3390/toxics5040029

de Oliveira JL, Campos EVR, Bakshi M, Abhilashi PC, Fraceto LF (2014) Applications of nanotechnology for encapsulation of botanical insecticides for sustainable agriculture and promises. Biotechnol Adv 23:1550–1561

Di Sia P (2017) Nanotechnology among innovation, health and risks. Procedia Soc Behav Sci 237:1076–1080. https://doi.org/10.1016/j.sbspro.2017.02.158

Ditta A, Arshad M (2016) Applications and perspectives of using nanomaterials for sustainable plant nutrition. Nanotechnol Rev 5:1–22. https://doi.org/10.1515/ntrev-2015-0060

Dudo A, Choi D-H, Scheufele DA (2011) Food nanotechnology in the news. Coverage patterns and thematic emphases during the last decade. Appetite 56(1):78–89. https://doi.org/10.1016/j.appet.2010.11.143

Ezz El-Din H, Manjaiah DH (2017) Internet of nano things and industrial internet of things. In: Acharjya D, Geetha M (eds) Internet of things: novel advances and envisioned applications. Studies in big data. Springer, Cham, pp 109–123. https://doi.org/10.1007/978-3319-53472-5_5

Feregrino-Perez AA, Magaña-López E, Guzmán C, Esquivel K (2018) A general overview of the benefits and possible negative effects of the nanotechnology in horticulture. Sci Hortic 238:126–137

Fraceto LF, Grillo R, de Medeiros GA, Scognamiglio V, Rea G, Bartolucci C (2016) Nanotechnology in agriculture: which innovation potential does it have? Front Environ Sci 4:20. https://doi.org/10.3389/fenvs.2016.00020

Glenn JC, Florescu E (2016) Millennium project team. 2015-16 state of the future. J Socialomics 5:1–6. https://doi.org/10.4172/2167-0358.1000168

Hajirostamlo B, Mirsaeedghazi N, Arefnia M, Shariati MA, Fard EA (2015) The role of research and development in agriculture and its dependent concepts in agriculture [Short Review]. Asian J Appl Sci Eng 4. https://doi.org/10.1016/j.cocis.2008.01.005

Handford CE, Dean M, Spence M, Henchion M, Elliott CT, Campbell K (2015) Awareness and attitudes towards the emerging use of nanotechnology in the agri-food sector. Food Control 57:24–23

Hans KB, Jana T (2018) Micronutrients in the life cycle: requirements and sufficient supply. NFS J 11:1–11

He X, Deng H, H-min H (2019) The current application of nanotechnology in food and agriculture. J Food Drug Anal 27(1):1–21. https://doi.org/10.1016/j.jfda.2018.12.002

Jayaseelan C, Rahuman AA, Roopan SM, Kirthia AV, Kim S-K, Iyappan M, Siva C (2011) Synthesis of pediculocidal and larvicidal silver nanoparticles by leaf extract from heartleaf moonseed plant, *Tinospora cordifolia* Miers. Parasitol Res 109(1):185–194. https://doi.org/10.1007/s00436-010-2242-y

Justo-Hanani R, Dayan T (2015) European risk governance of nanotechnology: explaining the emerging regulatory policy. Res Policy 44(8):1527–1536. https://doi.org/10.1016/j.respol.2015.05.001

Karpachev VV, Spiridonov JJ, Voropaeva NL, Tkachev AG, Shachnev NV, Figovsky OL (2016) Pre-sowing seed treatment nanotechnology with environment-friendly nanotube-based nanochips. Int Lett Nat Sci 58:29–34

Khan R, Fulekar MH (2016) Biosynthesis of titanium dioxide nanoparticles using *Bacillus amyloliquefaciens* culture and enhancement of its photocatalytic activity for the degradation of a sulfonated textile dye Reactive Red 31. J Colloid Interface Sci 475:184. https://doi.org/10.1016/j.jcis.2016.05.001

Kouhkan M, Ahangar P, Babaganjeh LA, Allahyari-Devin M (2019) Biosynthesis of copper oxide nanoparticles using *Lactobacillus casei* subsp. casei and its anticancer and antibacterial activities. Curr Nanosci 16:101. https://doi.org/10.2174/1573413715666190318155801

Kumar S, Bhanjana G, Sharma A, Sarita MC, Sidhu Dilbaghi N (2015) Herbicide loaded carboxymethyl cellulose nanocapsules as potential carrier in agri nanotechnology. Sci Adv Mater 7:1143–1148

Kumar S, Nehra M, Dilbaghi N, Marrazza G, Hassan AA, Kim K-H (2019) Nano-based smart pesticide formulations: emerging opportunities for agriculture. J Control Release 294:131–153. https://doi.org/10.1016/j.jconrel.2018.12.012

Kundu M, Krishnan P, Kotnala RK, Sumana G (2019) Recent developments in biosensors to combat agricultural challenges and their future prospects. Trends Food Sci Technol 88:157–178

Liu R, Lal R (2015) Potentials of engineered nanoparticles as fertilizers for increasing agronomic productions. Sci Total Environ 514(Suppl C):131–139. https://doi.org/10.1016/j.scitotenv.2015.01.104

Mabe FN, Talabi K, Danso-Abbeam G (2017) Awareness of health implications of agrochemical use: effects on maize production in Ejura-Sekyedumase municipality, Ghana. Adv Agric 11:7960964. https://doi.org/10.1155/2017/7960964

Mani PK, Mondal S (2016) Agri-nanotechniques for plant availability. In: Kole C, Sakthi Kumar D, Khodakovskaya MV (eds) Plant nanotechnology. Principles and practices. Springer, Cham, pp 263–303. https://doi.org/10.1007/978-3-319-42154-4_11

Martinho VJPD (2018) Interrelationships between renewable energy and agricultural economics: an overview. Energ Strat Rev 22:396–409. https://doi.org/10.1016/j.esr.2018.11.002

Milewska-Hendel A, Gawecki R, Zubko M, Stróż D, Kurczynska E (2016) Diverse influence of nanoparticles on plant growth with a particular emphasis on crop plants. Acta Agrobot 69 (4):1694. https://doi.org/10.5586/aa.1694

Monreal CM, DeRosa M, Mallubhotla SC, Bindraban PS, Dimkpa C (2016) Nanotechnologies for increasing the crop use efficiency of fertilizer-micronutrients. Biol Fertil Soils 52:423–437. https://doi.org/10.1007/s00374-015-1073-5

Morales-Díaz AB, Ortega-Ortíz H, Juárez-Maldonado A, Cadenas-Pliego G, GonzálezMorales S, Adalberto Benavides-Mendoza A (2017) Application of nanoelements in plant nutrition and its impact in ecosystems. Adv Nat Sci 8:1–13. https://doi.org/10.1088/2043-6254/8/1/013001

Mufamadi MS, Sekhejane PR (2017) Nanomaterial-based biosensors in agriculture application and accessibility in rural smallholding farms: food security. In: Prasad R, Kumar M, Kumar V (eds) Nanotechnology. Springer, Singapore, pp 263–278. https://doi.org/10.1007/978-981-10-4573-8_12

Nasrollahzadeh M, Sajadi SM, Sajjadi M, Issaabadi Z (2019) Applications of nanotechnology in daily life. Interface Sci Technol 28:113–143

Nima AZ, Lahiani MH, Watanabe FXY, Khodakovskaya MV, Biris AS (2014) Plasmonically active nanorods for delivery of bio-active agents and high-sensitivity SERS detection in planta. RSC Adv 4:64985–64993. https://doi.org/10.1039/C4RA10358K

Nuruzzaman M, Rahman MM, Liu Y, Naidu R (2016) Nanoencapsulation, nano-guard for pesticides: a new window for safe application. J Agric Food Chem 64:1447–1483. https://doi.org/10.1021/acs.jafc.5b05214

Ozdemir M, Kemerli T (2016) Innovative applications of micro and nanoencapsulation in food packaging. In: Lakkis JM (ed) Encapsulation and controlled release technologies in food systems. Wiley, Chichester

Pandey G (2018a) Challenges and future prospects of agri-nanotechnology for sustainable agriculture in India. Environ Technol Innov 11:299–307

Pandey G (2018b) Nanotechnology for achieving green-economy through sustainable energy. Rasayan J Chem 11(3):942–950

Pandey G (2018c) Prospects of nanobioremediation in environmental cleanup. Orient J Chem 34 (6):2838–2850. https://doi.org/10.13005/ojc/340622

Parisi C, Vigani M, Rodríguez-Cerezo E (2015) Agricultural nanotechnologies: what are the current possibilities? Nano Today 10:124–127

Patra JK, Baek K-H (2017) Antibacterial activity and synergistic antibacterial potential of biosynthesized silver nanoparticles against foodborne pathogenic bacteria along with its anticandidal and antioxidant effects. Front Microbiol 8:167. https://doi.org/10.3389/fmicb.2017.00167

Prasad R, Pandey R, Barman I (2016) Engineering tailored nanoparticles with microbes: quo vadis. WIREs Nanomed Nanobiotechnol 8:316–330. https://doi.org/10.1002/wnan.1363

Prasad R, Bhattacharyya A, Nguyen QD (2017) Nanotechnology in sustainable agriculture: recent developments, challenges, and perspectives. Front Microbiol 8:1014. https://doi.org/10.3389/fmicb.2017.01014

Quist-Jensen CA, Macedonio F, Drioli E (2015) Membrane technology for water production in agriculture: desalination and wastewater reuse. Desalination 364:17–32. https://doi.org/10.1016/j.desal.2015.03.001

Radad K, Al-Shraim M, Moldzio R, Rausch W-D (2012) Recent advances in benefits and hazards of engineered nanoparticles. Environ Toxicol Pharmacol 34(3):661–672. https://doi.org/10.1016/j.etap.2012.07.011

Rai PK, Kumar V, Lee SS, Raza N, Kim K-H, Ok YS, Tsang DCW (2018) Nanoparticle-plant interaction: implications in energy, environment, and agriculture. Environ Int 119:1–19. https://doi.org/10.1016/j.envint.2018.06.012

Raliya R, Tarafdar JC (2014) Biosynthesis and characterization of zinc, magnesium and titanium nanoparticles: an eco-friendly approach. Int Nano Lett 4:1. https://doi.org/10.1007/s40089-014-0093-8

Rienzie R, Ramanayaka S, Adassooriya NM (2019) Nanotechnology applications for the removal of environmental contaminants from pharmaceuticals and personal care products. In: Pharmaceuticals and personal care products: waste management and treatment technology, pp 279–296

Rossi M, Cubadda F, Dini L, Terranova ML, Aureli F, Sorbo A, Passeri D (2014) Scientific basis of nanotechnology, implications for the food sector and future trends. Trends Food Sci Technol 40:127–148

Saharan V, Mehrotra A, Khatik R, Rawal P, Sharma SS, Pal A (2013) Synthesis of chitosan based nanoparticles and their in vitro evaluation against phytopathogenic fungi. Int J Biol Macromol 62:677–683. https://doi.org/10.1016/j.ijbiomac.2013.10.012

Sayre PG, Steinhäuser KG, Teunenbroe T-V (2017) Methods and data for regulatory risk assessment of nanomaterials: questions for an expert consultation. Nano Impact 8:20–27. https://doi.org/10.1016/j.impact.2017.07.001

Schmid O, Stoeger T (2016) Surface area is the biologically most effective dose metric for acute nanoparticle toxicity in the lung. J Aerosol Sci 9:133–114

Sekhon BS (2014) Nanotechnology in agri-food production: an overview. Nanotechnol Sci Appl 7:31–53. https://doi.org/10.2147/NSA.S39406

Sertova NM (2015) Application of nanotechnology in detection of mycotoxins and in agricultural sector. J Cent Eur Agric 16:117–130. https://doi.org/10.5513/JCEA01/16.2.1597

Shweta TDK, Singh S, Singh S, Dubey NK, Chauhan DK (2016) Impact of nanoparticles on photosynthesis: challenges and opportunities. Mater Focus 5:405–411. https://doi.org/10.1166/mat.2016.1327

Siddiqui MH, Al-Whaibi MH, Firoz M, Al-Khaishany MY (2015) Role of nanoparticles in plants. In: Siddiqui MH, Al-Whaibi MH, Mohammad F (eds) Nanotechnology and plant sciences: nanoparticles and their impact on plants. Springer, Cham, pp 19–35. https://doi.org/10.1007/978-3-319-14502-0_2

Singh R, Glick BR, Rathore D (2018) Biosurfactants as a biological tool to increase micronutrient availability in soil: a review. Pedosphere 28(2):170–189

Sozer N, Kokini JL (2009) Nanotechnology and its applications in the food sector. Trends Biotechnol 27(2):82–89. https://doi.org/10.1016/j.tibtech.2008.10.010

Steinhäuser KG, Sayre PG (2017) Reliability of methods and data for regulatory assessment of nanomaterial risks. Nano Impact 7:66–74. https://doi.org/10.1016/j.impact.2017.06.001

Subramanyam SG, Siva K (2016) Bio-synthesis, characterization and application of titanium oxide nanoparticles by *Fusarium oxysporum*. Int J Life Sci Res 4:69–75

Sun-Waterhouse D, Waterhouse GIN (2016) Recent advances in the application of nanomaterials and nanotechnology in food research. In: Grumezescu A (ed) Novel approaches of nanotechnology in food. Elsevier, London, pp 21–66

Tarafdar A, Raliya R, Wang W-N, Biswas P, Tarafdar JC (2013) Green synthesis of TiO_2 nanoparticle using *Aspergillus tubingensis*. J Adv Sci Eng Med 5:943–949

Wakeil NE, Alkahtani S, Gaafar N (2017) Is nanotechnology a promising field for insect pest control in IPM programs? In: Grumezescu A (ed) New pesticides and soil sensors. Academic, London, pp 273–309

Wang L, Hu C, Shao L (2017) The antimicrobial activity of nanoparticles: present situation and prospects for the future. Int J Nanomed 12:1227–1249. https://doi.org/10.2147/IJN.S121956

Wang P, F-Jie Z, Kopittke PM (2019) Engineering crops without genome integration using nanotechnology. Trends Plant Sci 24(7):574–577. https://doi.org/10.1016/j.tplants.2019.05.004

Yan L, George C, Xiuli Y, Huining D (2019) Preparation of layer-by-layer nanofiltration membranes by dynamic deposition and crosslinking. Membranes 9:20. https://doi.org/10.3390/membranes9020020

Yang Y, Qin Z, Zeng W, Yang T, Cao Y, Me C, Kuang Y (2017) Toxicity assessment of nanoparticles in various systems and organs. Nanotechnol Rev 6(3):279–228

Zhang Q, Han L, Jing H, Blom DA, Lin Y, Xin HL, Wang H (2016) Facet control of gold nanorods. ACS Nano 10:2960–2974. https://doi.org/10.1021/acsnano.6b00258

Suitability of Fly Ash Amendment in Soil for Productivity of Agricultural Crops

12

Tanushree Bhattacharya and Shikha Kumari Pandey

Contents

Abstract

Technological progress has put enormous pressure on our valuable natural resources. The global energy need has increased with rapid industrialization, which to a large extent has been met by fossil fuels. Coal is an exhaustible energy source, which plays a crucial role in meeting the ever-increasing energy demands of countries around the world. Combustion of coal in thermal power stations produces a variety of residues—fly ash, bottom ash, flue gas desulfurization waste (scrubber sludge), fluidized bed boiler waste, and coal gasification ash. Over 225 million tons of coal is produced annually in India and over 100 thermal power stations generate more than 108 million tons of fly ash every year, which is expected to double within the next 5–6 years. So, it is immesnsely important to utilize fly ash for various purposes. One of the most viable option can be use as fertilizer in agriculture in suitable ammendments.

T. Bhattacharya (✉) · S. K. Pandey
Department of Civil & Environmental Engineering, Birla Institute of Technology, Mesra, Ranchi, Jharkhand, India
e-mail: tbhattacharya@bitmesra.ac.in

K. Bauddh et al. (eds.), *Ecological and Practical Applications for Sustainable Agriculture*, https://doi.org/10.1007/978-981-15-3372-3_12

Keywords

Amendments · Fly ash · Agriculture · Crop yield

12.1 Fly Ash Generation and Utilization

Coal is a fossil fuel formed by the residues of plants that died and lived around 100 to 400 million years ago, when the earth was covered with huge swampy forests. Coal is a nonrenewable form of energy as it takes millions of years for its formation. India is placed third in the world having 5.6% coal reserves of the total world coal reserves. Almost 70% of the total coal produced is used for electricity generation by thermal power plants (Gupta et al. 2007).

Coal-based power stations produce some by-products through the burning of coal. Coal is generally burnt into a pulverized furnace that produces two types of ashes as by-products—fly ash (FA) and the bottom ash (BA). The FA is recovered from the flue gas; it has fine texture and is the major fraction (90%) of ash produced. The residue gathered at the base of the furnace is called bottom ash; it is coarser compared to fly ash. Thirty percent of total electricity is produced worldwide from coal-fired thermal power stations. Consequently, a huge amount of fly ash is generated globally. It has been reported that more than 100 million tons of fly ash annual production is in the USA and EU, and about 13 million tons in Australia. Furthermore, coal consumption globally for thermal power production is expected to increase by 49% by the year 2030. As a result, fly ash generation will also increase. However, fly ash can be reused for various purposes like waste stabilization, building raw materials such as cement and bricks, and as fertilizer in agriculture with suitable soil amendment ratios. Utilization of ash in several countries are reported in the following order: Germany (85%) > Denmark (73%) > France and UK (60%) > Poland (50%) > USA (32%), Australia (31%) > China (25%) > India (15%) (Pandey and Bhattacharya 2019).

Coal used in Indian thermal power plants has ash content of 35 to 45% (Pandey and Bhattacharya 2019). At present in India, there are more than 100 coal-fired thermal power plants which produce around 170 million tons of fly ash every year. This would require about 4000 hectares of land for the construction of ash ponds (Dwivedi et al. 2007). The problem of fly ash disposal has increased to such an enormous extent in the country that the MoEF (Ministry of Environment and Forests) released a guideline on September 14, 1999, giving normative levels for sustainable utilization of fly ash. As per this guideline, it is compulsory for all coal-based thermal power plants (old or new) to consume total fly ash (100%) produced by them in a fixed time span. New thermal power plants are required to use 100% of the fly ash produced in the 9 years of commencing operation. The old ones, though, are required to attain 100 percent utilization target within 15 years from the date of regulation issue.

As per ASTM C618 (American Society of Testing Materials), fly ash is classified into two classes, "F" and "C." Class F fly ash is produced by burning of anthracite or bituminous coal and has low lime and high silica, iron oxide, and alumina content.

Class C fly ash is formed by the burning of lignite coal that has more lime content. The main distinction between these classes of fly ash is the quantity of silica, calcium, alumina, and iron in the ash. The chemical property is mainly influenced by the chemical composition of the parent coal.

12.2 Utilization of Fly Ash

Fly ash and bottom ash disposal are very critical issues worldwide as they have been considered as serious operational constraint in the utilization and processing of fly ash, bottom ash, and pond ash as raw materials. It is used in the preparation of bricks and ceramic products. For the first time, cordierite has been synthesized from fly ash. Several methods have been reported for the preparation of zeolite from fly ash. Fly ash was treated hydrothermally, and the performance of this material as cracking catalyst was investigated with heavy oil fraction as the cracking feedstock. On the other hand, there were many experimental analyses on fly ash to determine its basic compositional, physical, and chemical properties for technical studies and applications (Pandey et al. 2009).

However, the thrust areas of fly ash utilization are construction of roads and embankments, building components, hydraulic structures, agriculture-related studies, and applications for underground minefills (Bhattacharya 2007).

The technologies pertaining to these areas were identified for further development and technology demonstration projects are usually undertaken in several countries toward confidence building in all the thrust areas.

Over the past few decades, the utilization of fly ash in concrete has had a successful record. The benefits of using fly ash as part of concrete provide mechanical and strength properties of concrete. This has been widely researched and applied successfully in actual structures. At present, fly ash is used in more than 50% of all ready-mix concrete placed in the United States, yet many professionals persist to remain too restrictive when it comes to mixing fly ash in concrete. When fly ash was initially used in concrete in the 1970s, there were some reasons for restricting its use due to a dearth of research. Though, after widespread research and a number of decades of successful use of fly ash, it should be permitted to be used in concrete.

In India, various authors reported studies about the management of fly ash utilization and disposal (Kumar et al. 2002). It is being used in construction and cement industry, though the production rate is much more than the utilization. Leftover fly ash is disposed of into slurry ponds, landfills, and slag heaps. There is an environmental risk in the dumping of large quantities of ashes in slurry ponds and surface impoundments or its reuse in construction materials.

Fly ash dumps are not only creating land use problems but are also harmful for the environment. Fly ash dumping is considered the key cause of contamination due to the enrichment and holding of toxic elements with surface association in ash particles. During storage, transport, and disposal phases, the residues from coal combustion are subjected to leaching effects of rain and part of the toxic metals in the ashes pollute surface waters as well as groundwaters (Mishra et al. 2007). This fly ash can be leached in large amounts beyond the standards of drinking water and contaminate the sources of drinking water. Fly ash had trace amounts of toxic elements which can create an unfavorable condition for plants and human health. Fly ash dumping in the surface water like river bodies interrupts aquatic life, while heavy metals leach out and pollute water bodies. Hence, it is very essential to understand its leaching behavior to prevent the environmental impacts, particularly for the underwater environment. However, the chemical composition of fly ash provides an idea about the toxic element leaching through water, and it is important to conduct testing for leaching.

Health threats and environmental effects from coal burning thermal power plants are recognized as a result of the mobilization of toxic metals from fly ash. The huge amount of ash that accumulates in thermal power plants, its possible recycle, and the dispersion and mobilization of toxic elements require greater attention (Palit et al. 1991). Mobilization of different ash elements into the environment depends on climate, soil, indigenous vegetation, and agricultural practices . However, Goswami and Mahanta (2007) reported that utilization of fly ash and lime for the stabilization of lateritic soil was likely to have no significant impact on the environment, as most of the toxic metals present in the fly ash were within the threshold limits. The high pH induced by lime treatment of the mixes helped in keeping most of the metals within the stabilized soil matrix. The higher the acidity of the solution, the more leaching potential of heavy metals (Behera and Mishra 2012).

12.3 Properties of Fly Ash

The physical and chemical properties of fly ash mainly depend on the parent coal type and its combustion process (Chaudhary et al. 2011). The calcium oxide (CaO) content is high in fly ash of sub-bituminous or lignite coal and low in bituminous coal. The fly ash of coals of varying origins also vary in Si, Fe, Ca, and Mg oxide contents as well as reactive water-soluble and amorphous phases. The color of fly ash varies from tan to gray and to black, depending on the content of unburned carbon in the fly ash.

Fly ash usually consists of 49–67% silica (SiO_2), 16–29% alumina (Al_2O_3), 1–4% calcium oxide (CaO), 4–10% iron oxide (Fe_2O_3), and 0.2–2.0% magnesium oxide (MgO) as reported by Singh and Agrawal (2010). Fly ash may also contain sodium oxide (Na_2O), potassium oxide (K_2O), unburned carbon, and sulfate as reported recently. Fly ash has higher specific surface area, lower bulk density (1.0 to 1.8 g cm^{-1}), higher moisture retention capacity, higher electrical conductivity, and lower cation exchange capacity than normal soil. The pH of fly ash varies from 4.5 to 12.0 depending on constituents in the parent coal. The majority of various types of coal fly ash produced globally, including those in India, are alkaline in nature (Ram and Masto 2014).

The alkaline pH of fly ash may be due to the presence of Ca, Na, Mg, and OH along with other heavy metals. Mainly CaO, a major constituent of the fly ash, forms

Ca $(OH)_2$ with water and, thus, attributes toward alkalinity (Hodgson et al. 1982). Almost all metals present in soil are also found in fly ash, which include Al, B, Ca, Cd, Co, Cu, Fe, K, Mg, Mo, Mn, Na, Ni, Pb, Si, and Zn. Fly ash thus contains certain metals, essential for plant growth (Jala and Goyal 2006), and some metals of environmental concern (Ram and Masto 2014). Nitrogen present in the coal is volatilized during its combustion and thus fly ash may have nil or negligible nitrogen (Bradshaw and Chadwick 1980). The concentration of phosphorous is generally high (400–8000 mg kg^{-1}), but the form of P is not readily available to plants, due to interactions with Al, Ca, and Fe present in alkaline fly ash (Singh and Agrawal 2010).

12.4 Effects of Fly Ash Amendments in Agricultural Soil

Fly ash use as fertilizer source like B, Zn, P, Cu, and Zn is practiced in agriculture to overcome the fly ash disposal problem (Ram and Masto 2014). Alkaline and Ca-rich fly ash neutralizes acidic soils (Mishra and Shukla 1986). Fly ash lacks N and humus, which can be supplemented with the organic manure.

The fly ash amendment improves some physicochemical properties of soil bulk density, water-holding capacity, texture, pH, electrical conductivity, and particle size distribution (Ram and Masto 2014). The fly ash amendment is also known to reduce compaction of clay soils. The bulk density of soil declines with the addition of fly ash and thereby reduces porosity and increases water-holding capacity (Pandey et al. 2009). The electrical conductivity of the soil increases on application of fly ash as the levels of soluble major and minor inorganic constituents increase in soil (Jala and Goyal 2006).

12.4.1 Effects on Soil Properties

Soil quality governs the soil health, which leads to soil productivity and has pronounced consequences on the soil ecosystem. The health of the soil is dependent on the presence of inorganic and organic matter content and natural processes like salinization, erosion, and chemical contamination. These have a direct effect on groundwater quality, land use, and management practices (Acton and Gregorich 1995). Trace metals are present in fly ash, which easily infiltrate down from conventionally used earth-lined lagoons and contaminate groundwater. However, the solubility of these elements is $<10\%$ (Rohriman 1971). Natusch and Wallace (1974) observed that 5 to 30% of metals particularly Cd, Cu, and Pb are leachable. There are numerous reports on the occurrence of radionuclides in fly ash, especially of uranium and thorium series, but studies on their effect have been sporadic. The agronomic advantages of fly ash applications are chiefly related to amended physicochemical and biological properties of the soil. Because of the occurrence of Ca-Si minerals with a pozzolanic nature, along with the soil moisture, fly ash improves soil porosity, bulk density, and water-holding capacity. Several vital plant nutrients existing in fly ash also increase plant growth and crop yield (Ram and Masto 2014).

Agricultural application of fly ash has been suggested because of its significant content of K, Ca, Mg, S, and P. Fly ash application usually has a consistently promising impact on plant growth and nutrient uptake. Research in England reported that the creation of effective vegetative covers on fly ash deposits has shown that plant growth is habituated by the quantity of total soluble salts, pH, and available B in the ash and by the physical characteristics of the ash deposits. A general list of natural and crop species grown on fly ash has been obtained by Hodgson and Holliday (Hodgson and Townsend 1973). Of the natural species observed, *Funaria hygrometrica* and *Atriplex hastata* were among the first species to grow on fly ash. Crop species were classified into tolerant, semi-tolerant, and sensitive species with respect to their ability to withstand fly ash.

12.4.2 Effects on Soil Nutrient Status

In India, fly ash generated in thermal power plants is alkaline in nature; hence, their application increases the soil pH due to the rapid release of Ca^{2+}, Na^{+}, Al^{3+}, and OH^{-} ions (Jala and Goyal 2006). The acidic fly ashes generated in some countries can be used in reclaiming alkaline soils (Singh and Agrawal 2010). Amendment with fly ash at ratios from 25 to 100% in the garden soil increased the pH, particle density, porosity, and water-holding capacity (Pandey et al. 2009). The application of fly ash to the soil significantly increased the levels of soil nutrients, viz., Na, K, Ca, Mg, and Fe. The high boron level in fly ash restricts its utilization in crop production which can be overcome by properly weathering the fly ash before it is used in agriculture (Ram and Masto 2014).

Generally, fly ash amendment in soil increases plant growth and nutrient uptake (Aitken et al. 1984). Fly ash amendment of up to 50% improved the growth and yield of cucumber (*Cucumis sativus*) as reported by Ajaz and Tiyagi (2003). The photosynthetic pigments such as total chlorophyll and carotenoids in cucumber increased maximum at 25%. Root shoot length and dry matter of *Sesbania cannabina* and *Vigna radiata* grown at 10% and 25% fly ash amendment rates enhanced as compared to their controls.

The fly ash amendment in soil caused significant improvement in growth (shoot length, leaf area, and pigment composition) and yield attributes of rice (Mishra et al. 2007). Application of fly ash at and above 50% reduced photosynthetic pigments, protein content, and growth (plant height, root biomass, number of tillers, grain, and straw weight) in four cultivars of rice (Dwivedi et al. 2007), but lower concentrations (10–25%) of fly ash amendment in soil enhanced the plant growth (Dwivedi et al. 2007). Addition of 40% fly ash in soil increased dry matter accumulation and chlorophyll and protein content of *Brassica juncea* (Gautam et al. 2012). In a recent study, optimum growth (dry matter and seedling) of rice was recorded on application of 25% fly ash in soil (Panda and Tikadar 2014). The growth and reproductive ability of *B. juncea* enhanced on application of 20% to 40% fly ash in soil, but higher amounts of fly ash (60–100%) affected these growth parameters adversely (Pillai and Chaturvedi 2012).

12.4.3 Effects of Fly Ash on Crop Yield

Generally, fly ash amendment in soil increases plant growth and nutrient uptake (Aitken et al. 1984). Various studies have been reported on the amendment of fly ash in soil in different ratios for cultivation of crops (Table. 12.1). The fly ash amendment (50%) improved the growth and yield of cucumber (*Cucumis sativus*) as reported by Ajaz and Tiyagi (2003). The photosynthetic pigments such as total chlorophyll and carotenoids in cucumber increased maximum at 25% (Ajaz and Tiyagi 2003). Root shoot length and dry matter of *Sesbania cannabina* and *Vigna radiata* grown at 10 and 25% fly ash amendment rates enhanced as compared to their controls (Sinha and Gupta 2005). The fly ash amendment in soil caused significant improvement in growth (shoot length, leaf area, and pigment composition) and yield attributes of rice (Mishra et al. 2007).

Application of fly ash at and above 50% reduced photosynthetic pigments, protein content, and growth (plant height, root biomass, number of tillers, grain, and straw weight) in four cultivars of rice (Dwivedi et al. 2007), but lower concentrations (10–25%) of fly ash amendment in soil enhanced the plant growth. Addition of 40% fly ash in soil increased dry matter accumulation and chlorophyll and protein content of *Brassica juncea* (Gautam et al. 2012).

In a recent study, optimum growth (dry matter and seedling) of rice was recorded on application of 25% fly ash in soil. The growth and reproductive ability of *B. juncea* enhanced on application of 20% to 40% fly ash in soil, but higher amounts of fly ash (60%–100%) affected these growth parameters adversely (Pillai and Chaturvedi 2012).

Plants combat the stress caused by toxicants through enzymatic as well as nonenzymatic antioxidants. When grown in fly-ash-amended soil, antioxidant levels (carotenoids, cysteine, nonprotein thiol, free proline, and ascorbic acid) were higher in roots of *Sesbania cannabina* facing the stress directly than the shoots (Sinha and Gupta 2005). Several studies showed that fly ash application in soil significantly increased the yield of *Brassica oleracea* (Kim et al. 1997), clover (Summers et al. 1998), *Secale cereale* (Matsi and Keramidas 1999), cotton (Dunn and Stevens 2000), lettuce (Lau and Wong 2001), *Cynodon dactylon* (Adriano and Weber 2001), *Zea mays* (Tarkalson et al. 2005), *Oryza sativa* (Panda and Tikadar 2014), *B. campestris* (Jayasinghe and Tokashiki 2012), *B. juncea* (Gautam et al. 2012), *Solanum melongena* (Gond et al. 2013), and biofuel feedstock (Dzantor et al. 2013). The agronomic benefits of fly ash applications are primarily associated with improved physicochemical and biological characteristics of the soil. Because of the presence of Ca-Si minerals with a pozzolanic nature, along with the soil moisture, fly ash improves soil bulk density, porosity, and water-holding capacity.

Table 12.1 Reported studies on plant species grown on fly-ash-amended soil

Plant species	Amendments and doses	Findings	References
Cichorium intybus	3% FA dose and the soil was contaminated with Cd, Cu, Zn, and Ni	FA can produce alkalizing effects which can be used to decrease plant accumulation of heavy metals, mainly in acidic soils with less buffering capacity	Scotti et al. (1999)
Beta vulgaris L.	2%, 4%, and 8% FA amendment to soil	Application of low doses (2% FA, w/w) helped plant development and enhanced yield and carbohydrate content	Singh et al. (1994)
Brassica juncea L. var. Vaibhav	10%, 25%, 50%, 75%, and 100% FA (w/w) amendments to soil	*B. juncea* was reported as a good phytoextractor of metals, especially for Ni, when grown in FA-amended soil	Gupta and Sinha (2006)
Sesbania cannabina L.	10%, 25%, 50%, 75%, and 100% FA (w/w) amendments to soil	Maximum accumulation of Fe and less of Ni was reported. The concentration of metals was found to be reduced with an increase in FA application ratio. Antioxidant concentration in plant leaves was found to combat metal stress for all the exposure periods. So it was concluded that *S. cannabina* can be used for phytoremediation and revegetation of FA-contaminated sites	Sinha and Gupta (2005)
Glycine max, Cicer arietinum, Vigna radiate, Vigna mungo, Vigna unguiculata, Vicia faba	FA dose at 1 g/5 kg pulses	FA might be used as a post-harvest preserving material	Mendki et al. (2001)
Turf species, centipede grass [*Eremochloa ophiuroides* (Munro) hack.]	FA application at 0, 280, 560, and 1120 mg ha^{1}	Plant growth parameters like height and root depth were not adversely affected, as were the dry biomass yields throughout the study period and ash treatment did not affect the physical parameters of soil, but considerably improved water-holding capacity and plant-available water. Greater water-holding capacity enhanced the cohesion	Adriano and Weber (2001)
Phaseolus vulgaris	FA doses at 10%, 25%	Root to shoot translocation of metals was increased	Gupta et al. (2007)
Vicia faba L.	FA dose at 25, 50, 75, and 100%	FA amendment in the agricultural soil is beneficial for plant growth at a lower percentage	Rai et al. (2003)

References

Acton DF, Gregorich LJ (1995) The health of our soils: toward sustainable agriculture in Canada. Centre for Land and Biological Resources Research. Research Branch Agriculture and Agri-Food Canada, Ottawa, Canada

Adriano DC, Weber JT (2001) Influence of fly ash on soil physical properties and turf grass establishment. J Environ Qual 30:596–601

Aitken RL, Campbell DJ, Bell LC (1984) Properties of Australian fly ash relevant to their agronomic utilization. Aust J Soil Res 22:443–453

Ajaz S, Tiyagi S (2003) Effect of different concentrations of fly ash on the growth of cucumber plant, *Cucumis sativus*. Arch Agron Soil Sci 49:457–461

Behera B, Mishra MK (2012) Microstructure and leaching characteristics of Fly ash-mine overburden-lime mixtures. In: International conference on chemical, civil and environment engineering (ICCEE'2012), Dubai, March 24–25, pp 256–260

Bhattacharya T (2007) The effect of speciation of heavy metals on their uptake and growth of two wetland plants. PhD thesis. http://hdl.handle.net/10603/29859

Bradshaw AD, Chadwick MJ (1980) The restoration of land. Blackwell Scientific Publications, Oxford

Chaudhary SK, Rai UN, Mishra K, Huang HG, Yang XE, Inouhe M, Gupta DK (2011) Growth and metal accumulation potential of *Vigna radiata* L. grown under fly ash amendments. Ecol Eng 37:1583–1588

Dunn D, Stevens G (2000) Fly ash as a liming material for cotton: a rate study. In: Proceedings Cotton Conference, 2, 1402–1404

Dwivedi S, Tripathi RD, Srivastava S, Mishra S, Shukla MK, Tiwari KK, Singh R, Rai UN (2007) Growth performance and biochemical responses of three rice (*Oryza sativa* L.) cultivars grown in fly-ash amendment soil. Chemosphere 67:140–151

Dzantor EK, Pettigrew H, Adeleke E, Hui D (2013) Use of fly ash as soil amendment for biofuel feedstock production with concomitant disposal of waste accumulations. WOCA, World of Coal Ash Association, Lexington, KY, April 22–25, 2013. http://www.flyash.info

Gautam S, Singh A, Singh J, Shikha (2012) Effect of fly ash amended soil on growth and yield of Indian mustard (*Brassica Juncea*). Adv Biores 3:39–45

Gond DP, Singh S, Pal A, Tewary BK (2013) Growth, yield and metal residues in *Solanum melongena* grown in fly ash amended soils. J Environ Biol 34:539–544

Goswami RK, Mahanta C (2007) Leaching characteristics of residual lateritic soils stabilised with fly ash and lime for geotechnical application. Waste Manag J 27(4):466–481

Gupta AK, Sinha S (2006) Role of Brassica juncea (L.) Czern. (var. Vaibhav) in the phytoextraction of Ni from soil amended with fly ash: selection of extractant for metal bioavailability. J Hazard Mater 136:371–378

Gupta AK, Dwivedi S, Sinha S, Tripathi RD, Rai UN, Singh SN (2007) Metal accumulation and growth performance of *Phaseolus vulgaris* grown in fly-ash amended soil. Bioresour Technol 98:3404–3407

Hodgson DR, Townsend WN (1973) The amelioration and revegetation of pulverized fuel ash. In: Hutnik RJ, Davis G (eds) Ecology and reclamation of devastated land, vol 2. Gordon and Breech, London, pp 247–270

Hodgson DR, Dyer D, Brown DA (1982) Neutralization and dissolution of high-calcium fly ash. J Environ Qual 11:93–98

Jala S, Goyal D (2006) Fly ash as a soil ameliorant for improving crop production: a review. Bioresour Technol 97:1136–1147

Jayasinghe GY, Tokashiki Y (2012) Influence of coal fly ash pellet aggregates on the growth and nutrient composition of *Brassica campestris* and physico-chemical properties of grey soils in Okinawa. J Plant Nutr 35:47–45

Kim BJ, Back JH, Kim YS (1997) Effect of fly ash on the yield of Chinese cabbage and chemical properties of soil. Korean J Soil Sci Fertilizer 30:161–167

Kumar A, Vajpayee P, Ali MB, Tripathi RD, Singh N, Rai UN, Singh SN (2002) Biochemical responses of *Cassia siamea* Lamk grown on coal combustion residue (fly-ash). Bull Environ Contam Toxicol 68:675–683

Lau SSS, Wong JWC (2001) Toxicity evaluation of weathered coal fly ash amended manure compost. Water Air Soil Pollut 128:243–254

Matsi T, Keramidas VZ (1999) Fly ash application on two acid soils and its effect on soil salinity, pH, B, P, and on ryegrass growth and composition. Environ Pollut 104:107–112

Mendki PS, Maheshwari VL, Kothari RM (2001) Fly ash as a post-harvest preservative for five commonly utilized pulses. Crop Prot 20:241–245

Mishra LC, Shukla KN (1986) Effect of fly-ash deposition on growth, metabolism and dry matter production of maize and soybean. Environ Pollut Series A 42:1–13

Mishra M, Sahu RK, Padhy RN (2007) Growth, yield and elemental status of rice (*Oryza sativa*) grown in fly ash amended soils. Ecotoxicology 16:271–278

Natusch DSF, Wallace JR (1974) Urban aerosol toxicity: the influence of particle size. Science 186:695

Palit A, Gopal R, Dube SK, Mondal PK (1991) Characterization and utilization of coal ash in the context of super thermal power stations. In: Proceedings international conference, on environmental impact of coal utilization. IIT, Mumbai, pp 154–155

Panda D, Tikadar P (2014) Effect of fly ash incorporation in soil on germination and seedling characteristics of rice (*Oryza sativa* L.). Biolife 2:800–807

Pandey SK, Bhattacharya T (2019) Mobility, ecological risk and change in surface morphology during sequential chemical extraction of heavy metals in fly ash: a case study. Environ Technol Innov 13:373–382. https://doi.org/10.1016/j.eti.2016.10.004

Pandey VC, Abhilash PC, Singh N (2009) The Indian perspective of utilizing fly ash in phytoremediation, phytomanagement and biomass production. J Environ Manag 90:2943–2958

Pillai P, Chaturvedi A (2012) Effect of fly ash on reproductive biology of oil yielding plant *Brassica juncea* var. Pusa bold. Global J Biosci Biotechnol 1:175–177

Rai UN, Gupta DK, Akhtar M, Pal A (2003) Performance of seed germination and growth of *Vicia faba L.* in fly-ash amended soil. J Environ Biol 24:9–15

Ram LC, Masto RE (2014) Fly ash for soil amelioration: a review on the influence of ash blending with inorganic and organic amendments. Earth Sci Rev 128:52–74

Rohriman FA (1971) Analyzing the effects of fly ash on water pollution. Power 115:76

Scotti A, Silva S, Botteschi G (1999) Effect of fly ash on the availability of Zn, cu, Ni and cd to chicory. Agric Ecosyst Environ 72:159–163

Singh RP, Agrawal M (2010) Biochemical and physiological responses of rice (*Oryza sativa* L.) grown on different sewage sludge amendments rates. Bull Environ Contam Toxicol 84:606–612

Singh N, Singh SN, Yunus M, Ahmad KJ (1994) Growth response and element accumulation in *Beta vulgaris* L. raised in fly ash amended soils. Ecotoxicology 3:287–298

Sinha S, Gupta AK (2005) Translocation of metals from fly ash amended soil in the plant of *Sesbania cannabina* L. Ritz: effect on antioxidants. Chemosphere 61:1204–1214

Summers R, Clarke M, Pope T, O'Dea T (1998) Western Australian fly ash on sandy soils for clover production. Commun Soil Sci Plant Anal 29:2757–2767

Tarkalson DD, Hergert GW, Stevens WB, McCallister DL, Kachman SD (2005) Fly ash as a liming material for corn production. Soil Sci 170:386–398

Trichoderma: A Multifaceted Fungus for Sustainable Agriculture

13

Swati Sachdev and Rana Pratap Singh

Contents

Abstract

Sustainable agricultural practices are keys for food security of the world's burgeoning population. *Trichoderma* is a ubiquitous fungus that offers several avenues for sustainable agriculture. The panoply of mechanisms displayed by several species of *Trichoderma* makes them a better solution for conventional agricultural problems. Plant protection from unfavorable biotic and abiotic conditions under circumstances of changing global climatic scenario and promoting their growth in soil with limited or poor nutrient conditions are marvelous attributes of *Trichoderma*. Understanding the mechanisms such as the function of

S. Sachdev · R. P. Singh (✉)
Department of Environmental Science, School for Environmental Sciences, Babasaheb Bhimrao Ambedkar University, Lucknow, India

K. Bauddh et al. (eds.), *Ecological and Practical Applications for Sustainable Agriculture*, https://doi.org/10.1007/978-981-15-3372-3_13

secondary metabolites and cell wall degrading enzymes in mycoparasitism and antibiosis, pathways triggered for induced systemic resistance and enhanced nutrient use efficiency displayed by different strains of *Trichoderma* at the physiological, biochemical, and molecular level are essential for harnessing their potential efficiently. Gathering information on performances of *Trichoderma* spp. under variable environmental conditions and their vigilant amalgamation for selection of strains with multiple activity and/or development of consortia for formulation of successful product is the need for current agriculture scenario. The current chapter made an effort to compile information on the beneficial role and mechanisms involved by strains of *Trichoderma* at different levels to enhance knowledge for exploring future research opportunities.

Keywords

Antibiosis · Consortia · Induced systemic resistance · Mycoparasitism · Nutrient use efficiency · Sustainable agriculture

13.1 Introduction

Sustainable agriculture is an agricultural practice that maintains the world's food security to feed the growing population without disrupting ecological integrity. Since the green revolution, the use of agrochemicals has raised agricultural production manifold; however, as a consequence, it has deteriorated the environmental quality, depleted natural resources, caused genetic erosion, and created socioeconomic problems (Shiva 2016; Srivastava et al. 2016). Injudicious and excessive application of agrochemicals together with uncertainty in climatic behavior and other factors including resource depletion, land fragmentation and degradation, and amplified abiotic and biotic stress has challenged agricultural sustainability (Grover et al. 2011; Sachdev and Singh 2016a, b; Ehrlich and Ehrlich 2016; Kashyap et al. 2017). The abiotic and biotic factors driven by climate change have resulted in 1–5% reduction in agricultural production in the last 30 years (Newbery et al. 2016). A paradigmatic shift is obligatory to attain sustainability of the agricultural system for providing food security to increasing global population (9–10 million by 2050) by doubling food production (70–100%) (Godfray et al. 2010; Rockstrom et al. 2017). Sustainable intensification is an alternative approach that provides opportunity to increase food productivity in the same land area while minimizing ecological footprints (Rockstrom et al. 2017). Adoption of integrated pest and nutrient management methods, agroforestry, integrated waste management for livestock production, and use of agriculturally important microorganisms (AIMs) present opportunities for sustainable intensification with reduced negative environmental externalities (Godfray et al. 2010; Singh et al. 2016).

AIMs are beneficial microorganisms like free living fungi, endophytic fungi, arbuscular mycorrhizal fungi (AMF), and plant-growth-promoting rhizobacteria (PGPR) that promote growth and yield of plants by regulating growth of deleterious microflora, acquiring nutrients for plant growth, synthesizing phytohormones, and alleviating abiotic stress (Singh et al. 2016; Sachdev and Singh 2018a). Exploitation of AIMs for sustainable agriculture is an efficient and cost-effective strategy. Interaction of agriculturally important fungi with plant demonstrates paramount importance in sustainability of agriculture and ecosystem (Verbruggen et al. 2012; Ansari et al. 2013; Sachdev and Singh 2018a).

Trichoderma (*Hypocrea*: teleomorph) is a ubiquitous genus that belongs to the ascomyceta division of fungi (Bae et al. 2016). Till present 1100 strains of *Trichoderma* including teleomorphic and anamorphic stages are documented from 75 molecularly distinguished species and new strains are continuously being studied and recognized (Contreras-Cornejo et al. 2016). *Trichoderma* sustains its life under diverse climatic conditions and has wide geographical distribution ranging from tropical climate to polar (Hermosa et al. 2004). *Trichoderma* is a filamentous, saprophytic fungus that proliferates freely in soil and/or shows symbiotic relation with plant roots and foliar parts (Reino et al. 2008). *Trichoderma* spp. possess potential to act as biocontrol agents (Saber et al. 2017; Sachdev and Singh 2018b) and plant growth promoter (Hermosa et al. 2012; Sachdev et al. 2018) and therefore recognized as plant-growth-promoting fungi (PGPF) (Hyakumachi and Kubota 2003; Doni et al. 2014). The role of *Trichoderma* spp. in physiological stress mitigation (Bae et al. 2009; Mastouri et al. 2010; Shoresh et al. 2010) and soil bioremediation (Harman et al. 2004a; Caporale et al. 2014; Tripathi et al. 2017) is also documented in scientific literature by several workers.

The principal biocontrol attribute of *Trichoderma* is believed to involve the mechanism of mycoparasitism, antibiosis and niche exclusion and root colonization; however, other indirect mechanisms of induced systemic resistance and growth promotion are also acknowledged to play a crucial role in the biocontrol and alleviation of abiotic stresses (Harman 2011; Bae et al. 2016; Youssef et al. 2016). Production of antioxidants such as catalase (CAT), superoxide dismutase (SOD), and peroxidase (POD) that are active scavengers of reactive oxygen species (ROS) also enhances resistance toward different stress and is considered as another important mechanism for biocontrol and plant growth under unfavorable environmental cues (Shoresh et al. 2010; Chowdappa et al. 2013; Youssef et al. 2016).

Studies have demonstrated that the positive influence of biocontrol strains on plant is observed during unfavorable conditions (Nawrocka and Malolepsza 2013). Beneficial outcomes of inoculation of *Trichoderma harzianum* T-22 on tomato seedlings were more pronounced under the presence of the pathogen *Pythium ultimum*, abiotic stress, and/or physiological stress (Mastouri et al. 2010). Similarly, changes associated with resistance induced by *Trichoderma* in plants were observed by Sriram et al. (2009) and Moran-Diez et al. (2009) on inoculation with pathogens. Several studies support the fact that *Trichoderma* alleviate abiotic stress by enhancing root growth and nutrient uptake thereby demonstrating plant-growth-promoting attributes (Mastouri et al. 2010). The multifaceted mechanisms displayed by *Trichoderma* spp. that infer protection against unfavorable conditions and promote growth of agricultural crops are demonstrated in Fig. 13.1.

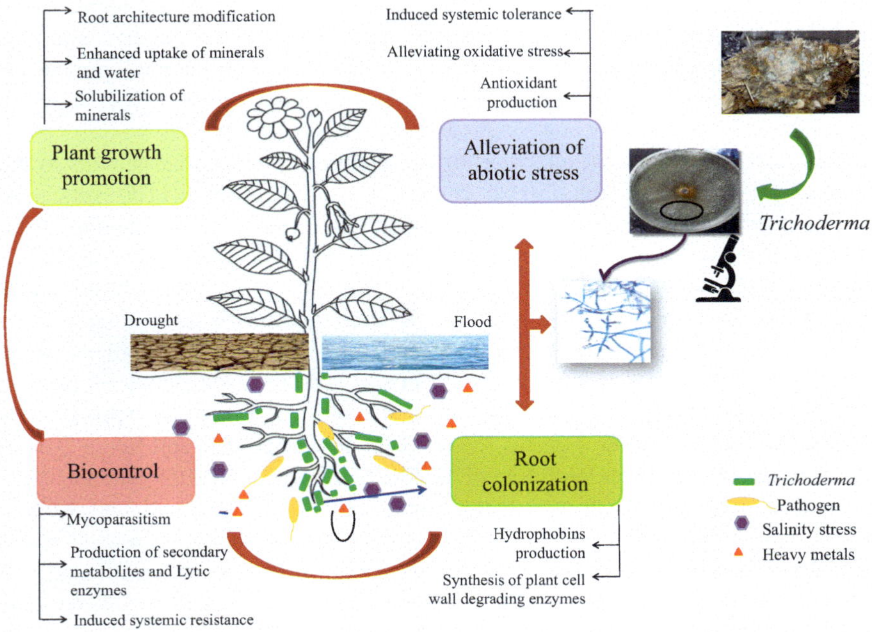

Fig. 13.1 The panoply of mechanisms displayed by *Trichoderma* spp. for sustainable growth of plants under diverse conditions

Trichoderma spp. are versatile fungi with multiple beneficial attributes that make them an appropriate player for establishing agricultural sustainability, and thus to exploit its canopy of benefits, it is essential to gain deeper knowledge on the mechanisms involved by this genus fungi at biochemical, physiological, and molecular levels and the amalgamation of these mechanisms displayed at the time of action. The present work makes an attempt to provide insight on the functioning of *Trichoderma* at different levels to increase understanding for effective use of potential of *Trichoderma* spp. for maintaining agricultural system sustainability.

13.2 *Trichoderma* as Efficient Root Colonizer

Trichoderma species is the most capable organism in soil to mobilize and utilize nutrients (Benitez et al. 2004). This trait is crucial for better root colonization and acquisition of nutrients, making *Trichoderma* an efficient substrate colonizer (Reino et al. 2008; Vieira et al. 2018). Root colonization is an initial mechanism that is prerequisite for displaying direct or indirect mechanisms of biocontrol, plant growth promotion, and alleviation of abiotic stress (Yedidia et al. 1999; Harman et al. 2004b; Benitez et al. 2004). The root exudates synthesized and released by plants attract *Trichoderma* spp. as a source of nutrition to flourish and establish themselves in the rhizosphere (Druzhinina et al. 2011; Hermosa et al. 2012). The carbon-rich nutrition such as sucrose derived from plant roots assists in the communication with soil microorganisms and set off root colonization (Vargas et al. 2009; Macías-Rodríguez et al. 2018). Vargas et al. (2009) and Hermosa et al. (2012) reported that the hydrolysis of sucrose upregulates the synthesis of elicitors that results in defense mechanism activation and increased photosynthesis rate in plant leaves. Similarly, the tomato-root-derived exudates, mainly sugars, acted as chief attractant for *T. atroviride* that improved growth of fungus and assisted root colonization (Macías-Rodríguez et al. 2018).

Root colonization by *Trichoderma* spp. is attained by recognizing and adhering to the root surface, penetrating in root, establishing communication with the plant via chemical signals (Harman 2011; Sachdev and Singh 2018a), and enduring the presence of toxicants produced by plants due to their invasion (Benitez et al. 2004; Hermosa et al. 2012). Colonization of periplasmic spaces in plant roots by *T. virens* was reported by Nogueira-Lopez et al. (2018). The adherence to root surface is mediated by expansin-like proteins and hydrophobins, peptides, and CWDEs (Moran-Diez et al. 2009; Hermosa et al. 2012). After successful colonization, *Trichoderma* grow along with the growth and expansion of plant roots and act as mycorrhiza (Harman 2011). *Trichoderma* produce compounds that induce local or systemic resistance in plant by altering the profile of phytohormones and secondary metabolites such as salicylic acid (SA) and jasmonic acid (JA) (Malinich et al. 2019), impede colonization by phytopathogens by competing for space and nutrition, release antibiotic compounds, improve water and mineral uptake, and acidify ambient environment via synthesis of organic acids ensuing solubilization of minerals and micronutrients for biofertilization (Benitez et al. 2004; Harman 2011; Contreras-Cornejo et al. 2016). Saravanakumar et al. (2016) in their work demonstrated that the production of cellulase after root colonization by *T. harzianum* induced systemic resistance in maize against pest *Curvularia lunata* responsible for leaf spot by upregulating the jasmonic/salicylic acid (JA/SA) pathway. Similarly, colonization of tomato roots by *Trichoderma* spp. induced systemic resistance by triggering the JA/SA-dependent signaling pathway against *Meloidogyne incognita*, which limited invasion, galling, and reproduction of nematode (Martínez-Medina et al. 2017). Malinich et al. (2019) reported that *T. virens* with the ability to colonize maize roots enhanced the expression of genes related to cell wall degradation enzymes, synthesis of phytohormones and secondary metabolites, and production of ROS and signal transduction which facilitates their entry into maize root and help them to survive in a new environmental setup. Macías-Rodríguez et al. (2018) observed successful root colonization by *T. atroviride* that incur bioprotection against the pathogen *Phytophthora cinnamomi* by competing for space and nutrition. The association of fungus with tomato root stimulated plant growth; however, this stimulation was dependent on the percentage or stages of root colonization.

The role of root colonization by *Trichoderma* spp. has also been evaluated in the management of pests feeding on plants. Muvea et al. (2014) provided evidences wherein *Trichoderma* spp. (*T. asperellum, T. atroviride,* and *Hypocrea lixii*) colonizing onion roots significantly lowered the feeding puncture by pest *Thrips tabaci* and also reduced the number of eggs laid. *T. longibrachiatum* MK1, with the ability to colonize tomato roots, promoted growth and development of plant and altered the level of volatile compounds that reduced infestation of the aphid *Macrosiphum euphorbiae* by recruiting its natural enemies *Aphidius ervi* (aphid parasitoid) and *Macrolophus pygmaeus* (aphid predator) by increasing attractiveness (Battaglia et al. 2013). Analogous to this, Contreras-Cornejo et al. (2018) reported enhanced recruitment of *Campoletis sonorensis*, a natural parasitoid of maize pest fall armyworm (*Spodoptera frugiperda*) by maize plants colonized by *T. atroviride*. This increased the susceptibility of pest toward its natural enemy was due to the production of metabolite 6PP (6-pentyl-2H-pyran-2-one) by *Trichoderma*.

13.3 *Trichoderma*: Biotic Stress Manager

The biocontrol potential of *Trichoderma* spp. is an attribute that involves mechanisms of mycoparasitism, competition for nutrition and ecological niche, modification in ambient environmental conditions, antibiosis, and induced systemic resistance (ISR) (Reino et al. 2008; Saravanakumar et al. 2016; Sharma and Gothalwal 2017). These mechanisms either act singularly or synergistically against pathogens after mutualistic association between *Trichoderma* and plant. The biocontrol activity of *Trichoderma* depends on its strain, host plant, pathogen, and prevailing environmental conditions including temperature, availability of nutrients, concentration of iron, and pH (Benitez et al. 2004). Activation of these mechanisms involves production of several metabolites and enzymes such as CWDE, siderophores, carbon, and nitrogen permeases (Sharma et al. 2012) and efficient root colonization (Harman 2011; Sachdev and Singh 2018a). Production of metabolites and root colonization by *Trichoderma* spp. bring changes in gene expression of the host plant, responsible for providing protection or responding toward stress (Marra et al. 2006; Bae et al. 2011).

The growth-promoting and biocontrol activity of *Trichoderma* spp. against phytopathogens increases on interaction with other beneficial biological agents. These enhanced results may be due to the synergistic effect of different or similar mechanisms occurring simultaneously. The production of a higher concentration of antimicrobial compounds on co-inoculation of *Trichoderma asperellum* GDFS1009 with *Bacillus amyloliquefaciens* was reported to significantly increase biocontrol activity against microbes as compared to individual microbial culture (Wu et al. 2018). Similarly, inoculation of *T. harzianum* with *Glomus versiforme* reduced the incidence of powdery mildew and improved growth in cowpea seedlings as compared to individual treatment (Omomowo et al. 2018). The biocontrol and plant-growth-promoting potential displayed by *Trichoderma* spp. is listed below in Table 13.1 with different activities demonstrated by them.

Table 13.1 Mechanisms displayed by *Trichoderma* spp. for biocontrol of phytopathogens and plant growth promotion

Trichoderma spp.	Pest and pathogen(s)	Crop(s)	Activity displayed	References
T. asperellum	*Fusarium graminearum* and *F. verticillioides*	Maize	*Trichoderma* reduced severity of stalk rot and ear rot by 49.83 and 39.63%; altered endophytic microbiome and reduced the accumulation of mycotoxins (deoxynivalenol and fumonisin B1) produced by pathogens	Anle et al. (2019)
Trichoderma spp.	*Fusarium oxysporum* f. sp. *cepae*	Onion	Antagonism	Bunbury-Blanchette and Walker (2019)
T. atroviride	*Meloidogyne javanica*	Tomato	Induced systemic resistance via enhanced expression of JA/SA-dependent genes and ROS-producing genes	Rubio et al. (2019)
T. longibrachiatum	*Valsa mali*	Apple tree	Antibiosis by production of secondary metabolites 1,2-benzenecarboxylic acid bis (2-methyl propyl) ester (DIBP) and 1,2-benzenecarboxylic acid mono(2-ethyl hexyl) ester (MEHP)	Zhang et al. (2018)
T. lixii and *T. brevicompactum*	*Fusarium oxysporum* f. sp. *lycopersici* and *Alternaria solani*	–	Antibiosis and mycoparasitism by production of CWDEs and secondary metabolites	Sachdev and Singh (2018b)
T. lixii	–	Spinach	Plant growth promotion	Sachdev et al. (2018)
T. harzianum	*Sclerotinia sclerotiorum*	Sunflower	Mycoparasitism	Mathews et al. (2019)
T. viride	*Fusarium oxysporum*	Wheat	Production of antioxidants and alleviation of oxidative stress	Mohapatra and Mittra (2017)

(continued)

Table 13.1 (continued)

Trichoderma spp.	Pest and pathogen(s)	Crop(s)	Activity displayed	References
T. harzianum/ T. atroviride	*Uncinula necator*	Grapes	The secondary metabolites harzianic acid and 6PP suppressed growth of pathogen in greenhouse and 6PP in field condition enhanced crop yield, antioxidant activity, and concentration of polyphenols	Pascale et al. (2017)
T. asperellum and *T. harzianum*	*Botrytis cinereal*	*Arabidopsis thaliana*, tomato	The volatile compounds triggered expression of transcription factor *MYB72* that activated iron uptake in root and primed leaves against pathogen by inducing systemic resistance via jasmonic acid (JA) signaling pathway	Martínez-Medina et al. (2017)
T. asperellum CCTCC-RW0014	*Fusarium oxysporum* f. sp. *cucumerinum*	Cucumber	Production of CWDE and secondary metabolites displayed antibiosis and mycoparasitism and resulted in 71.67% disease reduction with increased plant growth	Saravanakumar et al. (2016)
T. atroviride/ petersenii KACC 40557	*Phytophthora* spp.	Pepper and tomato	Metabolites showing antibiosis, inducing systemic resistance genes, and plant hormonal changes	Bae et al. (2016)
T. harzianum	*Rhizoctonia solani*	Tomato	Production of ROS-scavenging antioxidant enzymes provided 61.4% disease protection	Youssef et al. (2016)
T. longibrachiatum	*Fusarium oxysporum* f. sp. *cepae*	Onion	Enhanced plant growth and accumulation of important metabolites that provided protection via induced systemic resistance	Abdelrahman et al. (2016)

T. harzianum T-22	*Cucumber mosaic virus*	Tomato	Reduced disease symptoms due to elicitation of ISR via JA/ethylene and SA signaling pathway; and increase photosynthetic rate, chlorophyll content, plant growth, production of ROS-scavenging enzymes, and phytohormones	Vitti et al. (2016)
T. cremeum	*Fusarium oxysporum*, *R. solani*, and *B. cinereal*	Tomato	The metabolite cremenolide produced by *Trichoderma cremeum* showed antagonistic activity against pathogens under study and increased growth of tomato seedlings	Vinale et al. (2016)
T. atroviride	–	*Arabidopsis thaliana*	The volatile metabolites enhanced plant growth depending upon age of fungal culture and plant	Lee et al. (2015)
T. harzianum	*S. sclerotiorum* and *R. solani*	Tomato	The metabolites iso-harzianic acid (iso-HA) and harzianic acid inhibited mycelia growth of pathogens under *in vitro* condition and improved seed germination and root and shoot length and induced resistance to diseases in tomato in *in vivo* condition	Vinale et al. (2014b)
T. parareesei T6	*B. cinerea*	Tomato	Lateral root growth of seedlings increased and plant's defense system was primed by upregulation of expression of JA/ethylene-related *EIN2* and *LOX1* genes and SA-related *PR-1* gene	Rubio et al. (2014)
T. harzianum T-78	*B. cinereal*	Tomato	Elicitation of ISR mediated primarily by JA signaling pathways along with SA and ABA signaling that restricted pathogen growth and disease development	Martinez-Medina et al. (2013)

(continued)

Table 13.1 (continued)

Trichoderma spp.	Pest and pathogen(s)	Crop(s)	Activity displayed	References
T. harzianum OTPB3	*A. solani*; *Phytophthora infestans*	Tomato	Elicitation of ISR via enhanced production of defense-related enzymes, i.e., polyphenol oxidase, peroxidase, superoxide dismutase	Chowdappa et al. (2013)
T. virens PS1-7	*Burkholderia plantarii*	Rice	The active compound carot-4-en-9,10-diol represses production of tropolone by pathogen via quorum quenching	Wang et al. (2013)
T. harzianum ALL42	*R. solani* and *Macrophomina phaseolina*	–	*Trichoderma* was overgrown and disintegrated pathogen mycelia by coiling around hyphae producing appressoria and CWDE	Monteiro et al. (2010)
T. asperellum T-34	*Pseudomonas syringae* pv. *tomato; Hyaloperonospora parasitica; Plectosphaerella cucumerina*	*Arabidopsis thaliana*	T-34 triggered resistance in leaves against pathogens and primed leaves against pathogen invasion by increasing development of callose-containing papillae and jasmonic acid-responsive gene expression	Segarra et al. (2009)
T. harzianum	*R. solani*	–	Mycoparasitism was displayed via effective coiling around pathogen hyphae and production of hydrolytic enzymes chitinases, N-acetyl-b-D-glucosaminidase (NAGase), and b-1,3-glucanases	Almeida et al. (2007)
T. harzianum T39	*B. cinereal*	Tomato, tobacco, pepper, and lettuce	Activation of mechanism induced systemic resistance (ISR) resulted in 25–100% disease control	De Meyer et al. (1998)
T. harzianum	*B. cinereal*	–	*Trichoderma* initially displayed antibiosis and caused cell death which was proceeded by cell degradation by chitinolytic activity	Belanger et al. (1995)

13.3.1 Mycoparasitism

It is one of the major traits displayed by *Trichoderma* spp. against phytopathogens (Chet et al. 1981). This trait enables *Trichoderma* spp. to parasitize and kill host fungi (Mukherjee et al. 2012). The mycoparasitism is an intricate process that involves direct physical interaction of *Trichoderma* with pathogens followed by secretion of antibiotics and CWDEs such as β-glucanase, chitinase, and protease (Benitez et al. 2004; Druzhinina et al. 2011; Saravanakumar et al. 2016; Guzmán-Guzmán et al. 2018). The mycoparasitic activity of *Trichoderma* has been documented against several pathogens like *Fusarium* spp., *Alternaria alternata*, *Rhizoctonia solani*, *Sclerotinia sclerotiorum*, *Sclerotium rolfsii*, *Pythium* spp., and root knot nematodes (Howell 2003; Affokpon et al. 2011; Druzhinina et al. 2011; Troian et al. 2014; Contreras-Cornejo et al. 2016; Mathews et al. 2019).

The mycoparasitism mechanism takes into account several episodes that include recognition of pathogens, attack, penetration, and finally lysis of host pathogen. Mycoparasitism is not merely a contact mechanism; it is a host-specific process, wherein genes of *Trichoderma* responsible for the production of hydrolytic enzymes are transcribed after receiving signals from the host fungus (Zeilinger and Omann 2007). Once *Trichoderma* senses host pathogens, it coils around it and produces appressoria-like structure, which with the help of CWDEs disintegrate cell wall polysaccharides and penetrate inside the host cytoplasm, utilizes its cell content as a source of nutrition, and finally causes death of the pathogen (Kubicek et al. 2001; Viterbo et al. 2002; Gajera et al. 2012; Troian et al. 2014; Saravanakumar et al. 2016). The inhibition of fungal growth by *T. harzianum* was documented by Braun et al. (2018) who found attachment of *Trichoderma* on mycelia of *Aspergillus* resulting in enzymatic lysis of the latter.

The cell wall of fungi is generally comprised of chitin (Brozová 2004) and β-glucan (Troian et al. 2014), and during mycoparasitism, hydrolysis of pathogen fungal cell wall by *Trichoderma* takes places through expression of genes that encode for different CWDEs (chitinase, glucanase) (Lorito 2009; Sharma et al. 2012). Djonovic et al. (2007) reported the key role of β-1,3-gluacanase and β-1,6-glucanase in mycoparasitism by demonstrating the inhibitory action of *T. virens* transformant with coexpressed genes for β-1,3-gluacanase and β-1,6-glucanase against *Pythium ultimum* and *Rhizoctonia solani*. Antagonistic *T. harzianum* T20 was reported to show mycoparasitic activity against red rot pathogen *Colletotrichum falcatum* via upregulation of genes encoding CWDEs such as *chit33* (chitinase), *endo42* (endochitinase), *glu* (β-1,3 endoglucanase), *exc1* and *2* (exochitinase 1 and 2) during and after interaction with pathogen mycelium, and genes *prb1*, *pra1*, and *ssp* related to protease enzymes only during interaction (Elamathi et al. 2018). Analogously, the genome sequencing of *T. hamatum* GD12 revealed the presence of a unique genomic region which encodes for novel bioactive metabolites that participate in mycoparasitism against an array of soil pathogens (Studholme et al. 2013). The analysis of a unique genomic region of GD12 provides evidences for its involvement in secondary metabolism and production of antibiotics such as fengycin and surfactin (Wu et al. 2007). The mycoparasitism involves a highly conserved signaling pathway (Zeilinger and Omann 2007). The α-subunit of G proteins (G-α), a key component of cAMP and mitogen-activated protein (MAP) kinase signaling pathways, is found responsible for the production of extracellular enzymes, antibiotics, and coiling of *Trichoderma* around the host hyphae during mycoparasitism (Omero et al. 1999; Mukherjee et al. 2003; Silva et al. 2004; Almeida et al. 2007).

13.3.2 Competition

Competition for space and nutrition is an important mechanism for biocontrol of phytopathogens. *Trichoderma* are efficient competitors for limited nutrients that deprive other organisms from essential life-sustaining nutrients, leading to starvation resulting in death and elimination (Benitez et al. 2004). Competition for nutrients is reported as the best mechanism for growth suppression of fungal pathogen *Botrytis cinerea*, which is sensitive to low nutrient level (Benitez et al. 2004). Macías-Rodríguez et al. (2018) observed that in the presence of carbohydrate, *T. atroviride* growth was stimulated limiting the growth of the fungal pathogen *Phytophthora cinnamomi* on interaction by competing for space and nutrition.

Iron is a vital and indispensable micronutrient required for the optimum growth and viability of organisms (Eisendle et al. 2004). During limited availability of iron in soil, microorganisms secrete siderophores – an iron-chelating compound with low molecular weight that binds with unavailable forms of iron (III) and regulates its availability in the rhizosphere (Hider and Kong 2010; Vinale et al. 2013). The competition for iron-binding sites in the rhizosphere is determined by the siderophores' affinity for iron (Vinale et al. 2013). The iron complexes formed by chelation are recognized and absorbed by plants, which is important for satisfying the requirement of iron by plants, mainly in calcareous soil (Sharma and Johri 2003; Vinale et al. 2013). Some *Trichoderma* spp. are efficient siderophore producers that chelate iron and inhibit growth of other organisms (Benitez et al. 2004). *T. harzianum* produces harzianic acid that binds with Fe^{3+}, making it soluble and available for plants and influenced the growth of plant colonized by *T. harzianum* even under iron-deficient conditions (Vinale et al. 2013). Mukherjee et al. (2018) reported the role of intracellular siderophore ferricrocin synthesized by *T. virens* in inducing systemic resistance in maize thereby controlling the foliar pathogen *Cochliobolus heterostrophus*.

Trichoderma demonstrate good resistance against toxic metabolites secreted by other microorganisms and chemicals generally used in agriculture such as pesticides, fungicides, metals, and phenolic compounds and have the ability to proliferate in extreme competitive conditions. *Trichoderma* are fast-growing fungi that aggressively colonize on various substrates, compete for ecological niche, and eliminate pathogens (Bailey and Lumsden 1998). *Trichoderma*, when applied as soil treatment or coated on seeds, create a zone of protection around plants and provide resistance against pathogens. The spore of *T. harzianum* applied as soil treatment suppressed infection caused by *Fusarium oxysporum* f. sp. *melonis* and *Fusarium oxysporum*

f. sp. *vasinfectum* probably due to colonization of *Trichoderma* around plant developing roots suggesting competition for ecological niche (Ahmad and Baker 1987).

13.3.3 Antibiosis

Antibiosis involves low-molecular-weight diffusible nonvolatile and volatile metabolites possessing antibiotic properties that are toxic to other microorganisms and impede their growth (Vinale et al. 2014a). Many fungal species are well known to secrete secondary metabolites that may not be essentially involved in direct growth but play an imperative role in signaling and interaction with other organisms (Keller et al. 2005; Mukherjee et al. 2012) and check growth of other microorganisms by targeting their sporulation and hyphal elongation, while some of these metabolites directly alter growth and metabolisms of plants (Vinale et al. 2008, 2014a). The production of antibiotics by fungus depends on its growth stage, prevailing environmental conditions, nutritional status of colonizing substrate, and competition with other organisms for ecological niche and nutrition (De la Cruz-Quiroz et al. 2019).

Trichoderma species are able to secrete a variety of nonvolatile and volatile compounds such as nitrogen heterocyclic compounds (harzianic acid), peptaibols (alamethicin), tricholin, viridins (viridin, viridiol), diketopiperazines (gliovirin, gliotoxin), pyrones (6-pentyl-α-pyrone), terpenoids, trichothecene, and others that are capable of suppressing growth of various microbes (Sivasithamparam and Ghisalberti 1998; Reino et al. 2008; Vinale et al. 2014b; Contreras-Cornejo et al. 2016). The metabolites produced by *Trichoderma* help plants to withstand in the presence of pathogens and stimulate plant defense mechanism and promote plant growth (Vinale et al. 2008, 2012; Liu et al. 2016). The metabolites harzianic acid, harzianolide, T39 butenolide, T-22 azaphilone, harzianopyridone, and peptaibols produced by *T. harzianum* have shown marked inhibitory effect on mycelia growth of various fungal phytopathogens (*Gaeumannomyces graminis* var. *tritici*, *Pythium ultimum*, *Pythium irregular*, *Sclerotinia sclerotiorum*, *Rhizoctonia solani*, *Botrytis cinerea*, *Fusarium oxysporum* f. sp. *conglutinans*) (Vinale et al. 2006, 2009b, 2014b; Palyzová et al. 2019). The isocyano metabolites homothallins produced by *T. koningii* were reported to affect the morphology of *Phytophthora* spp. (Reino et al. 2008). The two volatile metabolites 6n-pentyl-α-pyrone (6PP) and 6n-pentenyl-α-pyrone produced by *T. harzianum* were found to possess an inhibitory effect against the pathogen *Rhizoctonia solani* responsible for damping off in lettuce seedlings (Claydon et al. 1987). Similarly, trichodermin, a trichothecene metabolite biosynthesized by *T. brevicompactum* (Shentu et al. 2014), and cerinolactone extracted from *T. sirenum* (Vinale et al. 2012) were observed to suppress the growth of *Botrytis cinerea* and *Rhizoctonia solani* significantly.

The metabolites produced by *Trichoderma* vary from isolates to isolates and even within species (Ghisalberti and Sivasithamparam 1991; Vinale et al. 2009a; Liu et al. 2016), and their activity is reported to have differential effects on different pathogens. The bioactive compounds showing detrimental effect on one pathogen might show no observable negative effect on the growth of another pathogen. For example, two diketopiperazine secondary metabolites of *T. virens*, gliotoxin and gliovirin, were found to have a differential inhibitory effect on phytopathogen *Rhizoctonia solani* and *Pythium ultimum*. Gliotoxin displayed higher suppressing effect on *R. solani* than *P. ultimum*, whereas gliovirin was effective only against *P. ultimum* and the growth of *R. solani* remained unaffected by its presence (Howell et al. 1993). Furthermore, the secondary metabolites of *T. asperellum* checked the growth of *Phytophthora capsici*; however, the growth of *Colletotrichum gloeosporioides* remained unaltered (De la Cruz-Quiroz et al. 2019). One of the reasons for inability of secondary metabolites to hinder the growth of pathogens may be because of the capability of some pathogenic fungi to biotransform these metabolites into nontoxic/less toxic form, as a strategy to sustain life under harsh environment (Zikmundova et al. 2002; Vinale et al. 2009a). This indicates that antibiotic activity of metabolites is dependent on *Trichoderma* strain, host pathogens, their viability, and rate of biotransformation (Vinale et al. 2009a).

The synthesis of secondary metabolites by *Trichoderma* is sometimes stimulated after its interaction with plant tissues. Inoculation of tomato tissues in growth medium with *T. harzianum* stimulated the biosynthesis of harzianic acid and its stereoisomer iso-harzianic acid. Biosynthesis of harzianic acid was stimulated positively; however, the production of iso-harzianic acid was negatively modulated (Vinale et al. 2014a). Similar observations were reported by Marfori et al. (2002) demonstrating production of antimicrobial compound trichosetin (*N*-desmethyl analog of equisetin) by *T. harzianum* cocultured with *Catharanthus roseus* callus, which was otherwise not found to be produced in the absence of either *T. harzianum* or *C. roseus* callus.

The secondary metabolites are often reported to act synergistically with CWDEs (Błaszczyk et al. 2014). The dual effect of lytic enzymes and antibiotics enhances the antagonistic activity of biocontrol agents suggesting synchronization between mycoparasitism mechanism and antibiosis in pathogen control (Li et al. 2016). *Trichoderma* is known to exploit hydrolytic enzymes and secondary metabolites simultaneously to antagonize target microorganisms. Synergism of lytic enzymes and secondary metabolites of *T. harzianum* has been reported against fungal pathogens (Schirmbock et al. 1994). Analogously, Li et al. (2016) identified significant synergistic effect of lytic enzymes and secondary metabolites synthesized by *T. asperellum* ZJSX5003 that enhanced the antagonistic activity of *Trichoderma* against *Fusarium graminearum* causing stalk rot in maize. The study on antagonistic effects of CWDEs and secondary metabolites corroborates with the fact that application of these enzymes and metabolites can provide significant avenues for disease control as they are commercially more acceptable and viable as compared to living microbes (Li et al. 2016).

Though metabolite productions by *Trichoderma* show significant benefits, few metabolites produced by some strains are also reported to be phytotoxic (Contreras-Cornejo et al. 2016). Bae et al. (2011) in their study observed metabolites, produced by *T. carribbaeum* var. *aequatoriale* DIS 320c, showing 100% growth inhibition of *Phytophthora capsici* that caused browning of pepper roots and damaged them.

Trichothecene produced by *T. brevicompactum* was reported to have strong inhibitory effect on tomato seed germination (Nielsen et al. 2005; Degenkolb et al. 2008; Tijerino et al. 2011a). Further, greenhouse assay on the effect of seed coating of *T. brevicompactum* to tomato seeds showed enhanced leaf necrosis by *B. cinerea* that rules out application of strains of *T. brevicompactum* as biocontrol agent (Tijerino et al. 2011b). Therefore, before use and commercialization of metabolites as biocontrol agents, it is important to understand the effect of their interaction with plants.

13.3.4 Modulation of Systemic Resistance (SR) in Plants by Trichoderma

On subsequent infection of pathogens, a systemic resistance (SR) is developed in plants to cope up with invasion. This natural phenomenon is termed as systemic acquired resistance (SAR). However, when nonpathogenic microbes invade or colonize plant roots, the systemic resistance developed in plants is termed as induced systemic resistance (ISR) due to involvement of different signaling pathways in SAR and ISR at biochemical and molecular levels (De Meyer et al. 1998; Nawrocka and Malolepsza 2013). Several biocontrol agents, such as rhizospheric fungi and bacteria, suppress plant diseases by reprogramming gene expressions (Vitti et al. 2016) which in turn release elicitor molecules that activate signaling pathways and trigger physiological as well as biochemical changes in plants (Mastouri et al. 2010; Harman et al. 2012; Nawrocka and Malolepsza 2013). Induced systemic resistance triggered by biocontrol rhizobacteria (*Pseudomonas* spp.) and *Trichoderma* spp. has shown similarity to much greater extent as demonstrated by Segarra et al. (2009), who reported involvement of MYB72 transcription factor for early signaling by both ISR displaying biocontrol agents.

Induced systemic resistance by biocontrol agents provides protection against infections occurring at distal parts of plants on subsequent infection of pathogens. De Meyer et al. (1998) in their study indicated elicitation of induced systemic resistance against *B. cinerea* by inoculating *T. harzianum* T39 at spatially separate site that rules out the possibility of involvement of other mechanisms of biocontrol. Similarly, Rubio et al. (2019) reported induction of systemic resistance in tomato plant by *T. atroviride* against nematode *Meloidogyne javanica* by inoculating on different spatial sites using the split root method. *T. atroviride* increased the expression of jasmonic acid/salicylic acid genes and ROS-producing genes. Another study investigated by Alizadeh et al. (2013) on the effectiveness of combined application of *T. harzianum* Tr6 with *Pseudomonas* sp. Ps14 on the efficacy of ISR in *Arabidopsis thaliana* and cucumber against *Fusarium oxysporum* f. sp. *radicis cucumerinum* and *B. cinerea*, respectively, depicted participation of signaling pathways in biocontrol. The results obtained revealed enhanced ISR in cucumber by use of consortium as compared to individual strain, whereas in *A. thaliana*, biocontrol potential of consortium was at par with individual treatment suggesting activation of two different signaling pathways in cucumber and a similar pathway in *A. thaliana* by two biocontrol agents.

Trichoderma spp. produce an arsenal of compounds and/or hormones that elicit defense in host plant against abiotic and biotic factors known as microbe-associated molecular patterns (MAMP) (Djonovic et al. 2006; Sriram et al. 2009; Shoresh et al. 2010; Druzhinina et al. 2011; Mathys et al. 2012; Alizadeh et al. 2013; Martinez-Medina et al. 2013). Plants recognize these MAMPs that trigger a cascade of signals eliciting resistance in plants (Hermosa et al. 2012). Phytohormones such as jasmonic acid (JA), ethylene (ET), salicylic acid (SA), and abscisic acid (ABA) act as primary signaling molecules that regulate signaling pathways activating local and systemic defense response in plants (Pieterse et al. 2009; Martinez-Medina et al. 2013). The jasmonic acid pathway has been recognized as a major signaling pathway participating in the activation of ISR by *Trichoderma* spp.; however, involvement of other signaling pathways (ET and SA) is strain dependent (Korolev et al. 2008; Mathys et al. 2012; Martinez-Medina et al. 2013). Inoculation of cucumber plants by *T. asperellum* T203 conferred protection against *Pseudomonas syringae* pv. *lachrymans* by eliciting induced resistance via jasmonic acid and ethylene signaling pathways (Shoresh et al. 2005). The infected plants inoculated with *T. asperellum* T203 showed higher expression of genes involved in induced resistance compared to plants only challenged with pathogen indicating involvement of *Trichoderma* spp. in resistance induction (Shoresh et al. 2005). Further, inhibition of JA/ET synthesis resulted in diminished protection against pathogens indicating the role of signaling molecules in microbe-mediated systemic resistance (Shoresh et al. 2005). Analogously, the study performed by Segarra et al. (2009) documented ISR triggered by *T. asperellum* T-34 on subsequent root colonization that developed resistance in leaves against several pathogens and primed leaf tissues for alteration in gene expression responsive for jasmonic acid signaling and development of callose-containing papillae on pathogen attack. Further, ISR was reported to be fully expressed in mutant *sid2* (salicylic acid impaired mutant) responsible for triggering SAR but disrupted in mutant *npr1* impaired in SAR and ISR.

13.4 *Trichoderma*-Mediated Tolerance to Abiotic Stress

In natural conditions, plants frequently encounter several abiotic stresses such as salinity, acidic condition, drought (water deficiency), flooding (anoxia), temperature fluctuation (heat/chilling stress), poor nutrition level in soil, light intensity, toxic levels of heavy metals which cause nutrient deficiency, oxidative stress, early senescence, impaired seed germination, plant dehydration, membrane dysfunction, photosynthesis inhibition, and disturbed ion and osmotic homeostasis (Essah et al. 2003; Wang et al. 2003; Chaves and Oliveira 2004; Agarwal and Grover 2006; Hirel et al. 2007; Katori et al. 2010; Contreras-Cornejo et al. 2014; Rawat et al. 2016; Meena et al. 2017). It has been reported that under the influence of abiotic stresses, plant growth and productivity are reduced by half as compared to that result in ideal growth condition (Boyer 1982; Brotman et al. 2013). Drought is the most severe stress followed by salinity, which are responsible for limiting crop productivity (Wani et al. 2016). Abiotic constraints have affected most of the global land area and thus unaffected land comprises only 3.5% of the world's land area (Meena et al. 2017). Around 7% of the world's land area has become saline and one-third of the irrigated land masses are under threat of induced salinization (Contreras-Cornejo et al. 2014). Around 64% of land area is subject to water-deficient conditions; 13% is affected by flood; 9% land area is nutrient deprived; acidic soil shares 15% and cold 57% (Mittler 2006; Cramer et al. 2011; Meena et al. 2017). Plants are gifted with intrinsic capabilities to cope with environmental stresses which work on the cost of their growth and productivity resulting in "trade-off" (Huot et al. 2014). The soil natural microbial biome supports plants to render effect of abiotic stresses (Turner et al. 2013) by evoking resistance in plants termed as induced systemic tolerance (IST) that improves metabolic activity (Nguyen et al. 2016; Meena et al. 2017), thereby clinching plants from reaching to a deteriorating level of trade-off.

Trichoderma species are well known to alleviate abiotic stresses and promote plant growth under stress conditions (Contreras-Cornejo et al. 2014; Ahmad et al. 2015; Vieira et al. 2018). Under the influence of abiotic stress condition (salt, suboptimal temperature, and osmotic stress), the rate of seed germination and uniformity was found to be increased on inoculation of seeds with *Trichoderma harzianum* T-22 (Mastouri et al. 2010). The exact mechanism for stress alleviations is not known but many studies have reported production of auxin-like and cytokinin-like compounds and ACC deaminase, upregulation of gene expression for induced resistance, increase in antioxidant capacity, production of phytoalexin and phenols, etc., by *Trichoderma* spp. implicated with stress tolerance in plants (Harman et al. 2004a; Rawat et al. 2011; Brotman et al. 2013; Contreras-Cornejo et al. 2014; Kashyap et al. 2017). In the present context, evidences reported suggest the ability of *Trichoderma* to enhance abiotic stress tolerance capacity in plants. The presence of salt in growth media is reported to inhibit root hair growth in plants (Halperin et al. 2003) and results in declined level of endogenous hormones (Afzal et al. 2006). Inoculation of *Arabidopsis* seedlings with *T. virens* or *T. atroviride* was found to increase growth of lateral roots and root hairs, suggesting root plasticity as a part of the mechanism displayed by *Trichoderma* spp. that may confer tolerance to salt stress (Contreras-Cornejo et al. 2014). The study further verified that auxin production by *Trichoderma* spp. maintains plant auxin homeostasis and normalized plant growth and development under salinity stress. Some of investigated *Trichoderma* spp. possessing ability to alleviate abiotic stress is listed in Table 13.2.

Apart from auxin synthesis, *Trichoderma* spp. enhance osmoprotectant and antioxidant production capacity in plant and demonstrated exclusion of Na^+ from roots under saline state conferring tolerance to stress condition (Yesilyurt et al. 2018; Anam et al. 2019). Proline is accumulated in plants when exposed to abiotic stress conditions and maintains osmotic level in cytosol providing protection against salt stress (Khomari et al. 2018). Malondialdehyde (MDA) concentration increase during abiotic stress condition indicates level of lipid peroxidation (Bernat et al. 2018). Wheat seed bioprimed with *T. harzianum* Th-14 were found with higher osmolyte concentration under salt stress condition that maintained cellular macromolecule

Table 13.2 Potential of *Trichoderma* spp. to alleviate abiotic stress

Trichoderma spp.	Stress	Crop(s)	Activity displayed	Reference
T. asperellum	Lead and salinity stress	*Suaeda salsa*	Bioaugmentation of plant with *Trichoderma* promoted plant growth and alleviated oxidative stress by reducing malondialdehyde concentration by 7–85% and peroxide level by 7–49%. It also decreased Pb, Na^+, and Ca^{+2} content and translocation in plant	Li et al. (2019)
T. asperellum RM-28	Sodic/saline–alkaline stress (red mud)	Sorghum–sudangrass	Inoculation of *Trichoderma* reduced pH and electrical conductivity of red mud, improved plant growth and chlorophyll content, and reduced oxidative stress	Anam et al. (2019)
T. harzianum	2,4-Dichloro-phenoxyacetic acid (2,4-D)	Wheat	Inoculation of wheat seedlings with *Trichoderma* in the presence of herbicide 2,4-D resulted in stimulated growth on increase in phosphatidylcholine and reduction in lipid peroxidation and oxylipins	Bernat et al. (2018)
T. citrinoviride	Salt stress	Maize	Seed biopriming increased osmoregulatory capacity of plant	Yesilyurt et al. (2018)
T. harzianum AK20G	Chilling stress	Tomato	*Trichoderma* improved plant growth rate and photosynthetic performance, increased relative water content and proline concentration, reduced electrolyte leakage and lipid peroxidation, and enhanced expression of genes encoding for metabolic proteins (*TAS14* and *P5CS*)	Ghorbanpour et al. (2018)

(continued)

Table 13.2 (continued)

Trichoderma spp.	Stress	Crop(s)	Activity displayed	Reference
T. harzianum Th-6	Hydroponic saline environment	Maize and rice	Improved rate of seed germination, plant growth, photosynthetic performance, relative water content, and antioxidant enzymes and inhibited uptake of Cl^- and Na^+ ions by plant leaves	Yasmeen and Siddiqui (2018)
T. asperellum	Saline–alkaline stress	Maize	K^+ and Ca^{2+} content increased and Na^+ decreased in seedlings. Concentration of osmolytes increased that enhanced water-absorbing capacity of cell and enzymatic and nonenzymatic antioxidant content and lowered accumulation of reactive oxygen species	Fu et al. (2017)
T. lixii ID11D	Salt stress	*Zea mays*	*Trichoderma* enhanced effective quantum yield and maximum quantum efficiency of PSII, photochemical quenching, electron transport rate, plant growth, proline, soluble protein, relative water content, and photosynthetic content while reducing non-photochemical quenching and reactive compounds	Pehlivan et al. (2017)
Trichoderma sp. M-35	Arsenic	Chickpea	Enhanced plant capability to tolerate arsenic stress and downregulated stress-responsive genes	Tripathi et al. (2017)
T. asperellum	Copper	Onion	Tolerance mediated by enhancing growth parameters resulted in lower accumulation, translocation, and toxicity	Vargas et al. (2017)

(continued)

Table 13.2 (continued)

Trichoderma spp.	Stress	Crop(s)	Activity displayed	Reference
T. harzianum Th-56	Drought stress	Rice	Tolerance to drought was improved in *Trichoderma*-treated plants that upregulated expression of aquaporin and dehydrin gene. In treated plants, content of proline, malondialdehyde, superoxide dismutase, and plant growth attributes was modulated	Pandey et al. (2016)
T. harzianum	Drought	Wheat	*Trichoderma* biopriming induced tolerance to drought by increasing water content, root vigor, osmoregulation, and photosynthetic activity, reducing ROS damage that delayed occurrence of stress response	Rawat et al. (2016)
T. harzianum	Salt stress	*Brassica juncea*	Oil content was improved; nutrient uptake efficiency, accumulation of osmoprotectant, and antioxidant were enhanced, whereas Na^+ uptake was reduced	Ahmad et al. (2015)
T. virens Tv29.8 and *T. atroviride* IMI 206040	Salinity stress	*Arabidopsis thaliana*	*Trichoderma* spp. via indole-3-acetic acid signaling mediated protection against salt stress and improved root architecture, i.e., lateral root growth and root hairs, under normal and saline conditions. Their inoculation also regulated the status of sodium (Na^+) ions in plant and increased antioxidant and osmoprotectant content	Contreras-Cornejo et al. (2014)

(continued)

Table 13.2 (continued)

Trichoderma spp.	Stress	Crop(s)	Activity displayed	Reference
T. parareesei T6	Salt stress	Tomato	Upregulation of expression of salt-tolerance-related *SOS1* gene	Rubio et al. (2014)
T. hamatum	Salinity	*Ochradenus baccatus*	Scavenging of ROS by antioxidant production	Hashem et al. (2014)
T. asperelloides T-203	Salt stress	*Arabidopsis thaliana* and cucumber	*Trichoderma* enhanced seed germination, plant growth, and resistance against stress by reducing ethylene concentration via activity of ACC deaminase and upregulated gene expressions for antioxidant production	Brotman et al. (2013)
T. harzianum T-22	Water deficit	Tomato	Colonized plants showed resistance by production of antioxidants that scavenged reactive oxygen species and recycled oxidized glutathione and ascorbate	Mastouri et al. (2012)
T. asperellum	Salinity	Cucumber	Alleviated salinity stress by improving plant growth, chlorophyll content, soluble protein and sugar content, efficiency of photosynthetic system, and antioxidant machinery and reducing lipid peroxidation	Qi and Zhao (2013)
T. harzianum Th-14	Salinity stress	Wheat	Biopriming of seeds with *Trichoderma* improved seed germination, plant height, chlorophyll content, and membrane stability index and increased proline and protein content	Rawat et al. (2011)

(continued)

Table 13.2 (continued)

Trichoderma spp.	Stress	Crop(s)	Activity displayed	Reference
T. hamatum DIS219b	Drought	*Theobroma cacao*	Colonization of seedling altered the expression of expression sequence tags (ESTs), reduced content of aspartic acid and glutamic acid, increased alanine and γ-aminobutyric acid concentrations, lowered wilting and delayed drought responses, and increased seedling root growth	Bae et al. (2009)

structure and function and lowered MDA content compared to other treatments and non-inoculated seeds (Rawat et al. 2011). Analogously, Alwhibi et al. (2017) reported alteration in plant growth regulators and osmolytes in tomato plants colonized with *T. harzianum* conferring tolerance to drought encompassing protection of membrane damage by oxidative stress and enhancing plant growth. The overexpression of genes encoding for membrane proteins aquaglyceroporins in *T. harzianum* that regulates movement of water and solutes through biomembranes was found to play an effective role in the biocontrol of *Fusarium oxysporum* and increased tolerance in bean plant against drought stress by enhancing root growth, leaf area, and dry weight (Vieira et al. 2018).

The ecological role of *Trichoderma* in the bioremediation of metal-contaminated sites widens their spectrum of commercial use. In reference to this context, many reports have been published citing the effectivity of different strains of *Trichoderma* spp. in remediation of toxic metals. *T. virens* PDR-8 have demonstrated potential of *Trichoderma* in efficiently removing heavy metals from liquid substrate and growth promotion of plants grown in metal-contaminated soil, suggesting their suitable application for bioremediation and biomass production as a source of biofuel (Babu et al. 2014). Simultaneously, Adams et al. (2007) documented phytostabilization of metal-contaminated sites and increase in plant growth attributes by *T. harzianum* T-22.

Genetic manipulation of plant genes could be another effective way to improve tolerance in plants to environmental constraints. *Trichoderma* strains possess the ability to express heat shock protein (HSP) gene under extreme environmental condition and their transgenic expression in plants incurs tolerance against abiotic stress. Transgenic expression of HSP gene (*T. harzianum hsp70*) in *Arabidopsis thaliana* has been reported to show resistance against heat, salt, osmotic, and oxidative stresses (Montero-Barrientos et al. 2010). Beneficial genes in *Trichoderma* with ability to serve for betterment of plant health can be transferred; however, it is a challenging task and use of transgenic plants is not considered economically or environmentally sound. Therefore, direct interaction of *Trichoderma* with plants is more useful and hence understanding this interaction more precisely is the need of present state of affairs.

During stress conditions (abiotic and biotic), abscisic acid (ABA) mediates adaptive responses in plants at physiological and molecular levels, which is generally referred to as "stress hormone" (Vishwakarma et al. 2017). ABA is well documented in literature to alleviate undesirable negative effects of water deficit by switching on water-saving mode through decreased stomatal opening and leaf expansion, thereby reducing water loss via transpiration, rendering vigorous root growth, modulating root architecture, and upregulating processes responsible for the synthesis of osmoprotectants and antioxidants (Chaves et al. 2003; Giuliani et al. 2005; Wani et al. 2016). Reduction of stomatal aperture and water loss from seedlings of *Arabidopsis thaliana* under saline condition on inoculation with *T. virens* Tv Gv29.8 or *T. atroviride* IMI 206040 was mediated through production of ABA (Contreras-Cornejo et al. 2015).

13.4.1 Oxidative Stress

Plants under stress (abiotic and biotic) conditions switch on the production of reactive oxygen species (ROS) such as hydroxyl ions (OH^-), hydrogen peroxide (H_2O_2), and superoxide radicals (O_2^-) (Singh et al. 2011). These ROS provide tolerance to unfavorable conditions and play a significant role in signal transduction (Jaspers and Kangasjärvi 2010). ROS limit growth of pathogens and trigger production of compounds such as phytoalexin, pathogenesis-related proteins, and other secondary metabolites, lignin formation, and callose deposition (Singh et al. 2011; Tartoura and Youssef 2011; Bernal-Vicente et al. 2015). Controlled production of ROS mediates signaling pathway and induces defense (Nanda et al. 2010); however, uncontrolled production during stress condition results in disruption of normal functioning of cells leading to oxidation of lipid, proteins, carbohydrates, DNA, or RNA, inactivation of vital cellular functions, and disintegration of cell wall (Singh et al. 2009, 2011). To protect plant's cellular machinery from oxidative stress, quenching of deleterious level of ROS is needed (Singh et al. 2011). To prevent oxidative damage and scavenge toxic level of ROS, plant constitutes an antioxidant machinery of several enzymatic compounds comprising of SOD, CAT, POD, ascorbate peroxidase (APX), glutathione reductase (GR), and nonenzymatic compounds such as carotenoids, tocopherols, etc. (Bernal-Vicente et al. 2015). Under stress conditions, for example, the antioxidant enzyme SOD catalyzes the superoxide free radical to less toxic H_2O_2 and oxygen molecules (Fridovich 1986), thereby reducing oxidative stress in cellular components.

Activation of plant's antioxidant machinery in the presence of stress is a short-lived phenomenon; however, several rhizospheric microorganisms have been reported to activate antioxidant machinery in plants that has prolonged effect

(Singh et al. 2011, 2013). The proteomic studies of plant roots have demonstrated that inoculation of plant with *Trichoderma* spp. improves their antioxidant enzyme activity (Segarra et al. 2007) and enhances tolerance toward abiotic stress via ROS detoxification and modulation of gene expression in related signaling pathways (Guler et al. 2016). The increase in antioxidant activity and ascorbic acid content in *Brassica rapa* plant treated with *T. harzianum* TM10 was reported by Gallo et al. (2013). Root colonization has been found to play an important role in the effective activation of the antioxidant machinery of plants by *Trichoderma* spp. Effective root colonization of maize seedling by *T. atroviride* ID20G was reported to alleviate drought stress and also found to reduce H_2O_2 concentration in seedlings (Guler et al. 2016). Similarly, Mastouri et al. (2012) in their work demonstrated the effect of colonization of tomato roots by *T. harzianum* T-22 on antioxidant production activity of plants. The gene expressions encoding for antioxidant enzymes (superoxide dismutase, ASC-GSH cycle enzymes) were found to be increased significantly that enhanced the redox buffering capacity and plant's ability to tolerate a wide range of environmental stresses such as water deficit, suboptimal temperature, and salinity.

Several other workers also demonstrated that the role of *Trichoderma* mediated antioxidant enzyme production in the alleviation of oxidative stress resulting from biotic stress. The inoculation of *T. harzianum* NBRI-1055 to sunflower plant revealed protection against *Rhizoctonia solani* by reprogramming oxidant and antioxidant compounds (Singh et al. 2011). Analogously, the amendment of *T. harzianum* T-78 with citrus compost was found to improve plant growth and suppressed the growth of pathogen responsible for Fusarium wilt in *Cucumis melo* L. (Lopez-Mondejar et al. 2010) and increased the activity of the enzymes monodehydroascorbate reductase (MDHAR) and dehydroascorbate reductase (DHAR) (Bernal-Vicente et al. 2015). The aforementioned studies describe that partial tolerance to biotic and abiotic stress is mediated through stimulation of antioxidant defense system in plants by colonization of *Trichoderma* (Mastouri et al. 2012; Bernal-Vicente et al. 2015).

13.5 *Trichoderma* as Plant Growth Promoter

Negatively stimulated plant growth results in deteriorated yield of agricultural produce qualitatively and quantitatively. The need of the present scenario is to increase agricultural produce to feed the growing global population. Many soil-dwelling microorganisms possess the ability to promote plant growth in numerous ways such as by facilitating nutrient uptake, enhancing root growth and development, and producing phytohormones. *Trichoderma* is one of the genera of soil-dwelling microbes that acquire the capacity to colonize the apoplast region of plant roots (Brotman et al. 2013) and after colonization grow along the root length with continuous growth, mediate positive alteration in root architecture, and provide anticipated beneficial effects including better anchorage and enhanced water and nutrient uptake/use efficiency (Contreras-Cornejo et al. 2009; Samolski et al. 2012; Brotman et al. 2013; Kashyap et al. 2017; Vinci et al. 2018).

13.5.1 Effect on Soil Productivity

Soil is a vital component for plant growth and maintaining its quality is essential for sustaining agricultural activity (Buragohain et al. 2019). Degradation of soil structure and chemistry severely affects its production capability. Poor organic matter, nutrient deficiency, and low microbial populations are the major problems that limit the plant's growth and agricultural produce. Desertification, nutrient leaching, and excessive conventional agricultural practices have resulted in nutrient-deficient conditions, erosion of organic matter, and reduced microbial diversity. Conventional agricultural practices such as monoculture, tillage, and repeated use of inorganic fertilizers impede the population of soil biodiversity by reducing macroaggregates that serve as microhabitat for microbial population in soil (Dick 1992). Inorganic fertilizers sometimes get fixed in soil components forming an insoluble complex and become unavailable for plants use. Nutrients like phosphorus, nitrogen, potassium, and other micronutrients (Fe, Mn, Cu, and Zn) are essential for balanced plant growth. Unavailability or inaccessibility to these nutrients leads to productivity loss. The large fraction of applied fertilizers is also leached into water bodies or volatilized in air and becomes an undesirable part of other ecosystems creating environmental havoc. Hence, involvement of biological methods for maintaining and improving soil productivity is needed for the sustainability of the agroecosystem.

Soil inhabiting microbes including *Trichoderma*, arbuscular mycorrhiza fungi (AMF), plant-growth-promoting rhizobacteria (PGPR), and other beneficial microbes have the capacity to fix nitrogen and solubilize and/or mineralize phosphate, potassium, and other elements, forming iron chelates and making them accessible for plant's uptake (Bhardwaj et al. 2014; Li et al. 2015). These microorganisms catalyze many chemical and biochemical reactions by altering soil pH (Li et al. 2015), increase pool of soil organic matter, enhance nutrient level, and modify microbial activity (Buragohain et al. 2019). *Trichoderma* modify plants' ambient environment by producing several compounds and alleviating toxic compounds which result in improved plant growth and increased soil microbial structure (Benitez et al. 2004; Chen et al. 2011). The degradation of toxic allelochemicals produced in the rhizosphere by cucumber plants by the *T. harzianum* SQR-T037 was reported by Chen et al. (2011).

Several other workers have also demonstrated the beneficial role of *Trichoderma* in improving soil health and plant growth promotion under suboptimal or adverse conditions. In an experiment, the application of *T. harzianum* T83 biofertilizer ameliorated the coastal saline soil and converted it into viable and productive soil by producing soluble sugars, amino acids, and organic acids and enhanced uptake of potassium and calcium ions. Further, it improved the growth of the plant *Suaeda salsa* L. by increasing the activity of the antioxidant machinery resulting in production of ROS enzymes such as CAT, POD, SOD, and polyphenol oxidase (PPO) that alleviated salinity stress (Chen et al. 2016). The effective potential of *Trichoderma* in reducing loss of nitrogen from soil via ammonia volatilization was demonstrated by Wang et al. (2018). The inoculation of *T. viride*-based biofertilizer in soil enhanced the process of nitrification and additionally increased the abundance of ammonia-oxidizing archaea (AOA) and ammonia-oxidizing bacteria (AOB) in soil, reduced *T. harzianum* soil pH, and altered microbial community composition and structure. These evidences clearly identify the effective role of different species of *Trichoderma* in maintaining and enhancing soil productivity without compromising its future production ability.

13.5.2 Effects on Plant Growth and Productivity

Trichoderma spp. are demonstrated as promising biofertilizers. Several studies have provided strong verification on the role of *Trichoderma* in plant growth via involvement of different mechanisms (Naseby et al. 2000). Plant hormones play a crucial role in plant growth promotion and development under normal and stressed environmental circumstances (Peleg and Blumwald 2011). Auxin, cytokinins, ethylene, and abscisic acid are major plant hormones that regulate the growth and development of plants (Mok and Mok 2001; Linkies et al. 2009; Raghavendra et al. 2010; Martínez-Medina et al. 2014). Interaction of biocontrol agents with plants also results in growth promotion, but studies on the involvement of phytohormones in plant growth promotion are limited. Though some evidences are lined in literature that indicate that production of phytohormones by several microorganisms leads to plant morphogenesis (Ortiz-castro et al. 2009; Martínez-Medina et al. 2014).

Some *Trichoderma* spp. are known to synthesize growth-regulating metabolites that alter the phytohormonal network of the host plant leading to growth promotion as well as resistance to diseases (Ortiz-castro et al. 2009; Hermosa et al. 2013; Martínez-Medina et al. 2014). Indole-3-acetic acid (IAA), the most abundantly occurring auxin compound, is an important player in plant growth and known to be responsible for lateral root growth and root hair development in plants. An increase in auxin (IAA) content and decrease in cytokinin and abscisic acid contents in shoots of melon were observed to be responsible for increased plant fitness (Martínez-Medina et al. 2014). Contreras-Cornejo et al. (2009) also detailed out the role of IAA synthesized by *T. viride* in enhanced growth of *A. thaliana*. The crude extracts of *Trichoderma* spp. showing the presence of IAA improved the growth potential of quinoa with grain yield, lettuce, and radish (Ortuno et al. 2017).

Some of the metabolites secreted by *Trichoderma* spp. have been investigated for their ability to promote plant growth. The volatile metabolites of several isolates of *Trichoderma* have demonstrated their active participation in inducing enhanced plant growth. The activation of Fe uptake in *Arabidopsis* and tomato by volatile metabolites of *T. harzianum* and *T. asperellum* via modulation of expression of transcription factor *MYB72* was reported by Martínez-Medina et al. (2017). A parallel study undertaken by Hung et al. (2013) determined volatile metabolites produced by *T. viride* increased biomass and chlorophyll content of *A. thaliana*. Concomitantly, growth of *Arabidopsis* was positively influenced by the volatiles produced by *T. virens* Gv29, *T. asperellum* LU1370, *T. atroviride* IMI206040, and *T.* sp. "*atroviride B*" LU132 (Nieto-Jacobo et al. 2017). Similarly, the metabolite chrysophanol of *T. harzianum* has shown ability to enhance the growth of cabbage by stimulating the expression of photosynthesis and sucrose-transport-related gene and elicit plant defense against the pathogen *Botrytis cinerea* (Liu et al. 2016).

13.5.3 Effect on Photosynthetic Pigments of Agricultural Crops

Trichoderma spp. systematically activates expression of photosynthetic genes by enhancing demand for photosynthates especially sucrose by sink tissues resulting in higher photosynthetic activity in plants (Vargas et al. 2009). The inoculation of maize leaves with *T. virens* displaying sucrolytic activity of invertase in the presence of plant-derived sucrose was found to stimulate the root sink activity, increase carbon partitioning toward roots, and increase photosynthetic rate (Vargas et al. 2009). In addition, the presence of sucrose-containing root exudates in tomato rhizosphere was reported to promote growth of *T. atroviride* resulting in successful root colonization and increased chlorophyll content in plant leaves even in the presence of a fungal pathogen (Macías-Rodríguez et al. 2018). Similarly, inoculation of tomato plant with *T. harzianum* T-22 was reported to increase the photosynthetic rate via improved redox status of plant (Vitti et al. 2016). The significant increase in photosynthetic rate and uptake of P in *Zea mays* on combined inoculation of *T. harzianum* OmG-08 and compost was reported by Vinci et al. (2018). *Trichoderma* spp. not only increase the photosynthetic activity of plants under normal environmental conditions but often reported to show improved rate of photosynthesis in stressed plants. Treatment of maize plant with *T. asperellum* growing under saline–alkaline stress resulted in improved photosynthesis due to increased photosynthetic pigments by reducing membrane lipid peroxidation, enhancingenhanced activity of reaction center of photosystem II and ATP synthetase enzymes in the chloroplast, and decreasing non-stomatal limitation (Fu et al. 2018). Analogously, the increase in photosynthetic pigment of spinach under suboptimal and normal water regime conditions on inoculation of *T. lixii* TvR1 was reported by Sachdev et al. (2018).

13.5.4 Commercial Role of *Trichoderma* for Sustainable Agriculture

The most widespread genus of fungi marketed as biopesticide is *Trichoderma*. It contributes more than 60% share to the registered biofungicides worldwide (Abbas et al. 2017). *Trichoderma*-formulated products have received worldwide acceptance and around 250 biofungicides are registered globally (Woo et al. 2014; Kashyap et al. 2017). For instance, Tellus (*T. gamsii* and *T. asperellum*), Binab (*T. polysporum* and *T. harzianum*), and TrichoPlus (*T. asperelloides*) were recently launched in the market as biofungicides by Syngenta agrochemical company, Switzerland; Bio-Innovation AB, Sweden; and BASF Corporation, Germany, respectively (Kashyap et al. 2017). Globally, *T. harzianum* is considered as the most potent BCA; however in India, most products based on *T. viride* are commercialized (Saxena et al. 2014). Lists of some commercialized *Trichoderma*-based bio-products are presented in Table 13.3.

Conidia-based *Trichoderma* bioformulations are applied for soil inoculation, soil drenching, seed biopriming, seedling dip, foliar application, and wound dressing (Oros and Naar 2017; Kashyap et al. 2017). The literature describing successful tales of bio-product application suggests seed treatment as the best application method and is widely accepted (Sharma et al. 2015; Rawat et al. 2016). The number of factors determines the *Trichoderma*-driven plant's benefits that include crop, plant genotype, persisting environmental and soil condition, method of bio-product application, inoculum size, plant growth stage during application, etc. (Kashyap et al. 2017).

Trichoderma is compatible with many agrochemical and other beneficial microbes, showing synergistic activity (Kashyap et al. 2017). Coupling of *Trichoderma* with other beneficial microorganisms and agrochemicals with different modes of actions improves their efficiency due to amalgamation of various mechanisms together. This unique property enhances their use in agricultural practices and is incorporated in integrated pest management (IPM) system, thereby reducing load of usage of chemical pesticides (Kashyap et al. 2017).

13.6 Concluding Remarks

Plants are sessile organisms that constantly encounter several biotic and abiotic stresses simultaneously in field conditions that challenge plant's survival and productivity. The concurrent occurrence of two or more stresses may detrimentally impede plant growth. For example, occurrence of phytopathogens and nutrient deficiency in soil simultaneously could severely effect plant growth and productivity and may result in total crop failure. To cope with such unfavorable conditions and increase crop productivity, use of chemical-based approaches are generally employed which are not sustainable and there is a need for an alternative approach that not only reduces impact of stress but maintains ecological integrity. In natural ecosystems, plants coexist with several beneficial microorganisms and show extensive communication with them. This beneficial communication confers protection against ecological stresses and positively modulates plant growth.

Trichoderma is one of the beneficial rhizospheric microbes that display a panoply of traits which unload plants from the debt of environmental stress. Several spp. of *Trichoderma* are commercially available in the market as biopesticides and biofertilizers, but lack their place in farmer's field. This is due to low consistency and efficacy in field conditions and deficiency in knowledge related to molecular mechanisms. During plant and *Trichoderma* interaction, plants undergo alterations at physiological and biochemical levels and reprogramming at molecular level to address environmental cues thereby improving plants fitness. Therefore, studies of plant–*Trichoderma* interaction at biochemical, physiological, and molecular levels can generate information of the mechanism modulating plant's activity under unfavorable environmental conditions. Thorough vigilant amalgamation of obtained

Table 13.3 *Trichoderma*-based commercial bio-products

Product	Active ingredient	Target	Formulation	Registered country (ies)	Manufacturer
Rootshield Plus	*T. harzianum* Rifai T-22 and *T. virens* G-41	Soilborne diseases including *Phytophthora*, *Pythium*, *Rhizoctonia*, *Fusarium*, *Thielaviopsis*, and *Cylindrocladium*	Wettable powders (WP) and granules (G)	California, USA	BioWorks Inc.
Eco-77	*T. atroviride* B77	Control foliar and wound diseases, wood rot pathogens (*Botrytis, Eutypa*)	WP	South Africa, Zambia, Kenya	Plant health Products (Pty) Ltd.
Trichopel	*T. harzianum* and *T. viride*	Phytopathogenic fungi (*Rhizoctonia, Phytophthora, Pythium*)	Granules	New Zealand, Australia	Agrimm Technologies Ltd., New Zealand
Trichodex	*T. harzianum* T39	Soilborne and foliar phytopathogenic fungi	WP	Australia, New Zealand, Vietnam, and South Africa	Valent Biosciences, Australia
TurfShield Plus	*T. harzianum* and *T. viride*	Turf diseases, enhance water and nutrient uptake, solubilize inorganic nutrients, and provide tolerance to various stresses	WP, granules	USA	BioWorks, Inc.
BioTrek 22 G	*T. harzianum*	Turf diseases	Granules	USA, Europe	Wilbur-Ellis
ECOSOM®-TV	*T. viride*	Seed- and soilborne fungi	WP	India	AgriLife, India
Supersivit	*T. harzianum*	Soilborne pathogens (*Pythium, Rhizoctonia, Fusarium*)	–	–	BioPlant, Denmark
Bio-Cure-F	*T. viride*	Soilborne fungi	Powder, liquid	European Union, India	T. Stanes and Company Ltd., Coimbatore, Tamil Nadu
Ecofit	*T. viride*	Pathogens causing root rot, seedling rot, collar rot, and damping off	Powder	India	Hoechst and Schering agrEvo. Ltd., Mumbai, India

(continued)

Table 13.3 (continued)

Product	Active ingredient	Target	Formulation	Registered country (ies)	Manufacturer
Bio-Shield	*T. viride*	Seed-borne fungal pathogens	WP	India	Ambika Biotech, India
Defence-SF	*T. viride*	Seed- and soilborne diseases	–	India	Wockhardt Life Science Ltd., Mumbai, India
Bioderma	*T. viride*	Fungal pathogens (soilborne and foliar)	WP and AS (aqueous suspension)	India	Biotech International Ltd., India
Unite	*Trichoderma* spp.	Controlling root disease-causing pathogens (*Fusarium*, *Pythium*, *Sclerotinia*, *Rhizoctonia*, *Phytophthora*)	WP	–	Agrimm Technologies Ltd., New Zealand
TrichoPlus	*T. asperelloides* JM41R	Soilborne diseases (*Fusarium*, *Pythium*, *Sclerotinia*, *Rhizoctonia*, *Phytophthora*)	WP	–	BASF Corporation, USA
Bio-Tam	*T. asperellum* and *T. gamsii*	Manage soilborne diseases and promote plant growth by enhancing water and nutrient uptake	WP	USA	AgroQuest
Remedier	*T. asperellum* ICC012 and *T. gamsii* ICC080	Soilborne diseases	WP	USA, European Union	Isagro Co., USA
Tenet	*T. asperellum* and *T. gamsii*	Control phytopathogenic fungi such as *Rhizoctonia*, *Botrytis*, *Phytophthora*, *Verticillium*, etc.	WP	–	Isagro Co., USA
Esquive	*T. atroviride*	Fungal pathogens (*Rhizoctonia*, *Phytophthora*, *Botrytis*)	WP	Europe	Agrauxine
Sentinel	*T. atroviride*	*Botrytis cinerea* affecting grapes and tomatoes	WP	New Zealand, Australia	Agrimm Technologies Ltd., New Zealand
Binab	*T. polysporum* and *T. harzianum*	*Botrytis cinerea* and *Chondrostereum purpureum*	WP	–	Bio-Innovation AB, Sweden

Trichotecin	Compounds produced by *Trichothecium roseum*	Seed phytopathogen	–	–	–
SoilGard	*T. virens/ Gliocladium virens strain* GL-21	Damping off and root rot pathogens (*Fusarium, Rhizoctonia, Pythium*)	Granules	USA	Certis, USA
T-22HC	*T. harzianum* Rifai strain T-22	Plant root pathogens (*Pythium, Rhizoctonia,* and *Fusarium*); enhances nutrient uptake	Powder	California (USA)	BioWorks Inc.
Bhoomika 1%WP	*T. viride*	Soilborne diseases	WP	India	Varsha Biosciences and Technology, Pvt. Ltd, India
Eco-T	*T. harzianum*	Root pathogens (*Pythium, Fusarium,* etc.); stimulate plant growth even under water-logging and drought conditions	WP	–	Plant Health Products (Pty) Ltd.
Hariz 1.15% WP	*T. harzianum*	Fungal pathogens (*Pencillium*, *Fusarium*, *Pythium*, *Botrytis*) and nematodes; mobilize phosphate and other micronutrients	WP	India	Varsha Biosciences and Technology Pvt. Ltd., India

Adopted and modified from Woo et al. (2014) and Abbas et al. (2017)

information could provide better insight to understand the mechanism at grassroot level for plants to survive under extreme environmental conditions and open new pathways for research to find more superior options for sustainable agriculture with global changing climatic scenario.

Acknowledgement The authors gratefully acknowledge University Grant Commission, New Delhi, India, for providing UGC-Senior Research Fellowship grant as financial support to Swati Sachdev.

Conflict of Interest The authors declare no conflict of interest.

References

Abbas A, Jiang D, Fu Y (2017) *Trichoderma* spp. as antagonist of *Rhizoctonia solani*. J Plant Pathol Microbiol 8:402. https://doi.org/10.4172/2157-7471.1000402

Abdelrahman M, Abdel-Motaal F, El-Sayed M, Jogaiah S, Shigyo M, Ito SI, Tran LSP (2016) Dissection of *Trichoderma longibrachiatum*-induced defense in onion (*Allium cepa* L.) against *Fusarium oxysporum* f. sp. *cepa* by target metabolite profiling. Plant Sci 246:128–138

Adams P, De-Leji FA, Lynch JM (2007) *Trichoderma harzianum* Rifai 1295-22 mediates growth promotion of crack willow (*Salix fragilis*) saplings in both clean and metal-contaminated soil. Microb Ecol 54(2):306–313

Affokpon A, Coyne DL, Htay CC, Agbede RD, Lawouin L, Coosemans J (2011) Biocontrol potential of native *Trichoderma* isolates against root-knot nematodes in west African vegetable production systems. Soil Biol Biochem 43:600–608

Afzal I, Basra SMA, Farooq M, Nawaz A (2006) Alleviation of salinity stress in spring wheat by hormonal priming with ABA, salicylic acid and ascorbic acid. Int J Agri Biol 8:23–28

Agarwal S, Grover A (2006) Molecular biology, biotechnology and genomics of flooding-associated low O_2 stress response in plants. Crit Rev Plant Sci 25:1–21. https://doi.org/10.1080/07352680500365232

Ahmad JS, Baker R (1987) Rhizosphere competence of *Trichoderma harzianum*. Phytopathology 77(2):182–189

Ahmad P, Hashem A, Abd-Allah EF, Alqarawi AA, John R, Egamberdieva D, Gucel S (2015) Role of *Trichoderma harzianum* in mitigating NaCl stress in Indian mustard (*Brassica juncea* L) through antioxidative defense system. Front Plant Sci 6:868. https://doi.org/10.3389/fpls.2015.00868

Alizadeh H, Behboudi K, Ahmadzadeh M, Javan-Nikkhah M, Zamioudis C, Pieterse CMJ, Bakker PAHM (2013) Induced systemic resistance in cucumber and *Arabidopsis thaliana* by the combination of *Trichoderma harzianum* Tr6 and *Pseudomonas* sp. Ps14. Biol Control 65:14–23

Almeida FB, Cerqueira FM, Silva RN, Ulhoa CJ, Lima AL (2007) Mycoparasitism studies of *Trichoderma harzianum* strains against Rhizoctonia solani: evaluation of coiling and hydrolytic enzyme production. Biotechnol Lett 29:1189–1193. https://doi.org/10.1007/s10529-007-9372-z

Alwhibi MS, Hashem A, Abd Allah EF, Egamberdieva D (2017) Increased resistance of drought by *Trichoderma harzianum* fungal treatment correlates with increased secondary metabolites and proline content. J Integr Agric 16(8):1751–1757. https://doi.org/10.1016/S2095-3119(17)61695-2

Anam GB, Reddy MS, Ahn YH (2019) Characterization of *Trichoderma asperellum* RM-28 for its sodic/saline-alkali tolerance and plant growth promoting activities to alleviate toxicity of red mud. Sci Total Environ 662:462–469

Anle H, Jianan S, Xinhua W, Liwen Z, Bo F, Jie C (2019) Reprogrammed endophytic microbial community in maize stalk induced by *Trichoderma asperellum* biocontrol agent against Fusarium diseases and mycotoxin accumulation. Fungal Biol 123:448–455

Ansari MW, Trivedi DK, Sahoo RK, Gill SS, Tuteja N (2013) A critical review on fungi mediated plant responses with special emphasis to *Piriformospora indica* on improved production and protection of crops. Plant Physiol Biochem 70:403–410

Babu AG, Shim J, Bang KS, Shea PJ, Oh BT (2014) *Trichoderma virens* PDR-28: a heavy metal-tolerant and plant growth-promoting fungus for remediation and bioenergy crop production on mine tailing soil. J Environ Manag 132:129–134. https://doi.org/10.1016/j.jenvman.2013.10.009

Bae H, Sicher RC, Kim MS, Kim SH, Strem MD, Melnick RL, Bailey BA (2009) The beneficial endophyte *Trichoderma hamatum* isolate DIS 219b promotes growth and delays the onset of the drought response in *Theobroma cacao*. J Exp Bot 60(11):3279–3295. https://doi.org/10.1093/jxb/erp165

Bae H, Roberts DP, Lim HS, Strem MD, Park SC, Ryu CM, Melnick RL, Bailey BA (2011) Endophytic *Trichoderma* isolates from tropical environments delay disease onset and induce resistance against *Phytophthora capsici* in hot pepper using multiple mechanisms. Mol Plant-Microbe Interact 24:336–351

Bae SJ, Mohanta TK, Chung JY, Ryu M, Park G, Shim S, Hong SB, Seo H, Bae DW, Bae I, Kim JJ (2016) *Trichoderma* metabolites as biological control agents against *Phytophthora* pathogens. Biol Control 92:128–138

Bailey D, Lumsden R (1998) Direct effects of *Trichoderma* and *Gliocladium* on plant growth and resistance to pathogens. In: Harman GE, Kubicek CP (eds) *Trichoderma* and *Gliocladium*: enzymes, biological control and commercial applications, vol 2. Taylor & Francis, London, pp 185–204

Battaglia D, Bossi S, Cascone P, Digilio MC, Duran Prieto J, Fanti P, Guerrieri E, Iodice L, Lingua G, Lorito M, Maffei ME, Massa N, Ruocco M, Sasso R, Trotta V (2013) Tomato below ground-above ground interactions: *Trichoderma longibrachiatum* affects the performance of *Macrosiphum euphorbiae* and its natural antagonists. Mol Plant-Microbe Interact 26:1249–1256

Belanger RR, Dufour N, Caron J, Benhamou N (1995) Chronological events associated with the antagonistic properties of *Trichoderma harzianum* against *Botrytis cinerea*: indirect evidence for sequential role of antibiosis and parasitism. Biocontrol Sci Tech 5(1):41–54

Benitez T, Rincon AM, Limón MC, Codon AC (2004) Biocontrol mechanisms of *Trichoderma* strains. Int Microbiol 7:249–260

Bernal-Vicente A, Pascual JA, Tittarelli F, Hernandezc JA, Diaz-Vivancos P (2015) *Trichoderma harzianum* T-78 supplementation of compost stimulates the antioxidant defence system in melon plants. J Sci Food Agric 95(11):2208–2214. https://doi.org/10.1002/jsfa.6936

Bernat P, Nykiel-Szymańska J, Gajewska E, Różalska S, Stolarek P, Dackowa J, Słaba M (2018) *Trichoderma harzianum* diminished oxidative stress caused by 2,4- dichlorophenoxyacetic acid (2,4-D) in wheat, with insights from lipidomics. J Plant Physiol 229:158–163. https://doi.org/10.1016/j.jplph.2018.07.010

Bhardwaj D, Ansari MW, Sahoo RK, Tuteja N (2014) Biofertilizers function as key player in sustainable agriculture by improving soil fertility, plant tolerance and crop productivity. Microb Cell Factories 13(1):66

Błaszczyk L, Siwulski M, Sobieralski K, Lisiecka J, Jedryczka M (2014) *Trichoderma* spp. – application and prospects for use in organic farming and industry. J Plant Prot Res 54 (4):309–317. https://doi.org/10.2478/jppr-2014-0047

Boyer JS (1982) Plant productivity and environment. Science 218:443–448

Braun H, Woitsch L, Hetzer B, Geisen R, Zange B, Schmidt-Heydt M (2018) *Trichoderma harzianum*: inhibition of mycotoxin producing fungi and toxin biosynthesis. Int J Food Microbiol 280:10–16

Brotman Y, Landau U, Cuadros-Inostroza A, Takayuki T, Fernie AR (2013) *Trichoderma*-plant root colonization: escaping early plant defense responses and activation of the antioxidant machinery for saline stress tolerance. PLoS Pathog 9(3):e1003221. https://doi.org/10.1371/journal.ppat.1003221

Brozová J (2004) Mycoparasitic fungi *Trichoderma* spp. in plant protection. A review. Plant Protect Sci 40:63–74

Bunbury-Blanchette AL, Walker AK (2019) *Trichoderma* species show biocontrol potential in dual culture and greenhouse bioassays against Fusarium basal rot of onion. Biol Control 130:127–135

Buragohain P, Sreedeep S, Lin P, Ni J, Garg A (2019) Influence of soil variability on single and competitive interaction of ammonium and potassium: experimental study on seven different soils. J Soils Sediments 19(1):186–197

Caporale AG, Sommella A, Lorito M, Lombardi N, Azam SMGG, Pigna M, Ruocco M (2014) *Trichoderma* spp. alleviate phytotoxicity in lettuce plants (*Lactuca sativa* L.) irrigated with arsenic-contaminated water. J Plant Physiol 171:1378–1384

Chaves MM, Oliveira MM (2004) Mechanisms underlying plant resilience to water deficits: prospects for water-saving agriculture. J Exp Bot 55:2365–2384. https://doi.org/10.1093/jxb/erh269

Chaves M, Manuela P, Pereira JSM (2003) Understanding plant responses to drought-from genes to the whole plant. Funct Plant Biol 30:239–264

Chen L, Yang X, Raza W, Li J, Liu Y, Qiu M, Zhang F, Shen Q (2011) *Trichoderma harzianum* SQR-T037 rapidly degrades allelochemicals in rhizospheres of continuously cropped cucumbers. Appl Microbiol Biotechnol 89(5):1653–1663

Chen T, Yuan F, Song J, Wang B (2016) Nitric oxide participates in waterlogging tolerance through enhanced adventitious root formation in the euhalophyte *Suaeda salsa*. Funct Plant Biol 43 (3):244–253

Chet I, Harman GE, Baker R (1981) *Trichoderma hamatum*: its hyphal interactions with *Rhizoctonia solani* and *Pythium* spp. Microb Ecol 7(1):29–38

Chowdappa P, Kumar SPM, Lakshmi MJ, Upreti KK (2013) Growth stimulation and induction of systemic resistance in tomato against early and late blight by *Bacillus subtilis* OTPB1 or *Trichoderma harzianum* OTPB3. Biol Control 65:109–117. https://doi.org/10.1016/j.biocontrol.2012.11.009

Claydon N, Allan M, Hanson JR, Avent AG (1987) Antifungal alkyl pyrones of *Trichoderma harzianum*. Trans Br Mycol Soc 88(4):503–513

Contreras-Cornejo HA, Macias-Rodriguez L, Cortes-Penagos C, Lopez-Bucio J (2009) *Trichoderma virens*, a plant beneficial fungus, enhances biomass production and promotes lateral root growth through an auxin-dependent mechanism in arabidopsis. Plant Physiol 149:1579–1592. https://doi.org/10.1104/pp.108.130369

Contreras-Cornejo HA, Macías-Rodríguez L, Alfaro-Cuevas R, Lopez-Bucio J (2014) *Trichoderma* spp. improve growth of *Arabidopsis* seedlings under salt stress through enhanced root development, osmolite production, and Na+ elimination through root exudates. Mol Plant-Microbe Interact 27(6):503–514. https://doi.org/10.1094/MPMI-09-13-0265-R

Contreras-Cornejo HA, Macías-Rodríguez L, Vergara AG, López-Bucio J (2015) *Trichoderma* modulates stomatal aperture and leaf transpiration through an abscisic acid-dependent mechanism in *Arabidopsis*. J Plant Growth Regul 34(2):425–432

Contreras-Cornejo HA, Macias-Rodriguez L, del-Val E, Larsen J (2016) Ecological functions of *Trichoderma* spp. and their secondary metabolites in the rhizosphere: interactions with plants. FEMS Microbiol Ecol 92(4):fiw036. https://doi.org/10.1093/femsec/fiw036

Contreras-Cornejo HA, del-Val E, Macías-Rodríguez L, Alarcón A, González-Esquivel CE, Larsen J (2018) *Trichoderma atroviride*, a maize root associated fungus, increases the parasitism rate of the fall armyworm *Spodoptera frugiperda* by its natural enemy *Campoletis sonorensis*. Soil Biol Biochem 122:196–202

Cramer GR, Urano K, Delrot S, Pezzotti M, Shinozaki K (2011) Effects of abiotic stress on plants: a systems biology perspective. BMC Plant Biol 11:163. https://doi.org/10.1186/1471-2229-11-163

De la Cruz-Quiroz R, Ascacio-Valdés JA, Rodríguez-Herrera R, Roussos S, Aguilar CN (2019) Phytopathogen biomass as inducer of antifungal compounds by *Trichoderma asperellum* under solid-state fermentation. In: Singh H, Keswani C, Reddy M, Sansinenea E, García-Estrada C (eds) Secondary metabolites of plant growth promoting rhizomicroorganisms. Springer, Singapore, pp 113–124

De Meyer G, Bigirimana J, Elad Y, Hofte M (1998) Induced systemic resistance in *Trichoderma harzianum* T39 biocontrol of *Botrytis cinerea*. Eur J Plant Pathol 104:279–286

Degenkolb T, Dieckmann R, Nielsen KF, Grafenhan T, Theis C, Zafari D, Chaverri P, Ismaiel A, Bruckner H, Von Dohren H, Thrane U, Petrini O, Samuels GJ (2008) The *Trichoderma brevicompactum* clade: a separate lineage with new species, new peptaibiotics, and mycotoxins. Mycol Prog 7:177–219. https://doi.org/10.1007/s11557-008-0563-3

Dick RP (1992) A review: long-term effects of agricultural systems on soil biochemical and microbial parameters. Agric Ecosyst Environ 40(1-4):25–36

Djonovic S, Pozo MJ, Dangott LJ, Howell CR, Kenerley CM (2006) Sm1, a proteinaceous elicitor secreted by the biocontrol fungus *Trichoderma virens* induces plant defense responses and systemic resistance. Mol Plant-Microbe Interact 19:838–853

Djonovic S, Vittone G, Mendoza-Herrera A, Kenerley CM (2007) Enhanced biocontrol activity of *Trichoderma virens* transformants constitutively coexpressing β-1, 3-and β-1, 6-glucanase genes. Mol Plant Pathol 8(4):469–480

Doni F, Isahak A, Zain CRCM, Yusoff WMW (2014) Physiological and growth response of rice plants (Oryza sativa L.) to *Trichoderma* spp. inoculants. AMB Express 4(1):45

Druzhinina IS, Seidl-Seiboth V, Herrera-Estrella A, Horwitz BA, Kenerley CM, Monte E, Mukherjee PK, Zeilinger S, Grigoriev IV, Kubicek CP (2011) *Trichoderma*: the genomics of opportunistic success. Nat Rev Microbiol 9:749–759

Ehrlich PR, Ehrlich AH (2016) Population, resources, and the faith-based economy: the situation in 2016. Biophys Econ Resour Qual 1:3. https://doi.org/10.1007/s41247-016-0003-y

Eisendle M, Oberegger H, Buttinger R, Illmer P, Haas H (2004) Biosynthesis and uptake of siderophores is controlled by the PacC-mediated ambient-pH regulatory system in *Aspergillus nidulans*. Euk Cell 3:561–563

Elamathi E, Malathi P, Viswanathan R, Ramesh Sundar A (2018) Expression analysis on mycoparasitism related genes during antagonism of *Trichoderma* with *Colletotrichum falcatum* causing red rot in sugarcane. J Plant Biochem Biot 27(3):351–361. https://doi.org/10.1007/s13562-018-0444-z

Essah PA, Davenport R, Tester M (2003) Sodium influx and accumulation in *Arabidopsis*. Plant Physiol 133:307–318

Fridovich I (1986) Superoxide dismutases. Adv Enzymol Relat Areas Mol Biol 58(6):61–97

Fu J, Liu Z, Li Z, Wang Y, Yang K (2017) Alleviation of the effects of saline-alkaline stress on maize seedlings by regulation of active oxygen metabolism by *Trichoderma asperellum*. PLoS One 12(6):e0179617

Fu J, Wang YF, Liu ZH, Li ZT, Yang KJ (2018) *Trichoderma asperellum* alleviates the effects of saline–alkaline stress on maize seedlings via the regulation of photosynthesis and nitrogen metabolism. Plant Growth Regul 85(3):363–374

Gajera HP, Bambharolia RP, Patel SV, Khatrani TJ, Goalkiya BA (2012) Antagonism of *Trichoderma* spp. against *Macrophomina phaseolina*: evaluation of coiling and cell wall degrading enzymatic activities. J Plant Pathol Microb 3:149. https://doi.org/10.4172/2157-7471.1000149

Gallo M, Esposito G, Ferracane R, Vinale F, Naviglio D (2013) Beneficial effects of *Trichoderma* genus microbes on qualitative parameters of *Brassica rapa* L. subsp. *sylvestris* L. Janch. Var. esculenta Hort. Eur Food Res Technol 236:1063–1071

Ghisalberti EL, Sivasithamparam K (1991) Antifungal antibiotics produced by *Trichoderma* spp. Soil Biol Biochem 23(11):1011–1020

Ghorbanpour A, Salimi A, Ghanbary MAT, Pirdashti H, Dehestani A (2018) The effect of *Trichoderma harzianum* in mitigating low temperature stress in tomato (*Solanum lycopersicum* L.) plants. Sci Hort 230:134–141

Giuliani S, Sanguineti MC, Tuberosa R, Bellotti M, Salvi S, Landi P (2005) Root-ABA1 a major constitutive QTL affects maize root architecture and leaf ABA concentration at different water regimes. J Exp Bot 56:3061–3070

Godfray HCJ, Beddington JR, Crute IR, Haddad L, Lawrence D, Muir JF, Pretty J, Robinson S, Thomas SM, Toulmin C (2010) Food security: the challenge of feeding 9 billion people. Science 327:812–818

Grover M, Ali Sk Z, Sandhya V, Rasul A, Venkateswarlu B (2011) Role of microorganisms in adaptation of agriculture crops to abiotic stresses. World J Microbiol Biotechnol 27:1231–1240. https://doi.org/10.1007/s11274-010-0572-7

Guler NS, Pehlivan N, Karaoglu SA, Guzel S, Bozdeveci A (2016) *Trichoderma atroviride* ID20G inoculation ameliorates drought stress-induced damages by improving antioxidant defence in maize seedlings. Acta Physiol Plant 38(6):132

Guzmán-Guzmán P, Porras-Troncoso MD, Olmedo-Monfil V, Herrera-Estrella A (2018) *Trichoderma* species: versatile plant symbionts. Phytopathology 109(1):6–16

Halperin SJ, Gilroy S, Lynch JP (2003) Sodium chloride reduces growth and cytosolic calcium, but does not affect cytosolic pH, in root hairs of *Arabidopsis thaliana* L. J Exp Bot 54:1269–1280

Harman GE (2011) *Trichoderma*—not just for biocontrol anymore. Phytoparasitica 39:103–108. https://doi.org/10.1007/s12600-011-0151-y

Harman GE, Lorito M, Lynch JM (2004a) Uses of *Trichoderma* spp. to alleviate or remediate soil and water pollution. Adv Appl Microbiol 56:313–330

Harman GE, Howell CR, Vitarbo A, Chet I, Lorito M (2004b) *Trichoderma* species –opportunistic, avirulent plant symbionts. Nat Rev Microbiol 2:43–56

Harman GE, Herrera-Estrella AH, Horwitz BA, Lorito M (2012) *Trichoderma* – from basic biology to biotechnology. Microbiology 158:1–2. (Special issue: 498)

Hashem A, Abd_Allah EF, Alqarawi AA, Al Huqail AA, Egamberdieva D (2014) Alleviation of abiotic salt stress in *Ochradenus baccatus* (Del.) by *Trichoderma hamatum* (Bonord.) Bainier. J Plant Interact 9(1):857–868

Hermosa MR, Keck E, Chamorro I, Rubio B, Sanz L, Vizcaino JA, Grondona I, Monte E (2004) Genetic diversity shown in *Trichoderma* biocontrol isolates. Mycol Res 108:897–906

Hermosa R, Viterbo A, Chet I, Monte E (2012) Plant-beneficial effects of *Trichoderma* and of its genes. Microbiology 158:17–25. https://doi.org/10.1099/mic.0.052274-0

Hermosa R, Rubio MB, Cardoza RE, Nicolas C, Monte E, Gutierrez S (2013) The contribution of *Trichoderma* to balancing the costs of plant growth and defense. Int Microbiol 16:69–80

Hider RC, Kong X (2010) Chemistry and biology of siderophores. Nat Prod Rep 27(5):637–657

Hirel B, Le Gouis J, Ney B, Gallais A (2007) The challenge of improving nitrogen use efficiency in crop plants: towards a more central role for genetic variability and quantitative genetics within integrated approaches. J Exp Bot 58:2369–2387. https://doi.org/10.1093/jxb/erm097

Howell CR (2003) Mechanisms employed by *Trichoderma* species in the biological control of plant diseases: the history and evolution of current concepts. Plant Dis 87:4–10

Howell CR, Stipanovic RD, Lumsden RD (1993) Antibiotic production by strains of *Gliocladium virens* and its relation to the biocontrol of cotton seedling diseases. Biocontrol Sci Tech 3 (4):435–441

Hung R, Lee S, Bennett JW (2013) *Arabidopsis thaliana* as a model system for testing the effects of *Trichoderma* volatile organic compounds. Fungal Ecol 6:19–26

Huot B, Yao J, Montgomery BL, He SY (2014) Growth–defense tradeoffs in plants: a balancing act to optimize fitness. Mol Plant 7:1267–1287

Hyakumachi M, Kubota M (2003) Fungi as plant growth promoter and disease suppressor. In: Arora DK (ed) Fungal biotechnology in agricultural, Food and environmental application. Marcel Dekker, New York, pp 101–110

Jaspers P, Kangasjärvi J (2010) Reactive oxygen species in abiotic stress signaling. Physiol Plant 138:405–413

Kashyap PL, Rai P, Srivastava AK, Kumar S (2017) *Trichoderma* for climate resilient agriculture. World J Microbiol Biotechnol 33(8):155. https://doi.org/10.1007/s11274-017-2319-1

Katori T, Ikeda A, Iuchi S, Kobayashi M, Shinozaki K, Maehashi K, Sakata Y, Tanaka S, Taji T (2010) Dissecting the genetic control of natural variation in salt tolerance of *Arabidopsis thaliana* accessions. J Exp Bot 61:1125–1138

Keller NP, Turner G, Bennett JW (2005) Fungal secondary metabolism – from biochemistry to genomics. Nat Rev Microbiol 3:937–947

Khomari S, Golshan-Doust S, Seyed-Sharifi R, Davari M (2018) Improvement of soybean seedling growth under salinity stress by biopriming of high-vigour seeds with salt-tolerant isolate of *Trichoderma harzianum*. New Zeal J Crop Hort 46(2):117–132

Korolev N, Rav David D, Elad Y (2008) The role of phytohormones in basal resistance and *Trichoderma*-induced systemic resistance to *Botrytis cinerea* in *Arabidopsis thaliana*. BioControl 53:667–683. https://doi.org/10.1007/s10526-007-9103-3

Kubicek CP, Mach RL, Peterbauer CK, Lorito M (2001) *Trichoderma*: from genes to biocontrol. J Plant Pathol 83:11–23

Lee S, Hung R, Yap M, Bennett JW (2015) Age matters: the effects of volatile organic compounds emitted by *Trichoderma atroviride* on plant growth. Arch Microbiol 197(5):723–727. https://doi.org/10.1007/s00203-015-1104-5

Li R-X, Cai F, Pang G, Shen Q-R, Li R, Chen W (2015) Solubilisation of phosphate and micronutrients by *Trichoderma harzianum* and its relationship with the promotion of tomato plant growth. PLoS One 10(6):e0130081. https://doi.org/10.1371/journal.pone.0130081

Li Y, Sun R, Yu J, Saravanakumar K, Chen J (2016) Antagonistic and biocontrol potential of *Trichoderma asperellum* ZJSX5003 against the maize stalk rot pathogen *Fusarium graminearum*. Indian J Microbiol 56(3):318–327. https://doi.org/10.1007/s12088-016-0581-9

Li X, Zhang X, Wang X, Yang X, Cui Z (2019) Bioaugmentation-assisted phytoremediation of lead and salinity co-contaminated soil by *Suaeda salsa* and *Trichoderma asperellum*. Chemosphere 224:716–725

Linkies A, Muller K, Morris K, Tureckova V, Wenk M, Cadman CSC, Corbineau F, Strnad M, Lynn JR, Finch-Savage WE, Leubner Metzger G (2009) Ethylene interacts with abscisic acid to regulate endosperm rupture during germination: a comparative approach using *Lepidium sativum* and *Arabidopsis thaliana*. Plant Cell 21:3803–3822

Liu SY, Liao CK, Lo CT, Yang HH, Lin KC, Peng KC (2016) Chrysophanol is involved in the biofertilization and biocontrol activities of *Trichoderma*. Physiol Mol Plant Pathol 96:1–7

Lopez-Mondejar R, Bernal-Vicente A, Ros M, Tittarelli F, Canali S, Intrigiolo F, Pascual JA (2010) Utilisation of citrus compost-based growing media amended with *Trichoderma harzianum* T-78 in *Cucumis melo* L. seedling production. Bioresour Technol 101:3718–3723

Lorito M (2009) Factors affecting the production of *Trichoderma harzianum* secondary metabolites during the interaction with different plant pathogens. Lett Appl Microbiol 48:705–711. https://doi.org/10.1111/j.1472-765X.2009.02599.x

Macías-Rodríguez L, Guzmán-Gómez A, García-Juárez P, Contreras-Cornejo HA (2018) *Trichoderma atroviride* promotes tomato development and alters the root exudation of carbohydrates, which stimulates fungal growth and the biocontrol of the phytopathogen *Phytophthora cinnamomi* in a tripartite interaction system. FEMS Microbiol Ecol 94(9): fiy137. https://doi.org/10.1093/femsec/fiy137

Malinich EA, Wang K, Mukherjee PK, Kolomiets M, Kenerley CM (2019) Differential expression analysis of *Trichoderma virens* RNA reveals a dynamic transcriptome during colonization of Zea mays roots. BMC Genomics 20(1):280. https://doi.org/10.1186/s12864-019-5651-z

Marfori EC, Kajiyama S, Fukusaki E, Kobayashi A (2002) Trichosetin, a novel tetramic acid antibiotic produced in dual culture of *Trichoderma harzianum* and *Catharanthus roseus* callus. Z Naturforsch C 57(5-6):465–470

Marra R, Ambrosino P, Carbone V, Vinale F, Woo SL, Ruocco M, Ciliento R, Lanzuise S, Ferraioli S, Soriente I, Gigante S, Turra D, Fogliano V, Scala F, Lorito M (2006) Study of the three-way interaction between *Trichoderma atroviride*, plant and fungal pathogens by using a proteomic approach. Curr Genet 50(5):307–321

Martinez-Medina A, Fernandez I, Sanchez-Guzman MJ, Jung SC, Pascual JA, Pozo MJ (2013) Deciphering the hormonal signalling network behind the systemic resistance induced by *Trichoderma harzianum* in tomato. Front Plant Sci 4:206. https://doi.org/10.3389/fpls.2013.00206

Martínez-Medina A, Alguacil MDM, Pascual JA, Van Wees SC (2014) Phytohormone profiles induced by *Trichoderma* isolates correspond with their biocontrol and plant growth-promoting activity on melon plants. J Chem Ecol 40:804–815. https://doi.org/10.1007/s10886-014-0478-1

Martínez-Medina A, Van Wees SC, Pieterse CMJ (2017) Airborne signals from *Trichoderma* fungi stimulate iron uptake responses in roots resulting in priming of jasmonic acid-dependent defences in shoots of *Arabidopsis thaliana* and *Solanum lycopersicum*. Plant Cell Environ 40 (11):2691–2705. https://doi.org/10.1111/pce.13016

Mastouri F, Bjorkman T, Harman GE (2010) Seed treatment with *Trichoderma harzianum* alleviates biotic, abiotic, and physiological stresses in germinating seeds and seedlings. Phytopathology 100:1213–1221

Mastouri F, Bjorkman T, Harman GE (2012) *Trichoderma harzianum* enhances antioxidant defense of tomato seedlings and resistance to water deficit. Mol Plant Microbe Interact 25:1264–1271

Mathews JR, Sivparsad BJ, Laing MD (2019) Greenhouse evaluation of *Trichoderma harzianum* for the control of Sclerotinia wilt (*Sclerotinia sclerotiorum*) of sunflower. S Afr J Plant Soil 36 (1):69–72

Mathys J, De Cremer K, Timmermans P, Van Kerckhove S, Lievens B, Vanhaecke M, Cammue BPA, De Coninck B (2012) Genome-wide characterization of ISR induced in *Arabidopsis thaliana* by *Trichoderma hamatum* T382 against *Botrytis cinerea* infection. Front Plant Sci 3:1–25

Meena KK, Sorty AM, Bitla UM, Choudhary K, Gupta P, Pareek A, Singh DP, Prabha R, Sahu PK, Gupta VK, Singh HB, Krishanani KK, Minhas PS (2017) Abiotic stress responses and microbe-mediated mitigation in plants: the omics strategies. Front Plant Sci 8:172. https://doi.org/10.3389/fpls.2017.00172

Mittler R (2006) Abiotic stress, the field environment and stress combination. Trends Plant Sci 11:15–19. https://doi.org/10.1016/j.tplants.2005.11.002

Mohapatra S, Mittra B (2017) Alleviation of *Fusarium oxysporum* induced oxidative stress in wheat by *Trichoderma viride*. Arch Phytopathol Plant Protect 50(1–2):4–96. https://doi.org/10.1080/03235408.2016.1263052

Mok DWS, Mok MC (2001) Cytokinin metabolism and action. Annu Rev Plant Physiol Plant Mol Biol 52:89–118

Monteiro VN, do Nascimento Silva R, Steindorff AS, Costa FT, Noronha EF, Ricart CA, de Sousa MV, Vainstein MH, Ulhoa CJ (2010) New insights in *Trichoderma harzianum* antagonism of fungal plant pathogens by secreted protein analysis. Curr Microbiol 61:298–305. https://doi.org/10.1007/s00284-010-9611-8

Montero-Barrientos M, Hermosa R, Cardoza RE, Gutierrez S, Nicola C, Monte E (2010) Transgenic expression of the *Trichoderma harzianum* hsp70 gene increases *Arabidopsis* resistance to heat and other abiotic stresses. J Plant Physiol 167:659–665

Moran-Diez E, Hermosa R, Ambrosino P, Cardoza RE, Gutierrez S, Lorito M, Monte E (2009) The ThPG1 endopolygalacturonase is required for the *Trichoderma harzianum*–plant beneficial interaction. Mol Plant-Microbe Interact 22:1021–1031

Mukherjee PK, Latha J, Hadar R, Horwitz BA (2003) TmkA, a mitogen-activated protein kinase of *Trichoderma virens*, is involved in biocontrol properties and repression of conidiation in the dark. Eukaryot Cell 2:446–455

Mukherjee PK, Horwitz BA, Charles MK (2012) Secondary metabolism in *Trichoderma* – a genomic perspective. Microbiology 158:35–45. https://doi.org/10.1099/mic.0.053629-0

Mukherjee PK, Hurley JF, Taylor JT, Puckhaber L, Lehner S, Druzhinina I, Schumacher R, Kenerley CM (2018) Ferricrocin, the intracellular siderophore of *Trichoderma virens*, is involved in growth, conidiation, gliotoxin biosynthesis and induction of systemic resistance in maize. Biochem Biophys Res Commun 505(2):606–611

Muvea AM, Meyhofer R, Subramanian S, Poehling HM, Ekesi S, Maniania NK (2014) Colonization of onion roots by endophytic fungi and their impacts on the biology of *Thrips tabaci*. PLoS One 9(9):e108242. https://doi.org/10.1371/journal.pone.0108242

Nanda AK, Andrio E, Marino D, Pauly N, Dunand C (2010) Reactive oxygen species during plant-microorganism early interactions. J Integr Plant Biol 52(2):195–204

Naseby DC, Pascual JA, Lynch JM (2000) Effect of biocontrol strains of *Trichoderma* on plant growth, *Pythium ultimum* populations, soil microbial communities and soil enzyme activities. J Appl Microbiol 88(1):161–169

Nawrocka J, Malolepsza U (2013) Diversity in plant systemic resistance induced by *Trichoderma*. Biol Control 67(2):149–156. https://doi.org/10.1016/j.biocontrol.2013.07.005

Newbery F, Qi A, Fitt BDL (2016) Modelling impacts of climate change on arable crop diseases: progress, challenges and applications. Curr Opin Plant Biol 32:101–109

Nguyen D, Rieu I, Mariani C, van Dam NM (2016) How plants handle multiple stresses: hormonal interactions underlying responses to abiotic stress and insect herbivory. Plant Mol Biol 91 (6):727–740

Nielsen KF, Gräfenhan T, Zafari D, Thrane U (2005) Trichothecene production by *Trichoderma brevicompactum*. J Agric Food Chem 53:8190–8196

Nieto-Jacobo MF, Steyaert JM, Salazar-Badillo FB, Nguyen DV, Rostas M, Braithwaite M, De Souza JT, Jimenez-Bremont JF, Ohkura M, Stewart A, Mendoza-Mendoza A (2017) Environmental growth conditions of *Trichoderma* spp. affects indole acetic acid derivatives, volatile organic compounds, and plant growth promotion. Front Plant Sci 8:102. https://doi.org/10.3389/fpls.2017.00102

Nogueira-Lopez G, Greenwood DR, Middleditch M, Winefield C, Eaton C, Steyaert JM, Mendoza-Mendoza A (2018) The apoplastic secretome of *Trichoderma virens* during interaction with maize roots shows an inhibition of plant defence and scavenging oxidative stress secreted proteins. Front Plant Sci 9:409

Omero C, Inbar J, Rocha-Ramirez V, Herrera-Estrela A, Chet I, Horwitz BA (1999) G proteins activators and cAMP promote mycoparasitic behavior in *Trichoderma harzianum*. Mycol Res 103(12):1637–1642

Omomowo IO, Fadiji AE, Omomowo OI (2018) Assessment of bio-efficacy of *Glomus versiforme* and *Trichoderma harzianum* in inhibiting powdery mildew disease and enhancing the growth of cowpea. Ann Agric Sci 63(1):9–17

Oros G, Naar Z (2017) Mycofungicide: *Trichoderma* based preparation for foliar applications. Am J Plant Sci 8(2):113–125

Ortiz-Castro R, Contreras-Cornejo HA, Macias-Rodriguez L, Lopez-Bucio J (2009) The role of microbial signals in plant growth and development. Plant Signal Behav 4(8):701–712. https://doi.org/10.4161/psb.4.8.9047

Ortuno N, Castillo A, Miranda C, Magnus MIC, Soto X (2017) The use of secondary metabolites extracted from *Trichoderma* for plant growth promotion in the Andean highlands. Renewable Agric Food Syst 32(4):366–375. https://doi.org/10.1017/S1742170516000302

Palyzová A, Svobodová K, Sokolová L, Novák J, Novotný Č (2019) Metabolic profiling of *Fusarium oxysporum* f. sp. *conglutinans* race 2 in dual cultures with biocontrol agents *Bacillus amyloliquefaciens, Pseudomonas aeruginosa*, and *Trichoderma harzianum*. Folia Microbiol 64:779–787. https://doi.org/10.1007/s12223-019-00690-7

Pandey V, Ansari MW, Tula S, Yadav S, Sahoo RK, Shukla N, Bains G, Badal S, Chandra S, Gaur AK, Kumar A, Shukla A, Kumar J, Tuteja N (2016) Dose-dependent response of *Trichoderma harzianum* in improving drought tolerance in rice genotypes. Planta 243:1251–1264. https://doi.org/10.1007/s00425-016-2482-x

Pascale A, Vinale F, Manganiello G, Nigro M, Lanzuise S, Ruocco M, Marra R, Lombardi N, Woo SL, Lorito M (2017) *Trichoderma* and its secondary metabolites improve yield and quality of grapes. Crop Prot 92:176–181

Pehlivan N, Yesilyurt AM, Durmus N, Karaglu SA (2017) *Trichoderma lixii* ID11D seed biopriming mitigates dose dependent salt toxicity in maize. Acta Physiol Plant 39:79. https://doi.org/10.1007/s11738-017-2375-z

Peleg Z, Blumwald E (2011) Hormone balance and abiotic stress tolerance in crop plants. Curr Opin Plant Biol 14:290–295

Pieterse CM, Leon-Reyes A, Van der Ent S, Van Wees SC (2009) Networking by small-molecule hormones in plant immunity. Nat Chem Biol 5(5):308–316

Qi W, Zhao L (2013) Study of the siderophore-producing *Trichoderma asperellum* Q1 on cucumber growth promotion under salt stress. J Basic Microbiol 53(4):355–364

Raghavendra AS, Gonugunta VK, Christmann A, Grill E (2010) ABA perception and signalling. Trends Plant Sci 15:395–401

Rawat L, Singh Y, Shukla N, Kumar J (2011) Alleviation of the adverse effects of salinity stress in wheat (*Triticum aestivum* L.) by seed biopriming with salinity tolerant isolates of *Trichoderma harzianum*. Plant Soil 347:387–400. https://doi.org/10.1007/s11104-011-0858-z

Rawat L, Bisht TS, Kukreti A, Prasad M (2016) Bioprospecting drought tolerant *Trichoderma harzianum* isolates promote growth and delay the onset of drought responses in wheat (*Triticum aestivum* L.). Mol Soil Biol 7(4):1–15. https://doi.org/10.5376/msb.2016.07.0004

Reino JL, Guerrero RF, Hernandez-Galan R, Collado IG (2008) Secondary metabolites from species of the biocontrol agent *Trichoderma*. Phytochem Rev 7:89–123. https://doi.org/10.1007/s11101-006-9032-2

Rockstrom J, Williams J, Daily G, Noble A, Matthewa N, Gordon L, Wetterstand H, DeClerck F, Shah M, Steduto P, De Fraiture C, Hatibu N, Unver O, Bird J, Sibanda L, Smith J (2017) Sustainable intensification of agriculture for human prosperity and global sustainability. Ambio 46(1):4–17

Rubio MB, Quijada NM, Pérez E, Domínguez S, Monte E, Hermosa R (2014) Identifying beneficial qualities of *Trichoderma parareesei* for plants. Appl Environ Microbiol 80(6):1864–1873. https://doi.org/10.1128/AEM.03375-13

Rubio MB, de Medeiros HA, Morán-Diez ME, Castillo P, Hermosa R, Monte E (2019) A split-root method to study systemic and heritable traits induced by *Trichoderma* in tomato plants. In: Reinhardt D, Sharma AK (eds) Methods in rhizosphere biology research. Springer, Singapore, pp 151–166

Saber WI, Ghoneem KM, Rashad YM, Al-Askar AA (2017) *Trichoderma harzianum* WKY1: an indole acetic acid producer for growth improvement and anthracnose disease control in sorghum. Biocontrol Sci Tech 27(5):654–676

Sachdev S, Singh RP (2016a) Studies on trends in use of pesticides and fertilizers for tomato cultivation in the vicinity of Lucknow, India. Int J Sci Technol Soc 2(1-2):49–54. https://doi.org/10.18091/ijsts.v2i1-2.7542

Sachdev S, Singh RP (2016b) Current challenges, constraints and future strategies for development of successful market for biopesticides. Clim Change Environ Sustain 4(2):129–136. https://doi.org/10.5958/2320-642X.2016.00014.4

Sachdev S, Singh RP (2018a) Root colonization: imperative mechanism for efficient plant protection and growth. MOJ Eco Environ Sci 3(4):240–242

Sachdev S, Singh RP (2018b) Isolation, characterisation and screening of native microbial isolates for biocontrol of fungal pathogens of tomato. Clim Change Environ Sustain 6(1):46–58

Sachdev S, Singh A, Singh RP (2018) Optimization of culture conditions for mass production and bio-formulation of *Trichoderma* using response surface methodology. 3 Biotech 8(8):360

Samolski I, Rincon AM, Pinzon LM, Viterbo A, Monte E (2012) The qid74 gene from *Trichoderma harzianum* has a role in root architecture and plant biofertilization. Microbiology 158:129–138

Saravanakumar K, Yu C, Dou K, Wang M, Li Y, Chen J (2016) Synergistic effect of *Trichoderma*-derived antifungal metabolites and cell wall degrading enzymes on enhanced biocontrol of *Fusarium oxysporum* f. sp. *cucumerinum*. Biol Control 94:37–46

Saxena A, Raghuwanshi R, Singh HB (2014) *Trichoderma* species mediated differential tolerance against biotic stress of phytopathogens in *Cicer arietinum* L. J Basic Microbiol 54:1–12. https://doi.org/10.1002/jobm.201400317

Schirmbock M, Lorito M, Wang YL, Hayes CK, ArisanAtac I, Scala F, Harman GE, Kubicek CP (1994) Parallel formation and synergism of hydrolytic enzymes and peptaibol antibiotics, molecular mechanisms involved in the antagonistic action of *Trichoderma harzianum* against phytopathogenic fungi. Appl Environ Microbiol 60(12):4364–4370

Segarra G, Casanova E, Bellido D, Odena MA, Oliveira E, Trillas I (2007) Proteome, salicylic acid, and jasmonic acid changes in cucumber plants inoculated with *Trichoderma asperellum* strain T34. Proteomics 7(21):3943–3952

Segarra G, Van der Ent S, Trillas I, Pieterse CMJ (2009) MYB72, a node of convergence in induced systemic resistance triggered by a fungal and a bacterial beneficial microbe. Plant Biol 11:90–96. https://doi.org/10.1111/j.1438-8677.2008.00162.x

Sharma PK, Gothalwal R (2017) *Trichoderma*: a potent fungus as biological control agent. In: Singh JS, Seneviratne G (eds) Agro-environmental sustainability: managing crop health, vol 1. Springer, Cham, pp 113–125

Sharma A, Johri BN (2003) Growth promoting influence of siderophore-producing *Pseudomonas* strains GRP3A and PRS9 in maize (*Zea mays* L.) under iron limiting conditions. Microbiol Res 158(3):243–248

Sharma R, Joshi A, Dhaker RC (2012) A brief review on mechanism of *Trichoderma* fungus use as biological control agents. Int J Innovovat Bio Sci 2(4):200–210

Sharma KK, Singh US, Sharma P, Kumar A, Sharma L (2015) Seed treatments for sustainable agriculture-a review. J Appl Nat Sci 7(1):521–539

Shentu X, Zhan X, Ma Z, Yu X, Zhang C (2014) Antifungal activity of metabolites of the endophytic fungus *Trichoderma brevicompactum* from garlic. Braz J Microbiol 45 (1):248–254. https://doi.org/10.1590/S1517-83822014005000036

Shiva V (2016) The violence of the green revolution: third world agriculture, ecology, and politics. University Press of Kentucky, Lexington, KY

Shoresh M, Yedidia I, Chet I (2005) Involvement of jasmonic acid/ethylene signaling pathway in the systemic resistance induced in cucumber by *Trichoderma asperellum* T203. Phytopathology 95(1):76–84. https://doi.org/10.1094/PHYTO-95-0076

Shoresh M, Harman GE, Mastouri F (2010) Induced systemic resistance and plant responses to fungal biocontrol agents. Annu Rev Phytopathol 48:21–43

Silva RN, Silva SP, Brandao RL, Ulhoa CJ (2004) Regulation of N-acetyl-b-D-glucosaminidase produced by *Trichoderma harzianum*: evidence that cAMP controls its expression. Res Microbiol 155:667–671

Singh BN, Singh BR, Sarma BK, Singh HB (2009) Potential chemoprevention of N-nitrosodiethylamine-induced hepatocarcinogenesis by polyphenolics from *Acacia nilotica* bark. Chem Biol Interact 181(1):20–28

Singh BN, Singh A, Singh SP, Singh HB (2011) *Trichoderma harzianum* mediated reprogramming of oxidative stress response in root apoplast of sunflower enhances defence against *Rhizoctonia solani*. Eur J Plant Pathol 131:121–134

Singh A, Sarma BK, Upadhyay RS, Singh HB (2013) Compatible rhizosphere microbes mediated alleviation of biotic stress in chickpea through enhanced antioxidant and phenylpropanoid activities. Microbiol Res 168(1):33–40

Singh JS, Abhilash PC, Gupta VK (2016) Agriculturally important microbes in sustainable food production. Trends Biotechnol 34:773–775. https://doi.org/10.1016/j.tibtech.2016.06.002

Sivasithamparam K, Ghisalberti EL (1998) Secondary metabolism in *Trichoderma* and *Gliocladium*. In: Harman GE, Kubicek CP (eds) *Trichoderma* and *Gliocladium*: basic biology taxonomy and genetics, vol 1. Taylor & Francis, London, pp 139–191

Sriram S, Manasa SB, Savitha MJ (2009) Potential use of elicitors from *Trichoderma* in induced systemic resistance for the management of *Phytophthora capsici* in red pepper. J Biol Control 23:449–456

Srivastava P, Singh R, Tripathi S, Raghubanshi AS (2016) An urgent need for sustainable thinking in agriculture—an Indian scenario. Ecol Indic 67:611–622

Studholme DJ, Harris B, Le Cocq K, Winsbury R, Perera V, Ryder L, Ward JL, Beale MH, Thornton CR, Grant M (2013) Investigating the beneficial traits of *Trichoderma hamatum* GD12 for sustainable agriculture—insights from genomics. Front Plant Sci 4:258. https://doi.org/10.3389/fpls.2013.00258

Tartoura KAH, Youssef SA (2011) Stimulation of ROS-scavenging systems in squash (*Cucurbita pepo* L.) plants by compost supplementation under normal and low temperature conditions. Sci Hort 130:862–868

Tijerino A, Hermosa R, Cardoza RE, Moraga J, Malmierca MG, Aleu J, Collado IG, Monte E, Gutierrez S (2011a) Overexpression of the *Trichoderma brevicompactum tri5* gene: effect on the expression of the trichodermin biosynthetic genes and on tomato seedlings. Toxins (Basel) 3 (9):1220–1232. https://doi.org/10.3390/toxins3091220

Tijerino A, Cardoza RE, Moraga J, Malmierca MG, Vicente F, Aleu J, Collado IG, Gutiérrez S, Monte E, Hermosa R (2011b) Overexpression of the trichodiene synthase gene *tri5* increases trichodermin production and antimicrobial activity in *Trichoderma brevicompactum*. Fungal Genet Biol 48:285–296

Tripathi P, Singh PC, Mishra A, Srivastava S, Chauhan R, Awasthi S, Mishra S, Dwivedi S, Tripathi P, Kalra A, Tripathi RD, Nautiyal CS (2017) Arsenic tolerant *Trichoderma* sp. reduces arsenic induced stress in chickpea (*Cicer arietinum*). Environ Pollut 223:137–145. https://doi.org/10.1016/j.envpol.2016.12.073

Troian RF, Steindorff AS, Ramada MHS, Arruda W, Ulhoa CJ (2014) Mycoparasitism studies of *Trichoderma harzianum* against *Sclerotinia sclerotiorum*: evaluation of antagonism and expression of cell wall-degrading enzymes genes. Biotechnol Lett 36(10):2095–2101

Turner TR, James EK, Poole PS (2013) The plant microbiome. Genome Biol 14(6):209

Vargas WA, Mandawe JC, Kenerley CM (2009) Plant-derived sucrose is a key element in the symbiotic association between *Trichoderma virens* and maize plants. Plant Physiol 151:792–808

Vargas JT, Rodriguez-Monroy M, Meyer ML, Montes-Belmont R, Sepulveda-Jimenez G (2017) *Trichoderma asperellum* ameliorates phytotoxic effects of copper in onion (*Allium cepa* L.). Environ Exp Bot 136:85–93. https://doi.org/10.1016/j.envexpbot.2017.01.009

Verbruggen E, Kiers ET, Bakelaar PN, Röling WF, van der Heijden MG (2012) Provision of contrasting ecosystem services by soil communities from different agricultural fields. Plant Soil 350(1-2):43–55

Vieira PM, Zeilinger S, Brandao RS, Vianna GR, Georg RC, Gruber S, Aragao FJL, Ulhoa CJ (2018) Overexpression of an aquaglyceroporin gene in the fungal biocontrol agent *Trichoderma harzianum* affects stress tolerance, pathogen antagonism and *Phaseolus vulgaris* development. Biol Control 126:185–191. https://doi.org/10.1016/j.biocontrol.2018.08.012

Vinale F, Marra R, Scala F, Ghisalberti EL, Lorito M, Sivasithamparam K (2006) Major secondary metabolites produced by two commercial *Trichoderma* strains active against different phytopathogens. Lett Appl Microbiol 43:143–148

Vinale F, Sivasithamparam K, Ghisalberti EL, Marra R, Barbetti MJ, Li H, Woo SL, Lorito M (2008) A novel role for *Trichoderma* secondary metabolites in the interactions with plants. Physiol Mol Plant Pathol 72:80–86

Vinale F, Ghisalberti EL, Sivasithamparam K, Marra R, Ritieni A, Ferracane R, Woo S, Lorito M (2009a) Factors affecting the production of *Trichoderma harzianum* secondary metabolites during the interaction with different plant pathogens. Lett Appl Microbiol 48(6):705–711

Vinale F, Flematti G, Sivasithamparam K, Lorito M, Marra R, Skelton BW, Ghisalberti EL (2009b) Harzianic acid, an antifungal and plant growth promoting metabolite from *Trichoderma harzianum*. J Nat Prod 72(11):2032–2035. https://doi.org/10.1021/np900548p

Vinale F, Arjona GI, Nigro M, Mazzei P, Piccolo A, Ruocco M, Woo S, Rosa DR, Herrera CL, Lorito M (2012) Cerinolactone, a hydroxylactone derivative from *Trichoderma cerinum*. J Nat Prod 75:103–106

Vinale F, Nigroa M, Sivasithamparam K, Flematti G, Ghisalberti EL, Ruocco M, Varlese R, Marra R, Lanzuise S, Eid A, Woo SL, Lorito M (2013) Harzianic acid: a novel siderophore from *Trichoderma harzianum*. FEMS Microbiol Lett 347(2):123–129. https://doi.org/10.1111/1574-6968.12231

Vinale F, Sivasithamparam K, Ghisalberti EL, Woo SL, Nigro M, Marra R, Lombardi N, Pascale A, Ruocco M, Lanzuise S, Manganiello G, Lorito M (2014a) *Trichoderma* secondary metabolites active on plants and fungal pathogens. The Open Mycology Journal 8(Suppl-1, M5):127–139

Vinale F, Manganiello G, Nigro M, Mazzei P, Piccolo A, Pascale A, Ruocco M, Marra R, Lombardi N, Lanzuise S, Varlese R, Cavallo P, Lorito M, Woo SL (2014b) A novel fungal metabolite with beneficial properties for agricultural applications. Molecules 19:9760–9772. https://doi.org/10.3390/molecules19079760

Vinale F, Strakowska J, Mazzei P, Piccolo A, Marra R, Lombardi N, Manganiello G, Pascale A, Woo SL, Lorito M (2016) Cremenolide, a new antifungal, 10-member lactone from *Trichoderma cremeum* with plant growth promotion activity. Nat Prod Res 30 (22):2575–2581. https://doi.org/10.1080/14786419.2015.1131985

Vinci G, Cozzolino V, Mazzei P, Monda H, Spaccini R, Piccolo A (2018) An alternative to mineral phosphorus fertilizers: the combined effects of *Trichoderma harzianum* and compost on *Zea mays*, as revealed by 1 H NMR and GC-MS metabolomics. PLoS One 13(12):e0209664. https://doi.org/10.1371/journal.pone.0209664

Vishwakarma K, Upadhyay N, Kumar N, Yadav G, Singh J, Mishra RK, Kumar V, Verma R, Upadhyay RG, Pandey M, Sharma S (2017) Abscisic acid signaling and abiotic stress tolerance in plants: a review on current knowledge and future prospects. Front Plant Sci 8:161. https://doi.org/10.3389/fpls.2017.00161

Viterbo A, Montero M, Ramot O, Friesem D, Monte E, Llobell A, Chet I (2002) Expression regulation of the endochitinase chit36 from *Trichoderma asperellum* (*T. harzianum* T-203). Curr Genet 42:114–122

Vitti A, Pellegrini E, Nali C, Lovelli S, Sofo A, Valerio M, Scopa A, Nuzzaci M (2016) *Trichoderma harzianum* T-22 induces systemic resistance in tomato infected by *cucumber mosaic virus*. Front Plant Sci 7:1520. https://doi.org/10.3389/fpls.2016 01520

Wang W, Vinocur B, Altman A (2003) Plant responses to drought, salinity and extreme temperatures: towards genetic engineering for stress tolerance. Planta 218:1–14. https://doi.org/10.1007/s00425-003-1105-5

Wang M, Hashimoto M, Hashidoko Y (2013) Repression of tropolone production and induction of a *Burkholderia plantarii* pseudo-biofilm by Carot-4-en-9,10-diol, a cell-to-cell signaling disrupter produced by *Trichoderma virens*. PLoS One 8(11):e78024. https://doi.org/10.1371/journal.pone.0078024

Wang X, Xu S, Wu S, Feng S, Bai Z, Zhuang G, Zhuang X (2018) Effect of *Trichoderma viride* biofertilizer on ammonia volatilization from an alkaline soil in northern China. J Environ Sci 66:199–207

Wani SH, Kumar V, Shriram V, Sah SK (2016) Phytohormones and their metabolic engineering for abiotic stress tolerance in crop plants. Crop J 4(3):162–176. https://doi.org/10.1016/j.cj.2016.01.010

Woo SL, Ruocco M, Vinale F, Nigro M, Marra R, Lombardi N, Pascale A, Lanzuise S, Manganiello G, Lorito M (2014) *Trichoderma*-based products and their widespread use in agriculture. The Open Mycology Journal 8(1):71–126

Wu CY, Chen CL, Lee YH, Cheng YC, Wu YC, Shu HY, Gotz F, Liu ST (2007) Nonribosomal synthesis of fengycin on an enzyme complex formed by fengycin synthetases. J Biol Chem 282:5608–5616. https://doi.org/10.1074/jbc.M609726200

Wu Q, Ni M, Dou K, Tang J, Ren J, Yu C, Chen J (2018) Co-culture of *Bacillus amyloliquefaciens* ACCC11060 and *Trichoderma asperellum* GDFS1009 enhanced pathogen-inhibition and amino acid yield. Microb Cell Factories 17(1):155

Yasmeen R, Siddiqui ZS (2018) Ameliorative effects of *Trichoderma harzianum* on monocot crops under hydroponic saline environment. Acta Physiol Plant 40(1):4

Yedidia I, Benhamou N, Chet I (1999) Induction of defense responses in cucumber plants (*Cucumis sativus* L.) by the biocontrol agent *Trichoderma harzianum*. Appl Environ Microbiol 65 (3):1061–1070

Yesilyurt AM, Pehlivan N, Durmus N, Karaoglu SA (2018) *Trichoderma citrinoviride*: a potent biopriming agent for the alleviation of salt stress in maize. Hacettepe J Biol Chem 46 (1):101–111

Youssef SA, Kamel A, Tartoura KA, Abdelraouf GA (2016) Evaluation of *Trichoderma harzianum* and *Serratia proteamaculans* effect on disease suppression, stimulation of ROS-scavenging enzymes and improving tomato growth infected by *Rhizoctonia solani*. Biol Control 100:79–86

Zeilinger S, Omann M (2007) *Trichoderma* biocontrol: signal transduction pathways involved in host sensing and mycoparasitism. Gene Regul Syst Bio 1:227–234

Zhang S, Xu B, Zhang J, Gan Y (2018) Identification of the antifungal activity of *Trichoderma longibrachiatum* T6 and assessment of bioactive substances in controlling phytopathogens. Pest Biochem Physiol 147:59–66

Zikmundova M, Drandarov K, Bigler L, Hesse M, Werner C (2002) Biotransformation of 2-benzoxazolinone and 2-hydroxy-1, 4-benzoxazin-3-one by endophytic fungi isolated from *Aphelandra tetragona*. Appl Environ Microbiol 68(10):4863–4870

Biochar Amendment in Agricultural Soil for Mitigation of Abiotic Stress

14

Khushbu Kumari, Zaira Khalid, Shahrukh Nawaj Alam, Sweta, Bhaskar Singh, Abhishek Guldhe, D. K. Shahi, and Kuldeep Bauddh

Contents

K. Kumari · Z. Khalid · S. N. Alam · Sweta · B. Singh · A. Guldhe · K. Bauddh (✉)
Department of Environmental Science, Central University of Jharkhand, Ranchi, India
e-mail: kuldeep.bauddh@cuj.ac.in

D. K. Shahi
Department of Soil Science and Agricultural Chemistry, Birsa Agricultural University, Ranchi, Jharkhand, India

K. Bauddh et al. (eds.), *Ecological and Practical Applications for Sustainable Agriculture*, https://doi.org/10.1007/978-981-15-3372-3_14

Abstract

Abiotic stresses like drought, cold, salinity, heat, oxidative stress and the presence of excess levels of heavy metal in the agroecosystems lead to a decrease in the growth and productivity of major crops worldwide. The majority of stresses are connected with each other and results in elevated adverse impacts on the plants as well as other important components of the environment. The intensity of stresses and associated adverse impacts are increasing substantially in the era of climate change that again triggers to produce abnormalities in the crops. To overcome the effects of abiotic stresses, a number of strategies have been investigated, such as developing and cultivate stress-tolerant varieties, use of organic fertilizers, and the application of high yielding varieties. Application of biochar to mitigate the impacts of major abiotic stresses especially drought, salinity, and heavy metal has been found very effective. Amendment of biochar in stress affected agroecosystems improves the soil physicochemical and biological features and thereby enhances the productivity of crops. In this chapter, efforts have been made to discuss about three major stresses, i.e. drought, salinity, and heavy metals, their impacts on soil as well as plant productivity. Further, the efficiency and mechanism of biochar in reducing the impacts of stresses when using as a soil amendment have also been discussed thoroughly.

Keywords

Abiotic stress · Biochar · Heavy metal · Drought · Salinity

14.1 Introduction

The projected world population by 2050 is from 7 billion (current estimated) to 8.9 billion approximately (Singh et al. 2011), and the food security to this burgeoning number is a paramount concern. The rapid increase in environmental contamination and global climate change is another important threat toward sustainable agriculture. Farmers are under pressure to use excess amounts of chemical fertilizers and pesticides to enhance the agricultural productivity. The application of synthetic fertilizers not only deteriorates the soil health but also enhances the greenhouse gases into the atmosphere as well as contaminates the surface and groundwater. Traditional cropping systems infuse a variety of toxicants to the soil and ultimately enter into the food chain. Abiotic stresses like salinity, acidity, drought, heavy metal, flood, high temperature, freezing, chilling, etc. also damage the agriecosystems, which results in decreased crop yield. These stresses decrease crop productivity by decreasing the photosynthetic pigments, protein denaturation, production of reactive oxygen species (ROS), lipid peroxidation, etc. (Fig. 14.1) (Chodak et al. 2015; Isayenkov and Maathuis 2019; Sallam et al. 2019).

Salinity, drought, and heavy metals are considered as major abiotic stress and reduce a significant portion of agricultural productivity. Plant growth is dependent on the rate and level of photosynthesis, hence environmental stress affecting

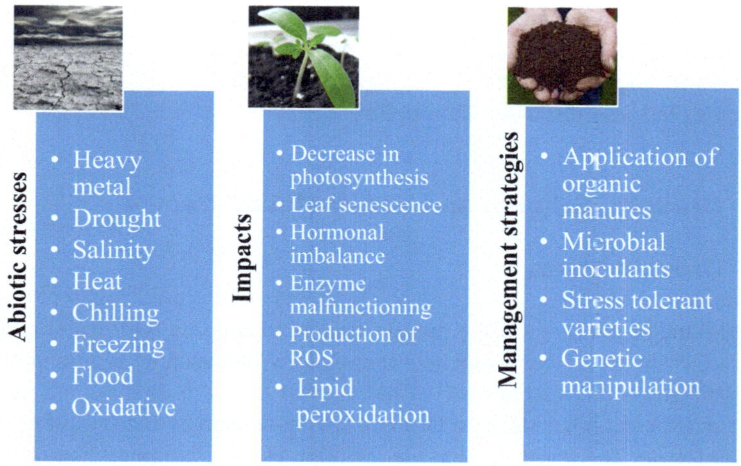

Fig. 14.1 Different abiotic stresses, their impacts and mitigation strategies (Chodak et al. 2015; Bauddh et al. 2016; Isayenkov and Maathuis 2019; Sallam et al. 2019)

photosynthesis directly impacts the growth of plant. Salt stress can also influence the synthesis of abscisic acid which when moves to the guard cells, closes the stomata. As a result of the closure of stomata, the rate of photosynthesis declines. The ability of germination and vigor is the most important and primary process in the production of any plant and the adverse impacts of abiotic stress on these parameters are well reported (Bhaskar et al. 2009; Kumar et al. 2009; Singh et al. 2010; Bauddh and Singh 2009, 2011, 2012). Germination being important in the reproduction often influenced by the environment created due to the stress induced by water-scarce conditions (Leila 2007). Water being absorbed by the seed is the first stage in the process of germination. The quantity of water absorbed, mainly dependent on the type, size, and chemical composition of the seed (Rahmani 2006). Generally, soil moisture level lesser than soil's field capacity is considered favorable for the germination of the seed but in the water-scarce conditions where there is less than desirable water absorption in these seeds are incomplete which can result in diminishing or even stopping the germination process (Leila 2007). Several strategies have been developed to mitigate the impacts of abiotic stresses, especially salinity, drought, and heavy metals. The tools like cultivation of stress-tolerant varieties, application of organic amendments, genetic manipulation, application of microbial inoculants, etc. have been found significantly effective to reduce the adverse impacts of majority of abiotic stresses. Application of biochar has been proven an efficient potential to reduce the adverse impacts of drought, salinity, and heavy metals. Biochar have several other beneficial properties like it retains moisture in soil, balance soil pH, use to make fuel, used in metal remediation from the soil, etc. (Yuan et al. 2019). Several studies have been found that application of biochar to the soil enhances the physicochemical and biological properties of soil and overcome the negative effects created by the stressed conditions.

In the present chapter, the efforts have been done to explore the adverse impacts of three major abiotic stresses, i.e., drought, salinity, and heavy metals. Further, the application of biochar to overcome the impacts of these stresses during the cultivation of crops has also discussed thoroughly.

14.2 Drought, Salinity, and Heavy Metal Stresses

14.2.1 Drought Stress

One of the major concerns of the present time is to cope up with the condition of water scarcity that is arising throughout the world. Water scarcity is among the most catastrophic stress that can severely affect human life both directly and indirectly arising from several other stresses. Drought can be defined as an event of prolonged shortage of water due to insufficient precipitation usually for more than one season (Trenberth et al. 2014). In most of the regions, the phenomenon of drought is a periodic condition and other places can remain in the state of drought for several years. Whenever an area experiences drought, it has a massive effect on life and activities in these areas. Drought also has an impact on hydration of soil, which slowly deteriorates soil quality and also severely affects the animals living in these regions. Some of the climatic factors that are precursors of drought include high wind speeds, high mercury concentrations, etc. Apart from the abovementioned drought, there are other types of drought such as meteorological drought, hydrological drought, and agricultural drought which are described below.

14.2.1.1 Meteorological Drought

It is defined as the drought condition arising due to absence of precipitation in a particular area for a very long time span. However, there is no common consensus among the researchers about a particular value that will define the drought condition like amount of rain or time span of absence of rain. Due to the varying climatic condition, even deserts have drought but their condition is different from the condition of a grassland's drought.

14.2.1.2 Hydrological Drought

Hydrological drought may be defined as the water deficiency in the surface and subsurface regions, which may affect the supply of water to the water bodies such as freshwater streams, lakes, ponds and most importantly confined water reservoirs and groundwater. Hydrological drought is not a seasonal phenomenon, it occurs and develops over an extended period of time.

14.2.1.3 Agricultural Drought

Agricultural drought is a condition when the water is in such scarce condition that it becomes difficult to sustain the grown crops. This condition usually arises when the expected rainfall is not achieved during essential growth period of crops (this period varies in different crops) and may result in the destruction of crops ranging from mild-to-absolute crop loss. This type of drought can develop very fast if expected rain does not arrive or there is an extensive heat wave, especially during essential growing periods of the crops.

It is a well-known fact that drought condition usually begins with the below-average rainfall in any particular region, although the scarcity of water supply is not the only reason for causing the drought situations. Any kind of activities that deplete the capacity of soil to hold water can be one of the main reasons in the commencement of water scarcity in and around any area. These activities may include over farming of a particular area and extensive deforestation. Both of these activities exploit and deplete the nutrients present in the soil and loosens the soil increasing the soil erosion and destroying the water holding capacity of the soil. This is the prime reason resulting in the desertification of that particular area. On top of that deforestation has been proven to be the reason for reduced rainfall, which adds up to the drought condition already prevailing. When there is a smaller number of trees around any area, vast area of land remains uncovered, which increases water loss directly from the land. All of this can lead to the complete destruction of any kind of forest on its own which further deteriorates the environmental conditions.

14.2.2 Salinity Stress

Salinity is simply the presence of salts such as sodium, magnesium and calcium and bicarbonates in the soil while sodicity refers specifically to the amount of sodium present in the soil. Saline soil may be defined as the soil with an electrical conductivity of at least 4 dS m^{-1} by measuring the saturated soil extract while sodic soil can be defined as the soil with an exchangeable-sodium percentage (ESP) which is more than 15 (United States Salinity Laboratory Staff 1954). Salinity is of two types; primary salinity and secondary salinity. The salting that results from anthropogenic activities, mainly land development and agriculture is known as secondary salinity. Unrestrained cutting of trees, overgrazing of economic land by animals, and the use of chemical fertilizers are some of the other contributing factors to soil salinity (Lakhdar et al. 2009). Salinity is not bound to rural areas only. Urban salinity occurs generally by the combination of both faulty irrigation-based salinity and due to dry land. Saline soils are mostly dominant in arid and semiarid regions throughout the world, and many productive lands are converted into salt-affected wastelands (Qadir et al. 2000). Globally, the extent of primary salinity is about 955 M ha, whereas the secondary salinization affects 77 M ha, where 58% of the area is affected by irrigation (Metternicht and Zinck 2003). The global estimate of the land affected by salinity is 400 million hectares, which makes up 6% of the total world land area (Arora et al. 2017). The distribution of saline lands is not uniform. The continent of Australia has the highest salinity surface area with 963 million hectares of saline land followed by Asia with about 923 million hectares of saline land (ICARDA 2002). Fifty percent of the arable land is expected to encounter a serious salinity problem by 2050 (Meng et al. 2016).

Table 14.1 Characteristics of salt-affected soil (Amini et al. 2015)

Soil type	pH	Electrical conductivity (EC) (mmohs/cm)	Exchangeable sodium % (ESP)	Sodium absorption ratio (SAR)
Saline	<8.5	>4.0	<15	<13
Sodic	>8.5	<4.0	>15	>13
Saline–sodic	<8.5	>4.0	>15	>13

Classification of salt-affected soil is done on the basis of pH, electrical conductivity (EC), sodium absorption ratio (SAR), and exchangeable sodium percentage (ESP) of the saturated extract (Bohn et al. 2001). Salt-affected soils are thus categorized into saline soil, sodic soil, and saline–sodic soil (Table 14.1).

14.2.3 Heavy Metal Stress

The rampant industrialization and urbanization along with biological and behavioral variations in human, lead to histrionic changes in the environment and climatic patterns. Haggag et al. (2015) discussed feed of devastating population and their food security increased the biotic and abiotic stresses on crop productivity and the environment. Minhas et al. (2017) emphasized on abiotic stresses like salinity, drought, flooding, metal toxicity, nutrient deficiency, high and low temperatures, UV-exposed photo inhibition, air pollution, wind, hail, and gaseous pollutants caused severe damage to the crop productivity and eventually reduced the crop yields. The global per capita food supply and sustainable agricultural food production are the major challenges (Tilman et al. 2002). The technological advancement used an excessive amount of chemical fertilizers such as new cultivars, mineral fertilizers, pesticides, synthetic fertilizers like DDT (1,1′-(2,2,2-trichloroethane-1,1-diyl) bis (4-chlorobenzene)) and polyaromatic hydrocarbons (PAH), etc. these activities increased the concentration of heavy metals in agricultural soil particularly Cd, Pb, As, Zn, Co, Ni, Mn, etc. (Atafar et al. 2010; Chibuike and Obiora 2014). According to Connor et al. (2018), a national soil quality survey in China showed about 19.4% of nations' agricultural land has been affected by heavy metal contaminations. Heavy metals are naturally present in the soil and geologic as a useful concentration but anthropogenic activities increased the concentration of these elements to amounts that are harmful to plants, animals, and living organisms (Chibuike and Obiora 2014). Plants growing on these heavy metal-containing soils show a reduction in growth, performance, decrease crop yield with quality, and metal accumulations in crops can lead to a potential threat to human health. Chromium toxicity in Bangladesh is increased by the continuous advancement in industrial activities cause cancers to humans through direct contact and other health problems such as respiratory tract and skin problems (Uddin et al. 2017). Some heavy metals are toxic even at low concentrations such as As, Cd, Pb, Cr, and Ni and also carcinogenic by nature (Jan et al. 2015; Mishra and Bharagava 2016 Mishra et al. 2018).

14.3 Adverse Impacts of Drought, Salinity, and Heavy Metal on Soil Quality

14.3.1 Effects of Drought

Drought is one of the worst abiotic stresses which vastly affects the fertility of the soil. It greatly affects the microbial activities, soil respiration, waste decomposition, etc. (Hoogmoed and Sadras 2016; Schmidt et al. 2016). Microbial community and their function in the soil ecosystem plays a very significant role in deciding the fertility and improving the structure of any soil (Wall et al. 2012; Arafat et al. 2017). For example, microbial communities generally mass-produce enzymes that are main reason for 90% of soil organic matter decomposition which in turn helps in C mineralization, this whole process is responsible for the cycling of the nutrients in the soil (Swift et al. 1979). In comparison to fungal properties, the bacterial population is more sensitive to the stress caused due to scarcity of the water and in turn, their reaction adversely impacts the fertility of the soil and soil structure (Bardgett and Wardle 2010).

14.3.1.1 Soil Enzymes

Soil enzymes and the related activities that occur are known to be the indicators of the soil health as they are directly dependent and can give information about microbial and physicochemical status of the soil (Baum et al. 2003). These enzymes are generally synthesized by fungi and bacteria present in the soil and may contain enzymes like ureases, pectinases, phosphatases, and proteases. All these enzymes are very important part and influence soil fertility. Drought is one of the major stresses that can directly and most frequently affect the soil enzymes. Soil enzyme is one of the first affected characteristics during water scarcity thus giving drought the control over nutrient availability and ultimately soil fertility (Sinsabaugh et al. 2013).

14.3.1.2 Microbial Activity

Water un-availability and its stress is one of the primary limiting factor in the survival of microbial community through induced osmotic stress, starvation, and resource competition, thus making the presence of water to be an important factor in the functioning of soil microbial communities and the effect of the stress can be very severe (Griffiths et al. 2003; Sowerby et al. 2005). It has been proven by the scientists from all over the world that soil fertility and its stability is greatly influenced by the microbial community present and they are very sensitive indicators of the environment to stresses like drought (Zornoza et al. 2007). Bacteria are considered to be a major part of any soil microbial community and very important to the fertility of the soil. During the conditions of drought, bacterial community is greatly affected thus limiting the decomposition capacity of the bacteria which may result in a change of structure of soil (Hueso et al. 2012). The availability of water also influences substrate availability and other soil characteristics like its physical characteristics that will affect the population and overall performance (Hueso et al. 2012)

14.3.1.3 Nutrient

Under drought conditions, nutrient uptake through roots as well as transport of nutrients from the roots to shoots is greatly affected mainly due to reduced transpiration rate and the impairment of permeability of membranes (Alam 1999). The drought condition also results in the reduction of the diffusion rate of the nutrients within the soil to the roots (Pinkerton and Simpson 1986).

Nitrogen is one such element which is among the most important mineral for plant growth and is required in subsequent amount as it is a common component of nucleic acid and amino acid. Therefore, the deficiency of nitrogen will immediately cause slow growth in plants. Various researchers have conducted study to establish a relationship between the availability of water and fertilizing property of nitrogen. It has been seen that under water-scarce condition nitrogen fertilizer does not yield good crops and at the same time, increasing water without sufficient supply of nitrogen fertilizer also inhibits growth (Smika et al. 1965). Scientists and researchers have also proven that under drought/water-scarce condition nitrogen-fixing property of leguminous plant is reduced (Streeter 2003). Phosphorus, another important element is a component of phospholipids, dinucleotides, nucleic acids, adenosine triphosphate, and phosphoproteins. Therefore, phosphorus is essential for processes like photosynthesis, transfer and storage of energy, transport of carbohydrates, and regulation of some enzymes. It has been proven that under drought conditions, there are more requirements for adding phosphorus and increased phosphorus uptake into the soil. It is a known fact that under drought or water-scarce condition, phosphorus uptake by plants is reduced (Pinkerton and Simpson 1986), i.e., the transfer and uptake of phosphorus and translocation of phosphorus to the shoots from roots are greatly affected during mildly dry soil conditions (Resnik 1970). In a study by Turner (1985) concluded that in drought-affected areas, phosphorus deficiency is one of the earliest symptoms observed in soil and plants grown in these soils. Therefore, the application of phosphorus fertilizers is required to increase the yield of crops under drought conditions (Garg et al. 2004). Potassium is essential in protein synthesis, photosynthesis, synthesis of glycolytic enzymes, maintaining turgor movements, and also cell expansion (Marschner 1995). Potassium becomes an essential element for the turgor pressure and turgor movement in plants in dry soil conditions. Kuchenbuch et al. (1986) in a study concluded that dry soil reduces the potassium uptake in onion plants. Calcium plays a very important role in maintaining several physiological changes that control growth and other responses to environmental stresses. Some of these physiological changes include solute and water and stomatal closure in a plant. Although calcium uptake decreases under drought conditions same as other elements, but overall calcium uptake by plants is only slightly decreased in comparison to other major elements like nitrogen, phosphorus, potassium, etc. There is not much work done for the effect of drought on elements like magnesium and sulfur. However, studies have shown that drought conditions reduce both magnesium and sulfur uptake (Scherer 2001).

14.3.2 Effect of Salinity on Different Properties of Soil

14.3.2.1 Physical Properties

High level of exchangeable sodium ion can structurally deteriorate the soil that results in the decline of pore volume and also in salt-affected soil it can disrupt the relation between soil–air and soil–water (Rengasamy and Olsson 1991). In sodic soil, slaking, dispersion, and clay swelling are the chief mechanisms that are involved in the breakdown of aggregates as sodicity in the soil affects soil hydraulic conductivity and its infiltration rate (Rengasamy and Sumner 1998). Slaking process causes permanent blockage of the pores even though it is a reversible process while dispersion is an irreversible process that can cause translocation of individual soil particles (Rengasamy and Sumner 1998). This mechanism of soil pore system and its structural stability can be explained by diffuse double layer (DDL) theory, which suggests that the attraction and repulsion between ions in soil particles is a result of intermolecular and electrostatic forces working on them (Quirk 1994). The increase of sodium ion on the exchange site of the clay or soil particles results in the increase of repulsive force which increases the inter-particulate distance turn the breakdown of soil aggregates affecting the soil structure (Oster and Shainberg 2001). Swelling and dispersion in sodium dominated soil bring about changes in the hydraulic properties of the soil.

Low salinity and high sodicity lead to a substantial reduction of hydraulic conductivity and infiltration because of the induced dispersion and swelling (Dikinya et al. 2006). Ensuing slaking and dispersion in the surface layer of sodic soils, a surface crust is formed, which is a thin layer formed after drying. This crust makes the surface soil prone to intense erosion as well as waterlogged conditions (Shainberg et al. 1992).

14.3.2.2 Chemical Properties

Saline and sodic soils have high values of SAR, ESP, EC, and pH. Saline soils normally suffer from deficiencies of NPK while their high pH has an adverse effect on the availability of micronutrients like Al, Fe, Zn, Cu, and Mn (Lakhdar et al. 2009). Salt toxicity, high osmotic pressure as well as soil degradation lead to decline in vegetation growth and decrease in carbon inputs in the soil further deteriorating the physicochemical properties of the soil (Wong et al. 2009). It is also observed that irrigation with sodium-rich water increases the soil sodicity, which results in the decrease of total C and N (Chander et al. 1997). The dispersion of soil aggregates uncovers the confined organic matter, thus increasing the decomposition rate by the soil microbes.

Increase in salinity decreases the mineralization of nitrogen and increases the loss of gaseous NH_3 (Paliwal and Gandhi 1976). It was further discovered that in the arid soils with increasing salinity and sodicity even with organic amendment leads to a decrease in the mineralization of C and N (Pathak and Rao 1998). The ability of plants to take up water is reduced by the high ion concentration, primarily Na and Cl ion. This, in turn, affects the plant growth and plant cell (Muneer and Oades 1989).

14.3.2.3 Biological Properties

The changes in soil chemistry negatively affect the soil microbial and biochemical processes, which play a crucial role in maintaining the soil's ecological functions (Rietz and Haynes 2003). Increasing soil salinity adversely impacts microbial growth and activity, as increased concentration of salt in the soil causes osmotic stress resulting in microbial cell dehydration (Wichern et al. 2006). Other factors, in addition to salt stress, that contributes to the decrease in microbial population and activities in salt-affected soils are Na^{+} toxicity; nutritional deficiency like Ca^{2+} deficiency; increase in the level of toxicity of ions such as bicarbonate, carbonate, and chloride; and also significant loss in organic matter resulted by structural degradation (Nelson and Oades 1998). A study was conducted in the arid region of southeast Spain to analyze the effect of salinity on the composition of microbial community in the soil. It was found that the increase in salinity had a negative impact on the microbial community of soil (Garcia et al. 1994). The enzymatic activity, as well as the rate of mineralization of C and N, decreased at high levels of salinity. In a study, the nitrification rate is increased by up to 83% with increase in salinity to 20 dS m^{-1}, which stimulates volatilization of ammonia (Wichern et al. 2006). Soil biochemical and microbial activities are negatively affected by salinity and sodicity induced by irrigation (Rietz and Haynes 2003). Many other studies have also reported that soil microbial biomass and enzyme activities significantly decrease with salinity and sodicity (Tripathi and Srivastava 2006). A shift in the microbial community has been observed with increasing salinity; an adaptive trait to reduce salt stress by lowering the metabolism is noticed (Yuan et al. 2007). Microbial and biochemical activities significantly affect soil ecological function. They play a central role in enhancing the structure of soil as well as in the stabilization of the soil aggregates (Six et al. 2005).

14.3.3 Impact of Heavy Metals

14.3.3.1 Impacts on Soil Physicochemical Quality

Metals exist either as separate entities or in combination with other soil components. Odueze et al. (2017) reported that the soil physicochemical properties affected by heavy metal concentrations present in different layers of soil locations and depth. According to a survey in China by Wang et al. (2018) emphasized on the effects of heavy metal contaminations in agricultural soils concerned to soil physicochemical properties and cultivation age of crop species. Similarly, in Zambia, the physicochemical characteristics of local forest soil were contaminated by the copper mining waste including heavy metals that decreased the pH level (5.05), affect the bulk density (1.24 g/cm^3), total organic carbon (2.24%), total nitrogen (0.07%), available

P (4.9 mg/kg), K (37.8 mg/kg), Na (49.0 mg/kg), Ca (99.0 mg/kg), and Mg (81.2 mg/kg) (Mutale et al. 2019). As discussed by Obasi et al. (2012), the physicochemical properties of soil are mainly influenced by the mobility and bioavailability of heavy metals. They emphasized that the physicochemical parameters of the soil help to support microbial diversity, metabolic changes, and lastly growth of plant species. The effects of heavy metal on soil physicochemical quality with different levels are shown in Table 14.2.

14.3.3.2 Impact on Soil Biology (Microbial and Enzymatic Activities)

Soil microbes such as bacteria, fungi, algae and protozoa have good metabolic activities for depolymerization and mineralization of organic matter in the forms of N, P, and S. Soil enzyme activities are good indicators of heavy metal toxicity in the soil for bioavailability assessment (Xian et al. 2015). Sobolev and Begonia (2008) reported the effects of Pb on soil microbial community in low (1 ppm) and high (500–2000) concentrations. Kouchou et al. (2017) observed heavy metal contaminations in alkaline soils of the region of Fez (Morocco) and showed adverse impact on actinomycetes and fungi. Atafar et al. (2010) indicated that the microbial biomass in the soil contaminated by Cu, Zn, Pb, and other heavy metals were inhibited drastically. Shi and Ma (2017) observed that Cd contamination affects the microbial activity like respiratory intensity, urease activity, and catalase activity in forest soil and garden soil.

Soil organic matter is the dominant factor contributing to the soil microorganisms and measurements of free soil enzyme activities that can serve as useful indicators of microbial metabolic potential as explained in Table 14.3 (Hagmanna et al. 2015). The research done by Jin et al. (2015) depicted that the heavy metal contaminations adversely affect the soil biological functions, including the size, activity, and diversity, of the soil microbial community and the activity of various enzymes involved in the transformation of C, N, P, and S. Some findings in the review done by Chibuike and Obiora (2014) showed the heavy metal affects the soil microorganism's quantity, diversity, density, and their activities. The metal toxicity on microorganisms is dependent on several aspects such as pH, soil temperature, organic matter, clay minerals, inorganic anions, and cations, and chemical forms of the metal (Wuana and Okieimen 2011; Chibuike and Obiora 2014). Therefore, the effects of the heavy metals on soil microbial and enzymatic activities are mainly dependent on the soil physiological properties, the stimulus of heavy metals on soil microbial activity, the end product on soil enzyme activity, and the composition of soil microbial community (Khan 2000). Similarly, Khan et al. (2010) found that heavy metals (Pb and Cd) have been an inhibitory impact on soil enzyme activities and as well as microbial community structure. Thus, heavy metals significantly impact on soil enzymes, they restrict the growth and reproduction of microorganisms, reduce the synthesis and metabolism of the microbial enzymes (Chu 2018). The effects of heavy metal on soil microbial diversity may be assessed more comprehensively by using methods such as microbial biomass, C and N mineralization, respiration, other enzymatic activities, etc. (Oijagbe et al. 2019).

Table 14.2 Effect of heavy metals on soil physicochemical quality

Heavy metals	Conc./level	Soil quality parameters	Remarks	References
Zn, Fe, Cu, Pb, Cd, Ni, and Cr	360.00–441, 169.60–547.20, 37.20–102.00, 18.80–80.00, 2.36–2.95, 11.00–19.20, and 18.00–42.20 μg/g	pH, organic carbon, and organic matter	The pH value increased from 7.07 to 7.69, organic carbon 0.46 to 1.18%, and organic matter 0.80% to 2.05% at 0–30 cm depth in the contaminated site	OlabimpeIyabo et al. (2017)
Mn, Zn, Pb, and Cd	182.69–697.06; 122.69–632.94; 19.38–158.50; and 0.25–1.63 mg/kg	pH, moisture content, bulk density, organic matter, and organic carbon	The changes in soil properties were reported such as pH values increased from 5.17 to 8.28, moisture content increased from 3.50 to 28.55%, bulk density increased from 0.78 to 2.29 g/ cm^3, organic matter increased from 0.09 to 16.01%, organic carbon increased from 0.02 to 8.48%	Olayinka et al. (2017)
Cr, Cu, Zn, and Ni	63.4, 201.2, 291.2, and 33.2 mg/kg	pH, EC, $CaCO_3$, and organic matter	The pH value increased from 8.4 to 8.5, EC decreased from 69.7 to 57.9 μS/ cm, $CaCO_3$ decreased from 35.9 to 34.3% and organic matter increased from 5.9 to 9.8	Kouchou et al. (2017)
Cr, Mn, Cd and Cu	0.17, 7.22, 0.02, and 0.80 mg/kg	pH, soil temperature, and organic matter	pH was slightly increased from 5.23 to 5.94, soil temperature significantly increased from 27.20 to 27.50 °C and organic matter increased from 2.42 to 3.93%	Eze et al. (2018)
As, Cd, Cr, Cu, and Pb	147.46, 2.31, 44.50, 4.10, and 13.01 mg/ kg	pH, EC, $CaCO_3$, and organic matter	Soil pH increased from 7.58 to 8.69, $CaCO_3$ increased from 0.45 to 8.1% and organic matter decreased 1.95% below the detection limit	Salman et al. (2019)

Table 14.3 Effects of heavy metals on soil biology (microbial and enzymatic activities)

Heavy metals	Conc./level	Soil microbial treatment	Remarks	References
Cu, Zn, Pb, Cd, Cr, Ni, Co, Mn, and Fe	73.7, 239.5, 40.4, 3.5, 36.2, 27.2,18.2, 224.5, and 27.0 mg/kg, respectively	Dehydrogenase, catalase, acid and neutral phosphatase, and sucrose	The soil enzymatic activities and microbial biomass carbon (MBC) and microbial biomass nitrogen (MBN) were mostly inhibited by heavy metal pollution and dehydrogenase activity was also affected by heavy metal pollution	Hu et al. (2014)
Vanadium (V)	254 and 1104 mg/kg and V contents ranged between 354 and 1104 mg/kg as incubation time increased	Impact of V on soil enzymatic activities, basal respiration (BR), microbial biomass carbon (MBC), and the microbial community structure	V in the soil had significant effects on soil dehydrogenase activity, BR. and MBC, while urease activity (UA) was less sensitive to V stress	Xiao et al. (2017)
Pb, As, and Cd	500, 50, and 1 mg/kg, respectively	Invertase, urease, aryl-sulfatase, catalase, alkalinephosphatase	Insignificant changes in soil urease, aryl-sulfatase, and alkaline phosphatase	Xian et al. (2015)
Cu	200, 500, and 1000 mg/kg Cu addition in soil	Microbial biomass carbon, soil microbial activity, and bacterial community	Changes in the microbial biomass carbon, microbial activity, and bacterial community composition between rhizosphere soil, bulk soil with the amendment of different Cu concentrations. Without any Cu addition, soil microbial biomass C in the rhizosphere of *Trifolium repens* and *Euphorbia splendens* were found to be higher	Wang et al. (2008)

(continued)

Table 14.3 (continued)

Heavy metals	Conc./level	Soil microbial treatment	Remarks	References
			than that of nonplanted soils	
Cu, Zn, Pb, and Cd	Cu (13.05–495.88), Zn (19.7–481.7), Pb (1.1–135.7), and Cd (3.7–59.4) mg/kg, respectively and available Cu (1.3–47.0), Zn (3.3–98.3), Pb (0.6–26.1), and Cd (1.1–13.2) mg/kg, respectively	Soil enzyme activities (sucrase, urease, and acid phosphatase), microbial biomass C, N, and P (MBC, MBN, and MBP), basal respiration, metabolic quotients, and N mineralization	Significantly decreased in microbial biomass, inhibited the enzymatic activities and adverse impact on metabolic activities and N mineralization	Zhang et al. (2010)

14.4 Effects of Drought, Salinity, and Heavy Metals on Growth and Productivity of Crops

14.4.1 Effects of Drought

Drought is a kind of stress that may affect the plants at individual level and even at whole crop level. This extreme stress can cause permanent damage to the plants and has an obvious symptom, e.g., wilting of the leaves. All plants require a particular amount of water for their survival throughout their lifecycle to give better productivity. Plants are well known to possess drought resistance. Other symptoms of drought can be seen by reduction in leaf area, increased leaf drop, reduced leaf growth, sunken stomatal development, yellowing of leaves, and wilting which all lead to the death of the plant (Nakayama et al. 2007).

Water stress inhibits cell enlargement more than cell division. It reduces plant growth by affecting various physiological and biochemical processes, such as photosynthesis, respiration, translocation, ion uptake, carbohydrates, nutrient metabolism, and growth promoters (Jaleel et al. 2008a, b, c, d, e; Farooq et al. 2008).

14.4.1.1 Effect of Drought on Seed Germination

Ability of seed germination and vigor is the most important and primary process in the production of any plant. Germination being important in the reproduction often influenced by environment created due to the stress induced by water-scarce condition (Leila 2007). Water being absorbed by the seed is the first stage in the process of germination. Under drought conditions, osmotic and matric potentials are dramatically decreased which does not allow the required amount of water to reach into plant tissues. The negative effect of drought on plant has been proven by several scientists again and again over time. Zareian (2004) stated that scarcity of water can significantly reduce the germination rate; however, different plant responds differently depending on the stage of germination. Shekari (2000) conducted a study on millet to assess the effect of drought on germination, root and shoot length and concluded that with increasing drought conditions the rate and percent increase in parameters observed reduces considerably. Drought affects many plant growth parameters affecting its overall germination, growth, yield, and production of dry matter.

14.4.1.2 Effects of Drought on Morphological Characters of Plants

Considering plant physiology, scarcity of water causes serious stress on plant growth, yield, and survival. Ghodsi et al. (1998) studied the effect of induced drought on plant and documented 30–50% decrease in yield due to drought and concluded that plant growth is severely affected as a result of high temperature, low humidity, high sun intensity, and high evapotranspiration. Similarly, Bagheri (2009) stated that high temperature due to high light intensity increases plant respiration, photosynthesis, and enzyme activity. Under drought condition, the increased light intensity from the sun increases the photoreaction processes such as photosynthesis and production of free oxygen radicals that ultimately leads to plant death.

Dryness causes the nutrient content of topsoil to be completely absorbed by the plants, which gets stored in fruits whereas salts and other ions get accumulated into the top soil due to increased drought conditions. The increase in ions and salts around the roots of the plants results in osmotic stress, which ultimately causes ion toxicity and death of plants. Under such conditions, the primary response of plants usually is biophysical responses. These responses after drought conditions include loosening and widening of cell walls, decrease in cell volume, decrease in cell pressure that also affects the potential of cell development and if this potential is reduced then growth rate also decreases (Bagheri 2009).

It has been proven by several researchers throughout the world that drought stress is among the worst stress which turns out to be an important limiting factor during the times of development of plants, especially during initial phases of germination (both expansion and elongation) and establishment (Anjum et al 2003; Kusaka et al. 2005; Bhatt and Rao 2005; Shao et al. 2008).

Most affected plant crops are usually submerged plant crops such as rice and are considered more susceptible to drought in comparison to other plants. In an experiment with soybean plants, stem length was found to be decreased under water-scarce conditions (Specht et al. 2001). In several studies concerning height of the plants, Wu et al. (2008) concluded that water-scarce conditions reduced the plant height of citrus seedlings by almost 25%. Heuer and Nadler (1995) stated that length of the stem is significantly affected due to drought conditions for potato plant and similar kind of results was observed for *Petroselinum crispum* (parsley) (Petropoulos et al. 2008), (soybean) (Zhang et al. 2004), *Vigna unguiculata* (Cowpea) (Manivannan et al. 2007), and *Abelmoschus esculentus* (Okra) (Sankar et al. 2007, 2008).

Under water-scarce conditions, cell growth and cell expansion are reduced due to extremely low turgor pressure whereas increase in osmotic regulation helps in maintaining cell turgor to help the plant grow under water-scarce conditions in plants such as pearl millet (Shao et al. 2008).

Bhatt and Rao (2005) conducted a study on the plant species *Abelmoschus esculentus* under dry conditions and concluded that reduction in the cell enlargement was responsible for the reduction of average plant height and also for the leaf area and size reduction. A similar study was conducted for assessing the leaf area change in the *Populus* plant under water-deficit condition by Wullschleger et al. (2005). It was concluded that water stress reduces the leaf growth thereby reducing leaf area in the plant and since the growth of leaf area is important for adequate photosynthesis and production/yield, the result was an overall reduction in the growth rate. Same results, i.e., reduction in growth was observed for *Glycine max* (soybean) (Zhang et al. 2004) and many other species (Farooq et al. 2009). The leaf growth was more sensitive to water stress in wheat than in maize (Sacks et al. 1997), *Vigna unguiculata* (Manivannan et al. 2007), and sunflower (Manivannan et al. 2007, 2008).

14.4.1.3 Effects of Drought on Yield

The main purpose of harvesting any crop in the world is to get a good amount of harvestable yield. Any specie of crop requires a minimum amount of water to sustain and produce a good harvestable yield. Plantings of sunflower with abundant water supply, the yield increase was associated with both an increase in grain number and in individual grain weight (Soriano et al. 2002). Water scarcity primarily affects the dry matter yield and is the most important factor in determining the yield (Petropoulos et al. 2008). Soriano et al. (2002) concluded that in the sunflower harvest water-deficient condition was an important factor in the decrease of the yield and production and that presence of adequate amount of water is important throughout the season from establishment to harvesting period (Soriano et al. 2002). In another study by Edward and Wright (2008) on the wheat plant, they concluded that grain size and number decreased dramatically under drought-induced conditions.

In a similar study on maize, water-deficit stress vastly decreased the final yield, which was attributed to the extent of defoliation caused by water scarcity mainly during the early growth period (Monneveux et al. 2006). Drought conditions decrease seed production and yield in soybean mainly as a response to water-scarce conditions, which generate fewer seeds and pods in an area (Specht et al. 2001). Whereas in a comparitive study between water-stressed soybean seeds and controlled seeds, the yield of stressed seeds was found to be much lesser in comparison to water-controlled plants (Specht et al. 2001).

Tahir and Mehid (2001) conducted a study to know the effect of drought on sunflower plant and found that water-deficit condition greatly reduced the head diameter, dry weight, and harvestable yield per plant per unit area. The study observed a negative correlation between head diameter and fresh root and shoots weight, whereas a positive correlation was observed between dry roots and shoot weight and achene yield per plant per unit area under water-scarce conditions. Water-scarce conditions prevailing for more than 12 days at flowering and grain filling stage was found to be the most damaging one in decreasing the achene yield in sunflower plant (Reddy et al. 2004), in common bean, and green gram (Webber et al. 2006), maize (Monneveux et al. 2006), and Parsley (Petropoulos et al. 2008).

14.4.1.4 Effects of Drought Stress on Pigment Composition

Photosynthetic pigments are important component mainly present in leaves of the plant and phytosynthetic bacteria. Chlorophyll is the most important pigment among several and drying of soil has a great effect on chlorophyll "a" and "b" which in turn affect the health of the plant (Farooq et al. 2009). However, during times of drought, carotenoids play a major role and help plants to withstand adverse effects of drying soil. Some of the changes that occur in the pigments during water-scarce conditions are given below.

14.4.1.4.1 Chlorophylls

Researchers from around the world have proven in a number of experiments that chlorophyll and carotenoid contents are reduced due to the drying of the land (Farooq et al. 2009). The decrease in chlorophyll and carotenoids have been greatly observed in cotton plants (Massacci et al. 2008), sunflower (Kiani et al. 2008), *Vaccinium myrtillus* (Bilberry) (Tahkokorpi et al. 2007), etc. It has also been proven that the overall foliar photosynthesis rate of plants decreases with the decrease in leaf water potential (Lawlor and Cornic 2002).

Apart from this, there is not a common consensus among researchers whether drought limits photosynthesis through stomatal closing or pigment or metabolic disfigurement (Anjum et al. 2003; Lawson et al. 2003). Some mixed results have also been observed in cases such as that of Ashraf et al. (1994) where they stated that chlorophyll b content was observed to increase in two lines of ladyfinger plants, whereas chlorophyll a was observed to have remained unaffected resulting in the reduction of chlorophyll a:b ratio in water-scarce conditions.

14.4.1.4.2 Carotenoids

Carotenoids are class of photosynthetic pigments mainly composed of yellow, orange, and also red pigments, including carotene that is mainly responsible for giving color to the plants and its parts such as autumn leaves and ripe tomatoes. They also protect chlorophyll pigments from photodamage. These pigments are mainly synthesized by all photosynthetic and some nonphotosynthetic organisms as well (Wilson et al. 2008).

The carotenoid play a more important role during times of drought by playing the protective role of β-carotene through direct quenching of triplet chlorophyll, preventing the production of any singlet oxygen thereby protecting plant tissue from oxidative damage occurring due to water-deficit conditions (Farooq et al. 2009).

14.4.1.5 Effect of Drought on Stomata

Stomata are minute pores usually found in the epidermis of leaves and stems whose main functions are water and gaseous exchange. Stomata are mainly found in the lower side of the leaves where generation of new stomata takes place continuously throughout the growth period of the leaf (Zhao et al. 2015). Active stomatal functioning decreases the rate of transpiration and is among the most important processes that maintain the water balance in the plant tissues. Stomatal functioning/ closure is also responsible for reducing cell growth and cell expansion, which affects the overall decrease in the rate of biomass production and yield (Rauf et al. 2015; Nemeskeri et al. 2015). It has been proven by many scientists that stomatal closure is among the primary response that a plant show in case of water scarcity to prevent or reduce water loss due to transpiration (Clauw et al. 2015; Nemeskeri et al. 2015). Closure of stomata is an active mechanism employed by plants to avoid dehydration (Osakabe et al. 2014; Clauw et al. 2015).

14.4.1.6 Drought-Associated Plant Diseases

Dry and hot conditions help the spreading of infections and diseases through pathogens and make plants susceptible to these diseases. Normally some diseases may be less harmful or lethal for the plant depending on the type. Some of the diseases that affect the plant during dry and water-scarce conditions are aflatoxin contamination, *Aspergillus* ear rot, etc. In general, certain diseases are prevalent while others are less severe or do not occur. With drought conditions, we anticipated yield losses due to soybean cyst nematode (SCN) and charcoal rot, as well as corn grain as a consequence of *Aspergillus* ear rot, charcoal rot, caused by *Macrophomina phaseolina,* etc. (Al-Kaisi et al. 2013).

14.4.2 Adverse Impacts of Salinity on Growth and Productivity of Crops

One of the most significant factors that limit the growth and production of crop is salinity. Salt stress generally affects the plants in two ways:

1. Osmotic stress, a high concentration of salt makes extraction of water arduous for roots.
2. Specific ion toxicity where the high levels of salt inside the plant pose harm to it (Munns and Tester 2008; Hussain et al. 2008).

Salinity has a number of damaging effects on physiological processes in plants like increase in the rate of respiration, ion toxicity, change in the metabolism of C and N, mineral distribution, membrane instability and permeability, reduced leaf area, dry mass, stomatal conductance as well as decreased efficiency of photosynthesis leads to the decrease in the productivity of plant (Sudhir and Murthy 2004; Kim et al. 2004; Hayat et al. 2012a, b).

14.4.2.1 Effect of Salt Stress on Plant Growth

Salinity stress has a significant impact on the growth and development of plants (Table 14.4). The effect of salt stress on processes such as germination, vegetative growth, and seedling growth and vigor has an adverse impact that ultimately leads to

Table 14.4 Effect of salinity on growth and productivity of crops

Salinity	Conc./level	Crop species	Growth parameters	Remarks	References
NaCl	4 dS/m, 6 dS/m, and 8 dS/m	Tomato	Number of leaves, leaf length, and yield	All the studied parameters were found significantly reduced	Rahman et al. (2018)
NaCl	250 mM	Mung beans	Root length and shoot length	Root length was reduced by 25.60–57.44% and shoot length was reduced by 24.85–30.17%	Alharby et al. (2019)
NaCl combined with Ca^{2+}:K^{+}:Mg^{2+} (10:12:3 mM) level	(30, 60, and 90 mM)	*Sorghum bicolor* L.	Plant growth	Significant decrease	Al-Amoudi and Rashed (2012)
NaCl	50, 100, 200, and 300 mmol/L	Lettuce	Plant growth	Significant reduction, very sensitive	Hniličková et al. (2019)
NaCl	(50–150 mM)	*Brassica juncea*	Germination and seedling growth	Both germination of seeds and seedling growth reduced with increased concentration of NaCl	Bhaskar et al. (2009)

poor plant growth (Sairam and Tyagi 2004). Osmotic phase inhibits the growth of young and new leaves while the ionic phase accelerates the senescence of mature leaf and causes them to fall off. In soil when there is an excess concentration of salt, not only is the plant growth inhibited but this salt stress even causes death. The connection between the increase in the concentration of sodium chloride and the decrease in the plant length is evident in the result of various studies (Houimli et al. 2008; Rui et al. 2009; Memon et al. 2010). The influence of salinity on plants is such that the number of leaves decreases with an increase in the concentration of salt (Gama et al. 2007; Ha et al. 2008). It is also observed that the leaf area gets negatively affected as well with the increase in the concentration of NaCl (Rui et al. 2009).

14.4.2.2 Effect of Salt Stress on Photosynthesis Process

The inhibition of photosynthesis takes place when high concentrations of Na^{+} and Cl^{-} are accumulated in the chlorophyll which is a major content for the photosynthesis to take place. This is directly correlated with the health of the plant (Zhang et al. 2008). It was observed that photosynthetic gas exchange parameters are reduced significantly due to salt stress conditions in *Solanum melongena* (Shaheen et al. 2012). Rate of photosynthesis, transpiration, and stomatal conductance is reduced significantly due to salt stress in *Cucumis sativus* as well as in *Lycopersicon esculentum* (Ahmad et al. 2012). It has been noted that the photosynthetic rate decreases with increase in salinity. This takes place due to certain factors (Iyengar and Reddy 1996):

1. Osmotic stress caused by high salt concentration dehydrates the cell membrane, which causes reduction in permeability to CO_2.
2. Toxicity caused by Na^+ and Cl^-. Cl^- inhibits N uptake by the roots, hence inhibiting the photosynthetic rate as well (Banuls et al. 1991).
3. The closure of stomata causes reduction in CO_2 supply restricting carboxylation reaction as well as the reduction in loss of water through transpiration affecting light-harvesting and energy conservation system (Brugnoli and Björkman 1992).
4. The alteration of the cytoplasmic structure causes a change in enzymatic activity.
5. Salinization enhances the leaf senescence rate and reduces sink activity.

14.4.3 Effect of Heavy Metals

14.4.3.1 Effect of Heavy Metals on Growth and Productivity of Crops

The heavy metals are available for plant uptake as soluble components in the soil solutions (Asati et al. 2016). At trace level some heavy metals like Zn, Ni, Cu, Co, etc. are beneficial for the plant growth. However, high concentrations of these metals in soil become toxic and reduce the growth and productivity of plants. Reduction in growth as a result of changes in physiological and biochemical processes such as decreased growth rate, chlorosis, necrosis, altered stomatal action, leaf rolling, reduced flow rate of cations, decreased water potential, alterations in membrane functions, inhibition of photosynthesis, respiration, altered metabolism, ultimately leads to food insecurity (Ashfaque et al. 2016; Hasanuzzaman et al. 2018). The direct effects of high metal concentration include inhibition of cytoplasmic enzymes and affect cell buildings due to oxidative stress, resulting in cellular damage (Dubey et al. 2018; Syed et al. 2018). Several researchers found that the accumulation of heavy metals in crop plants is mostly concerned with the probability of food contamination (Nazir et al. 2015). Though several heavy metals like Cd, As, Hg, etc. are not essential for plant growth, they are readily taken up and accumulated by plants. Arif et al. (2016) found the presence of heavy metals in both forms, i.e., essential and nonessential in the environment. Essential heavy metals such as magnesium (Mg), iron (Fe), manganese (Mn), and zinc (Zn) play a beneficial role in plant growth and development. A trace level of these beneficial elements improves the plant's nutritional level and also several mechanisms for the normal growth, productivity, and better yield of plants.

Connor et al. (2018) reported that the pH of paddy soils in southern/eastern China has decreased due to excessive use of fertilizers and consistently increased available heavy metal concentrations and thus, greater contaminant enrichment in crop production. The bioavailability determines contaminant availability to a receptor and the level of risk in land management. Toxic heavy metals are widespread throughout the world and cause acute and chronic toxic effects on plants grown in such contaminated soils (Yadav 2010). For example, Cu is a vital metal for the growth and development of plants; however, it becomes potentially toxic at higher concentrations (Asati et al. 2016). Su et al. (2014) reported when copper content in soil is more than 50 mg kg^{-1}, it affects citrus seedlings and if soil copper content is 200 mg kg^{-1}, wheat suffers from wilting. On the other hand, Zn is the main functional elements to produce plant chlorophyll and also helps in metabolic, growth and development, generation of oxidative properties (Arif et al. 2016; Andresen et al. 2018). A study conducted by Anwaar et al. (2015) indicated that exogenous Si application improved the growth and development of cotton crops suffering from Zn toxicity stress by restricting Zn bioavailability and oxidative damage. The high levels of Hg in agricultural lands may result in Hg toxification in the entire food chain. The Hg always exists in the forms of HgS, Hg^{2+}, Hg°, and methyl Hg. Therefore, the ionic form of Hg is always predominant in agricultural soil. Cr has been reported to cause oxidative stress that involves stimulation of lipid peroxidation in plants resulting in abrupt damage to cell membranes (Hayat et al. 2012a, b). The Pb uptake by plant roots reduced the crucial forms of precipitates with phosphates, sulfates, calcium, and chemicals in the rhizosphere (Fahr et al. 2013). Ni is also toxic to the plants, especially in ionic form like Ni^{+2} in soil and caused various physicochemical alterations like inhibition in growth and developments, photosynthesis rate, decreased chlorophyll content, and diverse toxicity symptoms such as leaf chlorosis, necrosis, and wilting in different plant species, including rice (Sachan and Lal 2017). Farid et al. (2015) examined the Cd, Pb, and Ni concentrations in soil 0.111 ppm, 0.87–8.97 ppm, and 0.017–1.72 ppm, respectively, at 0–15 cm while 0.88 ppm, 0.43–6.77 ppm, and 0.055–0.852 ppm, respectively, at 15–30 cm. Cd, Pb, and Ni concentrations in the plants ranged 0.00–2.25 ppm, 1.11–5.29 ppm, and 1.51–4.96, respectively, also transferred to the plant tissues ranged 0.00–2.25 ppm 1.11–5.29 ppm, and 1.51–4.96, respectively. Pb and Ni concentrations were found below permissible levels but Cd was found above the permissible levels in plants as well as in groundwater. Considerably, heavy metal toxicity varies from different plant species, concentration, definite metal, chemical forms, and soil solutions and other soil properties (Shah et al. 2010; Ackova 2018). Table 14.5 shows the impacts of heavy metals on different plant productivity and their growth.

Table 14.5 Effect of heavy metals on growth and productivity of crops

Heavy metals	Conc./level	Crop species	Growth parameters and other factors	Remarks	References
Cd and Pb	Both heavy metals were subjected to different concentrations (150, 300, 450, 600, 700, 900 μM, respectively)	Mustard (*B. juncea*)	Growth and biomass yield, chlorophyll and carotenoid contents	Root length decreased from 17% (with 300 μM Cd) to 54% (with 900 μM Cd), stem height declined from 4% (with 300 μM Cd) to 51% (with 900 μM Cd), at 900 μM of Pb decreased the chlorophyll "a" about 35% while compared with control, at 900 μM of Cd decreased the chlorophyll "b" from 0.74 mg/g to 0.15 mg/g fresh wt. and total chlorophyll content showed decreased higher concentration of both metals. Carotenoids decreased from 0.413 to 0.083 mg/g fresh wt. with 900 μM of Cd	John et al. (2009)
Cd	Applied concentration 0–5.0 mM ($CdCl_2$)	Mustard (*Brassica juncea* L.)	Carotenoids, chlorophyll, total sugars, and total protein	Cd toxicity in plants reported as decreased chlorophyll content, reduced biomass, and inhibited chlorophyll biosynthesis, and photosynthesis	Bauddh and Singh (2011)
Cd	100 mg Cd kg^{-1} soil	*Ricinus communis* and *B. juncea*	Biomass (fresh and dry wt.)	Significantly decreased in biomass production of both plant species	Bauddh and Singh (2012)
Cr	Roots 327.6 μg/g, stems 186.8 μg/g and leaves 116.7 μg/g of dry wt. at 0.4 mM Cr concentration	*Raphanus sativus* (Radish)	Growth (height, leaf area, and dry wt.), Hill reaction activity, chlorophyll, carotenoids, and catalase	Significant reduction in growth, chlorophyll content, Hill reaction activity, carotenoids, and catalase	Tiwari et al. (2013)
As	6–12 mg/kg	*Helianthus annuus* L. (Sunflower)	Growth and photosynthetic rate and chlorophyll pigments	Growth reduction, decreased photosynthetic rate and chlorophyll pigments, increased production of stress biomarkers	Yadav et al. (2014)

Cd	100 mg Cd kg^{-1} soil	Castor bean (*R. communis*) and Indian Mustard (*Brassica juncia* L.)	Biomass (fresh and dry wt.)	Significantly decreased in root and shoot fresh and dry wt. from 43.53 to 32.45% (root), 49.06 to 38.23% (shoot) and 53.84 to 26.58% (root), 45.33 to 33.84% (shoot), respectively	Bauddh et al. (2015)
Cd	17.50 mg Cd kg^{-1} soil	Castor bean (*R. communis*)	Plant growth, biomass (length, fresh and dry wt.) and yield	Decreased in plant growth, reduction in root and shoot length, fresh wt. and dry wt. from 4.78 to 2.77%, 2.25 to 2.11%, and 7.11 to 3.25%, respectively, and significantly decreased in seed yield (per plant)	Bauddh et al. (2016)
Cd, Pb, Ni, and Zn	Average concentrations 1.07, 17.22, 1.73, and 13.75 mg/kg in the plant's stem and 1.27, 12.32, 1.099, and 19.39 mg/kg in grain, respectively	Rice crop (*Oryza sativa*)	Plant growth	Metal concentration affects roots and some plant parts (stems and grains). Contained high concentrations of Cd and Pb that inhibited the growth of paddy plants	Rahimi et al. (2017)
Cu and Zn	Applied concentrations control, 5, 10, 25, 50, 100, 200, and 300 mg/l	Seeds of soybean (*Glycine max* (L.) Merr.)	Biomass content (above and below ground)	Maximum growth was observed at 5 mg/l and significant growth was observed at 200 mg/l concentration from control	Ganesh and Sundaramoorthy (2018)

14.5 Application of Biochar for Mitigation of Abiotic Stress

14.5.1 Mitigation of Drought Using Biochar

Restoring or improving soil water holding capacity is one of the only few options to improve soil productivity to achieve good crop production during the times of drought. Choosing a better management strategy among numerous others is very important as it can play a great role in restoring soil's water and nutrient holding capacity (Baronti et al. 2014). A recent advancement has been made in the field of biochar that can effectively achieve the task of improving soil characteristics such as its water holding capacity. Using biochar has been considered one of the most practical and sustainable way to achieve and increase fertility of the soil, including water holding capacity. Several advantages of biochar such as long residence time in the soil large surface area in comparison to other amendments to improve soil quality make it much more preferable than other options throughout the world (Lehmann et al. 2011). Biochar are usually prepared in the absence of oxygen and under high pressure and temperature (above 250 °C), where the end product is a carbonaceous material with very high stability mainly attributed to recalcitrant aromatic structures of biochar molecules (Wardle et al. 2008; Biederman and Harpole 2013). The biochar has been proven to increase the moisture-holding capacity of the soil mainly attributed to its larger surface area and to recalcitrant aromatic structures of biochar molecules which also makes it suitable for improving agronomic performance under various climatic conditions including dry conditions (Biederman and Harpole 2013). It also possesses smaller pore size that allows them to hold capillary water for longer period of time making it suitable for application in drought-affected areas. Biochar are considered to have high potential to change soil characteristics including pH, nutrient retention, and overall fertility; however, the capacity of change varies greatly depending on the physicochemical properties of biochar itself including feedstock material, pyrolysis temperature and pressure, specific surface area, and method of application in the field (Baronti et al. 2014). As mentioned previously, improving soil water holding capacity and moisture retention capacity are mainly attributed to the porosity and large surface area of the biochar and its application in the drought prone areas, there is a lack of common consensus among researchers of a typical type of biochar to be used in these areas as moisture retention capacity of applied biochar mainly depends upon the manufacturing procedure.

The effect of biochar also improves soil quality and makes it suitable for plant growth during various periods including germination to adulthood thereby increasing overall plant growth as it improves the accessibility of roots to soil water and air. The application also improves response of soil to soil moisture, size of aggregate, texture, soil matrix, dynamics, cation retention, and permeability making it a much more preferable option.

14.5.1.1 Application of Biochar with Microorganisms to Mitigate Drought Stress

Egamberdieva et al. (2017) reported an increase in biomass, growth, nodulation, and nutrient uptake (phosphorus and nitrogen) in lupin (*Lupinus angustifolius)* under water-scarce conditions when biochar is applied with microorganisms in comparison to solely applied microorganism's inoculation. The study also showed a considerable increase in relative water and chlorophyll content in comparison to the controlled ones. Similarly, Liu et al. (2017) observed that application of biochar with *Rhizophagus irregularis* decreased the water use efficiency, nutrient uptake, biomass, and overall growth as compared to control in potato. However, in contrary to the above findings, Pressler et al. (2017) concluded that application of wood-derived biochar with microorganisms had no significant effect on soil biota (including protozoa, bacteria, fungi, and nematodes) under low irrigation conditions.

Mickan et al. (2016) observed that inoculation of biochar with Arbuscular mycorrhizal fungi enhances drought tolerance of the plants by improving its physiological mechanisms such as nutrient uptake, and biochemical properties including osmotic adjustment, hormonal activities, and antioxidant systems. However, application of biochar with Arbuscular mycorrhizal fungi to the agricultural soil encouraged the growth of extra-radical hyphae and also increased mycorrhizal colonization of roots. Water potential was observed to be the same with and without biochar application.

Previous studies done by several researchers have proved that the application of biochar with microbial inoculation may prove itself to be far better helpful in mitigating and preventing drought stress in plants. However, there is no common consensus about the effectiveness and type of biochar to be used in such conditions, so there is a vast possibility of work in the near future and discover more advantages of biochar over other applications for drought mitigation.

14.5.2 Mitigation of Salinity Stress Using Biochar

Biochar has the potential to increase plant biomass and crop yield. Recent findings have shown that biochar treatment can increase crop yield by an average of 10% (Jeffery et al. 2011). Studies have also demonstrated that crop yield after several years of the application still show a significant increase after single treatment of biochar (Major et al. 2010). Salinity affects the total land area of about 7% of the world. It has been estimated that 30% of the irrigation land has been adversely affected by salinization (Chaves et al. 2009; Wicke et al. 2011). If no action is taken to prevent land degradation, up to 69% loss in revenue is estimated with time (Munns and Tester 2008). Charcoal and activated charcoal have been utilized in the industry for desalinization processes as it has high capacity to sorb a variety of salts (Bartell and Miller 1923; Zou et al. 2008). Decrease in biomass leads to lower carbon input that further deteriorates the soil. This is a major effect of salinity (Wong et al. 2009).

14.5.2.1 Effects of Biochar on Saline Soil Properties

Improvement of soil physicochemical and biological properties which is related to the removal of sodium such as leaching, absorption ratio as well as electrical conductivity has been observed with the application of biochar, which in turn assists in the reduction of salt stress (Sun et al. 2016; Drake et al. 2016). Soil enzymatic activity in saline soil varies with the rate of application of biochar, soil enzyme type as well as incubation time. The study shows that the soils' physicochemical properties have improved significantly (Wang et al. 2014; Bhaduri et al. 2016). It has also been observed that the application of biochar (30 g mm^{-2}) in salt-stressed soil though did not affect the pH of the soil, increased the electrical conductivity of the soil (Thomas et al. 2013). Application of furfural biochar in saline soil resulted in the decrease in pH on the other hand it increased the soil organic carbon as well as cation exchange capacity and available phosphorus in soil (Wu et al. 2014). It has been observed that composted biochar lead to the increase in cation exchange capacity and organic matter, while it decreased the exchangeable sodium and also the pH of the soil (Luo et al. 2017). The application of biochar has been observed to improve soil properties, such as increasing soil moisture, the Na binds with the biochar which results in the decrease in root sensitivity to osmotic stress.

14.5.2.2 Effect of Biochar on Plant Growth under Salinity Stress

As discussed in Table 14.6, many researchers have proven that amendment of biochar in fairly saline soils improves growth of plants, biomass as well as the rate of photosynthesis (Akhtar et al. 2015). In the tidal land soils containing high concentrations of soluble salt as well as exchangeable sodium, the application of rice hull biochar increased the growth of maize as well as its biomass (Kim et al. 2016). Application of biochar under saline irrigation (3.6 dS m^{-1}) increased the growth of tomato as well as increased the total biomass (Usman et al. 2016). A study conducted to estimate the effect of biochar derived from *Fagus grandifolia* on the growth and biomass production of two herbaceous species namely, *Prunella vulgaris* and *Abutilon theophrasti*, under saline condition. It was found that the biomass of both plants increased under salt stress. The effect on carbon profit through photosynthesis, water use efficiency, and chlorophyll fluorescence in both species did not have any significant impact under salt stress (Thomas et al. 2013). Amendment of biochar along with suitable microbial inoculation has been reported to improve growth and biomass production of the plant under saline conditions as compared to the control (Nadeem et al. 2013; Fazal and Bano 2016). The results obtained by Akhtar et al. (2015) depict that application of biochar increases leaf area, shoot, and root biomass, which increases with bacterial inoculation. The studies emphasize that the effect of biochar can be enhanced with the application of symbiotic microorganisms.

14.5.3 Mitigation of Heavy Metal Stress by Using Biochar

Several organic fertilizers like compost, vermicompost, manure, organic fertilizers, farmyard manure, etc. have been suggested to restore and improve soil quality and

Table 14.6 Effect of biochar on plant growth under salinity stress

Salinity	Biochar amount/dose	Crop species	Growth parameters	Remarks	References
Cd-contaminated saline soil	0.5% BC + 50 mM NaCl	*Triticum aestivum* L.	Plant growth	Growth increased while Cd and Na uptake decreased	Abbas et al. (2018)
Saline soil	2%	*Triticum aestivum* L.	Root and shoot length	Increase by up to 23%	Kanwal et al. (2018)
Saline soil	12 t ha^{-1} BPC—0.15 t ha^{-1} PS	*Triticum aestivum* L.	Yield	Increase in yield by 38%	Lashari et al. (2013)
Salt-stress condition	50 and 100 g kg^{-1}	*Glycine max* cv. M7	Nitrogen metabolism	Increased nodulation and nitrogen metabolism	Farhangi-Abriz and Torabian (2018a)
Salt stressed soil	10 and 20% BC	*Phaseolus vulgaris* L.	IAA content and growth	IAA content enhanced. Growth of root and shoot increased	Farhangi-Abriz and Torabian (2018b)
Saline-sodic soil	–	*Oryza sativa* L.	CEC and nutrient availability	28.8–29.0 $cmol_c$ kg^{-1} increase in CEC. P and K enhanced	Nguyen et al. (2016)

crop productivity (Mas-Carrió et al. 2018). Soil amelioration with biochar is the pre-eminent to promote sustainable agriculture because it has ability to nourish the soil and provide vital nutrients to the plants resulting in improved plant productivity. As described in Table 14.7, the application of biochar has capacity to reduce metal bioavailability and immobilized other contaminants also (Yuan et al. 2019). Similarly, several researchers explained that different types of biochar have potential and efficiency to adsorb the toxic agrochemicals and reduce the bioavailability of heavy metals and their uptake by plants (Abbas et al. 2017). Nie et al. (2018) assessed the bioavailability of Cd, Cu, and Pb and health of soil microbial population in the contaminated soil, by using sugarcane dry pulpy residues-derived biochar and they found that application of biochar enhanced growth of *Brassica chinesis* L. in terms of root and shoot development, control the bioavailability of metal and also enhanced the microbial population. Similarly, Zheng et al. (2015) investigated the effect of biochar (beanstalk and rice straw) on the bioavailability of Cd in contaminated soil and their accumulation into rice crop and they observed that biochar decreased the phytotoxicity of Cd (35–81 %). Hayyat et al. (2016) found that application of biochar in metal contaminated soil showed a significant decrease in metal toxicity and enhancement in the soil nutrient/fertility, plant growth, crop yield, carbon content, etc. Zheng et al. (2017) examined the effects of biochar (rice straw) on Cd-containing soil and their accumulation in lettuce plant. They observed

Table 14.7 Mitigation of heavy metal stress by using biochar

Heavy metals	Conc./level	Biochar	Amount/dose	Crop species	Growth parameters	Remarks	References
Pb, Cd, and Cr	200, 400, and 600 mg/kg, respectively	Agricultural residues (clean trees and burning green wastes)	Biochar at different rates 0.0, 1.0, 2.5, 5.0, and 10%	Maize plant (*Zea mays*)	Biomass content (shoot and root)	Biochar significantly decreased the phytotoxicity of Pb, Cd, and Cr and also enhanced biomass content and productivity of maize	Alaboudi et al. (2019)
Cd and Zn	43 mg/kg and 4340 mg/kg, respectively, concentration used	Coconut shells	Rates of biochar 5, 10, and 15% per mass of soil	*Salix smithiana*	Biomass yield	The effects of biochar showed a significant increase in biomass yield and also increased the remediation efficiency of metal	Břendová et al. (2015)
Cd, Cr, Ni, and Pb	0.752, 0.164, 2.03, and 0.087 mg/kg, respectively	Rice straw biochar with sewage sludge amendment	2.5, 5.0, 7.5 10, 15, 20 tonnes/hac.	Rice	Biomass yield	Biochar application increased the grain and straw yield of rice and increased bioavailability of essential plant nutrients	Jatav et al. (2016)
Cd, Cu, Zn, and Pb	1.4, 80, 1638, and 2463 mg/kg, respectively	Biomass materials of eucalyptus wood and poultry litter	2% w/w in soil	Rice	Biomass content (roots and shoots)	Modification of the surface of biochar material does not have any significant effects on the plant biomass (enhanced roots and shoots) with respect to control.	Lu et al. (2018)
Cu and Pb	600 and 21,000 mg/kg	Green waste compost and biochar (oak, ash, sycamore, and birch)	20 and 30%, respectively	Ryegrass (*Lolium perenne L. var. Cadix*)	Plant growth	Posed positive effect on plant growth and also enhanced a significant amount of nutrients to the soil	Karami et al. (2011)
Cd	(0.001, 1, and 5 mM	Rice husk	0.5% w/v	Seeds of alfalfa (*Medicago sativa* L.)	Growth and biomass (length, fresh and dry wt.)	Increased in germination rate, enhanced the biomass content (radical length, fresh and dry wt.) and restrict the bioavailability of metal	Zeid et al. (2018)

that the level of Cd reduced by 57 % while increasing biochar rates (0, 6, 12, 18 t ha^{-1}) increased shoot length and yield of lettuce plant. Khan et al. (2013) conducted an experiment to check the effects of biochar (derived from sewage sludge) on the rice plants and they found that the biochar increased the shoot biomass, grain yield, and the bioaccumulation of P and Na, decreased the bioaccumulation of N (excluding in grain) and K and decreased the bioavailability of metals (As, Cr, Co, Cu, Ni, and Pb); however, Cd and Zn concentrations were not affected in the plant parts by the application of biochar. Meng et al. (2018) found that the application of biochar (rice straw and swine manure) decrease the heavy metal bioavailability from contaminated soil in the order Pb > Cu > Zn > Cd. Kim et al. (2015) observed that the rice hull biochar significantly reduced the phytoavailability of metal present in contaminated soil and metal uptake by lettuce plant decreased with the increasing biochar application. Xu et al. (2018) found the adding of macadamia nutshell biochar (5% w/w) to the soil decreased Cd and Pb toxicity and increased total microbial phospholipid fatty acids (PLFAs), microbial respiration rate, biomass carbon and microbial availability.

14.6 Conclusions

Along with other conventional and biotechnological interventions, the application of organic manures may also be used to overcome the adverse impacts of abiotic stresses. The application of biochar as an organic amendment has proved to mitigate from drought, salinity and heavy metals. Application of biochar is an environment friendly and economically viable. Its application improves soil physicochemical and biological properties that help the plants to fight with adverse conditions of the environment. It has been found that biochar enhances water holding capacity, increase the nodulation, nutrient uptake, microbial diversity in the soil and thereby increases the growth and productivity of the plants. In the same way, biochar has potential, and efficiency to adsorbs the toxic agrochemicals and reduces their bioavailability, especially heavy metals and ultimately reduces their uptake by plants.

Acknowledgments Kuldeep Bauddh is thankful to the Science and Engineering Research Board (SERB) for the award of Research Grant (EEQ/2017/000476). Author Abhishek Guldhe is thankful to the Department of Biotechnology, Govt. of India for the Ramalingaswami Fellowship.

References

Abbas T, Rizwan M, Ali S, Zia-ur-Rehman M, Qayyum MF, Abbas F, Hannan F, Rinklebe J, Ok YS (2017) Effect of biochar on cadmium bioavailability and uptake in wheat (*Triticum aestivum* L.) grown in a soil with aged contamination. Ecotoxicol Environ Saf 140:37–47

Abbas AH, Sulaiman WR, Jaafar MZ, Aja AA (2018) Micelle formation of aerosol-OT surfactants in sea water salinity. Arab J Sci Eng 43:2515–2519

Ackova DG (2018) Heavy metals and their general toxicity on plants. Plant Sci 5(1):14–18

Ahmad M, Lee SS, Yang JE, Ro HM, Lee YH, Ok YS (2012) Effects of soil dilution and amendments (mussel shell, cow bone, and biochar) on Pb availability and phytotoxicity in military shooting range soil. Ecotoxicol Environ Saf 79:225–231

Akhtar SS, Andersen MN, Naveed M, Zahir ZA, Liu F (2015) Interactive effect of biochar and plant growth-promoting bacterial endophytes on ameliorating salinity stress in maize. Funct Plant Biol 42(8):770–781

Alaboudi KA, Ahmed B, Brodie G (2019) Effect of biochar on Pb, Cd and Cr availability and maize growth in artificial contaminated soil. Ann Agric Sci 4(1):95–102

Alam SM (1999) Nutrient uptake by plants under stress conditions. Handbook of plant and crop stress 19(2):285–313

AL-Amoudi OA, Rashed AA (2012) Effect of nutrient cations to improving salinity-tolerance responses in *Sorghum bicolor* L. Life Sci 50:77–81

Alharby HF, Al-Zahrani HS, Hakeem KR, Iqbal M (2019) Identification of physiological and biochemical markers for salt (NaCl) stress in the seedlings of mung bean [*Vigna radiata* (L.) Wilczek] genotypes. Saudi J Biol Sci 26(5):1053–1060

Al-Kaisi MM, Elmore RW, Guzman JG, Hanna HM, Hart CE, Helmers MJ, Hodgson EW, Lenssen AW, Mallarino AP, Robertson AE, Sawyer JE (2013) Drought impact on crop production and the soil environment: 2012 experiences from Iowa. J Soil Water Conserv 68(1):19A–24A

Amini S, Ghadiri H, Chen C, Marschner P (2015) Salt-affected soils, reclamation, carbon dynamics, and biochar: a review. J Soils Sediments 16(3):939–953

Andresen E, Peiter E, Kupper H (2018) Trace metal metabolism in plants. J Exp Bot 69:909–954

Anjum F, Yaseen M, Rasul E, Wahid A, Anjum S (2003) Water stress in barley (*Hordeum vulgare* L.) II. Effect on chemical composition and chlorophyll contents. Pak J Agric Sci 40:45–49

Anwaar SA, Ali S, Ali S, Ishaque W, Farid M, Farooq MA, Najeeb U, Abbas F, Sharif M (2015) Silicon (Si) alleviates cotton (Gossypium hirsutum L.) from zinc (Zn) toxicity stress by limiting Zn uptake and oxidative damage. Environ Sci Pollut Res Int 5:3441–3450

Arafat Y, Wei X, Jiang Y, Chen T, Saqib H, Lin S, Lin W (2017) Spatial distribution patterns of root-associated bacterial communities mediated by root exudates in different aged ratooning tea monoculture systems. Int J Mol Sci 18(8):1727

Arif N, Yadav V, Singh S, Singh S, Ahmad P, Mishra RK, Sharma S, Tripathi DK, Dubey NK, Chauhan DK (2016) Influence of high and low levels of plant-beneficial heavy metal ions on plant growth and development. Chemosphere 139:604–608

Arora N, Patel A, Sharma M, Mehtani J, Pruthi PA, Pruthi V, Poluri KM (2017) Insights into the enhanced lipid production characteristics of a fresh water microalga under high salinity conditions. Ind Eng Chem Res 56(25):7413–7421

Asati A, Pichhode M, Nikhil K (2016) Effect of heavy metals on plants: an overview. Int J Innov Eng Res Manag 5(3):56–66

Ashfaque F, Inam A, Sahay S, Iqbal S (2016) Influence of heavy metal toxicity on plant growth, metabolism and its alleviation by phytoremediation—a promising technology. J Agric Ecol Res Int 6(2):1–19

Ashraf MY, Azmi AR, Khan AH, Ala SA (1994) Effect of water stress on total phenols, peroxidase activity and chlorophyll content in wheat (*Triticum aestivum* L.) genotypes under soil water deficits. Acta Physiol Plant 16:185–191

Atafar Z, Mesdaghinia A, Nouri J, Homaee M, Yunesian M, Ahmadimoghaddam M, Mahvi AH (2010) Effect of fertilizer application on soil heavy metal concentration. Environ Monit Assess 160:83–89

Bagheri A (2009) Effects of drought stress on chlorophyll, proline and rates of photosynthesis and respiration and activity of superoxide dismutase and peroxidase in millet (*Panicum milenaceum* L.). National conference on water scarcity and drought management in agriculture. Islamic Azad University Arsanjan. p. 16

Banuls J, Legaz F, Primo-Millo E (1991) Salinity-calcium interactions on growth and ionic concentration of citrus plants. Plant Soil 133(1):39–46

Bardgett RD, Wardle DA (2010) Aboveground-belowground linkages: biotic interactions, ecosystem processes, and global change. Oxford University Press, Oxford

Baronti S, Vaccari FP, Miglietta F, Calzolari C, Lugato E, Orlandini S, Pini R, Zulian C, Genesio L (2014) Impact of biochar application on plant water relations in *Vitis vinifera* (L.). Eur J Agron 53:38–44

Bartell FE, Miller EJ (1923) Adsorption by activated sugar charcoal. II1, 2. J Am Chem Soc 45 (5):1106–1115

Bauddh K, Singh RP (2009) Genotypic differences in cadmium (Cd) toxicity in Indian mustard (*Brassica juncea* L.). Pollut Res 28(4):699–704

Bauddh K, Singh RP (2011) Differential toxicity of cadmium to mustard (*Brassica juncia* L.) genotypes under higher metal levels. J Environ Biol 32(3):355–362

Bauddh K, Singh RP (2012) Growth, tolerance efficiency and phytoremediation potential of *Ricinus communis* (L.) and *Brassica juncea* (L.) in salinity and drought affected cadmium contaminated soil. Ecotoxicol Environ Saf 85:13–22

Bauddh K, Kumar A, Srivastava S, Singh RP, Tripathi RD (2015) A study on the effect of cadmium on the antioxidative defense system and alteration in different functional groups in castor bean and Indian mustard. Arch Agron Soil Sci 62(6):877–891

Bauddh K, Singh K, Singh RP (2016) *Ricinus communis* L. a value added crop for remediation of cadmium contaminated soil. Bull Environ Contam Toxicol 96(2):265–269

Baum C, Leinweber P, Schlichting A (2003) Effects of chemical conditions in re-wetted peats on temporal variation in microbial biomass and acid phosphatase activity within the growing season. Appl Soil Ecol 22(2):167–174

Bhaduri D, Saha A, Desai D, Meena HN (2016) Restoration of carbon and microbial activity in salt-induced soil by application of peanut shell biochar during short-term incubation study. Chemosphere 148:86–98

Bhaskar P, Bauddh K, Singh RP (2009) Differential response of two high yielding cultivars of Indian mustard (*Brassica juncea* L.) to NaCl salinity during seed germination and early seedling growth. J Ecophysiol Occup Health 9:137–144

Bhatt RM, Rao NS (2005) Influence of pod load on response of okra to water stress. Indian J Plant Physiol 10(1):54

Biederman LA, Harpole WS (2013) Biochar and its effects on plant productivity and nutrient cycling: a meta-analysis. GCB Bioenergy 5(2):202–214

Bohn HL, McNeal BL, O'Connor GA (2001) Soil chemistry, 3rd edn. Wiley, New York

Břendová K, Tlustoš P, Száková J (2015) Biochar immobilizes cadmium and zinc and improves phytoextraction potential of willow plants on extremely contaminated soil. Plant Soil Environ 61(7):303–308

Brugnoli E, Björkman O (1992) Growth of cotton under continuous salinity stress: influence on allocation pattern, stomatal and non-stomatal components of photosynthesis and dissipation of excess light energy. Planta 187(3):335–347

Chander K, Goyal S, Mundra MC, Kapoor KK (1997) Organic matter, microbial biomass and enzyme activity of soils under different crop rotations in the tropics. Biol Fertil Soils 24 (3):306–310

Chaves MM, Flexas J, Pinheiro C (2009) Photosynthesis under drought and salt stress: regulation mechanisms from whole plant to cell. Ann Bot 103:551–560

Chibuike GU, Obiora SC (2014) Heavy metal polluted soils: effect on plants and bioremediation methods. Appl Environ Soil Sci 752708:1–12

Chodak M, Gołębiewski M, Morawska-Płoskonka J, Kuduk K, Niklińska M (2015) Soil chemical properties affect the reaction of forest soil bacteria to drought and rewetting stress. Ann Microbiol 65:1627–1637

Chu D (2018) Effects of heavy metals on soil microbial community. Earth Environ Sci 113:1–5

Clauw P, Coppens F, De Beuf K, Dhondt S, Van Daele T, Maleux K, Storme V, Clement L, Gonzalez N, Inzé D (2015) Leaf responses to mild drought stress in natural variants of Arabidopsis. Plant Physiol 167(3):800–816

Connor DO, Peng T, Zhang J, Tsang DCW, Alessi DS, Shen Z, Bolan NS, Hou D (2018) Biochar application for the remediation of heavy metal polluted land: a review of in situ field trials. Sci Total Environ 619–620:815–826
Dikinya O, Hinz C, Aylmore G (2006) Dispersion and re-deposition of fine particles and their effects on saturated hydraulic conductivity. Soil Res 44(1):47–56
Drake JA, Cavagnaro TR, Cunningham SC, Jackson WR, Patti AF (2016) Does biochar improve establishment of tree seedlings in saline sodic soils? Land Degrad Dev 27(1):52–59
Dubey S, Shri M, Gupta A, Rani V, Chakrabarty D (2018) Toxicity and detoxifcation of heavy metals during plant growth and metabolism. Environ Chem Lett 16(4):1169–1192
Edward D, Wright D (2008) The effects of winter water-logging and summer drought on the growth and yield of winter wheat (*Triticum aestivum* L.). Eur J Agron 28:234–244
Egamberdieva D, Reckling M, Wirth S (2017) Biochar-based Bradyrhizobium inoculum improves growth of lupin (*Lupinus angustifolius* L.) under drought stress. Eur J Soil Biol 78:38–42
Eze OC, Tukura BW, Atolaiye BO, Opaluwa OD (2018) Assessment of some physicochemical parameters of soil and heavy metals in vegetables cultivated on irrigated sites along the bank of Mpape River in FCT, Abuja, Nigeria. J Environ Sci Toxicol Food Technol 12(5):28–38
Fahr M, Laplaze L, Bendaou N, Hocher V, Mzibri ME, Bogusz D, Smouni A (2013) Effect of lead on root growth. Front Plant Sci 4:175. https://doi.org/10.3389/fpls.2013.00175
Farhangi-Abriz S, Torabian S (2018a) Biochar improved nodulation and nitrogen metabolism of soybean under salt stress. Symbiosis 74(3):215–223
Farhangi-Abriz S, Torabian S (2018b) Effect of biochar on growth and ion contents of bean plant under saline condition. Environ Sci Pollut R 25(12):11556–11564
Farid G, Sarwar N, Saifullah AA, Ghafoor A, Rehman M (2015) Heavy metals (Cd, Ni and Pb) contamination of soils, plants and waters in Madina town of Faisalabad metropolitan and preparation of GIS based maps. Adv Crop Sci Techn 4(1):1–7
Farooq M, Basra SM, Wahid A, Cheema ZA, Cheema MA, Khaliq A (2008) Physiological role of exogenously applied glycinebetaine to improve drought tolerance in fine grain aromatic rice (*Oryza sativa* L.). J Agron Crop Sci 194(5):325–333
Farooq M, Wahid A, Kobayashi N, Fujita D, Basra SM (2009) Plant drought stress: effects, mechanisms and management. In: Lichtfouse E, Navarrete M, Debaeke P, Véronique S, Alberola C (eds) Sustainable agriculture. Springer, Dordrecht, pp 153–188
Fazal A, Bano A (2016) Role of plant growth-promoting rhizobacteria (PGPR), biochar, and chemical fertilizer under salinity stress. Commun Soil Sci Plant Anal 47(17):1985–1993
Gama PB, Inanaga S, Tanaka K, Nakazawa R (2007) Physiological response of common bean (*Phaseolus vulgaris* L.) seedlings to salinity stress. Afr J Biotechnol 6(2):421–426
Ganesh SK, Sundaramoorthy P (2018) Copper and zinc induced changes in soybean (Glycine max (L.) Merr.). Innov Agri 1(1):9–12
Garcia C, Hernandez T, Costa F (1994) Microbial activity in soils under Mediterranean environmental conditions. Soil Biol Biochem 26(9):1185–1191
Garg BK, Burman U, Kathju S (2004) The influence of phosphorus nutrition on the physiological response of moth bean genotypes to drought. J Plant Nutr Soil Sci 167(4):503–508
Ghodsi M, Nazeri M, Zarea-Fizabady A (1998) The reaction of new cultivars and elite lines of spring wheat into drought stress. In Collection of abstract articles of 5th Iranian agronomy and plant breeding conference, Karaj, Iran. p. 252
Griffiths RI, Whiteley AS, O'Donnell AG, Bailey MJ (2003) Physiological and community responses of established grassland bacterial populations to water stress. Appl Environ Microbiol 69:6961–6968
Ha E, Ikhajiagba B, Bamidele JF, Ogic-Odia E (2008) Salinity effects on young healthy seedling of *Kyllingia peruviana* collected from escravos, Delta state. Glob J Environ Res 2(2):74–88
Haggag WM, Abouziena HF, Abd-El-Kreem F, El Habbasha S (2015) Agriculture biotechnology for management of multiple biotic and abiotic environmental stress in crops. J Chem Pharm Res 7(10):882–889

Hagmanna DF, Goodeya NM, Mathieua C, Evans J, Aronsonc MFJ, Gallagher F, Krumins JA (2015) Effect of metal contamination on microbial enzymatic activity in soil. Soil Biol Biochem 91:291–297

Hasanuzzaman M, Nahar K, Rahman A, Mahmud JA, Alharby HF, Fujita M (2018) Exogenous glutathione attenuates lead-induced oxidative stress in wheat by improving antioxidant defense and physiological mechanisms. J Plant Interact 13(1):203–212

Hayat S, Khalique G, Irfan M, Wani AS, Tripathi BN, Ahmad A (2012a) Physiological changes induced by chromium stress in plants: an overview. Protoplasma 249(3):599–561

Hayat S, Yadav S, Wani AS, Irfan M, Alyemini MN, Ahmad A (2012b) Impact of sodium nitroprusside on nitrate reductase, proline content, and antioxidant system in tomato under salinity stress. Hortic Environ Biotechnol 53(5):362–367

Hayyat A, Javed M, Rasheed I, Ali S, Shahid MJ, Rizwan M (2016) Role of biochar in remediating heavy metals in soil. Phytoremediation 3:420–437

Heuer B, Nadler A (1995) Growth and development of potatoes under salinity and water deficit. Aust J Agric Res 46(7):1477–1486

Hniličková H, Hnilička F, Orsák M, Hejnák V (2019) Effect of salt stress on growth, electrolyte leakage, Na^+ and K^+ content in selected plant species. Plant Soil Environ 65(2):90–96

Hoogmoed M, Sadras VO (2016) The importance of water-soluble carbohydrates in the theoretical framework for nitrogen dilution in shoot biomass of wheat. Field Crop Res 193:196–200

Houimli SI, Denden M, El Hadj SB (2008) Induction of salt tolerance in pepper (*Capsicum annuum*) by 24-epibrassinolide. Eur Asian J Bio Sci 2:83–90

Hu XF, Jiang Y, Shu Y, Hu X, Liu L, Luo F (2014) Effects of mining wastewater discharges on heavy metal pollution and soil enzyme activity of the paddy fields. J Geochem Explor 147:139–150

Hueso S, García C, Hernández T (2012) Severe drought conditions modify the microbial community structure, size and activity in amended and unamended soils. Soil Biol Biochem 50:167–173

Hussain TM, Hazara M, Sultan Z, Saleh BK, Gopal GR (2008) Recent advances in salt stress biology a review. Biotechnol Mol Biol Res 3(1):8–13

ICARDA (2002) Marginal quality water: a solution to fresh water scarcity. International Center for Agricultural Research in the Dry Areas, Aleppo, Syria

Isayenkov SV, Maathuis FJM (2019) Plant salinity stress: many unanswered questions remain. Front Plant Sci 10:80. https://doi.org/10.3389/fpls.2019.00080

Iyengar ERR, Reddy MP (1996) Photosynthesis in highly salt tolerant plants. In: Pessaraki M (ed) Handbook of photosynthesis. Marcel Dekker, New York, pp 897–909

Jaleel CA, Gopi R, Manivannan P, Gomathinayagam M, Sridharan R, Panneerselvam R (2008a) Antioxidant potential and indole alkaloid profile variations with water deficits along different parts of two varieties of *Catharanthus roseus*. Colloids Surf B Biointerfaces 62(2):312–318

Jaleel CA, Gopi R, Panneerselvam R (2008b) Growth and photosynthetic pigments responses of two varieties of *Catharanthus roseus* to triadimefon treatment. C R Biol 331(4):272–277

Jaleel CA, Gopi R, Sankar B, Gomathinayagam M, Panneerselvam R (2008c) Differential responses in water use efficiency in two varieties of *Catharanthus roseus* under drought stress. C R Biol 331(1):42–47

Jaleel CA, Manivannan P, Lakshmanan GM, Gomathinayagam M, Panneerselvam R (2008d) Alterations in morphological parameters and photosynthetic pigment responses of *Catharanthus roseus* under soil water deficits. Colloid Surface B Biointerfaces 61(2):298–303

Jaleel CA, Sankar B, Murali PV, Gomathinayagam M, Lakshmanan GM, Panneerselvam R (2008e) Water deficit stress effects on reactive oxygen metabolism in *Catharanthus roseus*; impacts on ajmalicine accumulation. Colloid Surface B Biointerfaces 62(1):105–111

Jan AT, Azam M, Siddiqui K, Ali A, Choi I, Mohd A, Haq R (2015) Heavy metals and human health, Mechanistic insight into toxicity and counter defense system of antioxidants. Int J Mol Sci 16(12):29592–29630

Jatav HS, Singh SK, Singh Y, Kumar O (2016) Biochar and sewage sludge application increases yield and micronutrient uptake in rice (*Oryza sativa* L.). Commun Soil Sci Plan 50 (14):1732–1742

Jeffery S, Verheijen FGA, van der Velde M, Bastos AC (2011) A quantitative review of the effects of biochar applications to soils on crop productivity using meta-analysis. Agric Ecosyst Environ 144:175–187

Jin Z, Li Z, Li Q, Hu Q, Yang R, Tang H, Li M, Huang B, Zhang J, Li G (2015) Canonical correspondence analysis of soil heavy metal pollution, microflora and enzyme activities in the Pb–Zn mine tailing dam collapse area of Sidi village, SW China. Environ Earth Sci 73 (1):267–274

John R, Ahmad P, Gadgila K, Sharma K (2009) Heavy metal toxicity: effect on plant growth, biochemical parameters and metal accumulation by *Brassica juncea* L. Int J Plant Prod 3 (3):65–67

Kanwal S, Ilyas N, Shabir S, Saeed M, Gul R, Zahoor M, Batool N, Mazhar R (2018) Application of biochar in mitigation of negative effects of salinity stress in wheat (*Triticum aestivum* L.). J Plant Nutr 41(4):526–538

Karami N, Clemente R, Moreno-Jiménez E, Lepp NW, Beesley L (2011) Efficiency of green waste compost and biochar soil amendments for reducing lead and copper mobility and uptake to ryegrass. J Hazard Mater 191(1-3):41–48

Khan M (2000) Effect of metals contamination on soil microbial diversity, enzymatic activity, organic matter decomposition and nitrogen mineralization (a review). Pak J Biol Sci 3 (11):1950–1956

Khan S, Hesham AEL, Qiao M, Rehman S, He JZ (2010) Effects of Cd and Pb on soil microbial community structure and activities. Environ Sci Pollut Res 17(2):288–296

Khan S, Chao C, Muhammad W, Arp HPH, Zhu YG (2013) Sewage sludge biochar influence upon rice (*Oryza sativa* L) yield, metal bioaccumulation and greenhouse gas emissions from acidic paddy soil. Environ Sci Technol 47(15):8624–8632

Kiani SP, Maury P, Sarrafi A, Grieu P (2008) QTL analysis of chlorophyll fluorescence parameters in sunflower (*Helianthus annuus* L.) under well-watered and water-stressed conditions. Plant Sci 175:565–573

Kim DI, Matsuyama Y, Nagasoe S, Yamaguchi M, Yoon YH, Oshima Y, Imada N, Honjo T (2004) Effects of temperature, salinity and irradiance on the growth of the harmful red tide dinoflagellate Cochlodinium polykrikoides Margalef (Dinophyceae). J Plankton Res 26(1):61–66

Kim HS, Kim KR, Kim HJ, Yoon JH, Yang JE, Ok YS, Owens G, Kim KH (2015) Effect of biochar on heavy metal immobilization and uptake by lettuce (*Lactuca sativa* L.) in agricultural soil. Environ Earth Sci 74:1249–1259

Kim BH, Ramanan R, Kang Z, Cho DH, Oh HM, Kim HS (2016) Chlorella sorokiniana HS1, a novel freshwater green algal strain, grows and hyper accumulates lipid droplets in seawater salinity. Biomass Bioenergy 85:300–305

Kouchou A, Rais N, Elsass F, Duplay J, Fahli N, Ghachtouli NEL (2017) Effects of long-term heavy metals contamination on soil microbial characteristics in calcareous agricultural lands (Saiss plain, North Morocco). J Mater Environ Sci 8(2):691–695

Kuchenbuch R, Claassen N, Jungk A (1986) Potassium availability in relation to soil moisture. Plant Soil 95(2):233–243

Kumar N, Bauddh K, Singh R, Anand K, Barman SC, Singh DP (2009) Phytotoxicity of trace metals (Cu and Cd) to gram (*Cicer aerientinum*) and mung (*Phasiolus mungo*). J Ecophysiol Occup Health 9(2):59–65

Kusaka M, Ohta M, Fujimura T (2005) Contribution of inorganic components to osmotic adjustment and leaf folding for drought tolerance in pearl millet. Physiol Plant 125(4):474–489

Lakhdar A, Rabhi M, Ghnaya T, Montemurro F, Jedidi N, Abdelly C (2009) Effectiveness of compost use in salt-affected soil. J Hazard Mater 171(1-3):29–37

Lashari MS, Liu Y, Li L, Pan W, Fu J, Pan G, Zheng J, Zheng J, Zhang X, Yu X (2013) Effects of amendment of biochar-manure compost in conjunction with pyroligneous solution on soil quality and wheat yield of a salt-stressed cropland from Central China Great Plain. Field Crop Res 144:113–118

Lawlor DW, Cornic G (2002) Photosynthetic carbon assimilation and associated metabolism in relation to water deficits in higher plants. Plant Cell Environ 25:275–294

Lawson T, Oxborough K, Morison JI, Baker NR (2003) The responses of guard and mesophyll cell photosynthesis to CO2, O2, light and water stress in a range of species are similar. J Exp Bot 54:1743–1752

Lehmann J, Rillig MC, Thies J, Masiello CA, Hockaday WC, Crowley D (2011) Biochar effects on soil biota–a review. Soil Biol Biochem 43(9):1812–1836

Leila R (2007) Response of Tunisian autochthonous peral millet to drought stress induced by polyethylene glycol 6000. Afr J Biotechnol 6:1102–1105

Liu C, Liu F, Ravnskov S, Rubæk GH, Sun Z, Andersen MN (2017) Impact of wood biochar and its interactions with mycorrhizal fungi, phosphorus fertilization and irrigation strategies on potato growth. J Agron Crop Sci 203:131–145

Lu HP, Li ZA, Gascó G, Méndez A, Shen Y, Paz-Ferreiro J (2018) Use of magnetic biochars for the immobilization of heavy metals in a multi-contaminated soil. Sci Total Environ 622-623:892–899

Luo JY, Zhang S, Peng J, Zhu XZ, Lv LM, Wang CY, Li CH, Zhou ZG, Cui JJ (2017) Effects of soil salinity on the expression of Bt toxin (Cry1Ac) and the control efficiency of *Helicoverpa armigera* in field-grown transgenic Bt cotton. PLoS One 12(1):0170379

Major J, Rondon M, Molina D, Riha SJ, Lehmann J (2010) Maize yield and nutrition after 4 years of doing biochar application to a Colombian savanna oxisol. Plant Soil 333:117–128

Manivannan P, Jaleel CA, Kishorekumar A, Sankar B, Somasundaram R, Sridharan R, Panneerselvam R (2007) Changes in antioxidant metabolism of *Vigna unguiculata* (L.) Walp. by propiconazole under water deficit stress. Colloids Surf B Biointerfaces 57(1):69–74

Manivannan P, Jaleel CA, Somasundaram R, Panneerselvam R (2008) Osmoregulation and antioxidant metabolism in drought-stressed *Helianthus annuus* under triadimefon drenching. C R Biol 331(6):418–425

Marschner H (1995) Mineral nutrition of higher plants. Academic, London, p 889

Mas-Carrió E, Dini-Andreote F, Brossi MJDL, Salles JF, Olff H (2018) Organic amendment under increasing agricultural intensification: effects on soil bacterial communities and plant productivity. Front Microbiol 9:2612

Massacci A, Nabiev SM, Pietrosanti L, Nematov SK, Chernikova TN, Thor K, Leipner J (2008) Response of the photosynthetic apparatus of cotton (*Gossypium hirsutum*) to the onset of drought stress under field conditions studied by gas-exchange analysis and chlorophyll fluorescence imaging. Plant Physiol Biochem 46(2):189–195

Memon SA, Hou X, Wang LJ (2010) Morphlogical analysis of salt stress response of Pak Choi. Electron J Environ Agric Food Chem 9(1):203–207

Meng L, Li W, Zhang S, Wu C, Jiang W, Sha C (2016) Effect of different extra carbon sources on nitrogen loss control and the change of bacterial populations in sewage sludge composting. Ecol Eng 94:238–243

Meng J, Tao M, Wang L, Liu X, Xu J (2018) Changes in heavy metal bioavailability and speciation from a Pb-Zn mining soil amended with biochars from co-pyrolysis of rice straw and swine manure. Sci Total Environ 633:300–307

Metternicht GI, Zinck JA (2003) Remote sensing of soil salinity: potentials and constraints. Remote Sens Environ 85(1):1–20

Mickan BS, Abbott LK, Stefanova K, Solaiman ZM (2016) Interactions between biochar and mycorrhizal fungi in a water-stressed agricultural soil. Mycorrhiza 26:565–574

Minhas PS, Rane J, Pasala RK (2017) Abiotic stresses in agriculture: an overview. In: Minhas P, Rane J, Pasala R (eds) Abiotic stress management for resilient agriculture. Springer, Singapore

Mishra S, Bharagava RN (2016) Toxic and geno-toxic effects of hexavalent chromium in environment and its bioremediation strategies. J Environ Sci Health C Environ Carcinog Ecotoxicol Rev 34(1):1–32

Mishra S, Bharagava RN, More N, Yadav A, Zainith S, Mani S, Chowdhary P (2018) Heavy metal contamination: an alarming threat to environment and human health. In: Sobti R, Arora N, Kothari R (eds) Environmental biotechnology: for sustainable future. Springer, Singapore, pp 103–125

Monneveux P, Sanchez C, Beck D, Edmeades GO (2006) Drought tolerance improvement in tropical maize source populations: evidence of progress. Crop Sci 46:180–191

Muneer M, Oades JM (1989) The role of Ca-organic interactions in soil aggregate stability. II. Field studies with 14C-labeled straw, $CaCO_3$ and $CaSO_42.H_2O$. Soil Res 27(2):401–409

Munns R, Tester M (2008) Mechanisms of salinity tolerance. Annu Rev Plant Biol 59:651–681

Mutale I, Franco IB, Jewette M (2019) Corporate sustainability performance: an approach to effective sustainable community development or not? A case study of the Luanshya copper mine in Zambia. Sustainability 11(20):5775

Nadeem SM, Zahir ZA, Naveed M, Nawaz S (2013) Mitigation of salinity-induced negative impact on the growth and yield of wheat by plant growth-promoting rhizobacteria in naturally saline conditions. Ann Microbiol 63(1):225–232

Nakayama NO, Saneoka HI, Moghaieb RE, Premachandra GS, Fujita KO (2007) Response of growth, photosynthetic gas exchange, translocation of 13C-labelled photosynthate and N accumulation in two soybean (Glycine max L. Merrill) cultivars to drought stress. Int J Agric Biol 9(5):669–674

Nazir R, Khan M, Masab M, Rehman HU, Rauf NU, Shahab S, Ameer N, Sajed M, Ullah M, Rafeeq M, Shaheen Z (2015) Accumulation of heavy metals (Ni, Cu, Cd, Cr, Pb, Zn, Fe) in the soil, water and plants and analysis of physico-chemical parameters of soil and water collected from Tanda Dam Kohat. J Pharm Sci Res 7(3):89–97

Nelson PN, Oades JM (1998) Organic matter, sodicity and soil structure. In: Sumner ME, Naidu RI (eds) Sodic soils, distribution, processes, management and environmental consequences. Oxford University Press, New York, pp 51–75

Nemeskeri E, Molnar K, Vigh R, Nagy J, Dobos A (2015) Relationships between stomatal behaviour, spectral traits and water use and productivity of green peas (*Pisum sativum* L.) in dry seasons. Acta Physiol Plant 37:1–16

Nguyen D, Rieu I, Mariani C, van Dam NM (2016) How plants handle multiple stresses: hormonal interactions underlying responses to abiotic stress and insect herbivory. Plant Mol Biol 91 (6):727–740

Nie C, Yang X, Niazi NK, Xu X, Wen Y, Rinklebei J, Ok YS, Xu S, Wang H (2018) Impact of sugarcane bagasse-derived biochar on heavy metal availability and microbial activity: a field study. Chemosphere 200:274–282

Obasi NA, Akubugwo EI, Ugbogu OC, Chinyere GC (2012) Heavy metals bioavailability and phyto-accumulation potentials of selected plants on burrow-pit dumpsites in Aba and Ntigha Dumpsite in Isiala Ngwa of Abia State, Nigeria. Nigerian J Biochem Mol Biol 27(1):27–45

Odueze JC, Nwakonobi TU, Itodo IN (2017) Influence of physico-chemical characteristics of soils on heavy metal contamination in Makurdi, Benue state. J Environ Sci Toxicol Food Technol 11 (5):84–92

Oijagbe IJ, Abubakar BY, Edogbanya PRO, Suleiman MO, Olorunmola JB (2019) Effects of heavy metals on soil microbial biomass carbon. J Biol Med 4(1):30–32

OlabimpeIyabo O, Oluwafemi JR, Grace AM, Oladele O (2017) Heavy metals and some physico-chemical parameters in soil of major domestic dumpsites in Akure township, ondo state, South-Western Nigeria. Int J Eng Res Manag Technol 4(10):26–32

Olayinka OO, Akande OO, Bamgbose K, Adetunji MT (2017) Physicochemical characteristics and heavy metal levels in soil samples obtained from selected anthropogenic sites in Abeokuta, Nigeria. J Appl Sci Environ Manag 21(5):883–891

Osakabe Y, Osakabe K, Shinozaki K, Tran LSP (2014) Response of plants to water stress. Front Plant Sci 5:1–8

Oster JD, Shainberg I (2001) Soil responses to sodicity and salinity: challenges and opportunities. Soil Res 39(6):1219–1224

Paliwal KV, Gandhi AP (1976) Effect of salinity, SAR, Ca: Mg ratio in irrigation water, and soil texture on the predictability of exchangeable sodium percentage. Soil Sci 122(2):85–90

Pathak H, Rao DLN (1998) Carbon and nitrogen mineralization from added organic matter in saline and alkali soils. Soil Biol Biochem 30(6):695–702

Petropoulos SA, Daferera D, Polissiou MG, Passam HC (2008) The effect of water deficit stress on the growth, yield and composition of essential oils of parsley. Sci Hortic 115(4):393–407

Pinkerton A, Simpson JR (1986) Interactions of surface drying and subsurface nutrients affecting plant growth on acidic soil profiles from an old pasture. Aust J Exp Agric 26(6):681–689

Pressler Y, Foster EJ, Moore JC, Cotrufo MF (2017) Coupled biochar amendment and limited irrigation strategies do not affect a degraded soil food web in a maize agroecosystem, compared to the native grassland. GCB Bioenergy 9(8):1344–1355

Qadir A, Sharif M, Ziad M (2000) Homotheties of cylindrically symmetric static manifolds and their global extension. Classical Quant Gravity 17(2):345

Quirk J (1994) Interparticle forces: a basis for the interpretation of soil physical behavior. Adv Agron 53:121–183

Rahimi G, Kolahchi Z, Charkhabi A (2017) Uptake and translocation of some heavy metals by rice crop (*Oryza sativa*) in paddy soils. Agriculture (Poľnohospodárstvo) 63(4):163–175

Rahman M, Sabir AA, Mukta JA, Khan MMA, Mohi-Ud-Din M, Miah MG, Islam MT (2018) Plant probiotic bacteria *Bacillus* and *Paraburkholderia* improve growth, yield and content of antioxidants in strawberry fruit. Sci Rep 8:2504–2510

Rahmani N (2006) Effect of irrigation and nitrogen application on the quantity and quality of medicinal plant Marigold (*Calendula officinalis* L.). Master Thesis Agronomy, University of Takestan Branch.

Rauf S, Al-Khayri JM, Zaharieva M, Monneveux P, Khalil F (2015) Breeding strategies to enhance drought tolerance in crops. In: Al-Khayri JM et al (eds) Advances in plant breeding strategies; agronomic, abiotic and biotic stress traits. Springer, Berlin, pp 1–70

Reddy AR, Chaitanya KV, Vivekanandan M (2004) Drought induced responses of photosynthesis and antioxidant metabolism in higher plants. J Plant Physiol 161:1189–1202

Rengasamy P, Olsson KA (1991) Sodicity and soil structure. Soil Res 29(6):935–952

Rengasamy P, Sumner ME (1998) Processes involved in sodic behaviour. In: Sumner ME, Naidu R (eds) Sodic soils: distribution management and environmental consequences. Oxford University Press, New York, pp 35–50

Resnik ME (1970) Effect of mannitol and polyethylene glycol on phosphorus uptake by maize plants. Ann Bot 34(3):497–504

Rietz DN, Haynes RJ (2003) Effects of irrigation-induced salinity and sodicity on soil microbial activity. Soil Biol Biochem 35(6):845–854

Rui L, Sun W, Chao M, Ji C, Wang M, Ye B (2009) Leaf anatomical changes of *Bruguiera gymnorrhiza* seedlings under salt stress. J Trop Subtropical Bot 17(2):169–175

Sachan P, Lal N (2017) An overview of nickel (Ni^{2+}) essentiality, toxicity and tolerance strategies in plants. Asian J Biol 2(4):1–15

Sacks MM, Silk WK, Burman P (1997) Effect of water stress on cortical cell division rates within the apical meristem of primary roots of maize. Plant Physiol 114(2):519–527

Sairam RK, Tyagi A (2004) Physiology and molecular biology of salinity stress tolerance in plants. Curr Sci 10:407–421

Sallam A, Alqudah AM, Dawood MFA, Baenziger PS, Börne A (2019) Drought stress tolerance in wheat and barley: advances in physiology, breeding and genetics research. Int J Mol Sci 20 (13):31–37

Salman SA, Zeid SAM, Seleem E-MM, Abdel-Hafiz MA (2019) Soil characterization and heavy metal pollution assessment in Orabi farms, El Obour, Egypt. Bull Nat Res Cent 43:42. https://doi.org/10.1186/s42269-019-0082-1

Sankar B, Jaleel CA, Manivannan P, Kishorekumar A, Somasundaram R, Panneerselvam R (2007) Effect of paclobutrazol on water stress amelioration through antioxidants and free radical scavenging enzymes in *Arachis hypogaea* L. Colloids Surf B Biointerfaces 60(2):229–235

Sankar B, Jaleel CA, Manivannan P, Kishorekumar A, Somasundaram R, Panneerselvam R (2008) Relative efficacy of water use in five varieties of *Abelmoschus esculentus* (L.) Moench. Under water-limited conditions. Colloids Surf B Biointerfaces 62(1):125–129
Scherer HW (2001) Sulphur in crop production. Eur J Agron 14(2):81–111
Schmidt R, Kunkowska AB, Schippers JH (2016) Role of reactive oxygen species during cell expansion in leaves. Plant Physiol 172(4):2098–2106
Shah FUR, Ahmad N, Masood KR, Peralta-Videa JR, Ahmad FD (2010) Heavy metal toxicity in plants. In: Ashraf M, Ozturk M, Ahmad M (eds) Plant adaptation and phytoremediation. Springer, Dordrecht
Shaheen HL, Shahbaz M, Ullah I, Iqbal MZ (2012) Morpho-physiological responses of cotton (*Gossypium hirsutum*) to salt stress. Int J Agric Biol 14(6):980–984
Shainberg I, Levy GJ, Rengasamy P, Frenkel H (1992) Aggregate stability and seal formation as affected by drops' impact energy and soil amendments. Soil Sci 154(2):113–119
Shao HB, Chu LY, Shao MA, Jaleel CA, Hong-Mei M (2008) Higher plant antioxidants and redox signaling under environmental stresses. C R Biol 331(6):433–441
Shekari F (2000) Effect of drought stress on phenology, water relations, growth, yield and quality canola, doctorate thesis in the field of Agriculture, University of Tabriz, p. 180.
Shi W, Ma X (2017) Effects of heavy metal Cd pollution on microbial activities in soil. Ann Agric Environ Med 24(4):722–725
Singh RP, Bauddh K, Sainger M, Sainger PA, Singh JS, Jaiwal PK (2010) Nitrogen use efficiency in higher plants under drought, high temperature, salinity and heavy metal contaminations. In: Jain V, Kumar PA (eds) Nitrogen use efficiency in higher plants. New India Publishing Agency (India) Pvt. Ltd., New Delhi, pp 99–123
Singh JS, Pandey VC, Singh DP (2011) Efficient soil microorganisms: a new dimension for sustainable agriculture and environmental development. Agric Ecosyst Environ 140:339–353
Sinsabaugh RL, Manzoni S, Moorhead DL, Richter A (2013) Carbon use efficiency of microbial communities: stoichiometry, methodology and modelling. Ecol Lett 16(7):930–939
Six D, Fily M, Blarel L, Goloub P (2005) First aerosol optical thickness measurements at dome C (East Antarctica), summer season 2003–2004. Atmos Environ 39(28):5041–5050
Smika DE, Haas HJ, Power JF (1965) Effects of moisture and nitrogen fertilizer on growth and water use by native grass 1. Agron J 57(5):483–486
Sobolev D, Begonia MF (2008) Effects of heavy metal contamination upon soil microbes: lead-induced changes in general and denitrifying microbial communities as evidenced by molecular markers. Int J Environ Res Public Health 5(5):450–456
Soriano MA, Villalobos FJ, Federes E (2002) Stress timing effects on sunflower harvest index. In: Villalobos FJ, Testi L (eds) VII Congress of the European Society for Agronomy. Sevilla, Consejería de Agricultural Pesca, Spain JA, pp 141–142
Sowerby A, Emmett B, Beier C, Tietema A, Penuelas J, Estiarte M, Freeman C (2005) Microbial community changes in heathland soil communities along a geographical gradient: interaction with climate change manipulations. Soil Biol Biochem 37(10):1805–1813
Specht JE, Chase K, Macrander M, Graef GL, Chung J, Markwell JP, Germann M, Orf JH, Lark KG (2001) Soybean response to water. A QTL analysis of drought tolerance. Crop Sci 41:493–509
Streeter JG (2003) Effects of drought on nitrogen fixation in soybean root nodules. Plant Cell Environ 26(8):1199–1204
Su C, Jiang L , Zhang W (2014) A review on heavy metal contamination in the soil worldwide: situation, impact and remediation techniques. Environ Skep Critic 3(2):24–27
Sudhir P, Murthy SD (2004) Effects of salt stress on basic processes of photosynthesis. Photosynthetica 42(2):481–486
Sun L, Wei P, Pu W, Wang B, Wu Y, Tan T (2016) The oil recovery enhancement by nitrogen foam in high-temperature and high-salinity environments. J Pet Sci Eng 147:485–494
Swift MJ, Heal OW, Anderson JM, Anderson JM (1979) Decomposition in terrestrial ecosystems. University of California Press, Berkeley

Syed R, Kapoor D, Bhat AA (2018) Heavy metal toxicity in plants: a review. Plant Arch 18 (2):1229–1238
Tahir MHN, Mehid SS (2001) Evaluation of open pollinated sunflower (*Helianthus annuus* L.) populations under water stress and normal conditions. Int J Agric Biol 3:236–238
Tahkokorpi M, Taulavuori K, Laine K, Taulavuori E (2007) After effects of drought-related winter stress in previous and current year stems of *Vaccinium myrtillus* L. Environ Exp Bot 61:85–93
Thomas S, Anand A, Chinnusamy V, Dahuja A, Basu S (2013) Magnetopriming circumvents the effect of salinity stress on germination in chickpea seeds. Acta Physiol Plant 35(12):3401–3411
Tilman D, Cassman KG, Matson PA, Naylor R, Polasky S (2002) Agricultural sustainability and intensive production practices. Nature 8(418):671–677
Tiwari KK, Singh NK, Rai UN (2013) Chromium phytotoxicity in radish (Raphanus sativus): effects on metabolism and nutrient uptake. Bull Environ Contam Toxicol 91:339–344
Trenberth KE, Dai A, Van Der Schrier G, Jones PD, Barichivich J, Briffa KR, Sheffield J (2014) Global warming and changes in drought. Nat Clim Chang 4(1):17–22
Tripathi VN, Srivastava S (2006) Ni^{2+}-uptake in *Pseudomonas putida* strain S4: a possible role of Mg^{2+}-uptake pump. J Biosci 31(1):61–67
Turner LB (1985) Changes in the phosphorus content of *Capsicum annuum* leaves during water stress. J Plant Physiol 121(5):429–439
Uddin SR, Farid M, Saeed R, Ali S, Md R, Md TH, SAH B (2017) Citric acid enhanced the antioxidant defense system and chromium uptake by *Lemna minor* L. grown in hydroponics under Cr stress. Environ Sci Pollut Res 24(21):17669–17678
United States Salinity Laboratory Staff (1954) Diagnosis and improvement of saline and alkali soils, Agriculture Handbook No. 60. United States Department of Agriculture, Washington, DC
Usman AR, Al-Wabel MI, Abdulaziz AH, Mahmoud WA, El-Naggar AH, Ahmad M, Abdulelah AF, Abdulrasoul AO (2016) Conocarpus biochar induces changes in soil nutrient availability and tomato growth under saline irrigation. Pedosphere 26(1):27–38
Wall DH, Ritz K, Six J, Strong DR, van der Putten WH (2012) Soil ecology and ecosystem services. Oxford University Press, Oxford
Wang YP, Li QB, Shi JY, Lin Q, Chen XC, Wu W, Chen YX (2008) Assessment of microbial activity and bacterial community composition in the rhizosphere of a copper accumulator and a non-accumulator. Soil Biol Biochem 40(5):1167–1177
Wang X, Yuan B, Chen Y, Li X, Ren Y (2014) Integration of micro-filtration into osmotic membrane bioreactors to prevent salinity build-up. Bioresour Technol 167:116–123
Wang S, Xu Y, Norbu N, Wang Z (2018) Remediation of biochar on heavy metal polluted soils. IOP Conference Series: Earth and Environmental Science 108(4):042113. IOP Publishing
Wardle DA, Nilsson MC, Zackrisson O (2008) Fire-derived charcoal causes loss of forest humus. Science 320(5876):629–632
Webber M, Barnett J, Finlayson B, Wang M (2006) Pricing China's irrigation water (working paper). School of Anthropology, Geography and Environmental Studies, The University of Melbourne, VIC, Australia
Wichern J, Wichern F, Joergensen RG (2006) Impact of salinity on soil microbial communities and the decomposition of maize in acidic soils. Geoderma 137(1-2):100–108
Wicke B, Smeets E, Dornburg V, Vashev B, Gaiser T, Turkenburg W, Faaij A (2011) The global technical and economic potential of bioenergy from salt affected soils. Energy Environ Sci 4:2669–2681
Wilson A, Punginelli C, Gall A, Bonetti C, Alexandre M, Routaboul JM, Kerfeld CA, Van Grondelle R, Robert B, Kennis JT, Kirilovsky D (2008) A photoactive carotenoid protein acting as light intensity sensor. PNAS 105(33):12075–12080
Wong VN, Dalal RC, Greene RS (2009) Carbon dynamics of sodic and saline soils following gypsum and organic material additions: a laboratory incubation. Appl Soil Ecol 41(1):29–40
Wu QS, Xia RX, Zou YN (2008) Improved soil structure and citrus growth after inoculation with three arbuscular mycorrhizal fungi under drought stress. Eur J Soil Biol 44(1):122–128

Wu W, Mhaimeed AS, Al-Shafie WM, Ziadat F, Dhehibi B, Nangia V, De Pauw E (2014) Mapping soil salinity changes using remote sensing in Central Iraq. Geoderma Reg 2:21–31
Wuana RA, Okieimen FE (2011) Heavy metals in contaminated soils: a review of sources, chemistry, risks and best available strategies for remediation. Int Sch Res Notices Ecol 402647:1–20
Wullschleger SD, Yin TM, DiFazio SP, Tschaplinski TJ, Gunter LE, Davis MF, Tuskan GA (2005) Phenotypic variation in growth and biomass distribution for two advanced-generation pedigrees of hybrid poplar. Can J For Res 35(8):1779–1789
Xian Y, Wang M, Chen W (2015) Quantitative assessment on soil enzyme activities of heavy metal contaminated soils with various soil properties. Chemosphere 139:604–608
Xiao X, Wang M, Zhu H, Guo Z, Han X, Zeng P (2017) Response of soil microbial activities and microbial community structure to vanadium stress. Ecotoxicol Environ Saf 142:200–206
Xu Y, Seshadri B, Sarkar B, Wang H, Rumpel C, Sparks D, Farrell M, Hall T, Yang X, Bolan N (2018) Biochar modulates heavy metal toxicity and improves microbial carbon use efficiency in soil. Sci Total Environ 621:148–159
Yadav SK (2010) Heavy metals toxicity in plants: an overview on the role of glutathione and phytochelatins in heavy metal stress tolerance of plants. Soil Biol Biochem 76(2):67–179
Yadav G, Srivastava PK, Singh VP, Prasad SM (2014) Light intensity alters the extent of arsenic toxicity in *Helianthus annuus* L. seedlings. Biol Trace Elem Res 158:410–421
Yuan B, Shen X, Li X, Xu C, Wang S (2007) Mitogen-activated protein kinase OsMPK6 negatively regulates rice disease resistance to bacterial pathogens. Planta 226(4):953–960
Yuan P, Wang J, Pan Y, Shen B, Wu C (2019) Review of biochar for the management of contaminated soil: preparation, application and prospect. Sci Total Environ 659:473–490
Zareian J (2004) Effects of drought on the different stages of growth, and growth traits, varieties of winter canola, MS Thesis Agronomy, Faculty of Agriculture, Islamic Azad University of Khorasgan
Zeid IM, Ghazi SM, Shedeed ZA, Nabawy DM (2018) Biochar mitigates cadmium stress on alfalfa seeds during germination. Int J Prog Sci Technol 6:251–261
Zhang H, Wang X, Wang S (2004) A study on stomatal traits of *Platanus acerifolia* under urban stress. J Fudan Univ 43:651–656
Zhang H, Kim MS, Sun Y, Dowd SE, Shi H, Paré PW (2008) Soil bacteria confer plant salt tolerance by tissue-specific regulation of the sodium transporter HKT1. Mol Plant-Microbe Interact 21(6):737–744
Zhang FP, Li CF, Tong LG, Yue LX, Li P, Ciren YC, Cao CG (2010) Response of microbial characteristics to heavy metal pollution of mining soils in Central Tibet, China. Appl Soil Ecol 45(3):144–151
Zhao W, Sun Y, Kjelgren R, Liu X (2015) Response of stomatal density and bound gas exchange in leaves of maize to soil water deficit. Acta Physiol Plant 37(1):1704. https://doi.org/10.1007/s11738-014-1704-8
Zheng R, Chen Z, Cai C, Tie B, Liu X, Reid BJ, Huang Q, Lei M, Sun G, Baltrėnaitė E (2015) Mitigating heavy metal accumulation into rice (*Oryza sativa* L.) using biochar amendment a field experiment in Hunan China. Environ Sci Pollut Res 22:11097–11108
Zheng R, Sun G, Li C, Reid BJ, Xie Z, Zhang B, Wang Q (2017) Mitigating cadmium accumulation in greenhouse lettuce production using biochar. Environ Sci Pollut Res 24(7):6532–6542
Zornoza R, Mataix-Solera J, Guerrero C, Arcenegui V, García-Orenes F, Mataix-Beneyto J, Morugán A (2007) Evaluation of soil quality using multiple lineal regression based on physical, chemical and biochemical properties. Sci Total Environ 378(1-2):233–237
Zou L, Morris G, Qi D (2008) Using activated carbon electrode in electrosorptive deionisation of brackish water. Desalination 225(1–3):329–340

Mitigation of Salinity Stress by Using the Vermicompost and Vermiwash

15

N. Ruiz-Lau, M. A. Oliva-Llaven, J. A. Montes-Molina, and F. A. Gutiérrez-Miceli

Contents

Abstract

Salinity in agricultural soils decreases productivity in a wide variety of crops. Vermicompost and vermiwash have humic substances that interact with the organic components of the soil and the roots of plants, where they can influence the tolerance to salinity. Plants grown in saline soils suffer from stress; however when adding vermicompost, the symptoms of the stress decrease. Also, an

N. Ruiz-Lau
CONACYT-Tecnológico Nacional de México/Instituto Tecnológico de Tuxtla Gutiérrez, Tuxtla Gutiérrez, Chiapas, Mexico
e-mail: nruizla@conacyt.mx

M. A. Oliva-Llaven
Facultad de Medicina Veterinaria y Zootecnia, Universidad Autónoma de Chiapas, Tuxtla Gutiérrez, Chiapas, Mexico

J. A. Montes-Molina · F. A. Gutiérrez-Miceli (✉)
Tecnológico Nacional de México/Instituto Tecnológico de Tuxtla Gutiérrez, Tuxtla Gutiérrez, Chiapas, Mexico

K. Bauddh et al. (eds.), *Ecological and Practical Applications for Sustainable Agriculture*, https://doi.org/10.1007/978-981-15-3372-3_15

increase in tolerance to salinity by application of vermiwash has been reported. The foliar and edaphic application of vermiwash and vermicompost improves the activity of antioxidant enzymes and reduces electrolyte leakage, oxidative stress, and accumulation of Na in the tissues suggesting the mechanisms by which the application of vermiwash and vermicompost can alleviate the damage caused by salinity. Plant-growth-promoting bacteria (PGPB) is another factor. Species of the genus Serratia, Pseudomonas, and Azospirillum and arbuscular mycorrhizal fungi can mitigate the effects of salinity in plants; however, research is still needed to explain the beneficial effects.

Keywords

Vermicompost · Vermiwash · Salinity stress · Humic substances

15.1 Introduction

In agriculture year after year, it is about improving and increasing the yield of crops. The magnitude of a yield indicates the level of efficiency of the corresponding combination of factors that influence the harvest. To achieve these specific characteristics, elements such as temperature, light, humidity, nutritional requirements, and climate must be met at the optimum level for each plant. If any of these factors is not met, the growth may be reduced, which is reflected in a decrease in yield. When environmental factors are appropriate, the plant must be properly fertilized to achieve optimum quality in agricultural production (Ceritoğlu et al. 2018). During the last few years in agriculture, the use of inorganic fertilizers has been more frequent than in later years. The indiscriminate use of high amounts of chemical fertilizers and the use of sewage has dramatically multiplied the surface area affected by salinity. Salinity in agricultural soils decreases crop productivity in a wide variety of plants, so that tolerance to this stress is an important characteristic in the improvement of plants. There are different mechanisms that provide tolerance; the development of succulent leaves, reabsorption and extrusion of sodium, and the accumulation of proline, both in leaf and root tissues, have been related to saline stress, indicating an essential role in tolerance. Because every year salinity in growing areas increases, it is sought to be implemented in agriculture strategies that allow to reduce the harm caused by salt stress in a more ecological way, hence the importance of using vermicompost and its derivatives to minimize the harm caused by salinity. Vermicomposting improves plant growth and soil structure (Ceritoğlu et al. 2018); therefore, the use of the vermicompost is an alternative to the use of agrochemicals to guarantee the yield and quality of the crops, reducing the cost and the deterioration of the environment. The vermicompost contains humic substances that interact with the organic components of the soil and the roots of plants, where they can influence fertility, the conservation of the same, and also the physiology of the plants (Rivera-Chávez et al. 2012). There are reports that mention that humic substances improve length and root biomass in plants, in addition to improving the acquisition of nutrients and increasing the concentration of chlorophyll in leaves (Ceritoğlu et al. 2018; Chiquito-Contreras et al. 2018; Yanan et al. 2018; Bidabadi et al. 2017).

15.2 Salinization of the Soils

Worldwide, salinity is a factor that limits the productivity of agricultural crops. Climate change, the increase in water loss, large-scale crops, and irrigation with groundwater have contributed to the salinization of agricultural soils (Costantini and Lorenzetti 2013; Daliakopoulos et al. 2016). Therefore, soluble salts can accumulate in soils in a natural or anthropogenic way, and this is known as salinization (Bojórquez-Quintal et al. 2012). The salts present in saline soils can be carbonates (CO_3); sulfates of magnesium ($MgSO_4$), calcium ($CaSO_4$), and sodium ($NaSO_4$); and chlorides (Cl^-) of sodium (NaCl), potassium (KCl), and magnesium ($MgCl_2$). Sodium chloride (NaCl) is one of the salts that cause greater problems due to its high solubility (Munns and Tester 2008).

Soil with an electrical conductivity (ECe) of 4 dS/m or more is considered saline. This corresponds to ~40 mM NaCl, which decreases the yields of many crops. NaCl is the main salt in saline soils, although other types of salts can also be found in solution, such as Na_2SO_4, $MgSO_4$, $CaSO_4$, $MgCl_2$, KCl, and Na_2CO_3 (Munns and Tester 2008). Therefore, the dominant cation is sodium and anion is chlorine in most saline soils, which increase the percentage of NaCl.

15.3 Effect of Salinity on Plants

Plants are sessile organisms; therefore, they live exposed to constant environmental changes that sometimes are negative for their development. Stress is defined as any abiotic factor that produces an adverse effect on the growth and development of plants, as well as on the quality and yield of production (Levitt 1980). Under natural conditions, combinations of two or more types of stress are common to most agricultural areas around the world and could impact crop productivity (Suzuki et al. 2014).

The excess of NaCl in the soil can produce osmotic, ionic, and oxidative stress. Together, they decrease the photosynthetic rate, lead to leaf necrosis and chlorosis, and cause changes in the architecture of the root. In the same way, it reduces the absorption and causes the excessive flow of water and ions from the cells, causing a decrease in the growth and yield of plants (Munns and Tester 2008; Bojórquez-Quintal et al. 2012).

The osmotic effect produces a water deficit due to the presence of high concentrations of solutes in the soil. Specific stress is caused by K^+ deficiency. An imbalance in the K^+/Na^+ ratio is caused by the increase in Na^+ influx. In the presence of salinity, the influx of Na^+ is carried out through the pathways that are for K^+ and the discrimination between the two ions is difficult. Due to this lack of selectivity, plants that develop in saline soils have toxicity due to excess Na^+ and K^+ deficiency.

The high concentrations of Na^+ in the cytoplasm should decrease to remain low in the cytosol and thus maintain a high K^+/Na^+ ratio, thanks to the mechanisms that reduce the influx of Na^+ towards the cells of the roots, to the compartmentalization of Na^+ in the vacuoles, and to an increase of Na^+ efflux. The osmotic adjustment and elimination of Na^+ by the cell are essential for salinity tolerance (Bojórquez-Quintal et al. 2012).

Saline stress induces serious alterations in the metabolism of the plant that affect its growth and development (Hasegawa et al. 2000). The increase in the intake of salts induces specific poisonings by ions such as Cl^-, Na^+, or sulfate (SO_4^{-2}) which minimize the uptake of essential minerals such as NPK (nitrogen, phosphorus, potassium) and calcium (Ca^{+2}) (Zhu 2001). The excess of Na^+ and Cl^- ions can cause harmful physiological effects for plants. The K^+ uptake is mainly affected by the presence of Na^+ and the stomatal closure is affected causing finally water loss; on the other hand, the Cl^- ion affects chlorophyll synthesis causing leaf chlorosis toxicity, Cl^- being more harmful than Na^+ (Tavakkoli et al. 2011). Photosynthesis is a process by which autotrophic organisms such as plants transform the light energy of the sun into chemical energy for its development. The accumulation of Na^+ and Cl^- in the chloroplasts of the cells can interrupt this process. Photosynthesis is directly correlated with the health of the plant. Under conditions of salinity, plants can present a loss of chlorophyll, so its concentration in the cell can be an indicator of stress (Chutipaijit et al. 2011).

15.4 Mechanisms of Tolerance to Salinity in Plants

Salinity tolerance is the ability of a plant to withstand a range of salt concentrations, the effects of high and low salt concentrations, with no or minimal reductions in growth or yield (Bojórquez-Quintal et al. 2012). This tolerance is mediated by several genes and in which anatomical, physiological, and biochemical components participate. In fact, the combination of several of them improves tolerance (Adem et al. 2014).

15.5 Compatible Solute

A mechanism to minimize the damage caused by salt stress is the synthesis of organic compounds called osmolytes, which do not cause damage in millimolar concentrations; these include amino acids (proline, glycine betaine), sugar (trehalose, mannitol, sucrose), urea, and others (Hayat et al. 2012; Yancey 2001). Organic osmolytes are typically called compatible solutes based on the fact that these solutes (except urea) do not react with other molecules in a harmful way; accordingly, they are regulated with a small impact on cellular functions (Yancey 2001, 2005).

An aspect of great relevance in the response of crops to salts is the osmotic adjustment, which is the net increase in the amount of osmotically active solutes in tissues. This osmotic adjustment is achieved through the accumulation of inorganic ions, or through the synthesis of organic solutes. Inorganic ions (compatible solutes) are determinants in the osmotic adjustment and accumulate in vacuoles, because they accumulate mainly in the cytoplasm, without causing inhibition of the enzymatic activity, and also prevent the dehydration of the cytosol by the reduction in water potential derived from the accumulation of salts in vacuoles (García and Medina 2003).

15.6 Transport of Sodium

Plants have mechanisms of influx (entry), efflux (exit), and/or compartmentalization of Na^+ in vacuoles that allow them to tolerate salinity through channels and/or transporters. Sodium moves from the solution of the soil to the root passively to the cytoplasm of the epidermal cells and the cortex. Once inside the epidermis, it is sent to the xylem of the root through the apoplastic or simplistic path. The transport of Na^+ in the xylem and from there to the whole plant is carried out through different mechanisms already reported as in plants such as NSCCs (nonselective cation channels), HKTs (high-affinity potassium transporters), and LTC1 (low-affinity cation transporter). The NSCCs allow the passive transport of broad ranges of ions by the plasma membrane. The families CNGCs (cyclic-nucleotide-gated channels) and GLRs (ionotropic glutamate receptor homologues) are suggested and involved in Na^+ uptake (Adams and Shin 2014). The compartmentalization of Na^+ in the vacuoles and the exit of it from the cytosol to the external environment are strategies to avoid the excessive accumulation of this cation in the cytosol (Keisham et al. 2018). Both processes are carried out by the CPA1 transporter family (cation/proton antiporter 1). It has been reported that in Arabidopsis, this family is divided into two groups: the NHX and SOS1 exchangers (Pires et al. 2013).

15.7 Vermicompost and Vermiwash

The non-thermophilic biodegradation of organic wastes such as vegetables, animals, industrial, and urban through the interaction of earthworms and microorganisms is known as vermicomposting (Gutiérrez-Miceli et al. 2008). In the vermicomposting process, two main products will be generated, vermicompost and vermiwash (also named as worm-bed leachate). In both products, although in different quantities, many useful elements have been found in addition to the plant nutrients such as vitamins, hormones, and humic substances (García-Gómez et al. 2008).

15.8 Use of Vermicompost to Mitigate Salt Stress in Crops

The vermicompost has positive effects on the growth of plants and the structure of the soil. One of the advantages of the production of this is its positive effect on the environment. Any material, such as vegetable, animal, industrial, and urban waste, can be transformed into beneficial fertilizers through the worm's digestive system through vermicomposting (Ceritoğlu et al. 2018). Vermicompost is an organic compound that can contain beneficial microorganisms and contains humic substances with effects similar to plant growth regulators (Beykkhormizi et al. 2018).

It can also play an effective role in reducing the damage caused by environmental factors in plants, since it has porosity and high levels of macro- and micronutrients and maintains humidity (Abdollah et al. 2015). The use of this as an organic fertilizer has increased in recent years as it is a friendly way to fertilize crops; however, it has also been reported that the application of these or their derivatives can help better tolerate salinity.

In a work carried out on five fennel breeds (*Foeniculum vulgare* Mill) (Urmia, Mashhad, Shiraz, Boushehr, Isfahan), it was demonstrated that the application of vermicompost extracts at 10% (VCE) improves the germination rate and the growth of the plants treated with NaCl (0, 40, 80, 120 mM). The length and dry weight of the roots and shoots of the local breeds of Shiraz, Urmia, Mashhad, and Boushehr treated with NaCl + VCE were higher compared to those that did not apply VCE. High concentrations of Na^+ can suppress K^+ uptake by the roots, being able to be a reason for the adverse effects caused by salinity stress, due to the importance of this ion in plants. When analyzing the Na^+ content in the root and shoots, this increase in the tissues and the K^+ content was reduced under salinity conditions; however, in those varieties to which the VCE treatment was applied, the K^+ content was higher (Beykkhormizi et al. 2018). The vermicompost interaction of sand (0, 100, 10:90, 25:75, 50:50, and 75:25 v/v) and NaCl (20, 40, 60, and 80 mmol L^{-1}) has also been studied on the concentration of minerals and some morphological and physiological characteristics in bean plants (*Phaseolus vulgaris* L. cv., Light Red Kidney). The use of vermicompost enhances the rate of photosynthesis and the content of K^+ and Ca^{2+} in tissues (leaves and roots). Both in conditions of stress and in their absence, it is recommended to use a volume of 10% vermicompost to increase the development of bean plants. Plants with vermicompost (10 and 75%) and with 80 mmol L^{-1} NaCl significantly improved the symptoms caused by salinity (Abdollah et al. 2015).

In studies on marjoram (*Origanum majorana*), Kassem et al. (2018) have shown that the application of vermicompost (0, 3, 4, and 5 ton fed^{-1}) supplemented with (foliar application) silicates (K_2SiO_4 [0, 2.5, 3.75, and 5 mM]) reduces the negative effects of a saline soil (EC = 11.7 dS m^{-1}). On the one hand, when applied separately, it decreases the concentration of Na^+ and proline in the shoots of the plant and improved the growth, the content of nutrients, the percentage of carbohydrates, and the yield of the essential oil. But the combination of both improves these parameters and the nutrient content (N, P, K, and Ca). The best treatment was 3–4 ton fed^{-1} and 3.75 mM K_2SiO_4. These results demonstrated that vermicomposting and the application of silicates could be implemented to damage saline soils and minimize damage due to stress on plants.

Reyes-Pérez et al. (2016) determined the impact of vermicompost humates (0, 1/60 v/v) on the growth and development of two varieties of basil (*Ocimum basilicum* L.), a tolerant (Napoletano) and a sensitive (sweet Genovese) variety to the salinity. These were subjected to three concentrations of NaCl (0, 50, and 100 mM). The results mention the Napoletano variety as the most tolerant with the application of the biostimulant, exhibiting the highest values in all the variables evaluated. The use of vermicompost humates stimulated the morphometric variables stem height and length of the root system, biomass of root, stem, leaf, and the leaf area of basil varieties in salinity conditions, allowing the tolerant variety to improve its emergence and growth and the sensitive variety to increase its tolerance to salt stress.

The benefits of the application of vermicompost and its derivatives in Solanaceae cultivation of economic importance, such as chili, tomato, and potato, have also been demonstrated. Skender et al. (2011) determined the influence of the composition of different substrates (peat moss Sphagnum [brand], vermicompost, and a mixture of peat-vermicompost [50 to 50%]) on the growth parameters of pepper seedlings (*Capsicum annuum* L. cv. cecil F1) during its development in nurseries and subsequent transplant to vermiculite under conditions of salt stress (0, 20, and 50 mM NaCl). The results of the relative growth rate (RGR), the net assimilation rate (NAR), the leaf area ratio (LAR), the specific leaf area (SLA), and the foliar weight fraction (LWF) indicated that the composition of the substrate showed a positive effect on the yield of seedlings in seedbeds and transplantation in the presence of salinity. The vermicompost turned out to be an appropriate substrate for the propagation of seedlings (alone or with the peat).

Although the growth of capsicum seedlings in the seedbeds was slower than those cultivated in peat, they had a greater development during their transplant, especially under salinity conditions.

The vermicompost humates improve the quality and productivity of crops and decrease the use of chemical fertilizers, promoting more sustainable cultural practices (Chiquito-Contreras et al. 2018). Humates are considered as a vegetal biostimulator that contains minerals (Ca, Mg, Na, P, K, N), organic compounds, and beneficial microorganisms. In studies in tomato, Chiquito-Contreras et al. (2018) evaluated the effect of vermicompost humates (1/10, 1/20, 1/30 v/v) on the growth of tomato (*S. lycopersicum*) seedlings grown in saline soils (3.50 dS m^{-1}). The growth parameters evaluated (height, weight of the aerial part, diameter of the stem, and length of the root) were higher in those that contained a higher ratio of vermicompost.

Solanum tuberosum L. (potato) is one of the most important crops in the world. For its cultivation, some farmers use hard water, which diminishes its yield (Pérez-Gómez et al. 2017). Pérez-Gómez et al. (2017) test the use of vermicompost and leachate verifying the effect of vermicompost (300, 580, and 860 g $plant^{-1}$) and its leachates (5, 10, and 15 mL $plant^{-1)}$ on the yield and growth of *Solanum tuberosum* plants and the characteristics of the tubers subjected to salinity stress (15, 20, and 25 mM NaCl) for 6 months. The vermicompost and leachate decrease the symptoms caused by salinity on the development and characteristics of the tubers. The treatments with 580 g $plant^{-1}$ of vermicompost and 15 mL $plant^{-1}$ of vermiwash promote an increase in the diameter of stems and in the height of the plants.

Similarly, the increase in NaCl concentration produced an increase in the content of the °Brix (total soluble sugars) in *S. tuberosum* plants supplemented with vermicompost. This response is associated with mechanisms of tolerance or adaptation of the plant to salinity; Chinsamy et al. report an increase in sucrose content in tomatoes under salinity conditions, which suggests an increase in the enzymatic activity of sucrose phosphate synthase.

The vermicompost that has been used to improve the quality of the fruit has also been studied. Yanan et al. (2018) evaluated the application of this on the richness of the soil and the development and quality of the fruit of strawberry plants (*Fragaria* × *ananassa* Duch cv 'Yanli'). The treatments evaluated were soil (100%, control) and vermicompost/soil in a range of 10%–50% (v/v). As a result, it produced an increase in biomass, height of the plant, and foliar area. In addition, the quality of the fruit was improved (increase in weight, vitamin C content, and soluble sugars). In the same way, the application of vermicompost improves the cation exchange capacity and the enzymatic and microbial activity of the soil. So they conclude that the vermicompost had a positive impact on the quality and development of the strawberry, which they attribute to the increase in the rate of photosynthesis, in the elimination of free radicals, and in the enzymatic activity of the soil. Their results showed that the incorporation of these substrates to the soil improved the growth parameters (increase in biomass) as well as the chlorophyll content under salinity conditions.

Oliva et al. (2008) investigated the used of the sheep manure vermicompost on the survival, development, and photosynthesis of tamarind exposed to salt stress under greenhouse conditions. The plants were cultivated in peat moss added with NaCl (0, 20, 40, 60, and 80 mM) and supplemented with or without of vermicompost (10% [v/v]). The results showed that in the absence of vermicompost, the survival of the plants was 20% at a concentration of 80 mM NaCl and in those supplemented with vermicompost was 85%. In general, in the supplemented treatments, vermicompost did not reduce growth or photosynthetic activity, so in tamarind plants, it limits the symptoms caused by salinity.

15.9 Use of Vermiwash to Mitigate Salt Stress in Crops

Studies using vermiwash show that they have beneficial effects on plant growth. Bidabadi et al. (2017) report an increase in tolerance to salinity (0, 30, and 60 mM NaCl) in pomegranate (*Punica granatum* cv 'Rabab' and 'Malas') by application of vermiwash (foliar). The dry and fresh weight as well as the leaf area of pomegranate plants treated with vermiwash showed a significant increase in both the plants that were in saline conditions or not. The decrease in chlorophyll and photosynthesis caused by NaCl is lower with vermiwash. The Na^+ content increased in the seedlings when increasing the salinity in the treatments without vermiwash. Similarly, the foliar application of vermiwash improves the activity of antioxidant enzymes and reduces electrolyte leakage, oxidative stress, and accumulation of Na^+ in the tissue, suggesting that vermiwash can alleviate the damage caused by the salinity. In the same sense, vermiwash has been used to improve tolerance to salinity in tomato seedlings (*Solanum lycopersicum*, cv Ailsa Craig). Benazzouk et al. (2018)

evaluated the effect of 150 mM NaCl for 7 days, applying 6 mL L^{-1} vermiwash prior to exposure to stress and others during the presence of stress. The application of the vermiwash increased the concentration of Na^+ in the root system (16.9%) and reduced in the aerial part (leaves 21.4%). The previous application of vermiwash followed by salt stress was more efficient than applying it during the stress period, since this way the net photosynthesis, the osmotic adjustment, and the K^+/Na^+ ratio at the end of the treatments were maintained.

Several reports have shown the positive effects of humic acid on crop under salt stress conditions. The exogenous application of humic acid improves plant growth under salt stress conditions because it improves the accumulation of essential nutrients, for example, in bean (*Phaseolus vulgaris* L) and pepper (*Capsicum annum* L) (Cimrin et al. 2010). The use of these was able to increase the vegetative characteristics, yield of pods and P, Ca, K, total proteins, and total soluble phenols in green pods in bean plants (Ahmed et al. 2018).

In a study made with two maize cultivars in a greenhouse experiment, different NaCl concentrations were applied with irrigation water. The one of humic acid (100 mg L^{-1}) was applied by means of a previous treatment to the seeds (imbibition) and of foliar way in the plants. The NaCl (100 mM) caused a reduction in biomass (fresh and dry) of both maize cultivars. Foliar application and seed imbibition of humic acid enhanced the fresh and dry biomass; however, foliar application was more efficient as compared to pretreatment in seed. To explain these results, it was found that electrolyte leakage and therefore the membrane integrity are improved with humic acid application. Also other parameters such as chlorophyll, proline, and Na^+ content in leaf and enzyme activities involved in oxidative stress are factors that could be influencing (Kaya et al. 2018). Other factors in vermicomposting include microorganisms. Many plant-growth-promoting bacteria (PGPB) can mitigate the effects of salinity in plants, for example, species of the genus Serratia, Pseudomonas (Zahir et al. 2009), and Azospirillum (Pereyra et al. 2012) and arbuscular mycorrhizal fungi (Kaya et al. 2009). Therefore, the application of PGPB and humic acid together can mitigate the negative effect of salinity stress (Bacilio et al. 2016).

15.10 Conclusion

Salinity is one of the main types of stress that causes adverse effects on the growth and productivity of crops, taking into account several physiological and metabolic processes. Currently, land for agricultural use is negatively affected by the concentration of salts that accumulate in the soil due to current agricultural practices. This problem can be greater in the drier and hot regions, where the drought periods are very long. An alternative to minimize the damage or improve the tolerance to salinity in crops of agricultural importance is the use of vermicompost and vermiwash; its application in the soil improves the productivity of crops in the presence of salinity. As the vermicompost is obtained by the composting of organic materials using earthworms and because worldwide large amounts of organic waste are produced as animal excreta, domestic, and/or agricultural waste, this could be an option to increase its use in agriculture or in the restoration of degraded soil. Whit this, organic waste would be used in a sustainable way.

References

Abdollah BK, Parvaneh A, Ganjeali A, Parsa M (2015) Effect of vermicompost on some morphological, physiological and biochemical traits of bean (*Phaseolus vulgaris* L.) under salinity stress. J Plant Nutr 39(6):883–893. https://doi.org/10.1080/01904167.2015.1109104

Adams E, Shin R (2014) Transport, signaling, and homeostasis of potassium and sodium in plants. J Integr Plant Biol 56:231–249

Adem GD, Roy SJ, Zhou M, Bowman J, Shabala S (2014) Evaluating contribution of ionic, osmotic and oxidative stress components towards salinity tolerance in barley. BMC Plant Biol 14 (1):113. https://doi.org/10.1186/1471-2229-14-113

Ahmed AHH, Mohamed HFY, Orabiand IOA, EL-Henfy AM (2018) Influence of gamma rays, humic acid and sodium nitroprusside on growth, chemical constituents and fruit quality of snap bean plants under different soil salinity levels. Biosci Res 15(2):575–588

Bacilio M, Moreno M, Bashan Y (2016) Mitigation of negative effects of progressive soil salinity gradients by application of humic acids and inoculation with *Pseudomonas stutzeri* in a salt-tolerant and a salt-susceptible pepper. Appl Soil Ecol 107:394–404

Benazzouk S, Zahr-Eddine D, Stanley L (2018) Assessment of the preventive effect of vermicompost on salinity resistance in tomato (*Solanum lycopersicum* cv. Ailsa Craig). Acta Physiol Plant 40:121

Beykkhormizi A, Siavash HS, Mohammad R, Sarafraz A, Seyed MM, Seyed MMK (2018) Alleviation of salinity stress by vermicompost extract: a comparative study on five fennel landraces. Commun Soil Sci Plant Anal 49(17):1–8. https://doi.org/10.1080/00103624.2018.1499109

Bidabadi SS, Safoora DG, Wright C (2017) Vermicompost leachate reduces some negative effects of salt stress in pomegranate. Int J Recycl Org Waste Agric 6:255–263

Bojórquez-Quintal JE, Echevarría-Machado I, Medina-Lara F, Martínez-Estévez M (2012) Plants challenges in a salinized world: the case of capsicum. Afr J Biotechnol 11:13614–13626

Ceritoğlu M, Sezer Ş, Murat E (2018) Effects of vermicompost on plant growth and soil structure. Selcuk J Agr Food Sci 32(3):607–615

Chiquito-Contreras RG, Reyes-Pérez JJ, Troyo-Diéguez E, Rueda-Puente EO, Torres-Rodríguez JA, Murillo-Amador B (2018) Growth of tomato seedlings (*Solanum lycopersicum* L.) treated with vermicompost humate. Revista Mexicana de Ciencias Agrícolas 20:4187–4197

Chutipaijit S, Cha-um S, Sompornpailin K (2011) High contents of proline and anthocyanin increase protective response to salinity in *Oryza sativa* L. spp. indica. Aust J Crop Sci 5:1191–1198

Cimrin KM, Türkmen Ö, Turan M, Tuncer B (2010) Phosphorus and humic acid application alleviate salinity stress of pepper seedling. Afr J Biotechnol 9:5845–5851

Costantini ACE, Lorenzetti R (2013) Soil degradation processes in the Italian agricultural and forest ecosystems. Ital J Agron 8(4):28

Daliakopoulos IN, Tsanis IK, Koutroulis A, Kourgialas NN, Varouchakis AE, Karatzas GP (2016) The threat of soil salinity: a European scale review. Sci Total Environ 573:727–739

García M, Medina E (2003) Crecimiento y acumulación de prolina en dos genotipos de caña de azúcar sometidos a salinización con cloruro de sodio. Rev Fac Agron 20:168–179

García-Gómez RC, Dendooven L, Gutiérrez-Miceli FA (2008) Vermicomposting leachate (worm tea) as liquid fertilizer for maize (*Zea mays*) forage production. Asian J Plant Sci 7(4):360–367

Gutiérrez-Miceli FA, Moguel-Zamudio M, Abud-Archila M, Gutiérrez-Oliva VF, Dendooven L (2008) Sheep manure vermicompost supplemented with a native diazotrophic bacteria and mycorrhizas for maize cultivation. Bioresour Technol 99:7020–7026

Hasegawa PM, Bressan R, Zhu J, Bohmert H (2000) Plant cellular and molecular responses to high salinity. Annu Rev Plant Physiol Plant Mol Biol 51:463–499

Hayat S, Hayat Q, Alyemeni MN, Wani A, Pichtel S, Ahmad AJ (2012) Role of proline under changing environments: a review. Plant Signal Behav 7(11):1446

Kassem AS, Mohammed KH, Abd El-Rheem M, Elsawy AM, Essa EM (2018) Effect of vermicompost supplemented by foliar application of silicate on marjoram plants grown in saline soil. Int J Adv Sci, Eng Inf Technol 8(4):1029–1035

Kaya C, Ashraf M, Sonmez O, Aydemir S, Tuna AL, Cuilu MA (2009) The influence of arbuscular mycorrhizal colonization on key growth parameters and fruit yield of pepper plants grown at high salinity. Sci Hortic 121:1–6

Kaya C, Akram NA, Ashraf M, Sonmez O (2018) Exogenous application of humic acid mitigates salinity stress in maize (*Zea mays* L.) plants by improving some key physico-biochemical attributes. Cereal Res Attrib 46(1):67–78

Keisham M, Mukherjee S, Bhatla SC (2018) Mechanisms of sodium transport in plants – progresses and challenges. Int J Mol Sci 19(3):647

Levitt J (1980) Responses of plants to environmental stresses: chilling, freezing and high temperature stresses. Academic Publisher, New York, NY

Munns R, Tester M (2008) Mechanisms of salinity tolerance. Annu Rev Plant Biol 59:651–681

Oliva MA, Rincón R, Zenteno E, Pinto A, Dendooven L, Gutiérrez F (2008) Vermicompost role against sodium chloride stress in the growth and photosynthesis in tamarind plantlets (*Tamarindus indica* l.). Gayana Botanica 65(1):10–17

Pereyra NA, García P, Colabelli MN, Barassi CA, Creos CM (2012) A better water status in wheat seedlings induced by Azospirillum under osmotic stress is related to morphological changes in xylem vessels of the coleoptile. App Soil Ecol 53:94–97

Pérez-Gómez JJ, Abud-Archila M, Villalobos-Maldonado JJ, Enciso-Saenz S, Hernández de León H, Ruiz-Valdiviezo VM, Gutiérrez-Miceli FA (2017) Vermicompost and vermiwash minimized the influence of salinity stress on growth parameters in potato plants. Compos Sci Util 25(4):1–6. https://doi.org/10.1080/1065657X.2017.1333932

Pires IS, Negrão S, Pentony MM, Abreu IA, Oliveira MM, Purugganan MD (2013) Different evolutionary histories of two cation/proton exchanger gene families in plants. BMC Plant Biol 13(1):97. https://doi.org/10.1186/1471-2229-13-97

Reyes-Pérez JJ, Murillo-Amador B, Nieto-Garibay A, Troyo-Diéguez E, Rueda-Puente EO, Hernández-Montiel LG, Preciado RP, Beltrán-Morales P, Rodríguez-Félix F, López-Bustamante RJ (2016) Use of humates of vermicompost to reduce the effect of salinity on growth and development of basil (*Ocimum basilicum* L.). Revista Mexicana de Ciencias Agrícolas 7(6):1375–1387

Rivera-Chávez FH, Vázquez-Gálvez G, Castillejo-Álvarez LE, Angoa-Pérez MV, Oyoque-Salcedo G, Mena-Violante HG (2012) Effect of arbuscular mycorrhizal fungi and vermicompost liquid extract on quality of strawberry. Ra Ximhai-Revista de Sociedad, Cultura y Desarrollo Sustentable 8(3):119–130

Skender K, Ismet B, Sallaku G, Balliu A (2011) The influence of vermicompost on plant growth characteristics and stand establishment rate of pepper (*Capsicum annuum* L.) seedlings under saline conditions. J Food Agric Environ 9(1):488–490

Suzuki N, Rivero RM, Shulaev V, Blumwald E, Mittler R (2014) Abiotic and biotic stress combinations. New Phytol 203:32–43

Tavakkoli E, Fatehi F, Conventry S, Rengasamy P, McDonald GK (2011) Additive effects of Na^+ and Cl^- ions on barley growth under salinity stress. J Exp Bot 62:2189–2203

Yanan Z, Junxiang Z, Rui Z, Hongyan D, Zhihong Z (2018) Application of vermicompost improves strawberry growth and quality through increased photosynthesis rate, free radical scavenging and soil enzymatic activity. Sci Hortic 233:132–140

Yancey PH (2001) Water stress, osmolytes and proteins. Amer Zool 41:699–709

Yancey PH (2005) Organic osmolytes as compatible, metabolic, and counteracting cytoprotectants in high osmolarity and other stresses. J Exp Biol 208:2819–2830

Zahir ZA, Ghani U, Naveed M, Nadeem SM, Asghar HN (2009) Comparative effectiveness of pseudomonas and Serratia sp. containing ACC-deaminase for improving growth and yield of wheat (*Triticum aestivum* L.) under salt-stressed conditions. Arch Microbiol 191:415–424
Zhu JK (2001) Plant salt tolerance. Trend Plant Sci 6:66–71

Case Studies on Cultural Eutrophication on Watersheds Around Lakes that Contribute to Toxic Blue-Green Algal Blooms

16

Mary Claire Cooperrider, Lydia Davenport, Sydney Goodwin, Landon Ryden, Nathan Way, and John Korstad

Contents

Abstract

Over the years, farmers have sought fertilizers containing increasing amounts of phosphorus and nitrogen to enhance crop yields. Excess nutrients, particularly phosphorus and nitrogen, are known to stimulate excess growth of blue-green algae (cyanobacteria). Cyanobacteria blooms are also named cyanobacteria harmful algal blooms (CHABS). These harmful blooms are detrimental to people,

M. C. Cooperrider · L. Davenport · S. Goodwin · L. Ryden · N. Way
Oral Roberts University, Tulsa, OK, USA

J. Korstad (✉)
Department of Biology and Renewable Energy, Oral Roberts University, Tulsa, OK, USA
e-mail: jkorstad@oru.edu

K. Bauddh et al. (eds.), *Ecological and Practical Applications for Sustainable Agriculture*, https://doi.org/10.1007/978-981-15-3372-3_16

profits, and aquatic ecosystems and are increasing in abundance. This chapter summarizes the process of cultural eutrophication, the connection to farming, and sustainable means of remediation.

Keywords

Sustainable farming · Nutrient pollution · Lake eutrophication · Lake Erie

16.1 Introduction

Eutrophication is the natural aging process of aquatic ecosystems over time. Cultural eutrophication is the human-caused acceleration of aging, usually resulting from increased nutrient input. This results in algal blooms, which in marine environments may cause red tides and dead zones. Cultural eutrophication often results from nutrient input from agriculture practices, sewage effluent, erosion, industrial waste, and other human factors. Cultural eutrophication leads to poor biotic diversity. Nutrients can enter a body of water in two ways: nonpoint source and point source. A nonpoint source is where pollution enters into water from many small sources such as residential lawns or cattle in ranches that excrete in streams. An identifiable source of water pollution is the point source, which is mainly pipes and drains that come from an industrialized area (ResponSEAble 2019). Cultural eutrophication often leads to low dissolved oxygen in the hypolimnion (deepest water) of lakes. People are directly affected through their health, recreational activities, and food sources. Recently, there has been increased attention to practical ways to prevent and remediate cultural eutrophication.

16.2 Cyanobacteria

Cyanobacteria, also called blue-green algae, resemble other bacteria as they lack membrane-bound organelles, but they differ in having photosynthetic pigments (chlorophyll a, carotenoids, phycobilin, and, in some species, phycoerythrin), making them autotrophs. Many scientists believe this indicates that algae may have evolved from cyanobacteria. In the presence of increased phosphorus and nitrogen, cyanobacteria proliferate and are useful as bioindicators of cultural eutrophication. Certain species of cyanobacteria can proliferate even without an influx of nitrogen in the water through nitrogen fixation (Sharma et al. 2013). Phosphorus, then, is the nutrient that enables bacterial growth. Cyanobacteria are the subject of large bodies of research due to their adaptability to diverse environments that can range from freshwater to saltwater. Ideally, cyanobacteria exist at a pH of 4–5; however, they are capable of survival in acidic solutions as low as pH 2.9. As of 2012, 202 species of cyanobacteria had been identified, and their genomes sequenced (Sharma et al. 2013).

16.3 How Blooms of Cyanobacteria Affect Humans

Although cyanobacteria produce toxins that can negatively affect humans, more research is being done on the possible benefits of some of these cyanotoxins. Sharma et al. (2013) explained that cyanobacteria blooms may release biologically active compounds produced during primary and secondary metabolism that hamper growth of other species of algae, fungi, bacteria, viruses, larvae, protozoa, and helminths in the water in a process called allelopathy. The chemicals can also negatively affect cell mitotic division, cancer growth, coagulation, and hemagglutination properties. However, during an algal bloom, recreational activities in lakes can be limited or prohibited due to the harmful effects and overall unpleasantness of chemicals such as geosmin (*trans*-1,10-dimethyl-*trans*-9-decalol) and 2-methylisoborneol that result in repugnant, musky odor and taste in the water. Some of the more harmful and damaging toxins produced by certain species of cyanobacteria include Microcystin-LR, which can cause abdominal pain, sore throat, vomiting and nausea, diarrhea, blistering around the mouth, and pneumonia; cylindrospermopsin, which causes fever, headache, vomiting, and bloody diarrhea; and anatoxin-a group, which can lead to tingling and burning sensation, numbness, drowsiness, incoherent speech, and salivation (Bennett 2017). Cyanotoxins pollute freshwater reservoirs that humans rely on, and algal blooms result in these water sources being unusable. Cyanotoxins in the water directly affect fish and can be harmful to humans if they consume the affected fish or even if they come into contact with the water. When fish in stock or production ponds and in recreational lakes are affected, people are unable to fish and business profits decrease.

16.4 How Do Blooms of Cyanobacteria Affect Ecosystems?

Sharma et al. (2013) give four main effects of CHABs: "... occurrence of tastes and odors, production of toxins, depletion of dissolved oxygen, and loss of water clarity." After the blue-green algae have bloomed, they begin to die and sink to the lake bottom where they are decomposed by other bacteria. This process of degradation leads to hypoxia, exhausting and depleting the oxygen concentration in the water. This is adverse for aquatic fauna. Cyanobacteria float on the water's surface using gas vesicles, inhibiting sunlight transmittance into the water, which impairs other photosynthetic phytoplankton. All phytoplankton require nitrogen, so the ability of cyanobacteria to conduct nitrogen fixation allows them to outcompete all other algae that are nonnitrogen fixers (Dolman et al. 2012). This drastically changes the species diversity of freshwater ecosystems. Overall, the more that can be accomplished in decreasing CHABs, the more ecosystems will have the opportunity to prosper.

16.5 Current Farming Practices

According to the Census of Agriculture, approximately 940 million acres of farmland existed in the United States in 2002 (https://oceanservice.noaa.gov/education/tutorial_pollution/06operations.html). Over the years, human population growth has led farmers to add more fertilizers to increase their crop yield. Farmers and ranchers in the meat industry have started to put more feed out for their livestock, causing a greater need for higher crop yields.

16.5.1 Fertilizers

The use of fertilizers is one of the main practices leading to cultural eutrophication. This is because fertilizer is used not just for agriculture but also for home lawns, golf courses, and other large areas of land not used for farming. Fertilizers are mainly used to increase nutrients in the soil to increase crop yield. With the increased demand for produce and a lack of space for farming, farmers try to increase their crop yield per square acre of land. To achieve this, they increase the amount of fertilizer they use in their fields. This practice has the desired effect of increased plant biomass but leaves excess amount of nutrients in the ground. When it rains, the runoff water is saturated with the excess nutrients and washes into lakes and rivers in the watershed. The main nutrient in the runoff from fertilized fields is nitrates because they are water soluble (Muir 2012). About 20% of the nitrogen put in fertilizer is lost due to surface runoff (Sources of Eutrophication 2019). Phosphorus is not water soluble, but it does move with the soil; if soil particles are carried into a body of water, then the phosphorus will tag along.

16.5.2 Animal Waste and CAFOs

The same concept of increased population growth leading to higher crop yields can be applied to livestock. To increase the price and amount of product in livestock, farmers and ranchers increase the amount of food they give to their livestock. The leftover food is broken down and taken into the soil. Excess nitrogen and phosphorus runoff again enters aquatic ecosystems. Farms that use the process of overfeeding animal stock to increase production, known as concentrated animal feeding operations (CAFOs), are increasingly more common (Madaan 2016). Even if a farmer is not overfeeding their animals, the waste, if not handled properly, can lead to nutrient runoff and eutrophication. It is just the increased abundance of nutrients that comes from overfeeding along with the built-in side effect of increased amounts of waste that makes overfeeding a major source of cultural eutrophication. Many farmers use animal waste as organic fertilizers. However, the same issues arise with organic fertilizers as with inorganic fertilizers.

16.6 Some Case Studies

16.6.1 Lake Erie Case Study

One prominent example of cultural eutrophication devastating a body of water is Lake Erie in the United States. With the surrounding land used primarily for urban or agricultural purposes, the Lake Erie watershed has experienced severe pollution in recent history. Among the Laurentian Great Lakes, Erie is the most susceptible to algal blooms for a number of reasons. The first is geography. Being the shallowest of the Great Lakes, at an average depth of only 20 m, Lake Erie is the smallest by water volume. It also lies the farthest south among the lakes. The lake has three basins, with the Western Basin having an average depth of only 8 m (United States Environmental Protection Agency 2019). Naturally, these factors result in a higher average water temperature that is ideal for organisms to flourish.

The second reason is land use. The Lake Erie watershed is the most populated at 12.6 million people spanning both the United States and Canada, with 17 population centers larger than 50,000 people (United States Environmental Protection Agency 2019). Outside of urban areas, the region is heavily cultivated for corn and soybean, with the largest percentage of agricultural land among the Great Lakes. As discussed in the previous sections, the ensuing runoff from these developmental practices contributes to nutrients such as nitrogen and phosphorus entering the lake, leading to algal blooms.

The third reason is commercial industry. The Lake Erie region is historically known for its heavy industry and manufacturing, particularly nearby Detroit, Michigan, and Cleveland, Ohio (Karkkainen 2019). The lake also hosts the largest commercial fishing industry, particularly in the Canadian portion. Major shipping ports and areas of tourism further play into making Lake Erie a prime location for cultural eutrophication to occur.

Historically, Lake Erie (Fig. 16.1) has been at the center of discussion in the United States regarding water contamination. Since the arrival of European settlers, the water quality of Lake Erie has fluctuated because of eutrophication and invasive species. Large algal blooms began to form in the 1940s (Andrews 2010). However, cyanobacteria have plagued the area long before. Major point and nonpoint sources of phosphorus resulted in extensive hypoxic areas, especially in the West–Central Basin, in the 1960s and 1970s (Andrews 2010). Industrial waste and sewage from nearby cities were discharged into the tributaries and rivers of Lake Erie. Dead fish began to litter the coast due to the dead zones, causing some sensational news pundits to exclaim "Lake Erie is dead" (Rotman 2019). While the fishing industry was greatly affected by these events, it would not be until the Cuyahoga River Fire of 1969 that the public would become involved (Michigan Environmental Council 2011). This was not the first time Cleveland's river would catch fire. In fact, it had caught fire ten times since 1868, but it was the fire of 1969 that moved the hearts of the nation to combat industrial water pollution (When our Rivers Caught Fire 2011). This event, coupled with the 1969 Santa Barbara oil spill, pushed the then President Richard Nixon to form the Environmental Protection Agency in 1970 (Andrews 2010). Legislation pushed by the EPA such as the Clean Water Act of 1972 helped to

Fig. 16.1 Lake Erie (Mosley and Glassner-Shwayder 2015)

initiate the change people were demanding. Canada and the United States further acted by implementing the Great Lakes Water Quality Agreement (GLWQA) in 1972, which limited the amount of point source phosphorus and solid waste that could be discharged into the Great Lakes. This agreement acknowledged that lowering the discharge of phosphorus was key to controlling excess algal growth (Dolan 1992). As a result, during the late 1970s and 1980s, phosphorus levels in the Great Lakes declined substantially (Reavie and Allinger 2014).

The GLWQA was a success in many regards, to the point that in 1978 the agreement was broadened to target all waste entering the Great Lakes to promote healthier ecosystems (Reavie and Allinger 2014). This agreement became a model for binational agreements and a testament to what two nations can accomplish toward preserving the environment. However, in the late 1990s, blue-green algae began to reappear (Dybas 2019). Many researchers have attributed this rise to increased agricultural runoff from soluble reactive phosphorus, whereas algal blooms in the past were more likely caused by industrial runoff. In recent years, algal blooms in the Western Basin of Lake Erie have become annual occurrences (Dybas 2019). In the summer of 2011, Lake Erie saw its largest algal bloom on record, followed by a severe dead zone in early 2012 (Record Setting Algal Bloom in Lake Erie 2013). In addition to harming the natural ecosystems, these blooms are also harmful for the local urban populations. In 2014, Toledo, Ohio, experienced water shortages when Lake Erie became too contaminated to drink, endangering 500 000 people in the area (Graddy 2018). These stories are becoming all too common, leading many to question what courses of action are necessary to counter this cultural eutrophication.

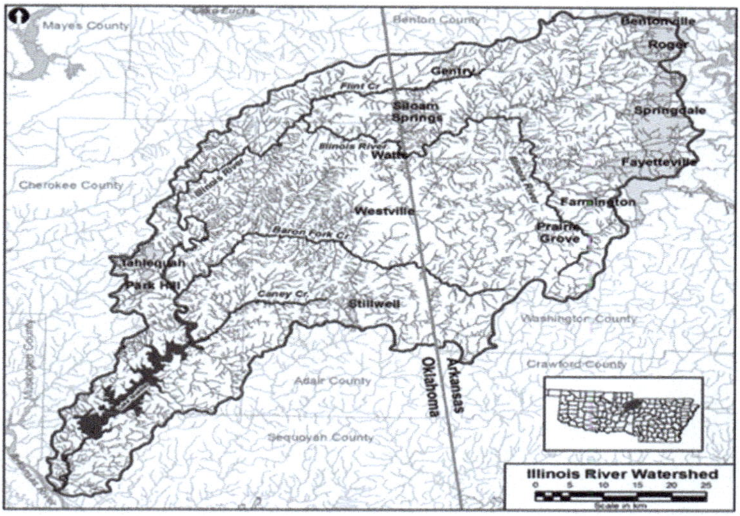

Fig. 16.2 Illinois River watershed (Engel et al. 2013)

Some researchers have argued that these blooms are from the fact that farmers are largely exempt from the Clean Water Act. The 1972 Act only enforces point-source discharges of phosphorus as a violation, whereas the nonpoint-source runoff from agriculture is not enforced (Graddy 2018). Many argue that farmers will require an economic incentive to reform agricultural practices. Farmers in this region receive billions in taxpayer dollars for farming and insurance subsidies. This gives the federal government leverage to instigate change in farming practices. Lake Erie remains a major ecological concern as cultural eutrophication occurs in more bodies of water across the United States.

16.6.2 Illinois River Case Study

Another example of the effects of eutrophication can be found in the Illinois River watershed, which occupies 4 257 km^2 in Oklahoma and Arkansas (Engel et al. 2013). Land use in the area is primarily agricultural, raising poultry and beef cattle, and forest. The northeast portion is also urban. The Illinois River, known as a "Wild and Scenic River", feeds Tenkiller Ferry Reservoir (Lake Tenkiller). The Illinois River is a popular tourist attraction, with people coming for camping, kayaking/ canoeing, hunting, fishing, hiking, horseback riding, wildlife viewing, sightseeing, and more. However, Lake Tenkiller and the Illinois River have experienced diminishing water quality due to eutrophication, causing concerns for many different people, and the Illinois River watershed (Fig. 16.2) has become an example of eutrophication (ResponSEAble 2019). Lake Tenkiller eutrophication is due to an excess of nutrients put into the water, primarily phosphorus (Engel et al. 2013). This addition of phosphorus into the Illinois River comes from both point-source and nonpoint-source pollution. Point-source pollution comes directly from a known source, such as outflow from wastewater treatment plants, septic systems, industries, sewer overflows, end drainage pipes, and more (Illinois River Watershed Partnership 2014). Nonpoint-source pollution can be more difficult to control and usually occurs through rain flow or groundwater, as the pollution eventually makes its way into lakes, rivers, wetlands, and groundwater. Pollutants come from urban and rural sources and include excess fertilizer, herbicides, and insecticides, as well as oil, grease, toxic chemicals from urban runoff, sediment from improperly managed construction sites, and more (Illinois River Watershed Partnership 2014). Poultry waste is another major source of pollution, which is a significant contributor to the increasing phosphorus levels in the Illinois River Watershed. Between 317,500 and 454,000 tons of poultry litter are produced every year, with high concentrations of phosphorus (2% of weight). The preferred method of using this waste is to apply it to the land, and runoff with excess manure that has been applied to the land enters the Illinois River, which empties into Lake Tenkiller. The average amount of phosphorus that enters Lake Tenkiller is 262,000 kg per year (Engel et al. 2013). Similar studies conducted in other agricultural watersheds, such as the Lake Mendota watershed in Wisconsin and watersheds in North Carolina, also reveal animal waste as a significant contributor to phosphorus levels. In the Illinois River watershed, poultry production contributes over 76% of the net annual additions of phosphorus and has been a significant contributor since 1964 (Engel et al. 2013).

Addressing the problem of eutrophication in the watershed would have to include reducing the amount of phosphorus that comes from poultry litter. The impact of phosphorus in the Illinois River watershed not only reaches beyond simply the watershed itself but also impacts the Arkansas River Basin and eventually the Gulf of Mexico (Lamp 2019).

Different solutions have been offered to attempt to reduce phosphorus inputs, specifically from poultry litter. Disputes have arisen between Oklahoma and Arkansas, who have differing economic interests, and a complex legal history has developed (Lamp 2019). In Oklahoma, the economy depends mostly on tourism, along with some poultry and cattle agriculture. An estimated $12 million per year is generated from tourism associated with the Illinois River. Lake Tenkiller produces another $30 million through tourism every year. In Arkansas, poultry production is the major economic source, producing $2 billion from chickens and $297 million from eggs in 2003. A lawsuit was submitted by the State of Oklahoma against the State of Arkansas to the Supreme Court in 1992, for allegedly polluting the Illinois River. In 1997, an agreement was reached to reduce target phosphorus by 40%. In 2002, Oklahoma was displeased with the progress and made a numerical standard of phosphorus levels to be 0.037 mg/L by 2012. In 2005, a lawsuit was filed against 14 different poultry corporations as varying interests continued to clash. In 2007, the city of Tulsa sued Tyson Foods, Inc. and other producers, and Arkansas later filed a lawsuit against Oklahoma, which halted the lawsuit against Tyson Foods, Inc. The

Cherokee Nation became involved and the various stakeholders continue to clash (Lamp 2019).

There are many stakeholders in the Illinois River watershed, each with their own interests. Focht (2007) studied the different stakeholders and came up with five different groups. The "reservationist" perspective is concerned with the threat to water quality and upholding ecological, cultural, and spiritual values. The "stewardship" perspective held by governments and recreationists desires responsible use of the watershed so they can enjoy the available resources. The "traditionalist" perspective pertains to the poultry farmers and ranchers who do not believe that their practices are responsible for any harm and view all of the conflict as a threat to their traditional way of life and their livelihood. The "pessimistic" perspective includes people concerned about the unseen threats to the watershed. Finally, the "conservationists" are concerned for recreation, depending on it for business. All of these views, except the traditionalist view, favor restricting the land use. This issue is very complex, and each perspective has blamed the others.

There are several socioeconomic impacts resulting from taking action or from the lack of action. If no action is taken, the water quality will continue to deteriorate. If the lawsuit is successful, the price of poultry would probably rise, or poultry producers would turn to Mexico for their production, and litigation itself is an expensive process. The outcome of the lawsuit is still pending and could cause significant implications regarding the water quality of the watershed. The role of stakeholders is important, and the public must become involved in the effort if true change will come. Several strategies have been suggested to reduce the pollution of poultry litter into the Illinois River. These include composting, applying chemical treatment, using free range or organic methods, treating litter with alum to reduce phosphorus, finding additional uses for the litter, soil testing, creating buffer zones, finding waste storage, converting cropland to forest, making a riparian establishment, exporting the waste to other areas such as fields that need fertilizer applications (Lamp 2019), and biogas production facilities (https://www.epa.gov/agstar/learning-about-biogas-recovery).

Formal collaboration between the states has taken place and began in 2003 with the first Joint Statement of Principles and Action goal of limiting phosphorus to 1.0 mg/L, and Arkansas enacted legislation regarding poultry litter. The second statement came in 2013 and established a 3-year extension for the goal. A comprehensive watershed plan was established in 2010 (Mitchell et al. 2018). The Recommended "Best Management Practices" developed by the watershed management plan in 2012 included conservation of existing forests, implementing riparian buffers, stream bank fencing, alternative water source, farm ponds, litter transfer, and utilization (Watershed Management Plan 2012: Recommended Best Management Practices 2012). Recently, conservation organizations have been disappointed with the memorandum of agreement that was established in 2018, claiming that it has fallen short of establishing pollution controls and does not provide actual measures of cleanup but only forms new bureaucracies and studies and calls for another watershed plan (STIR, Inc. 2018). This is a complex environmental and social problem involving many different people with many diverse opinions. While

Oklahoma's lawsuit against poultry production companies may not achieve results, it can still motivate the public to act in support of water quality. The most effective method will be a cooperative effort of the various stakeholders.

The Illinois River watershed presents a complex problem, as phosphorus levels continue to rise and diverse groups have separate interests. This issue must be addressed holistically as it affects sustainability—planet, people, and profit. People must come together to improve water quality and health in the watershed, pursuing sustainability by protecting people that live in the region, the profit of various industries involved, and water quality.

16.7 Addressing the Eutrophication Problem

The eutrophication problem is one that will not change without key management systems and treatment methods. With phosphates being the primary nutrient controlling algal growth, there are both internal and external measures that can be taken for the reduction of this nutrient and, thus, for cyanobacterial control. Achieving standard target values and understanding the carrying capacities for total phosphorus in water bodies is imperative to determining which measures of action should be taken. As discussed through the case studies on Lake Erie and the Illinois River, every watershed has its own set of contributing nutrients, landform susceptibility, and location of sensitivity. Unfortunately, due to these differences, there are no universal solutions to eutrophication, and each body of water will require a unique combination of management plans for proper stewardship.

16.7.1 Target Nutrient Values

16.7.1.1 Phosphorus

Phosphorus is one of the essential nutrients for every form of life; however, excess phosphorus overstimulates the growth of algae. When the algae die, bacterial decomposition consumes much of the watershed's dissolved oxygen, limiting the oxygen supply to other aquatic animals. Low-oxygen levels, called hypoxia, can result in "dead zones", where most animal life is void. The total phosphorus levels in eutrophic lakes must be reduced to a level where phosphate is "limiting". According to Chorus and Bartram (1999), the ideal ratio of phosphorus to support cyanobacteria in lakes is 1–100 μg of organic substances. The results from their study can serve as a rough reference for estimating the target nutrient values of phosphorus in other water bodies. Phosphorus, in relation to nitrogen, is consumed by biomass of just under one-seventh the amount consumed from nitrogen. Even so, phosphorus is the most frequent limiting factor for phytoplankton growth in aquatic environments. There is a storage mechanism in cyanobacteria, and other phytoplankton have developed to store up to 75 times the amount of phosphorus needed for normal growth (enough for 3–4 cell divisions) and recycle the phosphate molecules within 5–100 min (Chorus and Bartram 1999). This means that one cell is able to exponentially multiply without the addition of phosphorus. The carrying capacity for phosphorus is determined when the soluble-reactive phosphate is found to be over the detection limits, and surplus is created over the requirements of the cyanobacteria and algae.

16.7.1.2 Nitrogen

Nitrate is constantly being used and reused within natural ecosystems by a variety of plants, including algae. However, an excess of this nutrient in aquatic ecosystems can also lead to algal blooms, hypoxia, and dead zones. The EPA recommends appropriate reference and the target value for nitrate concentrations of 0.001–0.075 mg/L in aquatic ecosystems (Brouwer and Roelofs 2002). Unfortunately, according to the same study, about 60% of the tested streams have concentrations of 2 mg/L or over. Natural sources of ammonia in aquatic ecosystems include fertilizer runoff, industrial wastewater, decomposition of organic matter, and animal (including human) excretion. Ammonia is oxidized to nitrate by bacteria, therefore increasing the nitrate levels. While there are no national criteria for phosphorus or nitrate levels, there is a national criterion established for the concentration of unionized ammonia in surface water. Due to ammonia toxicity to fish, the target values designated by the EPA of ammonia fall between 0.07 and 2.1 mg/L of total ammonia for waters with a pH of 6.5–9.0 and temperatures ranging from 0 to 30 °C (Waite 2008). While phosphorus is the most frequent limiting nutrient for phytoplankton growth in aquatic environments, nitrogen in arid continental regions is the primary limiting nutrient for phytoplankton growth. To determine an aquatic body's carrying capacity for nitrogen, the Redfield ratio can be applied using the concentrations of dissolved inorganic nitrogen.

16.8 Seeking Ways to Sustainably Manage Cultural Eutrophication of Lakes

Possible remediation of cultural eutrophication of lakes is complex due to the various influences ranging from point and nonpoint sources of contributing nutrients in different watersheds that cover only local areas to vast areas that extend hundreds of km. Additional influences include regional soil types, geology, geography, weather, and human activities in the watersheds such as agriculture, industries, logging (forestry), politics, and economy. Various methods to control cultural eutrophication of lakes are discussed below. The reversal of eutrophication is called "oligotrophication" and is understood by limnologists to be analogous to the adage that "an ounce of prevention is worth a pound of cure". In other words, it is more time-consuming, more labor-intensive, and costlier to remediate than the process that led to the current condition. This is not saying that we should give up on lake remediation; that is important, but we should be more preemptive than neglectful.

16.8.1 Nutrient Reduction

Lowering nutrients entering lakes is by far the most important variable that can be controlled to lessen the impact of cultural eutrophication. Nutrient reduction in agricultural watersheds can be reduced by applying only the minimum amount of nutrients (especially nitrogen and phosphorus) needed by the plants (crops) for near-optimum growth so that excess nutrients do not run off into area streams. Nutrients in runoff water can also be lowered by contouring slopes and maintaining plants all year round along the periphery of agricultural land and bordering streams (riparian zone) and lakes (terrestrial area above the littoral zone).

16.8.2 Other Methods to Decrease Algal Blooms (Including Blue-Green Algae)

16.8.2.1 Biomanipulation

Biomanipulation is the intentional exploitation of the lake biota to reduce algal biomass (Shapiro et al. 1975). With biomanipulation, the primary goal is to increase larger herbivorous zooplankton, such as *Daphnia*, which consume more phytoplankton than smaller zooplankton. This is a "top-down" method where piscivorous fish such as pike are added so that planktivorous (zooplankton-feeding) fish such as bluegill sunfish decrease, which increases the percentage of large zooplankton, which overall lowers algal biomass.

16.8.2.2 Flushing

Flushing is the process of cleansing an aquatic environment with high volumes of water. For eutrophication, the flushing process removes some of the dissolved in-lake phosphorus as well as cyanobacteria, assuming the use of high volumes of "clean" water. A side effect that must be considered is the understanding that the phosphorus, while being flushed out, is being flushed into another waterway.

16.8.2.3 Algicides

Algicides can be used to kill prevailing algal blooms. Unfortunately, algicides poison through lysine can pass through drinking water and should only be used as a last resort. Copper sulfate is one of the most effective algicides and is commonly used even with the ecological risks of copper deposits accumulating in sediments and killing beneficial microfauna. Due to this accumulation, harmful species that are resistant to the effects of copper will be introduced and disrupt the naturally residing species. Using algicides is an extreme and effective method of killing algae, yet this is a surface-level solution to a deeper problem.

16.8.2.4 Barley Straw

Recently investigated is the idea of using rotting barley straw to control the growth of cyanobacteria. Barley straw has been found to effectively prevent the growth of certain cyanobacteria such as Microcystis, Anabaena, and Aphanizomenon (Rajabi et al. 2010). Their study ultimately showed that barley straw can be an effective and cost-efficient method for managing the occurrence of cyanobacterial blooms.

16.8.2.5 Sediment Removal

Removal of iron-depleted sediment is a highly effective short-term solution to eutrophication. Unfortunately, this process is not solely capable of transforming a eutrophied ecosystem. Removing the sediment within a water body that does not have a strong buffer system will decrease the overall nutrient levels as well as the base saturation in the sediment (Smolders et al. 2006). Acidification and re-acidification can occur if the water is rich with acid or ammonium and could require a countermeasure of controlling the supply of calcareous groundwater (Rajabi et al. 2010). Ultimately, sediment removal can be costly in process and disposal as it is considered chemical waste due to its accumulation of toxic compounds.

16.8.2.6 Phytoremediation

Phytoremediation is defined as the "the engineered use of green plants to remove, or render harmless, various environmental contaminants such as inorganic and organic compounds" (Ansari et al. 2013). For long-term sustainability, it is necessary to reintroduce native plant systems along lakeshores and rivers (littoral and riparian habitats, respectively) as they protect aquatic environments from eutrophication. Some of the most effective plants include *Eichornia crassipes*, *Salvinia auriculata*, *Typha*, *Glyceria*, and many other species as they remove excess nutrients and stabilize the soil (Ansari et al. 2013). Each plant species and its ratio of success is slightly dependent on the surrounding factors such as season, temperature, pH, species diversity, light intensity, etc. However, with the optimum flora for a specific aquatic system, phytoremediation can be an effective, long-term solution to eutrophication.

16.8.2.7 Fertilizer Requirement

Fertilizers tend to be one of the largest contributors to eutrophication. Minimal or alternative use of fertilizers (especially containing phosphorus and nitrogen) will protect aquatic systems, although it is difficult to control and convince its necessity to those who use it. However, if soil is tested, it is often found to be oversaturated with high P and N and can survive without extra fertilizer application for months at a time. Therefore, testing soil nutrients and responding to their deficit or abundance can allow insight for ideal fertilizer amounts that optimize crop productivity while also minimizing algal growth (i.e., eutrophication).

16.8.2.8 Domestic Wastewater

Domestic wastewater is often ignored as a determining factor of the levels of nutrients in aquatic ecosystems; however, it is commonly the most destructive. Most developed countries have some sort of wastewater management system for their population. In locations that have a seasonal increase in population or communities with a weak waste management system, untreated wastewater (sewage, laundry detergents, and other contaminants in the effluent) greatly contributes to the level of nutrients in aquatic ecosystems. Treatment can include following the World Health Organization's guidelines for wastewater in agriculture and aquaculture (WHO 2006). In protecting aquatic communities from eutrophication, it is important to reduce the import of nutrients through domestic wastewater.

16.8.2.9 Herbivorous Fish

In some situations, herbivorous fish such as tilapia (*Oreochromis*) and Silver Carp (*Hypophthalmichthys*) have been added to lakes to consume and, thus, control blue-green algae. Grass carp (often nonreproducing "triploid" types) can also be introduced to control aquatic macrophyte (plant) growth in the littoral zone.

16.9 Conclusion

The significance of cultural eutrophication is irrefutable. Government agencies, communities, and businesses recognize the problem, and some are implementing plans of action. National Oceanic and Atmospheric Administration (NOAA) is again predicting a harmful blue-green algal bloom for Western Lake Erie in future summers. Cultural eutrophication is antithetical to sustainability in that it does not support the balanced "people, planet, profit" model. People are deprived of freshwater resources and recreational opportunities; ecosystems are directly altered and degraded, and businesses and cities lose income due to expensive water treatment and decreased tourism. Efforts toward avoiding cultural eutrophication result in aiding people in enjoying outdoor activities, the planet prospering and persisting, and businesses continuing to prosper into the future.

References

Andrews RNL (2010) The EPA at 40: an historical perspective. Duke Envtl Law Pol Rev 21:223–258

Ansari AA, Gill SS, Khan FA, Naeem M (2013) Phytoremediation systems for the recovery of nutrients from eutrophic waters. In: Ansari A, Gill A, Singh S (eds) Eutrophication: causes, consequences and control. Springer, Dordrecht

Bennett L (2017) Algae, cyanobacteria blooms and climate change. Climate Change. http://climate.org/wp-content/uploads/2017/05/bennett_algalblooms-1.pdf. Accessed 12 Apr 2019

Brouwer E, Roelofs JGM (2002) Oligotrophication of acidified, nitrogen-saturated soft water lakes after dredging and controlled supply of alkaline water. Arch Hydrobiol 155:83–97

Chorus I, Bartram J (1999) Toxic cyanobacteria in water: a guide to their public health consequences, monitoring, and management. E & FN Spon, London

Dolan D (1992) Elsevier B. V. https://www.sciencedirect.com/science/article/pii/S0380133093712125

Dolman AM, Rücker J, Pick FR, Fastner J, Rohrlack T, Mischke U, Wiedner C (2012) Cyanobacteria and cyanotoxins: the influence of nitrogen versus phosphorus. PLoS One 7(6): e38757

Dybas C (2019). Lake Erie's toxic algal blooms: why is the water turning green? National Science Foundation. https://phys.org/news/2019-04-lake-erie-toxic-algae-blooms.html. Accessed 9 Sept 2019

Engel B, Smith M, Fisher JB, Olsen R, Ahiablame L (2013) Phosphorus mass balance of the Illinois River watershed in Arkansas and Oklahoma. J Water Res Protect 5(6):591–603

Focht W (2007) Assessment and management of policy conflict in the Illinois river watershed in Oklahoma: an application of Q methodology. Int J Pub Admin 25(11):1311–1349

Graddy S (2018) 4 Years since Toledo water crisis, toxic algal blooms on rise across U.S. Environmental Working Group. https://www.ewg.org/release/4-years-toledo-water-crisis-toxic-algal-blooms-rise-across-us. Accessed 14 Sept 2019

Illinois River Watershed Partnership (2012) Watershed management plan 2012: Recommended best management practices. http://irwp.org/assets/PDF/NewFolder/WMP-BMPs-4-categories.pdf. Accessed 20 Apr 2019

Illinois River Watershed Partnership (2014) About IRWP. http://irwp.org/about-irwp/. Accessed 20 Apr 2019

Karkkainen B (2019) The Great Lakes water resources compact and agreement: a model for transboundary governance at subnational scales? Sea Grant Law Policy Journal 9(3):1–20

Lamp J (2019) Poultry litter and water quality in the Illinois River. Personal interview

Madaan S (2016) Causes and effects of eutrophication. Earth Eclipse. www.eartheclipse.com/ecosystem/causes-and-effects-of-eutrophication.html. Accessed 4 Sept 2019

Michalak A, Anderson E, Beletsky D, Boland S, Bosch N, Bridgeman T, Chaffin J, Kyunghwa C, Confesor R, Daloglu I, Depinto J, Evans M, Fahnenstiel G, He L, Ho J, Jenkins L, Johengen T, Kuo K, Laporte E, Zagorski M (2013) Record setting algal bloom in Lake Erie. National Center for Biotechnology Information, U.S. National Library of Medicine. https://www.ncbi.nlm.nih.gov/pmc/articles/PMC3631662/#r5. Accessed 9 Sept 2019

Michigan Environmental Council (2011) When our rivers caught fire. https://www.environmentalcouncil.org/when_our_rivers_caught_fire. Accessed 4 Sept 2019

Mitchell W, Selig G, Woodyard PLLC (2018). Arkansas/Oklahoma memorandum of agreement announced: Illinois River watershed collaboration. JD Supra. https://www.jdsupra.com/legalnews/arkansas-oklahoma-memorandum-of-42044/. Accessed 20 Apr 2019

Mosley C and Glassner-Shwayder K (2015) Tracking harmful algal blooms in Lake Erie. National Oceanic and Atmospheric Administration. https://research.noaa.gov/article/ArtMID/587/ArticleID/635/Tracking-harmful-algal-blooms-in-Lake-Erie. Accessed 4 Sept 2019

Muir P (2012) Human impacts on ecosystems. 1. Eutrophication, Ohio State University. http://people.oregonstate.edu/~muirp/eutrophi.htm. Accessed 4 Sept 2019

Rajabi H, Filizadeh Y, Soltani M, Fotokian MH (2010) The use of barley straw for controlling of cyanobacteria under field application. J Fish Aquat Sci 5:394–401

Reavie E and Allinger L (2014) Research summary: Lake Erie's ecological history as recorded by Phytoplakton. Fondriest Environmental Inc. https://www.lakescientist.com/research-summary-lake-eries-ecological-history-recorded-by-phytoplankton/. Accessed 9 Sept 2019

ResponSEAble (2019) Eutrophication and agriculture – ResponSEAble – protecting the ocean. http://www.responseable.eu/key-stories/eutrophication-and-agriculture. Accessed 20 July 2019

Rotman M (2019) Lake Erie. Cleveland Historical, Center for Public History. https://clevelandhistorical.org/items/show/58. Accessed 4 Sept 2019

Shapiro P, Lamarra V, Lynch M (1975) Biomanipulation: an ecosystem approach to lake restoration. In: Brezonik PL, Fox JL (eds) Water quality management through biological control. University of Florida, Gainesville, FL, pp 85–96

Sharma NK, Rai AK, Stal LJ (2013) Cyanobacteria: an economic perspective. Wiley. https://ebookcentral.proquest.com/lib/oru-ebooks/detail.action?docID=1569020. Accessed 12 Apr 2019

Smolders AJ, Lamers LP, Lucassen EC, Van Der Velde G, Roelofs JG (2006) Internal eutrophication: how it works and what to do about it – a review. Chem Ecol 22(2):93–111. 93-107

STIR, Inc. (2018) Conservation groups object to Illinois River agreement with Arkansas. Save The Illinois River, Inc. https://www.illinoisriver.org/articles/page/250/conservation-groups-object-to-illin ois-river-agreement-with-arkansas. Accessed 20 Apr 2019

United States Environmental Protection Agency (2019) Lake Erie. Environmental Protection Agency. https://www.epa.gov/greatlakes/lake-erie. Accessed 4 Sept 2019

Waite IR (2008) Effects of urbanization on stream ecosystems in the Williamette River basin and surrounding area, Oregon and Washington. U.S. Dept. of the Interior, U.S. Geological Survey, Reston, VA

World Health Organization (2006). Guidelines for the safe use of wastewater, excreta, and greywater. https://www.who.int/water_sanitation_health/wastewater/wwuvol2intro.pdf. Accessed 25 Aug 2019

World Resources Institute (2019) Sources of eutrophication. World Resources Institute. www.wri.org/our-work/project/eutrophication-and-hypoxia/sources-eutrophication#agriculture. Accessed 4 Sept 2019

Agricultural Practices Contributing to Aquatic Dead Zones

17

Anna Bailey, Lauren Meyer, Neilson Pettingell, Madeline Macie, and John Korstad

Contents

Abstract

Aquatic dead zones are areas of decreased biodiversity due to hypoxic conditions. Runoff from farms often contains high levels of nutrients, particularly nitrogen and phosphorus, which flows into larger bodies of water and eventually the ocean and stimulates algal growth. When the algae die, they settle to the bottom and are decomposed by bacteria. The amount of oxygen produced by photosynthesis is exceeded by the amount of oxygen consumed during decomposition, thus leading to hypoxia in the bottom waters. This low oxygen starves organisms such as fish and shellfish, leading to decreased biodiversity or extinction of animal life in the area. Dead zones are linked to agricultural activity and are growing, which poses a threat to both aquatic organisms and humans. In this chapter, the case of the Gulf of Mexico dead zone is elaborated and sustainable solutions are posed to help mitigate this growing problem. The US "baby boom", a marked population increase, directly impacted the farming industry, and larger industrial farms have since been used to supply the buyers' needs. These farms use practices that are not sustainable

A. Bailey · L. Meyer · N. Pettingell · M. Macie
Oral Roberts University, Tulsa, OK, USA

J. Korstad (✉)
Department of Biology and Renewable Energy, Oral Roberts University, Tulsa, OK, USA
e-mail: jkorstad@oru.edu

K. Bauddh et al. (eds.), *Ecological and Practical Applications for Sustainable Agriculture*, https://doi.org/10.1007/978-981-15-3372-3_17

including planting monoculture crops that have poor nutrient absorption, planting crops that alter the soil composition causing increase flooding, and using chemical fertilizers. By identifying the source of the nutrients that cause dead zones, more sustainable solutions have been identified to stop the growing areas of hypoxia. Larger industrial farms are directly influencing the size and continual growth of the dead zones. Sustainable farming practices benefit both the consumers and aquatic ecosystems, and this chapter gives practical solutions for sustainable agriculture.

Keywords

Farming · Nutrients · Dean zones · Sustainable practises

17.1 Introduction

The creation of dead zones by hypoxia, a lack of oxygen in the water, is a growing ecological concern, and preventing their formation takes knowledge and effort from numerous parties. This chapter discusses the nature of dead zones and the sources of the nutrient input, with a primary focus on the influx of nitrogen and phosphorus from plant agriculture. The case study that is further analyzed is the dead zone in the Gulf of Mexico, where the various causes and negative effects of the occurrence are analyzed. Finally, this chapter will look at the ways farmers in the Mississippi River and other watersheds contribute to dead zones and can act more sustainably to prevent dead zone formation.

While the passage of the Clean Water Act in 1972 has made vast improvements in preventing the pollution of toxic materials into the rivers and lakes, two pollutants remain ubiquitous in America's and other countries' bodies of water, that is, nitrogen and phosphorus (Manuel 2014). These "pollutants", are not like those traditionally thought of in the sense they are not inherently toxic. In fact, the presence of nitrogen and phosphorus in ecosystems is essential. Nitrogen and phosphorus are needed to make cellular components such as proteins, DNA, cell membranes, and ATP, the energy currency of the cell. Without access to these, life would cease to exist. However, from an ecological perspective, the overabundance of these nutrients can be disastrous. As levels of nitrogen and phosphorus increase in aquatic systems, excessive algal growth called "blooms" occurs. The presence of algae is normally not negative; these photosynthetic microscopic organisms play an important role, that is, as the base, of aquatic food chains. However, too much of a good thing can become bad.

As the number of algae dramatically increases due to an increased provision of nutrients, termed by scientists as eutrophication, a whole host of problems begin to occur (Simmon 2019). Together, these problems form ecological "dead zones", areas where aquatic life and biodiversity are greatly reduced. The primary reason dead zones occur is that as algal levels increase, oxygen levels dramatically decrease, creating a condition called hypoxia in which nearly all aquatic life is incapable of tolerating (Manuel 2014). There are several ways excessive algal presence leads to a depletion of oxygen. For one, the higher rates of algae reproduction create higher rates of death and decomposition of algae. When this happens, dead algae

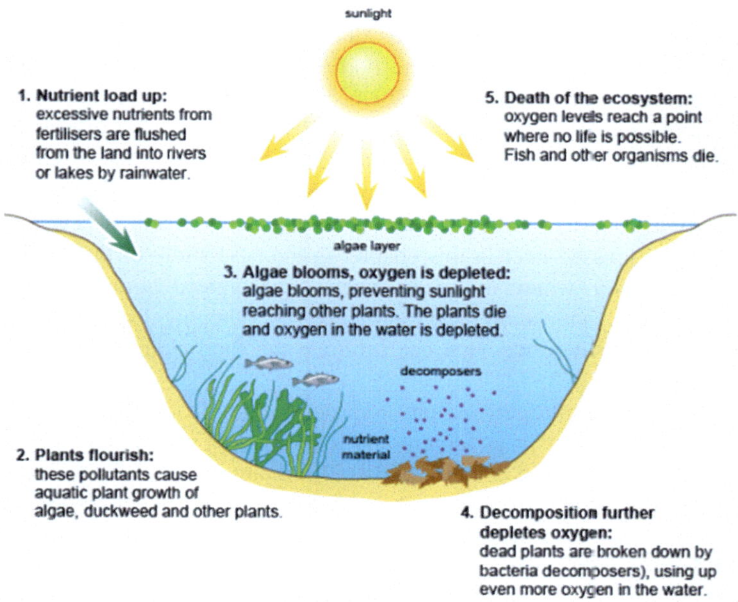

Fig. 17.1 Eutrophication (https://www.mrgscience.com/ess-topic-44-water-pollution.html)

accumulate at the bottom. Bacteria and fungi begin the decomposition process. As decomposition occurs, oxygen is utilized by the heterotrophic decomposers and carbon dioxide is released, depleting the available oxygen for other organisms (Rutledge et al. 2011). Additionally, the scum algae can form on the surface of the water and block out the sunlight necessary for the algae at depth, preventing photosynthesis and the release of oxygen. These plants are often what contribute significant amounts of oxygen to the water (Rutledge et al. 2011). A simplified diagram and explanation of this process is shown in Fig. 17.1.

While the accumulation of algae in general is harmful, there are certain kinds of algae that cause additional problems beyond just contributing to hypoxia. The major algal groups are cyanobacteria, green algae, dinoflagellates, coccolithophores, and diatoms. Increases in phosphorus and nitrogen input affect especially cyanobacteria, which are not as commonly consumed by other marine organisms (Rea and Patel 2017). Because of anthropogenic cultural eutrophication, more than 40% of the world's lakes and reservoirs in America, Asia, and Europe now host favorable conditions for the mass development of cyanobacteria. These organisms can cause a host of problems because "many cyanobacterial species are capable of synthesizing a wide range of odors, noxious compounds, or potent toxins" (Bláha et al. 2009). The toxins that are produced by 25–75% of the cyanobacterial algae pose serious threats to aquatic and human health (Bláha et al. 2009). Cyanobacteria are primarily known for causing problems in freshwater systems; however, certain species of cyanobacteria can thrive in saltwater environments and cause problems.

Dinoflagellate overgrowth in marine bodies of water experiencing dead zones from nutrient input can also pose a problem. Dinoflagellate blooms, or "red tides", have an abundance of one kind of dinoflagellate, *Karenia brevis*, that turns the surface of the water red due to its red pigment (Hall 2018). Like cyanobacteria, these organisms produce toxins that can harm the marine and human life that comes in contact with them (Hall 2018). More of the harmful effects of these specific organisms are discussed later in this chapter in the case study of the dead zone in the Gulf of Mexico, where these organisms have created numerous ecological problems: besides cyanobacteria and dinoflagellates, other algal species are capable of producing toxic substances that can infiltrate the food chain and endanger the surrounding life.

So where is all the nitrogen and phosphorous coming from? There are plenty of natural sources; the nutrients are commonly found in soil, water, and, in the case of nitrogen, the air we breathe (Manuel 2014). There are also anthropogenic sources that are what have been the main culprits behind dead zones. In agriculture, nitrogen and phosphorus runoff comes from chemical fertilizers and manure from animal farms (EPA 2017). Nitrogen input can also come from legume crops planted near watersheds due to the nitrogen fixation that occurs in their root nodules. In urban areas, nutrient pollution can be the result of wastewater treatment plants, failed septic tanks, storm water draining, pet waste, and fertilized home gardens (EPA 2015). Industrial waste and air pollution can also contribute to nutrient problems in dead zones.

17.2 Global Occurrences of Dead Zones

Dead zones are a global phenomenon; according to a 2008 study, there are over 400 dead zone occurrences around the world (Simmon 2019). Some of the most prominent ones reside in the Chesapeake Bay, the Baltic Sea, and the Gulf of Mexico.

The Chesapeake Bay was one of the first dead zones ever discovered and has contained a dead zone since the 1970s (Rutledge et al. 2011). Chesapeake Bay dead zone is caused by excessive levels of nitrogen stemming from urbanization west of the bay and agriculture toward the east. Many factories located in the well-developed area west of Chesapeake eject copious amounts of nitrogen pollution into the air. This type of pollution is responsible for about one-third of all the nitrogen found in the bay. The east side of Chesapeake Bay is a hub for poultry farming, which generates significant amounts of nitrogen-rich manure. Despite numerous efforts by the Chesapeake Bay Foundation to rid the bay of pollutants, Chesapeake Bay dead zone still exists today (Erickson 2018). As of the summer of 2018, researchers from the University of Michigan and the University of Maryland Center for Environmental Science expected the size of Chesapeake Bay dead zones to cover a volume of 7.9 cubic kilometers, with about 22% of that volume predicted to be anoxic, that is, severely deprived of oxygen (Erickson 2018). The study echoes the findings of Rutledge and her team, arguing that Chesapeake Bay dead zone is primarily caused by excess nutrients that flow into the bay. According to Don Scavia, an aquatic ecologist and professor emeritus of sustainability at the University of Michigan, the dead zone in the Chesapeake Bay is significantly larger than the goal set by the Chesapeake Bay Total Maximum Daily Load agreement. A 10-year study of the Chesapeake Bay's fishes by researchers at the Virginia Institute of Marine Science found a drastic decline in species richness, species diversity, and catch rate under the dead zone's low-oxygen conditions. This was especially true for the demersal fish that lived near the bay's floor (Buchheister et al. 2013).

As troubling as the Chesapeake Bay dead zone is, hypoxic conditions in the Baltic Sea are proving to be even more catastrophic. As of the summer of 2018, the dead zone in the Baltic Sea covers an area of about 70,000 square kilometers (Davis 2018), which constitutes about 19% of the entire Baltic Sea. According to scientists from Finland and Germany, the Baltic Sea's dissolved oxygen levels have hit a 1500-year low after a notable decrease in marine oxygen over the last century (Davis 2018). These researchers used sediment cores obtained from the Archipelago Sea, a portion of the Baltic Sea off the coast of Finland, and tested the cores using chemical proxies, as historical oxygen levels cannot be directly measured (Duncombe 2018).

Data collected from the sediment cores revealed that dead zones existed during the Middle Ages, specifically during the Medieval Warm Period from 900 to 1350 AD. Warm water retains less oxygen compared to colder water. Despite hotter than average temperatures during the period, evidence of hypoxia on the coasts was much less pronounced than it is today, indicating that modern-day dead zones are not merely a result of natural causes or even global warming. Human activity in agriculture and industry appears to contribute significantly to hypoxic conditions in the Baltic Sea today.

The study also found evidence of dead zones near the beginning of the twentieth century, and the sediment record shows that the Baltic Sea's oxygen levels have consistently decreased ever since. Even more worrisome is the fact that species that inhabit the seafloor are completely disappearing as a result of the declining oxygen levels. Joonas Virtasalo, a member of the research team from the Geological Survey of Finland, describes an alarming finding: "The seafloor is completely devoid of macrofauna for the first time in our studied record." This discovery has been confirmed through field observations and careful monitoring of the research site. While neighboring pockets of the Archipelago Sea still contain enough oxygen to sustain life, the study warns that oxygen is quickly depleting in coastal waters, particularly near the seafloor. The scientists conclude that human activity, particularly excess nutrient runoff, has magnified the size and severity of dead zones, which are already growing because of climate change.

Despite the study's discouraging findings, efforts are continually being made to reduce fertilizer runoff and combat the rampant growth of dead zones in the Baltic Sea. One such effort in Stockholm, Sweden, to reduce nutrient pollution has proved successful in reversing the dead zones humans have created in the surrounding waters. However, it is clear that much more needs to be accomplished before any

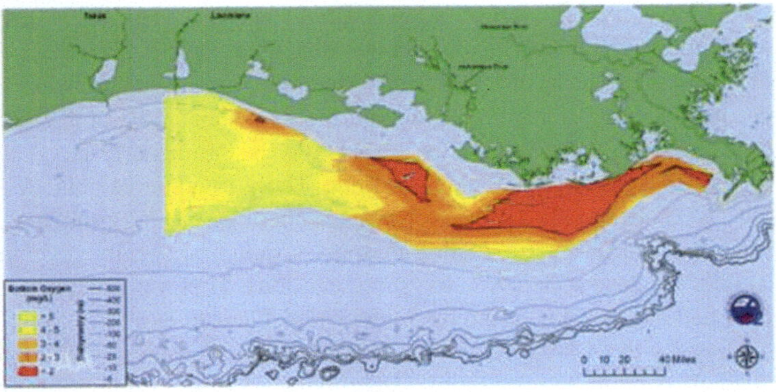

Fig. 17.2 Extent of the Gulf dead zone (USEPA 2019)

change in the dead zone size or severity is achieved. Sami Jokinen, the lead author of the Biogeosciences study, calls for an even greater level of action: "To achieve good ecological status in coastal areas under the projected global warming, the required reduction in nutrient input might be higher than previously thought" (Duncombe 2018).

17.2.1 The Gulf of Mexico Dead Zone

One of the world's largest and most costly dead zones has been the Gulf of Mexico dead zone (Fig. 17.2). This dead zone was first noted by shrimp trappers in the 1950s; however, it was not until the 1970s that scientists began investigating and documenting the phenomenon (Rea and Patel 2017). Today, the Gulf of Mexico has the largest dead zone in the United States. In the summer of 2017, it spanned over 22500 km^2, roughly the size of New Jersey (Belva 2017). Satellite imagery shows a great span of the zone because the nutrients are often carried in the sediments that are dumped into the Gulf, which is another issue by itself (Fig. 17.3). Figure 17.4 depicts where the Neuse River, part of the Mississippi River watershed, meets the ocean waters and shows the severity of the nutrient-containing sediment from the upstream entering the ocean (Chohan 2007). The Gulf of Mexico dead zone occurs every summer due to nutrient pollution from the Mississippi River watershed (EPA 2019a, b). The zone has varied in size throughout history. For instance, in 1988, a time of severe drought, the zone seemed to be nearly gone, but in 1933, a year of intense flooding, the zone expanded nearly 7800 km^2 (Rea and Patel 2017).

Nutrient loading from the Mississippi into the Gulf is a major factor, which increases the number of nitrates and phosphates in the water. A trend can be drawn between the influx of these nutrients and the seasonal increase of phytoplankton production, which is quite large (Courtney and Courtney 2015). The fertilizers used for everything from beets to wheat and corn in Minnesota, Illinois, Kansas, and Iowa are not entirely absorbed by the soil for usage in farming, the excess washes

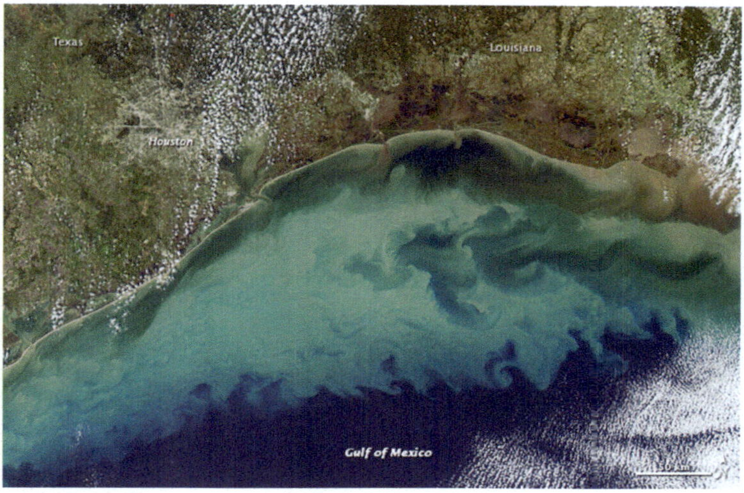

Fig. 17.3 Satellite image of the Gulf of Mexico dead zone in 2009 (NASA Earth Observatory (2019)

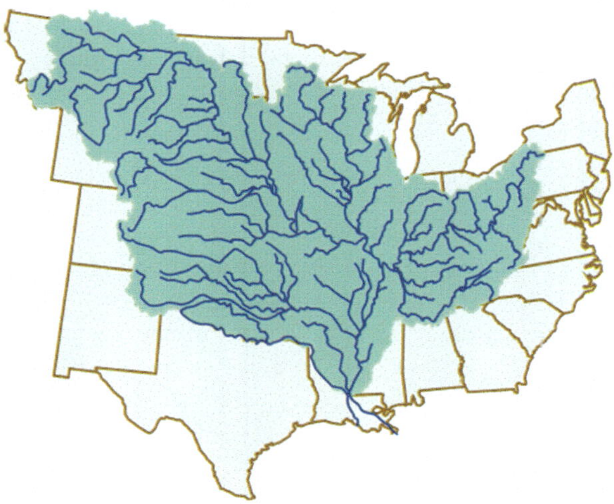

Fig. 17.4 Expanse of the Mississippi River watershed (US Geological Survey 2000)

away with rain and irrigation (Davis 2017). It then flows into a nearby body of water that ultimately flows into the Gulf of Mexico. Warmer temperatures of other bodies of water, including the Great Lakes, have led to prolongation of warmer seasons and increased stratification in the water (Salt Water Sportsman 2019).

The dead zone in the Gulf of Mexico has arisen from an excessive influx of nutrients, and the algae which thrive from these nutrients have 'bloomed', and the increased rate of bacterial decomposition of the algal biomass uses up the oxygen in the water. The leaves deoxygenated, eventually forming hypoxic areas where fish suffocate and die. In addition, after the phytoplankton bloom in response to the nutrient overloading, they die and sink to the bottom of the body of water. The algal blooms on the surface also decrease the light penetration to the Gulf bed and deprive the lower dwelling plants of light. The decomposition of bacteria consumes oxygen and generates carbon dioxide that further stifle the fish (Davis 2017). Hypoxic zones are somewhat mitigated by the tropical storms in the summer in areas such as the Louisiana–Texas shelf, which mix the warmer surface layers of water that are well-oxygenated and stratified with the cooler oxygen-starved bottom layers (Courtney and Courtney 2015). This natural mitigation, however, can only do so much in terms of reduction of the dead zone because of the increase influx of nutrients.

The expanse of the Mississippi River watershed is greater than what many people think. Figure 17.4 shows the expanse of the watershed and how the deleterious effects of nitrogen and phosphorus often occur hundreds or thousands of kilometers from where the nutrients originate (Manuel 2014). These areas contain some of the states with the highest corn, soy, and potato production. Major contributors to this Mississippi River basin (MRB) include the Ohio and Upper Mississippi River basin watersheds that all lead into the Gulf of Mexico. Within the Mississippi River watershed, there are three main sources of nutrient input: urban, natural and agricultural runoff. For nitrogen contributions in this watershed, about 14% are from urban sources, 26% from natural sources, and 60% from agriculture. About 49% of phosphorus originates from agriculture, 29% from urban sources, and the remainder from natural sources (Fig. 17.5).

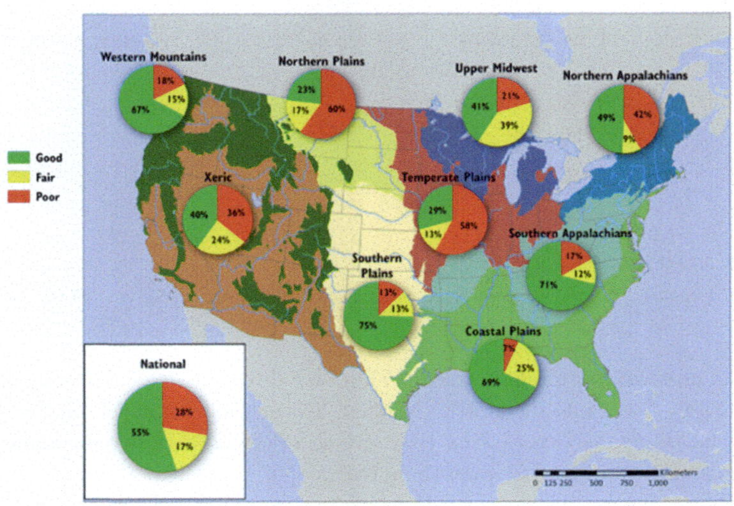

Fig. 17.5 Nitrogen pollution in US rivers and streams (Manuel 2014)

As shown in previous statistics, agriculture in particular has played a major role in nutrient pollution in the Mississippi River watershed, which dumps directly into the dead zone in the Gulf of Mexico. In the mid-to-late 1900s, there was a major population boom in the United States, and with this came an increased demand for crop production and yield (Davis 2017). Due to developments in technology as well as land use changes in agriculture in the United States, agricultural production value has nearly doubled since the baby-boom era (Talis 30). With this increase in value and demand have come new agricultural practices that contribute to nutrient pollution and dead zone formation.

An interesting trend to note in Fig. 17.6 includes the change in farm composition from the 1850s to 2017. There was a large peak that occurred around 1910–1954, which coincides with the population increase (USDA 2015). In addition, the overall land dedicated to farms since the 1930s has stayed roughly the same, if not decreased. What has changed is the average farm size. The data published by the USDA shows the transition from a large quantity of smaller farms to a smaller amount of larger, or industrial, farms. Corporations which run and fund these are not always as environmentally conscious.

One contributing factor to the worsening conditions of nutrient pollution and runoff in the Mississippi River watershed has been the conversion of natural prairie

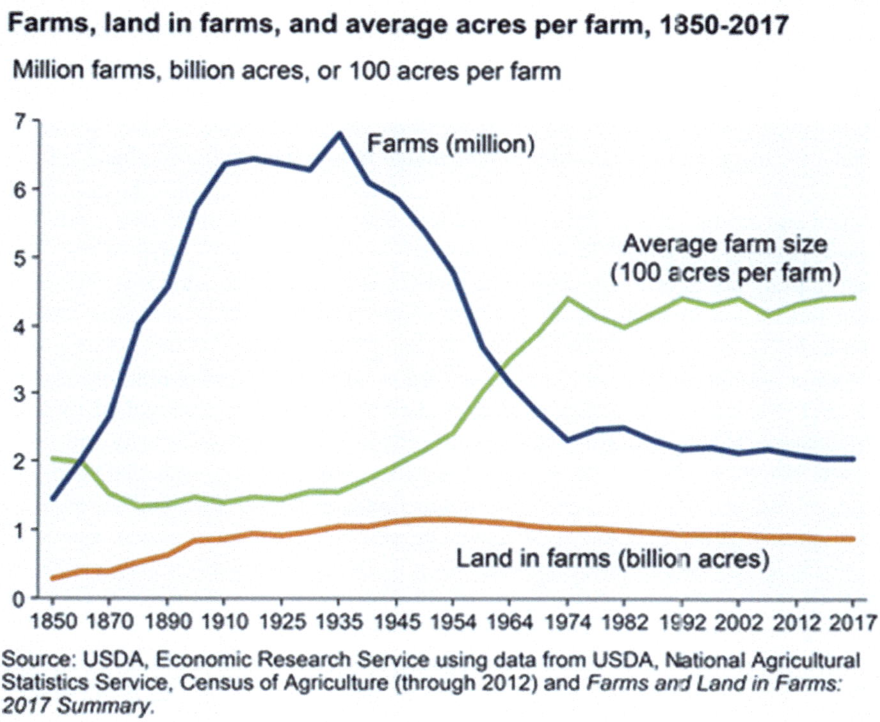

Fig. 17.6 Average acres per farm in the USA (USDA 2015)

lands and forests surrounding the basin into farmland and urban centers. Inorganic farming, pesticides and other chemical fertilizer usage, and the transition from diverse and runoff-adapted polycultures (areas of plant diversity) to more monocultures (areas where one plant (i.e., corn) dominates) are all factors affecting these basins. Agricultural lands, which predominantly influence the basin, almost entirely rely on inorganic farming. Inorganic farming, a practice that causes more nutrient runoff, relies on mineral fertilizers (MF) over biosolids to catalyze the growth of plants (Wallace et al. 2013). MF includes a larger concentration of nutrients than biosolids that are primarily sourced from natural causes. A study done by the University of Reims Champagne-Ardenne tested the effects of switching from inorganic farming to organic, and the results were in favor of organic farming. The study focused on the use of MF or biosolids and the conclusion drawn was that MF contributed to larger nutrient loads despite runoff-preventing irrigation methods being used (Morvan et al. 2018). The study promotes the agricultural practice of organic fertilizer, yet farmers continue to use conventional fertilizer. As fertilizer use has increased sevenfold due to this increasing demand on agriculture, nitrate runoff has tripled, which ends up in the Mississippi River basin (MRB) (Tallis et al. 2019).

Additionally, the root systems of annual agriculture products such as corn and soy are significantly less extensive than those of the prairie grass and trees that previously dominated. A lot of current agricultural land in the United States was prairie land filled with deep-rooted tall grass, now replaced with primarily corn and soybeans. The roots of corn and soybean are significantly shallower and are not present for large portions of the year after harvesting. However, the original prairie grasses and trees maintain their deeper and stronger root systems throughout the year (Gallegos 2017). Larger roots such as those of the native prairie grass have the ability to take up more water and a higher percentage of nutrients into the plants for the process of photosynthesis, leaving less water remaining in the topsoil. The properties of corn and soybeans can inhibit rainfall and water sources from leaching into the ground because of their shallow roots and only having growth for parts of the year, leading to higher rates of flooding. The higher the occurrence of flooding, the more often nutrients will be found in water supplies because of the increased runoffs (Davis 2017).

Newer agricultural processes have also led to soil erosion and overall damage to the land (Morvan et al. 2018). Soil degradation involves the removal of topsoil, which causes a decline in soil quality on the chemical and biological level and leads to soil being transported to undesirable areas. Topsoil loss, especially in the Midwest, is an alarming issue with severe implications for the Gulf of Mexico dead zone. Figures 17.3 and 17.4 show the current degree of sediment dumping occurring in the Gulf. Conventional farming methods that focus on crop yield and short-term cost-effectiveness rather than soil preservation have harmed not only the future of agriculture in America but also the vital Gulf ecosystem.

Increased production of certain agricultural products has also led to rapid nutrient inflow into the Mississippi River basin. Corn in particular has been one major culprit because it allows more nitrate to come off the fields than any other crops, which is why it is termed a “leaky” crop (Davis 2017). Between the 1940s and 1990s, corn production nearly tripled (Davis 2017). One reason for the increase is the multipurpose abilities of corn. Corn can be eaten plain, converted to a sweet corn syrup, or used for a source of food in the meat industry. In the past 70 years, there has been an increase of popularity for corn syrup in the American diet due to its sweetening capabilities. Additionally, increased meat production has indirectly led to increased corn production. In the early 1900s, farmers discovered that feeding cows corn, soy, and grain, rather than grazing on pastureland, can create greater yields and allow a greater supply of meat for the food market (Davis 2017). Instead of land being used solely for crops consumed by humans, an extra sum of land began to be used for crops consumed by cattle. Cassidy et al. (2013) estimates that nearly 36% of farmland today is utilized for feed crop. Ground beef companies such as ConAgra, Iowa Beef Processors, and Cargill, with headquarters located along the Mississippi River watershed, have adapted these practices to produce high quantities of beef. The companies even allocate land near the cattle farms for growing of cattle feed (Davis 2017). In more recent history, an alcohol derived from corn that is added to gasoline, that is, ethanol, has increasingly gained demand, through government subsidies. This has created a demand not just for corn to feed animals and humans but also for fuel.

However, corn is just one of the many crops that have seen a dramatic increase of production due to American demands. As the land in the Mississippi River watershed has been slowly converted from natural ecosystems to massive monocultures laden with chemical fertilizers, pesticides, and shallow-rooted plants, we have seen unprecedented levels of nitrogen and phosphorus in our rivers, streams, and now oceans. But just how harmful have these dead zones been to those near and dependent on the Gulf of Mexico?

The economic effects of hypoxic zones in the Gulf of Mexico are still largely unknown, but Smith et al. (2017) have provided valuable insight into the connection between dead zones and the economy, particularly concerning the shrimp industry. The study was done with the help of the National Oceanic and Atmospheric Administration (NOAA) and received funding from the National Centers for Coastal Ocean Science (NCCOS), a subgroup of the NOAA (2017). Lead author Martin D. Smith, an environmental economics professor at Duke University, explains the purpose of the study:

> Many studies have documented the ecological impacts of hypoxia, but establishing a clear causal link to economic losses in affected fisheries has been elusive. Our study does this by showing how seasonal hypoxia off the Louisiana and Mississippi coasts drives monthly fluctuations in market prices in the Gulf brown shrimp fishery, a major fishery that was once the most valuable in America.

Smith notes that finding a causal relationship between dead zones and fishery profits has proved particularly challenging because of the many feedbacks given by both human and environmental systems that interfere with the control areas, which makes it tough to establish any causality (Smith et al. 2017). The NCCOS study overcomes feedback contamination by using a market counterfactual, in which time series models are used to construct a model of what the shrimp fishery industry would look like if hypoxic conditions were not present in the Gulf of Mexico. Thus, price fluctuations in resource markets such as shrimp are used to measure the impact dead zones have on the economy.

Smith and his team created a counterfactual market by pairing the ecological effects of hypoxic zones with a multi-market model that utilizes shrimp size to determine pricing. When using a naive model, the authors found no evidence that hypoxia causes price changes in shrimp markets because feedbacks in the human and ecological systems contaminate the control areas and obstruct any attempt at establishing causality. However, time series pricing models remove feedback pollution by implementing a counterfactual so causality can be easily measured using correlation coefficients.

As Gulf waters lose oxygen, large shrimp become more expensive as compared with their smaller counterparts. Even small changes in price of shrimp can produce massive ripple effects throughout the Gulf fishing industry, as shrimp accounted for $455,363,744 in 2017 alone, which constituted over 51% of all the revenue generated by the Gulf of Mexico fisheries (NOAA 2017). Dead zones pose not only ecological threats but also economic ones.

Decreased shrimp production is not the only sector harmed by nutrient pollution and dead zones. As mentioned before, nutrient input can lead to the blooms of multiple species of algae that can produce harmful toxins. Red tides formed by dinoflagellates in particular have caused problems in the Gulf of Mexico, especially near the Florida coast. *Kerenia brevis*, one of the algae responsible for red tides that is commonly found near the Florida coast of the Gulf of Mexico, produces a neurotoxin called brevetoxin that disrupts the firing of nerve cells. These toxins bioaccumulate up the food chain and harm many species including fish, seabirds, and marine mammals. For instance, during a 1996 bloom, 149 endangered manatees died off the coast of Florida, and during another bloom from 1987 to 1988, over 740 bottlenose dolphins died after eating contaminated menhaden fish that consume the algae (Hall 2018). One study done later collected many of these deceased animals and found abnormally high levels of brevetoxin in their tissues (Flewelling et al. 2005). The accumulation of these toxins reduces biodiversity, which greatly harms the health and stability of any ecosystem.

These toxins can also harm people. Humans are exposed to algal toxins through food, water, or air. The most common way individuals fall ill is through consuming shellfish, filter feeders that easily accumulate toxins (Fleming et al. 2011). Consuming shellfish with harmful dinoflagellate exposure can lead to neurotoxic shellfish poisoning (NSP) in humans. This severe acute disease often requires emergency room and intensive care during the first hours to prevent respiratory failure (Fleming et al. 2011). It is also concerning that there are currently no studies showing the long-term health effects on low levels of exposure that do not lead to NSP. Additionally, people can acquire respiratory problems from simply breathing near these tides. There have been numerous studies in asthmatics and non-asthmatics linking aerosol exposure to health issues (Fleming et al. 2011). For example, one study conducted in the Gulf area found that lifeguards who were exposed to the tides experienced a statistically significant higher rate of respiratory health problems (Cheng et al. 2005).

Because of these harms to human health, there are many economic implications from having a dead zone present in the Gulf of Mexico. One of the major industries this affects is tourism. These organisms not only frighten beachgoers with illness but also produce an unpleasant odor. This can cause a large strain on coastal industries that depend on visitors coming to the area. Toxic algal blooms can also disrupt the fishing industry. As mentioned, contaminated shellfish can pose a significant health threat, causing fishers to be cautious about catching these creatures, especially during the summer months when the dead zone in the Gulf of Mexico gets bad. One alarming report found that "At present, harmful algal blooms cause about $82 million in economic losses to the seafood, restaurant, and tourism industries each year" (NOAA 2019). As shown, these dead zones are not just areas with unfavorable conditions that make "tree huggers" groan in discontentment: they are areas with serious and widespread harm to both aquatic and human life. They are areas where human health, recreation, seafood supplies, and biodiversity are all jeopardized.

17.3 Sustainable Solutions to Dead Zones

When analyzing how to mitigate dead zones and other ecological concerns, it is important to consider sustainability. In 1987, the UN World Commission declared that "sustainable development is development that meets the needs of the present without compromising the ability of future generations to meet their own needs". Another often used and simplified definition encompasses meeting the needs for the people, planet, and profit. At the heart, sustainability involves finding a balance between each of these factors and will inevitably require give and take from the profit and planet side. Finding sustainable solutions to the Gulf of Mexico and other dead zones will not be a simple process. Nutrient pollution can come from natural, agricultural, municipal, and industrial sources, and tracking the source is not always easy. However, the process of eutrophication and dead zone formation is becoming an increasingly researched topic with more and more data and insight being released each year. There is hope for the Gulf and other dead zones, but it is going to take the input of all parties, including those involved in agriculture, to begin to see change.

One potential method for reducing nutrient pollution, especially in the Gulf, is the practice of sustainable farming. Sustainable farming aims at reducing fertilizer, using minimal tillage, and keeping the land as viable as possible. Two methods farmers use to reduce fertilizer and nutrients in the soil are alternating crops and using leaf biomass assisting composting both on and in the soil (Enger and Smith 2016). There are several benefits to using crop rotation, including improved soil tilth, nutrient cycling, improved soil physical properties, and enhanced weed control (Al-Khaisi 2001). Al-Khaisi, professor of Soil Science at Iowa State University, highlights the importance of crop rotation in nitrogen management, stating how "crop rotation can play a major role in minimizing the potential risk of nitrate leaching to surface and groundwater by enhancing soil N availability by reducing the amount of N fertilizer applied, and by minimizing the potential risk of N leaching". This is because leguminous plants such as soybeans and alfalfa have symbiotic relationships with nitrogen-fixing bacteria in their roots, that is, after harvesting increased amounts of natural nitrogen will be available in the soil the next year. By rotating these sorts of crops with corn and grains, farmers can maximize nitrogen availability in their soil and reduce the need for mineral fertilizer and excess runoff (Al-Khaisi 2001).

Farmers can also use biomass as an organic alternative to chemical fertilizer to minimize nutrient pollution levels. Biomass is simply plant life that can be obtained from outside sources, such as nearby foliage or compost sites. Biomass contains a lot of nutrients essential for crop growth, so when it is distributed on the field, the crops can yield higher numbers (Enger and Smith 2016). In Brussels, Belgium, a man named Sidi Toure struggled to maintain his four hectare farmland and was able to only plant every other year because of lack of soil quality. He was losing significant profit, so he decided to utilize the practice of natural regeneration. Similarly, to the use of leaf biomass, he used the natural plant life of trees on his farmland to maintain the health of the soil. The foliage from the trees that he previously cut down could now grow and be utilized for soil regeneration. Sidi now has an annual rise in yield (State News Service 2019).

While obtaining adequate biomass to make an impact may be difficult for some farmers, there is one natural source of nitrogen and phosphorous that is readily available and abundant in the Midwest, that is, animal manure. In 2015, the US Department of Agriculture reported that animals at feeding operations produce 335 million tons of manure annually in the United States, and that is just in dry weight. This is compared to the 7 million tons Americans produce in dry weight each year. Main (2015) states how, before modern farming, "The manure these animals produced was typically used to fertilize crops grown at the same or nearby farms, which were then used to feed the animals, creating a semi-sustainable feedback loop". This changed with the rising popularity of chemical fertilizers due to greater ease of transporting and handling and their smelling better. Today, almost all manure sits in lagoons to fester, occasionally leaking out and contaminating groundwater and local watersheds. Rather than letting this nutrient-rich substance stagnate in plastic-lined lagoons, agricultural workers could possibly begin to use naturally available manure in their soils (Main 2015). Upon analyzing the effects of manure fertilizer on soil quality, crop growth, and erosion, the findings showed that manure has a significant value in reducing runoff and erosion and has a residual benefit for runoff and erosion that would persist for at least the next three cropping seasons. Additional soil quality benefits were observed regarding soil bulk density, soil organic matter, and pH (Koelsh 2017).

While there is increased demand for more organic farming methods, some farmers will inevitably use inorganic fertilizer due to economic constraints. In this case, a new solution has to be devised using additional farming techniques in order to control the nutrient pollution they will produce. The land could survive with the use of applied fertilizer as long as there is a focus on maintaining the soil well and using land structure techniques to reduce runoff into local watersheds. As soil degrades and erodes due to poor soil management, fertilizer loses its ability to further percolate in the soil and begins to move freely through the land into the waterways. The techniques farmers could use to better maintain their soil include using organic matter to mulch soil, reducing tillage, and making specific landscaping to control runoff (Enger and Smith 2016).

The best solution to decrease soil degradation is to reduce the practice of tilling or to eliminate it altogether. Tillage involves the mechanical agitation and overturning of soil to prepare it for seeding (Enger and Smith 2016). Tilling is normally used for weed suppression, turning over cover crops and forages, seedbed preparation, soil aeration, burying heavy crop residue, leveling the soil, incorporating fertilizer into the root zone, and activating pesticides (Al-Khaisi 2004). While tilling does offer some immediate benefits to farmers, in the long run, it can lead to the breakdown of soil structure and increase erosion. As aforementioned, the significant soil erosion and the dumping of nitrogen- and phosphorus-rich sediment have detrimental effects on the Gulf of Mexico dead zone. However, it can also have a negative impact on farmers, too. Al-Khaisi, professor of soil management at Iowa State University, elaborates on these harms:

> There may even be some initial loss of productivity with moderate levels of erosion. With frequent tillage sustained over a couple years, topsoil loss begins to exceed replacement. In time, the soil is in danger of yield setbacks due to organic matter and nutrient loss as well as the damage done to the soil's physical properties (Al-Khaisi 2004).

Conventional tillage can cause soil erosion, lowering of soil quality by fracturing, and disrupting soil structure (Al-Khaisi 2004). Reducing tillage not only is in the best interest of the Gulf's ecosystems and communities but also is vital for the long-term yields and soil quality in agriculture. According to one UN official, most of the world's topsoil could be eroded within the next 60 years with the current rates of erosion (Arsenault 2014). The USDA's Natural Resources Conservation Service reports that Iowa has lost an average of 17.272 cm of topsoil since 1850 (Smith 2017). This may not seem significant, but to give more perspective, it takes nearly 1000 years to generate just 3 cm of topsoil (Arsenault 2014). A study from the Environmental Working Group by Cox et al. (2011) states how a great deal of this eroded soil in the Midwest, choked with full of N and P from chemical fertilizers, "flows downstream to the Gulf of Mexico, contributing to the largest dead zone in U.S. coastal waters and the second largest in the world." Clay Mitchell of Buckingham, IA, grower and Harvard biomedical engineer estimates that "American farmland has lost half of its topsoil and organic matter, and tremendous productive potential" (Smith 2017). This leads to significant yield decline and a loss of profits for farmers, with some estimations saying that the 17 cm loss of topsoil has led to 10 bushels per acre less of corn product (Smith 2017).

There is also some evidence to show conventional tillage has negative effects on beneficial soil microbe life. In the Appalachian Mountains, several plant and soil scientists studied the effects of tilling versus no tilling on cropland to determine which was better for soil microorganisms. The study revealed that the microbial counts were higher and more diverse in the areas with no tillage because they had healthier soil to thrive in. The experiment included the importance of microbe communities.

> Maintaining the complexity and diversity of soil microorganisms is critical to sustain soil fertility, because soil microbes mediate the biogeochemical cycles of C and N, as well as serve as an important reservoir for plant nutrients (Wang 2017).

Microbes are necessary for land sustainability by acting as natural nutrient producers. The study included that the practice of no tilling reduced the runoff ratio by 64.9% when compared to conventional tilling (Wang 2017).

The structure of the land can also reduce water flow significantly. Two specific styles of farming that control runoff include contour and strip farming. Contour farming tills the land at right angles and creates ridges that act as small dams to prevent water from running down hills (Enger and Smith 2016). Strip farming does not change the land slope and instead focuses on different crop types. In this method, strips of different kinds of crops are alternated because some plants take up more water and nutrients than others. A plant such as corn could be paired with soybeans to retard the flow of water, reducing erosion and allowing more water to sink into the ground (Enger and Smith 2016).

These methods are certainly more sustainable and beneficial for the environment, but there are economic considerations. Many reasons farmers chose inorganic methods and high tillage techniques are for convenience and to maximize profits. When you go to a grocery store and see higher prices for organic foods, it is because organic farming typically costs more. There are several reasons for this, including the costs of more labor, higher fertilizer prices, higher prices to transport organic fertilizer, getting USDA certified, and less government subsidies (Fox News 2012).

Farmers often do not choose to switch to organic farming because it is more expensive than their conventional practices, but according to studies done on comparing the two, this common-held belief may not be entirely accurate. Uri (2000) explains that after analyzing multiple experiments through the compilation of the data, the cost of conventional farming is not any different than conservation farming. He analyzed the yield, fertilizer costs, pesticide costs, seed costs, production costs, and machine costs and came to the conclusion that the costs is dependent on the type of conservation practices and type of crop. Despite the data, he infers that farmers should still utilize conservation practices due to the benefits they have on long-term soil quality. Uri explains "if agricultural output prices are increasing or are expected to increase over time, a more soil-conserving practice will be adopted, all other things equal, since the value of returns associated with soil loss will be higher". He sees the future economic loss if the issue of soil degradation is not addressed. Farmers not motivated to reduce nutrient pollution for the sake of the cause can be encouraged to switch to conservation methods due to the alarming depletion rates of soil.

However, in the long term, there can be benefits to converting to more sustainable farming methods. Like all markets, the more mass-produced organic farming becomes, the lower the costs will be. As environmental concern continues growing in America, government subsidies for low-pollution farming could very well begin to take place. In the long term, conservation tillage could be of significant economic benefit. It is widely known that the rich topsoil in the Midwest is getting rapidly depleted, and as the fruitful soil erodes and dumps into the rivers, profits will follow suit (Enger and Smith 2016).

While government action concerning the dead zone in the Gulf of Mexico has largely been absent, the Environmental Protection Agency (EPA) and the United States Department of Agriculture (USDA 2015) have created numerous initiatives to address both point and nonpoint sources of nutrient pollution. Under the 1972 Clean Water Act, those who release nitrogen and phosphorus from point sources must have received a National Pollutant Discharge Elimination System (NPDES) permit from either the state or the EPA (Fox et al. 2017). This permit places limits on the quantity and concentration to which nutrient pollutants can be released into surface waters. Thus, the NPDES permit ensures that water bodies are clean enough to meet water quality benchmarks by limiting the amount of point source nutrients that slip into water basins.

Compared to point sources, nonpoint sources are much harder to regulate because the nutrient pollution cannot be attributed to a specific location. However, the EPA and USDA are still organizing efforts to limit nutrient pollution from nonpoint sources by providing monetary incentives to farmers and states. The USDA has created several programs to combat nutrient pollution through incentives, including the Conservation Stewardship Program (CSP) and the Environmental Quality Incentives Program (EQIP). Additionally, as a part of the Clean Water Act, the EPA was awarded a $160 million annual grant, which is used to aid pollution reduction efforts at the state level. Despite efforts by the EPA and USDA to reduce nutrient pollution, their programs also create major problems, especially economically. In particular, stricter regulations on the amount of wastewater that can be released drive up the price of utilities, which has implications for taxpayers and may not be worth the added cost (Fox et al. 2017).

Consumers can also play a role in mitigating the effects of dead zones. While it is easy to step back and critique today's agricultural workers and their practices, one must keep in mind that agricultural supply is ultimately tailored to meet the demand of the consumers. While all the people in America need food, our patterns in what we choose to consume and our patterns with consuming ultimately dictate the supply of food. There are many ways the consumers can begin to encourage better farming practices that help our planet, from the fields in Iowa to the depths of the Gulf of Mexico. One way this can be achieved is by choosing to waste less food. According to the United States Department of Agriculture, 31% of food is wasted at the retail and consumer level, the equivalent of $131 billion in food (USDA 2015). Even more food waste occurs at the manufacturing level. They state how "the land, water, labor, energy and other inputs used in producing, processing, transporting, preparing, storing, and disposing of discarded food…generate impacts on the environment that may endanger the long-run health of the planet" (USDA 2015). Another way to promote more sustainable agriculture is to choose to eat food that is organically grown with soil conservation in mind. Creating a greater market demand for food that releases less nitrogen and phosphorous into rivers and causing dead zones will ultimately lead to a greater supply of those foods. If practices are refined and mass-produced, the profit from growing organic can possibly become comparable to inorganic farming. Consuming less meat and avoiding the use of corn-to-ethanol biofuel is another way to mitigate the environmental impacts of our current agricultural system. Currently, 36% of the calories produced by the world's crops are used for animal feed, and only 12% of those feed animals will ultimately contribute to the human diet (Cassidy et al. 2013). Another 4% is used for biofuels such as ethanol. The Institute on the Environment states that even "small shifts in our allocation of crops to animal feed and biofuels could significantly increase global food availability". This increase in food supply could allow farming areas to focus more on environmental quality when growing produce and less on maintaining extremely high yields to meet the American demand. Finally, in America anyone can choose to be an activist and use their voice and voting power to promote more sustainable and environmentally friendly policies and practices.

Dead zones are multifaceted ecological phenomena with widespread implications. Combatting dead zones in the Gulf of Mexico requires input from many parties. Federal legislation may aid in a quicker turnaround for the major contributors to the dead zone. Involvement from fertilizer companies is also required to ensure proper farming practices. A community effort is needed to combat the hypoxic zones along coastal areas, including the mentioned dead zone in the Gulf of Mexico. Population growth requires increased food production, but this does not justify harmful practices. For example, with the wide range of corn products used, from corn syrup to cow feed, this crop and others should be treated with sustainable practices in mind. In addition, extra precaution should be taken in the tributaries that feed into the Mississippi River and, ultimately, the Gulf of Mexico to ensure that green belts are maintained to enhance soil retention and nutrient absorption. All the people living and all the farming and industry groups operating around the Mississippi River watershed should be properly informed about needed sustainable practices.

17.4 Conclusion

The excessive influx of nitrogen and phosphorous from plant agriculture is a contributing factor toward the development of a dead zone. In time, the increased algal blooms decrease the oxygen available in the water and lead to hypoxia that decreases the biodiversity of the affected body of water. These nutrients are shown to carry into areas such as the Gulf of Mexico and the Chesapeake Bay from tributaries that drain runoff containing fertilizer and manure, which can be linked to non-sustainable farming practices. Due to the increase in agriculture since the baby boom, these dead zones have grown, and action needs to be taken to halt this growth. There are many alternative farming practices that can help mitigate this problem such as the use of natural fertilizer instead of chemical, varying the plant type used, crop rotation, the reduction of tilling, and contour and strip farming. In addition, the less food wasted by consumers will impact the suppliers trying to meet the demands. Federal legislation and farmer education can also be influential in the change to more sustainable farming practices. Overall, the dead zones affect human health and quality of life due to toxins leaching into the ground and draining into surface water bodies where they are bioaccumulated by fish. The solution is to change to more sustainable farming methods.

References

Al-Khaisi M (2001) Value of crop rotation in nitrogen management. https://crops.extension.iastate.edu/value-crop-rotation-nitrogen-management. Accessed 21 Apr 2019

Al-Khaisi M (2004) Frequent tillage and its impact on soil quality. https://crops.extension.iastate.edu/frequent-tillage-and-its-impact-soil-quality. Accessed 21 Apr 2019

Arsenault C (2014) Only 60 years of farming left if soil degradation continues. https://www.reuters.com/article/us-food-soil-farming/only-60-years-of-farming-left-if-soil-degradation-continues-idUSKCN0JJ1R920141205. Accessed 21 Apr 2019

Belva K (2017) Gulf of Mexico 'dead zone' is the largest ever measured. http://www.noaa.gov/media-release/gulf-of-mexico-dead-zone-is-largest-ever-measured. Accessed 21 Apr 2019

Bláha L, Babica P, Maršálek B (2009) Toxins produced in cyanobacterial water blooms – toxicity and risks. J Interdiscip Toxicol 2(2):36–41

Buchheister A, Bonzek CF, Gartland J et al (2013) Patterns and drivers of the demersal fish community of Chesapeake Bay. Mar Ecol Prog Ser 481:161–180

Cassidy ES, West PC, Gerber JS et al (2013) Redefining agricultural yields: from tonnes to people nourished per hectare. Environ Res Lett 8(2013):034015. (8p). https://doi.org/10.1088/1748-9326/8/3/034015

Cheng YS, Zhou Y, Irvin CM et al (2005) Characterization of marine aerosol for assessment of human exposure to brevetoxins. Environ Health Perspect 113(5):638–643

Chohan R (2007) Mississippi dead zone. https://www.nasa.gov/vision/earth/environment/dead_zone.html. Accessed 21 Apr 2019

Courtney MW, Courtney JM (2015) Predictions wrong again on dead zone area – Gulf of Mexico gaining resistance to nutrient loading. https://arxiv.org/abs/1307.8064. Accessed 21 Apr 2019

Cox C, Hug A, Bruzelius N (2011) Losing ground. https://static.ewg.org/reports/2010/losingground/pdf/losingground_report.pdf. Accessed 21 Apr 2019

Davis JE (2017) Booms, blooms, and doom: the life of the Gulf of Mexico dead zone. Ala Rev 7 (2):156–170

Davis N (2018) Baltic sea oxygen levels at '1,500-year low due to human activity'. http://theguardian.com/environment/2018/jul/05/baltic-sea-oxygen-levels-at-1500-year-low-due-to-human-activity. Accessed 21 Apr 2019

Duncombe J (2018) Just how anomalous is the vast Baltic Sea dead zone. http://eos.org/articles/just-how-anomalous-is-the-vast-baltic-sea-dead-zone. Accessed 21 Apr 2019

Enger DE, Smith FB (2016) Environmental science a study of interrelationships. McGraw-Hill, New York

EPA (2015) Mississippi river/Gulf of Mexico watershed nutrient task force 2015 report to congress. https://www.epa.gov/sites/production/files/2015-10/documents/htf_report_to_congress_final_-_10.1.15.pdf. Accessed 21 Apr 2019

EPA (2017) Polluted runoff: nonpoint source (NPS) pollution. https://www.epa.gov/nps/basic-information-about-nonpoint-source-nps-pollution. Accessed 21 Apr 2019

EPA (2019a) Nutrient pollution: Sources and solutions. https://www.epa.gov/nutrientpollution/sources-and-solutions. Accessed 21 Apr 2019

EPA (2019b) The effects: dead zones and harmful algal blooms. https://www.epa.gov/nutrientpollution/effects-dead-zones-and-harmful-algal-blooms. Accessed 21 Apr 2019

Erickson J (2018) Chesapeake Bay: larger-than-average summer 'dead zone' forecast for 2018 after wet spring. http://news.umich.edu/chesapeake-bay-larger-than-average-summer-dead-zone-forecast-for-2018-after-wet-spring/. Accessed 21 Apr 2019

Fleming LE, Kirkpatrick B, Backer LC et al (2011) Review of Florida red tide and human health effects. Harmful Algae 10(2):224–233

Flewelling LJ, Naar JP, Abbott JP et al (2005) Brevetoxicosis: red tides and marine mammal mortalities. Nature 435(7043):755–756

Fox R, Krantz A, O'Neill E (2017) Addressing nutrient pollution in our nation's waters: the role of a statewide utility. US Water Alliance. http://uswateralliance.org/sites/uswateralliance.org/files/publications/Environmental_Utility_FINAL.pdf. Accessed 21 Apr 2019

Fox News (2012) 10 reasons organic food is so expensive. https://www.foxnews.com/food-drink/10-reasons-organic-food-is-so-expensive. Accessed 21 Apr 2019

Gallegos J (2017) The Gulf of Mexico dead zone is larger than ever. Here's what to do about it. http://link.galegroup.com/apps/doc/A499860292/BIC?u=oru&sid=BIC&xid=f793bf91. Accessed 7 Apr 2019

Hall D (2018) What exactly is a red tide? https://ocean.si.edu/ocean-life/plants-algae/what-exactly-red-tide. Accessed 7 Apr 2019

Koelsh R (2017) Manure impact on erosion and runoff. https://water.unl.edu/article/animal-manure-management/manure-impact-erosion-and-runoff-0. Accessed 7 Apr 2019

Main D (2015) Two numbers: animal manure a growing headache in America. https://www.newsweek.com/2015/12/18/two-numbers-animal-manure-growing-headache-america-402205.html. Accessed 7 Apr 2019

Manuel J (2014) Nutrient pollution: a persistent threat to waterways. Environ Health Perspect 122 (11):A304–A309

Morvan X, Verbeke S, Laratte AR et al (2018) Impact of recent conversion to organic farming on physical properties and their consequences on runoff, erosion and crusting in a silty soil. Catena 165:398–407

NASA Earth Observatory (2019) Sediment in the Gulf of Mexico. https://earthobservatory.nasa.gov/images/41237/sediment-in-the-gulf-of-mexico. Accessed 7 Apr 2019

NOAA (2017) Annual commercial landing statistics. www.st.nmfs.noaa.gov/commercial-fisheries/commercial-landings/annual-landings/index. Accessed 21 Apr 2019

NOAA (2019) Why do harmful algal blooms occur? https://oceanservice.noaa.gov/facts/why_habs.html. Accessed 7 Apr 2019

Rea WJ, Patel KD (2017) Reversibility of chronic disease and hypersensitivity: treatment options of chemical sensitivity, 7th edn. CRC Press, London

Rutledge K, McDaniel M, Boudreau D et al (2011) Dead zone. Nat Geo. www.nationalgeographic.org/encyclopedia/dead-zone/. Accessed 7 Apr 2019

Salt Water Sportsman (2019) Restored wetlands reduce the Gulf of Mexico dead zone. Salt Water Sportsman 80(2):17. https://www.zinio.com/articlepaywall. Accessed 9 Aug 2019

Simmon R (2019) What causes ocean dead zones? Sci Am. https://www.scientificamerican.com/article/ocean-dead-zones/. Accessed 7 Apr 2019

Smith R (2017) Iowa's black gold is washing away. https://iowastartingline.com/2017/08/14/iowas-black-gold-washing-away/. Accessed 7 Apr 2019

Smith MD, Oglend A, Kirkpatrick J et al (2017) Seafood prices reveal impacts of a major ecological disturbance. PNAS 114(7):1512–1517

State News Service (2019) Surprising benefits of an age-old land regeneration technique. http://link.galegroup.com/apps/doc/A580215438/BIC?u=oru&sid=BIC&xid=9e122fc8. Accessed 20 Apr 2019

Tallis H, Polasky S, Hellmann J et al (2019) Five financial incentives to revive the Gulf of Mexico dead zone and Mississippi basin soils. J Environ Manage 223:30–38

Uri ND (2000) An evaluation of the economic benefits and costs of conservation tillage. Environ Geol 39(3/4):238–248

US Geological Survey (2000) Restoring life to the dead zone: addressing gulf hypoxia, a national problem. https://www.nwrc.usgs.gov/factshts/016-00/016-00.htm. Accessed 21 Apr 2019

USDA (2015) Food waste. https://www.usda.gov/oce/foodwaste/faqs.htm. Accessed 7 Apr 2019

USEPA (2019) Northern Gulf of Mexico hypoxic zone. https://www.epa.gov/ms-htf/northern-gulf-mexico-hypoxic-zone. Accessed 4 Aug 2019

Wallace CB, Burton MG, Hefner SG et al (2013) Effect of preceding rainfall on sediment, nutrients, and bacteria in runoff from biosolids and mineral fertilizer applied to a hayfield in a mountainous region. Agric Water Manag 130:113–118

Wang Y (2017) Long-term no-tillage and organic input management enhanced the diversity and stability of soil microbial community. Sci Total Environ 609:341–347

Mining, Agriculture Change and Resilience: Reflections from Indigenous Knowledge in the Anthropocene

18

Rajanikant Pandey

Contents

Abstract

This chapter highlights the impact of iron mining on the subsistence-based agricultural practices of Ho people in Noamundi, Jharkhand. Historically, the Ho's relationship with natural environment has been shaped by their specific existence as subsistence cultivators and has been termed locally as 'Ho Honko of Hodesum' (the Ho people of Ho country). The unregulated and rapid pace of mining today has transformed this territory into a capitalistic 'resource frontier'. Iron mining has irreversibly damaged the subsistence economy, as well as the eco-cosmological associations of people with the natural environment. The impact of mining on local agricultural economy is being negotiated on an everyday basis. The chapter describes the Ho's subsistence-based agricultural activities to delineate the indigenous endurance mechanism within mining-

R. Pandey (✉)
Department of Tribal Studies, Central University of Jharkhand, Ranchi, Jharkhand, India
e-mail: rajanikant.pandey@cuj.ac.in

K. Bauddh et al. (eds.), *Ecological and Practical Applications for Sustainable Agriculture*, https://doi.org/10.1007/978-981-15-3372-3_18

induced environmental constraints. The chapter narrates the case of mining and agricultural coexistence in the major iron mining regions of India and presents an ethnographic understanding of people's engagements with the mining-driven ecological changes. This ethnographic study attempts to critically outline how the Ho people have shown resilience and capacity to survive amidst socioecological perturbations.

Keywords

Agriculture · The Anthropocene · Iron mining · Indigenous · Resilience

18.1 Introduction

The central concern of this chapter is that subsistence agriculture by local indigenous communities has come into serious competition with the oldest means of industrial pursuit called mining. The conflicts over access and control of natural resources such as land and water are predominant in the mining areas, owing to the adverse impacts of extractive industry on local agricultural practices. Both the mining and agriculture economically provide essential pathways for the development and growth of any developing economies. The pro-community researchers have emphasised on the critical role of agriculture in upholding indigenous well-being and argued that subsistence agriculture provides direct mechanism to local community for their survival. At the same time, mining is essential to promote industrial development and to generate revenues for governments. The states use royalties provided by mining companies for modern infrastructure development and social welfare work. Mining companies through its Corporate Social Responsibility (CSR) further contribute to the community development programmes, build schools and hospitals and make other infrastructural investments in the remote inaccessible tribal territories.

However, the compatibility of mining and peasantry in the tribal areas is becoming controversial, mainly owing to the rise of the idea of the Anthropocene. In this new geological time, 'anthropos has become an ambivalent figure, possessed of an agency scaled up to embrace—and endanger—the whole planet' (Kirksey and Helmreich 2010, p. 549). Human beings can reach every corner of the earth and drive 'climate change, mass extinctions and the large-scale destruction of ecological communities' (Kirksey and Helmreich 2010, p. 549). The rapid pace and overexploitation of the earth's crust through mechanised mining is the biggest example of ecological disruptions happening in the 'Anthropocene'. Mining generates significant 'externalities' such as air, soil and water pollution, which seems to have a direct impact on agricultural cycle. Mining-induced environmental impacts can be perpetual and can make community resources such as water bodies, fertile lands and local vegetation unusable. The mining projects acquire large amounts of land that have been otherwise engaged in cultivation by local tribes. The number of tribal peasants displaced by mining has increased in India and other parts of the world, which has perpetuated extreme poverty, and the peasants are forced to take up limited and unsafe unskilled mining works. The supporters of extractive industry argue that both mining and peasantry are not necessarily incompatible and can coexist. But there is no practical evidence of such coupling, particularly in the developing countries such as India, where the mining industry has weakened the local agricultural development. The Anthropocene questions the role of human agency in damaging the planetary equilibrium.

In this context, the study attempts to discuss a particular case of longitudinal impact of iron ore mining on the indigenous agricultural practices of the Ho community in a predominant tribal district of Jharkhand, a state of India. The identity of West Singhbhum district has emerged as iron belt because it has a century-old history of iron industry in the region. The chapter discusses the indigenous mode of agricultural practices in a tribal village and the ways through which the mining-based industrial developments have challenged the traditional ecology and economy. The much talked about corporate practices of environmental management have failed to identify with the local indigenous agricultural knowledge as well as the environmental changes and subsequent economic disruptions. This chapter ethnographically narrates the tribal wisdom and demonstrates the local understanding of mining-induced changes and ongoing community resilience for their survival amidst industrial chaos.

18.2 Methodology and Sources of Data

The study is a qualitative research based on tools and techniques from anthropological, environmental and historical modes of inquiries. To narrate this theoretical and methodological underpinning, the ethnographic fieldwork was conducted at the Noamundi Block of West Singhbhum district of Jharkhand West Singhbhum district is the largest district of Jharkhand (7224 km^2) with 17% of the total forest and 99% of iron reserve of the state. Majority of this district is covered by the Saranda forest that covers 211,840 acres of land. The forest region itself has 25% iron of the country and 90% iron ore reserve of the Jharkhand State. This iron reserve is of good quality with 58–67% iron content. The low sulphur and phosphorus in ore body makes it the best quality iron for steel manufacturing (Dungdung 2015). Mining operation in Noamundi began in 1917, and it is estimated that since mine prospecting in 1907 to mine operation in 1926, around 237 acres of cultivated *raiyati* lands in Noamundi were converted into extensive opencast iron mining areas (Kumar 2017). Both the mining and logging activities are going on in this region, which led to penetration in the large chunk of forest and agricultural lands of the tribal regions.

The study was conducted in three panchayats, that is, Noamundi Basti, Balijharan and Mahudi of Noamundi Block. Noamundi Basti and Gitillore village were the major villages of my field research (Figs. 18.1 and 18.2). The other villages such as Mahudi, Balijharan, Toretoppa, Gundijoda and Duccasai were also visited during the fieldwork in 2016–2017. In total, around 100 people were interviewed from the Ho village; these people were affected by the mining projects. The other caste communities such as Gupta, Burman, Majhi, Tanti, Karuwa, Brahmins, Gop,

Fig. 18.1 The *bera* land rice field in Gitillor village

Fig. 18.2 The *fertile bera* land ploughing in Noamundi

Gwala, Muslim, Poddar, Soy, Gond, Oraon, Bengali, Odhiya and Luhara were also interviewed. Some important people such as community leaders, forest department officials NGO representatives, etc., were selected purposefully to gather answers for specific queries that emerged during village visits. Many published reports, print and virtual media and the Tata Steel's website provided useful secondary data, which directed the interviews and subjects of informal discussion with the village indigenes.

18.3 Indigenous Mode of Agriculture

The economic life of the Ho people revolves around the occupation of gathering and peasant mode of subsistence cultivation. Almost all the households depend solely on rice cultivation; hence, both the land and forest are the most important assets in their economic life. Subsistence agriculture is the core of the Ho people; hence, their cultural existence can be understood in an ecological landscape. The primary subsistence economy defines the ways of the Ho society's engagement with nature and appropriation of natural resources available to them. The Ho economy has been discussed by scholars (Majumdar 1950; Misra 1987; Das Gupta 1978) in the past; however, as a backdrop for this information, a direct ethnographic inquiry is presented in this section to understand their primary occupation in the present and its coexistence with the recent mining boom in the area.

18.3.1 Indigenous Classification of Land

The undulated topography of the place has created a system of indigenous classification of land according to altitude, fertility and cultivation of different kinds of crops. There are three classes of agricultural land in the Ho village. The lowland is called *bera*, which is considered very fertile and used for paddy cultivation. This lowland or wetland has sufficient moisture and water supply for crops. The lands below the stream bed and usually the lands near the perennial stream bed come under this category. The Ho people have evolved some technical know-how, such as making embankments and providing enough water supplies in the field. Other than the paddy crop, many seasonal vegetables are grown in this kind of land; hence, there is prevalence of multi-cropping. *Pir kundi* is the term for the middle land where moisture content is not enough. The *pir kundi* land is basically rain-fed and therefore has rare facility of irrigation. It is told that if they have good source of water channel nearby the *kundi* land, it would be possible to lift some water to the field through the process of making embankments at different places down the stream. The land is used for cultivating an early variety of rice, and during summers, they grow some vegetables and other crops such as chilli, sugar cane, etc. *Gora* is the highland on the hill slopes, where water is scarce. In the *gora*, pulses such as black gram (*ramba*), yellow gram (*rahi*), red gram (*masuri*) and many other varieties such as khesari (*kansari*), grom gram (*moroy*), green gram (*mungi*), horse gram (*kudti*), etc., and oil seeds such as flax (*tisi*) are grown. The Ho cultivates those varieties of plants in the upland because these plants need less water. In the village, one can see kitchen gardens near every house that are rich in varieties of edible plants and vegetables. They have one or two edible fruits that are planted near their house. Vegetables such as brinjal (*benga*), greens (*aya*) and beans (*simili*) and fruits such as papaya (*bindi daru*), tomato (*biliti*), custard (*mandal*), lemon (*nimbu*), mango (*uli*), tamarind (*jojo*), etc., are found abundantly in the village.

Rice is the most important crop raised by the Ho community, as it is their staple food. There are no irrigation facilities, and crop cycle mostly depends on rain.

Though the rainfall for the last few years has not been good, this region in general gets abundant rainfall during the monsoon. For the production of paddy crop particularly in the lowlands, there is a need for little irrigation and manure. The Ho peasants can provide a detailed description of the major activities involved in agriculture. There are different methods of paddy cultivation; generally, transplantation *(rowai)* and broadcasting (*herr*) method of cultivation are practised in the Ho village. In the *bera* land where paddy is grown, transplanting is preferred. In some cases, broadcasting is done. The lower *bera* land can hold water for longer period, where transplanting is invariably done annually. In the *pir kundi*, mostly broadcasting is practised for rice cultivation. Ploughing starts in the months of April–May, and generally, the *bera* land is ploughed after *pir kundi*.

18.3.2 Pattern of Labour in Agriculture

Manual labour is the primary requirement for agriculture. All the adult male and female members of a family are involved in one or other activity related to land, but women members are exempted from ploughing. There are strict customary taboos on women touching the plough. The seed is spread with hands using a small basket. As the seedling grows up to a foot high, the field is again ploughed. After about 2 weeks, weeding is required in the field. The male and female members do weeding by hand. The land is ploughed three to four times so that the soil becomes very fine. During the ploughing of the field, the villagers would help one another by engaging their ploughs and bullocks in others field. They sought help from those whom they have helped. The owner of the plot whose land is jointly ploughed provides goat meat and rice beer to those who provide him/her assistance. The meat is distributed according to the number of bullocks and men, which have been supplied for helping by any family. Mostly close kin from lineage living nearby are requested for assistance, but sometimes relatives from neighbouring village may come to support. The cash transaction is less accepted in the village, but sometimes people may demand cash if they have urgent need. Those peasants having bigger chunk of land may hire people on the basis of cash transaction. While ploughing the land, some cow/buffalo dung is spread, and leveling is done with a wooden leveler. After some good rain, seedlings from the seeding beds are transplanted in the *bera* land. The harvesting of the paddy is done in November–December. While the paddy is half ripe, the Hos do some multi-cropping by disseminating some pulses and oil seeds in the paddy field. These latter crops will be ready for harvesting around February. In the *pir kundi*, the agricultural cycle is one month earlier than that of the *bera* land. The broadcasting occurs in June and harvesting takes place in October. Multi-cropping of pulses and oil seeds is also done in the *pir kundi* land. Broadcasting in the *gora* is done in the summer months, that is, April–May, and the paddy is ready for harvesting by September. The villagers have recently started complaining about climate change and groundwater depletion because of excessive mining, which is causing difficulty in paddy cultivation in the *gora* land. In the *gora* land, pulses such as green gram with yellow gram or green gram and milo are mixed and sown along with the paddy.

The pulses mature and ripen before harvesting of the paddy, and these pulses are manually removed. Many of the Ho peasants use the *gora* land for cultivating pulses and oil seeds exclusively. A lot of agricultural work such as transplanting, weeding, harvesting, etc., is assigned to women, and they work in a team of six to eight members. While the crop is ready for harvest, it is removed from the field mostly by women. There is a ritual of keeping few bunches of paddies at various spots; every labourer collects a bunch from stalks that are left after work is over. The bunch is generally a token labour cost, which may equal to about 1 kg of paddy.

The paddy stalks are gathered from the field and piled on the threshing ground (*kolom*). They carry the paddy stalks in bundles on their head from the field to the *kolom*. Women are more involved in this work than men. There are few households that keep bullock carts for this work, whereas others carry paddy manually or on yoke. The bullock cart is borrowed if required from others in the village. The threshing ground is generally on a high ground, which is chosen after conducting some rituals, and it is freshly plastered with cow dung before keeping the paddy. After keeping the paddy in the *kolom*, the village men will sleep in the *kolom* with their bows and arrows and guard the crop from animals and malevolent spirits. They will prepare some temporary shed near any tree to look after the crop till the threshing ends, and grains are kept in a house. The Ho people scatter the paddy stalks in the *kolom*, and bullocks or buffaloes are allowed to move above that in circle. They borrow cattle for the purpose of threshing. The members of the *Gop* caste also assist along with their cattle. The rice beer is offered twice or thrice during threshing, and sometimes meat and rice is also offered to all working in the *kolom*. Winnowing is required to remove dust and other grass from the grains. A basket is used for this, and when air is blown, the grain falls at a little distance away from the dust and gets segregated. The grains are stored with joy in big-size straw baskets (*pura*).

Agriculture requires lots of manual labour and sustained attention of family members. The village members have to cooperate with each other while working in the field because agricultural works such as transporting, threshing, harvesting, etc., require more members. Therefore, strong community sentiment is visible in the different spheres of agricultural life. The cultural core of subsistence agriculture and simple technology determine the cooperative behaviour patterns amongst the Ho people (Steward 1955). Agricultural tools such as ploughshare (*panl*), sickles (*datrom*), axe (*hanke*), spade (*kullam*), etc., are usually purchased in cash from the local market in Noamundi. In the nearby villages, there are blacksmiths (*luhara*) or carpenters (*kamar*) who also help in producing the tools, but their primary service is to sharpen the agriculture tools of the villages as required in exchange of yearly payment in paddy. Generally, the carpenters visit the village, field and *kolom* of clients to provide their service on the spot. Other household articles and utensils made of wood, bamboo, cane, brass, aluminum and bell metal may also get the attention of the carpenters or blacksmiths.

Generally, every household in the village has some amount of cultivable land; majority of them possess four to 8 *bighas* (approximately 1600 yard^2) of land. However, they do not keep track of the exact measurements of their lands and refer to their possessed lands as one field, two fields, etc. Further, the landless Ho family can temporarily use others' land by taking lease or mortgage. However, there is no information regarding the selling and purchasing of lands in the village. The practice of lease and mortgage now involves monetary payment. In leasing 1 *bigha* of *bera* land, the usual amount per year varies from Rs. 5000 to 10,000 depending on the quality of land. One *bigha* of *pir kundi* land is leased at Rs. 5000 to 8000. The tenure of lease is usually 5years, after which the land is automatically returned to the owner. During the lease period, the tenant cultivates the land and enjoys the product. Generally, all these transactions are oral, and everyone in the village knows the nature of such transactions, and community sanction is imposed automatically. However, nowadays, these transactions are made in writing with revenue stamps in the presence of five or six witnesses. In case money is taken by the owner of the land on loan with some interest, he keeps his land with the creditor peasant as mortgage. The land is returned to the owner as soon as the money is recovered. To minimise the pressure on land due to population growth, the Hos have started to increase the agricultural production by using chemical fertilisers provided from the block office. Only affluent peasants can afford to reclaim wasteland considering that the forest departments keep a close watch on those lands. Moreover, there is a tendency of working rigorously on the *gora* land for removing the top earth to level it as *bera* land. There is rising pressure on agricultural land in the village. The Gitillore village started with some five families, and within a span of 60 years, there are presently 47 families. The land and people ratio is decreasing, and new wastelands are being converted into agricultural lands. However, strict forest rules do not allow such opportunities beyond a point. The limited arable land in the village cannot sustain the increasing population properly. The surplus labour has to be provided with other means of livelihood and income. To minimise the pressure on land due to population growth, the Hos have started to increase the agricultural production by using chemical fertilisers and improved seeds provided from the block office.

The agricultural production is mostly dependent upon rain, and there are limited opportunities for irrigation in the village. When there is good rain, 2 *puras* or approximately 3 quintals of paddy can be obtained from about a *bigha* of *bera* land. In 1 *bigha pir kundi* land, around 1 *pura* paddy is produced. The *gora* land is less fertile with limited yield, which has also stopped these days. Land cannot provide full employment to the working population. The surplus paddy production is now becoming difficult, and people have to depend on annual crops to survive. It is also observed that owing to the lack of surplus, they cannot afford basic necessities of life.

Villagers with comparatively little land venture into subsidiary income such as rice beer (*diyang*) selling, manual or construction labour, MGNREGA work, tailoring, cycle repairing service, etc., which may be available in the Noamundi market. The rearing of the lac and silkworm is a very old institution amongst the Hos of Singhbhum. The local businessmen of other castes dominate the business of lac and silkworm, which they cheaply purchase from the villagers. Nowadays, this trade is also abandoned because of mining-induced environmental changes, deforestation and shrinking markets for such products. The people look at mining as an opportunity for employment and income; however, the jobs in mining are limited. Few educated youths are working in banks, schools and shops on daily wages. Very few members have moved out of the village to urban and industrial towns for employment. Some of the Ho youths also engage in poultry, vegetable gardening, goat breeding and lac and silkworm rearing. By selling these products, they improve their income to some extent. The Noamundi mining town at least offers good marketing facilities for the inhabitants of the adjacent tribal villages that supply the town people with different commodities such as *mahua*, rice beer, vegetables, milk, earthen pots, woods, etc. There is great demand for vegetables in the township, and many villages including the Ho have taken up vegetable gardening, as it requires relatively less labour and attention.

18.4 Mining-Induced Changes in Agricultural Lifeworld

Mining started in Noamundi in 1917, and even after more than a century of mining activities in this resource frontier, agriculture is still the basic occupation of the Ho people. However, agriculture today hardly provides enough to meet the basic monetary requirements to fulfill other purposes. The increasing population and decreasing size of land holding are also posing threat to the dependence on agriculture. It is also observed that women are involved in the agricultural work, and they consider agriculture as the main or the only occupation beyond household chores. The Hos believe that agricultural production has steeply decreased because of the changing rain pattern and other environmental fluctuations induced by rampant mining. The religious spirits (*Bongas*) are also unhappy with rampant mining in the area. A fairly educated mining worker who also cultivates his land remarked, 'In olden days, *Bongas* used to hear our prayers and gave us abundant rain but now-a-day due to all sorts of vices, He must be angry with use and do not pay heed to our prayers. For the last 5 years the production has suffered a lot in this area due to less rain, and the cultivators have to go through a tough time'. Nevertheless, even after several years of mining in the area, a Ho still considers land as the chief source of their survival because outsiders (*dikus*) control mining, and very limited opportunity is available to them. They believe that mining will stop and they will have no occupation. Therefore, they feel afraid to switch solely to mining jobs and try to preserve their hold on agricultural and forest lands. The villagers living in the neighbourhood of Noamundi mines have observed that the agricultural production in the villages within a radius of about 10 km has gone down because of the damage caused by the red oxide dust emitted from the mines. This is also one of the reasons for diversion of attention from the fields, which has caused considerable loss in agriculture.

The respectful employment opportunity in the mining is limited and needs sufficient skilled training certificates from educational institutions. Owing to the lack of skill training and degrees, there is little desire amongst younger generation to move into the unskilled mining labour and face daily abuses of labour contractors.

Many Ho men and women were employed during the early years of manual mining, but the mechanisation of mines has made those manual labour works redundant. Huge occupational changes have occurred in the mining town owing to the impact of industrialisation; the migrants from neighbouring states such as Odisha, West Bengal and Bihar have reaped the benefits of mining jobs and associated opportunities in transportation, etc. However, the indigenous villagers continue to survive mainly on agriculture and by combining the traditional occupation with new opportunities emanating from mining and resultant urbanisation.

The mining town bears the outlook of urban dwelling and quite a modern lifestyle, but the desired modernity and urbanity are invisible in the local tribe lifeworld. The Noamundi has been declared census town against the wishes of tribal leaders, as they know that they will have to pay township like taxes without actually experiencing any benefits of town life. However, the market has reached close to the tribal villages where they get many modern goods and amenities. Traditionally, the tribal weekly *haat* was a chief place of economic exchange, and now, the local market in Noamundi is open all 7 days. The reciprocity-based weekly markets are still relevant for tribal people, though the mode of exchange is mostly monetary transactions. The system of barter or reciprocal exchange is limited within village affairs, and money is gradually taking precedence in village transactions. Sometimes salt sellers, ornaments sellers and cloth merchants exchange their products for grains, oil seeds and agricultural products of the Ho community. The Ho indigenes also collect dry branches of trees and make bundles of firewood and sell them in the market amidst restrictions from forest departments.

Rice is the staple food of the Ho community. They eat unhusked rice along with fried leaves of various kinds; they also consume onion, garlic, chilies and a pinch of salt with it. The pulses are rarely used, and vegetables are boiled with turmeric and chilli. Some educated women know how to prepare rice cakes. The Ho people eat meats of fish, sheep, goat, hare, deer, fowl, pigeon, duck, goose, etc. with great contentment. Eating beef and pork is not a taboo, but they generally avoid eating these meats. The most important food items are homemade unfiltered rice beer (*diyang*) and clear-filtered rice beer (*rassi*). It provides instant energy and is consumed by everybody irrespective of their sex and age. For the fermentation of the rice beer, the roots of jungle shrubs and herbs called *ranu* are used. The *diyang* not only has food value but also has religious significance since it is used in almost every ritual of the Ho community. The *mahua* leaf is also used for making drinks. The growth of mining town and associated market has triggered some changes in food habits, as they now have access to different kinds of condiments, oil, beverages, etc.. The availability of market-based liquor has affected tribal labourers, as they have started to consume more alcohol. Mining has several negative effects on the local subsistence agriculture and communities. The local Ho people strongly believe that 'mine eat us, agriculture feed us' (*Khadan humko khata hai, kheti humko khilata hai*). Upon inquiring about their unwillingness to work in mines, one peasant stated that 'mining is for time-being, what we will do when ore ends? The cultivation of rice can feed our generations, so we prefer agriculture'. The tribals describe the impact of iron mining on local agriculture, environment and resources as it is locally conceived, conceptualised and negotiated to uphold the indigenous mode of agricultural existence. The discussion below attempts to sum up the concerns of local peasants on mining-induced changes in their lifeworld.

18.4.1 Loss of Land

A precise estimate of the amount of agricultural land lost to mining in Noamundi is not readily available. The mining requires land for mining site, roads, railways and ropeways for mineral transport, townships for housing miners and manager, infrastructure for administrative purposes, land for stockyard and preliminary processing operations, etc. The land required for the direct engagement in mining is very less compared to the total land affected by mining. The actual land utilised for mining activities is many times larger than the leased area. People assert that iron and other mining and allied activities going on in the Noamundi have caused severe damage to the land resources of the area. It is observed that majority of the land area in Noamundi was covered by dense vegetation, which was being eliminated; hence, deforestation is obvious side effect of mining. Many unregulated mines are abandoned without being covered and stabilised, and one can see open pits in the forest areas of Noamundi Block. Illegal mining starts after big companies abandon mines, which is more dangerous as the left-out poor-quality ores are illegally and unscientifically removed. These ores contain less mineral and more sludge that ultimately pollute the land. The people think that with end of land, their life will also end someday.

Mining exploration in Noamundi has been detrimental to the fertility of the land and has caused great soil erosion. The soil profile was also altered because of opencast mining operations. The extractive operations like shifting of mining overburden and reject dumps have caused severe soil erosion and degraded the productive capacity of the lands in the area. Excessive mining in the region has rendered the tribal lands unsafe for habitation, agriculture and grazing. It has also affected the fertility of the land, and growing of foreign variety grass is not good for the local vegetation. Many of the villagers have countered the claims made by mining companies about the reclamation of lands and abandoned mines. They retort that reclamation is false claim, as land once lost will never be as fertile as it used to be.

Many peasants working in the land are aware of the environmental impact of iron mining and ore washeries. The excess red water coming from mines, which is washed away in the lowland, causes a layer of 4–5 inches over the topsoil, thus making the *bera* land highly infertile. The Ho peasants believe that the *gora* lands have become completely useless owing to the excess presence of iron dust. The mining also impacts the topography and soil cover of the area. Due to digging of opencast mines, fertile topsoil is lost forever, and fertility of the soil cannot be scientifically achieved in the near future. The villagers are mainly worried about the excessive dumping of overburden rock mass in the form of large heaps without any permit of the *Gram Sabha* or consent of the people. The soils removed for the mining and dumped elsewhere are of no use for the villagers, as they cannot even utilise it for other purposes without the company's permission.

18.4.2 Air and Water Contamination and Scarcity

Air pollution is the most obvious impact of iron mining. The pervasiveness of red oxide dust is posing serious challenges to the quality of air available to the local population, including the tribal people, for breathing. People have been informed over the years about the serious side effects of red oxide dust typical to iron ore mine area. The local narratives about air pollution are filled with critical claims about the increasing problems of lung and respiratory diseases. The people in Noamundi substantiate their claim not through any scientific medical data but through their 'social memory' of pristine past (see Jacka 2015, p.163). Many village respondents told the researcher that during the initial years of limited amount of mining, serious air pollution was never experienced, but in recent years, the reckless pace of mining has enormously magnified the amount of dust in the air. The frequent movement of vehicles has further enhanced air pollution.

The villagers are more concerned about the pollution of water bodies that are the only source for their daily water use. The impacts of mining on water bodies are obviously easily perceived as the taste and colour of the water bodies have drastically changed. They are convinced about the fact that water is not as fresh as it used to be, even though the local company keeps on maintaining that water pollution is scientifically 'within the prescribed limit'. Water issues are completely linked with agriculture, as local irrigation is dependent on either rain or some small water channels that are available in the vicinity of village. The major challenge according to the community leader, Laxmi Suren, is the decreasing groundwater level. She stated that 'local wells are drying just because of deep mining and digging of deep water channels near mines'. Mining wastes from washeries contaminate the major local streams such as Balijhore Nallah and Murga Nallah. The major mining corporation established a water park, which has decelerated the natural flow of water downstream where villagers have access to water. The red oxide dust in the air settles down on open natural water bodies and makes the water further polluted. The company ignored the villager's claim about the negative impacts of building the park, since it is obvious to the villagers living along the water stream. A Ho youth from Duccasai specified that 'building of rainwater harvesting unit and park in the way of natural flow of the Balijhore Nallah in the mining area has affected the downstream movement of water that reaches the village. The Nallah has become just a *Nali* (water pipe) and the dust and toxic substances carried by rainwater enter the Nallah, which in turn affect the only source of drainage in the villages'. Mining plants have washeries that are claimed to have zero-discharge systems, but according to the local community leaders, lots of effluents and suspended particles flow down the water stream that reaches the only source of water consumed by the villagers. According to the villagers, dirty water in the nallah is adversely affecting fishes, thereby affecting the occasional dependence of the villagers on fishing. It is believed that a perennial well in the vegetable market area dried down because of the deep boring by the company. The Baitarani River is the major source of water for washeries, which are used for washing ores, and enormous amount of surface water is used daily in the mining plants. However, the people do not get adequate supply and access to the primary source of water.

18.4.3 Forest and Loss of Biodiversity

The large-scale exploitation of the forest region due to the extensive development of mines and related infrastructural developments along with the illegal cutting of trees in the forests for timber have severe negative consequences for tribal ecology and economy. The operation of the heavy mining machines, blasting and excess traffic generates lots of noise in the Noamundi that affects the local fauna. Many birds, reptiles and snakes are disappearing from the local forests. Numerous heavily loaded vehicles and big trucks going through forest patches disturb the wild animals as well as the vegetation. The forest in this area is home to elephants, flying squirrel, four-horned antelope, sloth bear, leopard and deer; unfortunately, the populations of these animals are gradually decreasing.

The chief characteristic of central Indian forest belt including the Saranda forests is the presence of a large number of elephants. The elephants have cohabited with the human population peacefully in this forest without any major conflicts. But in recent times, the matter of elephant–man conflict has become a regular feature in the Noamundi region. The elephants used to follow their traditional travel route for moving into the jungle to have food and fodder, but now they are forced to change their usual route due to the fortification of mining townships and project areas with high walls. Their natural abode has been disturbed, and elephants have no options but to move further afield in search of food. They frequently enter the villages at night and cause damage to the mud houses and standing crops. Sometimes, elephants get angry and even kill human beings during their village raids. Ironically, the forest department officials place the logic that the Ho people are chopping down the local trees for timber and fuel, which is often annoying the elephants. The local geologist argues that high iron ore contents in the water streams make the water unfit for drinking, and the noise of the trucks carrying ore scares elephants and irritates them (Priyadarshi 2012). More and more clearances to iron ore mining projects are resulting in continuous encroachments of animal's habitat, which is the only reason behind interspecies conflict.

Many smaller plants and creepers that grow in association with other plant species are vanishing owing to the impact of red oxide pollution retarding the growth of plants. Those plants that rely on other plant species for pollination are also affected. The company propagates the idea of reclamation of the mined land after closure, but the locals believe that making artificial park and garden cannot compensate the originally created natural forest by the supreme deity *Singbonga*. The Ho people laugh at the idea of botanical garden and ethnomedicine-workshop-like activities conducted by the mining company. After the loss of forest and natural animal and plant species, the synthetic beautification cannot fill their hungry stomachs. Indigenous people affiliate aesthetically with forest and the smell of different plants and their flowers in different seasons. The people strongly believe that *Bongas* (spirits) residing in the local environment are becoming powerless due to the persistent deforestation for mining in Noamundi.

18.5 Community Resilience and Survival

There has been scholarly contestation over the quixotic view of the tribe as conservationist of the local ecology. Redford (1990) questioned the concept of 'ecologically noble savage' and contended that as indigenous people come in contact with the Western world, they also depict the capacity, desire and perhaps need to overexploit their environment. Edgerton (1992) proposed that many societies living in proximity with nature have taken maladaptive strategy and driven themselves into extinction and debunked the myth of primitive harmony with nature. However, the supporters of tribal wisdom such as Sponsel (1995, p. 283) argued that Amazonian indigenous people have developed and used land and resources on a sustained basis without major irreversible environmental degradation and destruction for millennia. Parajuli (1998) used the term ecological ethnicities for 'those people who have developed a respectful use of the natural resources and consequently a commitment towards creating and preserving a technology that interacts with local ecosystems in a sustainable manner, and that can include peasants, indigenous peoples, rural inhabitants, fisher-folk, forest dwellers, nomadic shepherds, and a host of people marginalized by development projects and the programs of environmental modernization' (17). These indigenous ethnicities represent 'viable, functioning, ecological alternatives to existing models of modernization and environmental destruction' (Little 1999). The present study depicts that the Ho community in Noamundi is a typical example of such ethnicities that have relied on their traditional wisdom and agricultural mode of living even after centuries of mining and subsequent adverse changes in local ecology and economy.

The mining in Noamundi resource frontier has been taking place for 100 years now. Six generations of the Ho indigenes have witnessed the ravaging of forest, plundering of mountains and removing of earth from their nature. After a century of experiences with mining and industrialisation, the tribal world view is still limited to the idea of ecological sustenance through agricultural economy. The earlier generations of Ho people have seen physical dislocation and went through social factionalism, thereby experiencing emotional trauma owing to the loss of ancestral and sacred place. Moreover, they have developed resilience towards the onslaught of extractive modernity. Jacka (2015) used the concept of resilience to describe the capacity of indigenous people to survive in a mining zone. People have learnt to negotiate their everyday existence in terms of informed awareness towards political developments in the region. There is immediate belief in the traditional mode of agricultural existence and dependence upon nature and spirits (*Bongas*) residing in them. In Noamundi, the indigenous Ho people have demonstrated ample resilience amidst several challenges to the agricultural ecology. People are attracted towards modernity and aspire for prosperity such as mainstream living in mining camps. They know that a job in a mining company can provide monetary support to afford education for their kids, but there is no immediate urge to get that money by hook or by crook. In fact, there is intrinsic tendency to look beyond money and have resonance within the traditional agricultural lifestyle. The dialectics that exists in the academic world about the development versus environment is interestingly visible in the life of the local Ho people. They fear that natural resources and agricultural economy will come to an end because of mining activities, but at the same time, there is attraction towards mine-driven economic opportunities.

18.6 Conclusion

In the Anthropocene, the extraction of resources such as minerals, oils and gases is discussed in media, academics and at various public forums. The debates on sustainable mining have created increased awareness and 'socially informed connections' amongst the local indigenous people and promoted them to engage proactively with mining companies for exploiting natural resources with unprecedented pace. The indigenous communities such as the Ho have started to come forward and voice their concern against the challenges of global extractive capitalism, which they believe is threatening the local ecology, agricultural and small-scale economies and culture (Padel et al. 2013). Additionally, mining areas such as Western Singhbhum are flooded with multiple companies, both public and private, which are operating in one resource territory to take advantage of the availability of resources. The collective impact of such a rapid pace of mining in the multiple mines present in the Ho territory has unprecedented effects on the local forest, agriculture, land and water.

The mining companies should take concrete steps to address these indigenous concerns. The environmental impact assessment (EIA) process for mining projects needs to be significantly revamped to make it unbiased and independent. At present, EIA agencies rely on the data provided by mining companies, which are mostly not scrutinised by independent scientific bodies that are represented by local communities. Thus, extractive corporations have the capacity and mechanisms to create environmental optimism and downplay the harmful impacts of the mining projects on traditional agricultural land. In Noamundi, the indigenous peasant communities have severe lack of trust in corporate-centric environmental surveys and reports. Strategic planning on land management is another key step that should be taken into consideration to nullify the impacts of mining on agriculture in the tribal territory. There is no clear policy guideline to compensate for the continuous impacts of mining on agricultural production, which might occur due to the ongoing land expansion for mining-based infrastructural developments. The one-time payment for land acquisition may not suffice the sustainable livelihood dependence of indigenous communities on land and related natural resources. The prior demarcation of land for mining and agricultural use in consultation with the local tribal communities would help in reducing future conflicts before the extension of lease area for mining projects and infrastructure developments.

It may be emphasised that mining industry in the Anthropocene needs to ensure that local agricultural practices are more sustainable. For reconciling mining with agriculture in India, particularly in the tribal state such as Jharkhand, the needs of the tribal communities should be aligned with the existing mining policy of the national and local governments. It is recommended that mining companies should also look into the challenges associated with agricultural developments as part of mining and related industrial developments. It is only then we can actually achieve sustainable future for resources that are present both above and below the ground and reply to the environmental challenges being posed in the Anthropocene.

References

Das Gupta PK (1978) Impact of industrialization on tribal life of South Bihar. Anthropological Survey of India, Calcutta

Dungdung G (2015) Mission Saranda: a war for natural resources in India. Deshraj Prakasha, Ranchi

Edgerton R (1992) Sick societies: challenging the myth of primitive harmony. Free Press, New York

Jacka JK (2015) Alchemy in the rain forest: politics, ecology, and resilience in a New Guinea mining area. Duke University Press, Durham

Kirksey SE, Helmreich S (2010) The emergence of multispecies ethnography. Cult Anthropol 25 (4):545–576

Kumar U (2017) Colonial law making, mining lease instrument, and displaced rights of Ho Adivasis. J Adivasi Indig Stud 6(1): 17–40. http://joais.in/Default.aspx. Accessed 11 Aug 2017

Little PE (1999) Environments and environmentalism in anthropological research. Annu Rev Anthropol 28:253–284

Majumdar DN (1950) The affairs of a tribe: a study in tribal dynamics. Universal Publishers, Lucknow

Misra KK (1987) Social structure and change among the Ho of Orissa. Gian Publishing House, Delhi

Padel F, Dandekar A, Unni J (2013) Ecology, economy: quest for a socially informed connection. Orient Blackswan, New Delhi

Parajuli P (1998) How can four trees make a jungle? Terra Nova 3(3):15–31

Priyadarshi N (2012) Effects of mining on environment in the state of Jharkhand, India. http://nitishpriyadarshi.blogspot.in/2012/05/effects-of-mining-on-environment-in.html. Accessed 10 Aug 2017

Redford KH (1990) The ecologically noble savage. Orion 9(3):22–37

Sponsel LE (1995) Relationships among the world system, indigenous peoples, and ecological anthropology in the endangered Amazon. In: Sponsel LE (ed) Indigenous peoples and the future of Amazonia. University of Arizona Press, Tucson, pp 264–293

Steward JH (1955) Theory of culture change: the methodology of multilinear evolution. University of Illinois Press, Urbana, IL

Role of Indian Seed Industry in Promoting Food and Nutritional Security and Agricultural Sustainability

19

R. P. Singh and R. C. Agrawal

Contents

Abstract

The greatest challenge before Indian agriculture is to achieve sustainable agricultural growth instead of volatile one, which has been the hallmark of the decade. Increased crop productivity is required to boost agriculture growth for poverty reduction and enhance yield stability under extreme weather conditions due to climate change (CC) effects. Development, deployment and diffusion of climate-resilient, efficient, resource-utilizing and high-yielding varieties (HYVs) are required to attain the desired growth as the genetic gain made by plant breeding could only be translated through obtaining higher productivity. The seed and varietal replacement rates are still low, and ultimately, investments incurred to develop new varieties are not realized due to poor adoption and diffusion, and majority of marginal and small farmers (M&SFs) remain deprived of quality seed, which ultimately leaves huge gaps. Attention needs to be paid to the development and deployment of varieties amongst the traditional crops having local importance and specific adaptation. More efforts from the seed industry are required to

R. P. Singh
Birsa Agricultural University, Ranchi, Jharkhand, India

R. C. Agrawal (✉)
Protection of Plant Varieties and Farmers' Rights Authority, Ministry of Agriculture and Farmers Welfare, New Delhi, India

K. Bauddh et al. (eds.), *Ecological and Practical Applications for Sustainable Agriculture*, https://doi.org/10.1007/978-981-15-3372-3_19

resolve the seed issues amongst high-volume low-value crops as they are mainly grown in low input.

Keywords

Climate change · Food and nutritional security · Crop breeding · Seed systems · Production and productivity

19.1 Introduction

From the point of view of sustained agricultural growths, marginal environments are extremely in need of not only food but also nutritional security . To accelerate varietal adoption and diffusion, change in breeding philosophy of wide adaptation to local adaptation is required even with specific crop breeding for specific cropping pattern and for crops which grown in different seasons need to be prioritised. Indian soils are hungry as well as thirsty, and to cope up with this situation, the drought-tolerant, resource-efficient, short-duration and climate-compliant varieties are needed with the combined efforts of public and private sectors and farmers, that is, formal and informal seed systems together. Greater emphasis needs to be placed on breed varieties with higher nutritional contents under high CO_2 concentration condition to protect people from macro- and micronutrient deficiencies. The commercialisation and integration of the legislatively protected (through PPV & FR Act, 2001) climate-resilient farmers' varieties (FVs) through formal or semiformal seed sector are of paramount importance in enhancing resiliency, and for this, the provisions need to be incorporated through regulatory mechanism. Similarly, policy changes are required with respect to regulatory approval process for faster clearance and releasing of genetically modified (GM) varieties in order to reduce vulnerability, variability in yield and yield gaps vis-a-vis to promote sustainable agriculture through enhanced adaptation and mitigation of the adverse conditions due to climate change (CC). However, the seed industry should take heed of the lesson learnt with regard to species diversity loss due to Bt cotton; nevertheless, the fifth-ranked globally and well-established Indian seed industry is capable enough to provide leadership role not only to generate resource-efficient better technologies but also to make them available in a cost-effective manner to M&SFs to promote sustainable agriculture as well as sustained growth for nation building.

Since the 1970s, the world population has more than doubled (from 3.7 to 7.6 billion) and is likely to be around 9.5 billion by mid century. Edible crop production tripled since the 1970s, and to meet the demand of the ever bulging population, $\geq$70% more production would be required (IPBES 2019). Globally, more than 820 million people remain undernourished, and more than two billion people are micronutrient deficient (Willett et al. 2019). The Millennium Development Goals (MDGs), after attaining the desired success in achieving the goals of reducing poverty by half, child mortality, hunger, etc. in 2015, were replaced by the Sustainable Development Goals (SDGs), which were adopted globally with the mandate to end poverty and hunger, achieve food and nutritional security, promote sustainable agriculture and enhance adaptation to climate change (CC). As per the recent Intergovernmental Panel on Climate Change's (IPCC's) Special Report on Global Warming of 1.5 °C, human-induced warming has reached approximately 1 °C (0.8–1.2 °C) above pre-industrial levels in 2017, but it is increasing by 0.2 °C (0.1–0.3 °C) per decade; nevertheless, IPCCs report released on 8 October 2018 states that global food demand is expected to increase by 14% per decade until 2050, whereas global crop yields are expected to decrease by 1–2% per decade (Dunphy 2019). Similarly, the Paris Climate Change (COP 21) Agreement (2015), unequivocally adopted resolution to pose restriction on temperature increase to less than 20 °C limit along with the emphasis to allow temperature rise only to 1.5 °C above pre-industrial levels by adoption and implementation of the appropriate adaptation strategies (FAO and others 2018). India is the most vulnerable country in the world in terms of facing extreme weather with high climate risk index (CRI), a 'warning sign' (Eckstein et al. 2017). With respect to food and nutritional security, anaemia is prevalent amongst 51.4% of Indian women of reproductive age, and 38.4% of children below 5 years of age are stunting (FAO and others 2017). India, although ranked amongst the world's top producers of wheat, rice, pulses, oilseeds, cotton, sugarcane, etc., the productivity of these crops remained below the global average; therefore, the central challenge of Indian agriculture is low productivity coupled with high variability in yield and yield gaps. The decadal volatile agricultural growth in India seriously questions the sustainability of agriculture. The Economic Survey (2015–2016) categorically stressed the fact by making a statement that 'Indian agriculture, is in a way, a victim of its own success, which over time is posing to be a major threat'. The food demand is expected to double by 2050 due to population increase, which is likely to touch 1.65 billion in 2050 from the current level of 1.36 billion. In India, according to the 70th Round of National Sample Survey, at the national level, on an average basis, out of the total expenditure incurred on various input factors on crop production by each cultivating agricultural household, only 11% is incurred to seed in comparison to the 24% to fertilisers and manures and 21% to human labour (NSSO 2016). Thus, quality seed is known to have the lowest input cost in India but enhancing 20–25% productivity. Therefore, the role and responsibilities of Indian formal seed sector increase tremendously not only for production and productivity enhancement on a sustainable basis as being the basic input provider but also to increase resiliency while reducing vulnerability posed by the increased frequencies of extreme events due to climate change effects. In this chapter, issues related to the present agricultural scenario and the challenges and opportunities before Indian seed industry will be discussed.

19.2 Agricultural Scenario

Since the beginning of the new century, the agricultural growth fluctuated from as high as 5.8% (2005–2006) to 0.4% (2009–2010) down to the lowest of −0.2% (2014–2015) and 1.2% (2015–2016), increased up to 6.3% (2016–2017), reduced to 5% (2017–2018) and further reduced to 2.9% (2018–2019), reflecting high degree of volatility in growth. According to the Agriculture Census, 2015–2016, the number of operational holdings has increased to 146 million in 2015–2016 from 138 million in 2010–11 (a 5.3% increase). The area operated by the marginal and small farmer (M&SF) holdings increased from 38.9% in 2000–2001 to 47.4% in 2015–2016, while that of the large holdings decreased from 37.2 to 20% during this period (Economic Survey 2018–2019). India has ~17% of the world human and almost same proportion of cattle population, while land share is only 2.4% and water share just 4%; these do signify extraordinary human as well as animal pressure, in terms of land resources, such as food, fiber, fuel and shelter. In the country, more than 250 double-cropping systems of primary, secondary and tertiary importance do exist, while 30 are of primary importance (Singh et al. 2009). Farmers grow crops of different kinds of varieties, depending on the season and resource availability. The local varieties still cover significant area and virtually grown in all regions and seasons in India. There are adoption and diffusion of one kind (improved HYVs versus local/farmers' varieties) of variety in a particular season and the total absence of another kind in the same region or state, or they even grow FVs for their own consumption, while modern varieties (MVs) are grown only for market purposes. Therefore, highly asymmetrical adoption pattern of MVs in different seasons in the same region and in different regions in the same season does exist (Singh et al. 2016, 2017, 2018). About 70% of Indian soils lack organic carbon (<1%) and are micronutrient deficient (Singh 2017a). Also fertiliser response ratio decreased nearly four times within a span of 40 years, that is, from 13.4 kg of grains/kg nutrient fertiliser in the 1970s to merely 3.2 kg of grains/kg nutrient fertiliser in 2010. Nitrogen-use efficiency (NUE) is as low as 30–50%. The use efficiency of other nutrient elements is 15–20% for P, 60–70% for K, 8–10% for S and 1–5% for micronutrients, which is a major cause for concern (NAAS 2018). The remaining nutrients leached down to ocean and are responsible for dead zones, ecosystem degradation and biodiversity loss too. In India, almost 89% of groundwater extracted is for irrigation, and by 2050, India will be in the global hot spot for 'water insecurity' (Economic Survey 2018–2019).

19.3 Food and Nutritional Concerns Due To Climate Change

Micronutrient deficiencies cause a much larger burden of disease than food insecurity Globally, people derive most of their nutrients from plants, including 63% of the total dietary protein, 68% of zinc and 81% of iron. When grown under elevated CO_2 concentrations (546–586 ppm), many edible crops, such as, wheat, rice, barley and soybeans, have lower concentrations of nutrients, such as iron, zinc and protein, which are important for overall health. In India, due to higher CO_2 levels, an estimated 53 million protein-deficient and 48 million zinc-deficient people have been added, thus increasing health burden (Smith et al. 2018). Greenhouse gases (GHGs) affect the quantity and nutritional quality of harvested produce, especially the major cereal crops, by reducing protein and mineral concentrations up to 5–15% and B vitamins up to 30%, which is due to the increased concentrations of CO_2

(Ebi and Loladze 2019). Projected increases in the atmospheric CO_2 will decrease growth in the global availability of nutrients, such as protein (19.5%), iron (14.4%) and zinc (14.6%), by 2050 (Beachet et al. 2019). Presently, breeding programmes for improving nutritional value of crops are not of high priority, and major transnational corporations (TNCs) even underline the relevance of vegetables in delivering nutrition, while only a small fraction of the global seed companies collaborate to develop field crop varieties with improved nutritional value (Anonymous 2019).

19.4 Challenges

The formal seed sector in India accounts for about 30–35% of the total seed distributed in the country, while the informal seed sector comprising mainly of farm-saved seed (FSS) accounts for the remaining 65–70% (DAC & FW 2016a). With regard to the use of quality seed, out of 138.11 million operational holdings, only 39.41% used certified seeds while 26.96% used seeds of notified varieties and only 9.84% used hybrid seeds (DAC & FW 2016b). These figures clearly indicate that the formal seed system produces and distributes the quality seeds of only few major crops and varieties having wide adaptation for more profit motives (Howard 2015). In this process, a large number of crops species such as small millets, minor pulses, traditional tubers, vegetables, etc. with local significance, including local varieties (LVs) having specific adaptation and usually covering relatively small areas, have not attracted the attention of the organised formal seed sector and ultimately got converted into underutilised or orphan crops (van Bueren et al. 2018). Due to the input and labour-intensive and productive industrial agriculture, these crops are disappearing, are even lost or became extinct (Guhakar 2015). In India, during the onset of the Great Depression (GR) era, conventionally developed HYVs were simply replaced with FVs, but did not vanished completely, and some of them are under cultivation even today due to their superior quality and/or resilient nature and constitute a major portion of agro-biodiversity, which is extremely important not only for adaptation and mitigation of climate change but also for future use in crop improvement for food and nutritional security of the country (Singh 2018a).

Cotton represents the classic example of cyclic rise and fall of diversity. The Indian monopoly on cotton muslins for more than *three millenniums ended in less than three decades* after the British consolidated their power since 1818, when it became the raw material for the British textile industry in Manchester and Lancashire. During this period, the New World Cottons were introduced in India (Kotak 2002), and the diversity of species that has the world's best quality fiber was eroded in the first wave. In the second wave, biodiversity was eroded from the farmers' field as genetically modified (GM) Bt cotton covered almost all area under the crop. By 2012, more than 1100 *Bt* hybrid varieties were planted on 95% of the total cotton area (Gutierrez 2018), and cotton diversity in the form of two Old-World diploid climate-resilient cotton species, as well as intraspecific hybrids, was eroded, which used to cover $\geq$25% of the total cotton area of the country before the introduction of transgenic Bt-cotton hybrid technology. The diversity of cotton was compromised at gene level (as there was no genetic variation in maturity period), species level (almost complete replacement and displacement of the old-world diploid cottons) and agro-ecosystem level by jeopardising various ecosystem services, such as provisioning (cotton-species depletion) and regulatory (pollinators, climate resiliency of diploid cotton, biological pest control nutrient recycling, etc.) aspects, which will ultimately lead to the ecosystem services degradation. However, in the recent past, pink bollworm has become the dreaded cotton pest affecting cotton production and productivity, and in this regard, Sigaard (2019) remarked that 'the bio-hegemonic pillar of discursive power has been maintained despite the pink bollworm issue'. With respect to biodiversity loss, according to the Intergovernmental Science-Policy Platform on Biodiversity and Ecosystem Services (IPBES) Chair, Sir Robert Watson, 'The loss of biodiversity is therefore shown to be not only an environmental issue, but also a developmental, economic, security, social and moral issue as well' (IPBES 2019).

Also, in almost all crops, some of the old varieties still occupy the prominent position by virtue of achieving the mega variety status (Singh 2015, 2018b). Through inadequate and slow varietal replacement, the benefit of investment on varietal development hampers the agricultural growth seriously, and therefore, poor seed and varietal replacement rates are considered twin perennial bottlenecks that ail the Indian seed sector (Singh et al. 2017; Singh 2018a); the cultivation of a few genetically uniform varieties has increased vulnerability to pests and diseases, which causes a huge yield gap between actual and potential yields (Atlin et al. 2017). The important rainfed crops, from food and nutritional security's point of view, are groundnut, soybean, chickpea and upland rice; in these crops, high seed rate and low seed multiplication ratio increase seed cost. Moreover, in case of monsoon failure or drought after germination, nonavailability of the second batch of quality seeds for gap filling and/or resowing causes low yield and high variability in yield amongst these and other rainfed-grown crops. Varietal mismatch, poor seed conversion ratio, poor seed multiplication rates and nonavailability of suitable and locally adapted varieties of crops for specific season, which are being grown in different seasons in the same year in a particular region such as rice, maize, sorghum, groundnut, sunflower, black gram, green gram, etc., pose serious challenges before the formal seed systems in the country and warrant solution on a priority basis.

Intercropping in India is practised since ancient times in order to reduce vulnerability and temporal and spatial variability in yield and to increase yield stability and security against various abiotic and biotic stresses. Lack of development and deployment of suitable varieties, which are adapted specifically to the intercropping (IC) environments, is a serious constraint and responsible for lower yield (Litrico and Violle 2015; Saxena et al. 2018). Commoditisation of seed worldwide through stringent IPRs, which increased corporate concentration through mergers and acquisitions, further complicates the situation by eroding diversity and thereby promoting monocultures with more and more profit including market control (Howard 2015; Kotschi and Horneburg 2018). Adoption of the breeding philosophy for wide adaptation, while ignoring the local and/or specific adaptation (Baranski 2015; Ceccarelli 2015; Singh et al. 2015), is also responsible for the erosion of agro-biodiversity.

19.5 Way Forward

Ceccarelli (2016) argued that five global issues, namely, decline of agro-biodiversity, climate change, hunger and malnutrition, poverty and water and seed, are central to all the issues, and it can also be the solution to all these issues. The Indian seed market has reached a value of US$3.6 billion in 2017, exhibiting a compound annual growth (CAGR) of around 17% during 2010–2017, and is further expected to grow at a CAGR of 14.3% from 2018– to 2023, reaching a value of more than US$8 billion by 2023 (Anonymous 2018). Therefore, it is in a position to invest more to overcome the challenges mentioned earlier. The development and deployment of suitable varieties, which are adapted specifically to the IC environments, need to be prioritised, which require emphasis on breeding crop varieties using different multiple traits in comparison to breeding crop varieties using simple traits to be grown as mono-crops (Brooker et al. 2015; Saxena et al. 2018). Efforts are needed to develop crop varieties specifically suited to low-input marginal environments having specific traits such as increased resource-use efficiency, higher photosynthetic efficiency and abiotic stress tolerance with special traits such as better root characteristics, stay green traits and extreme temperature tolerance particularly at anthesis stage. Study on genetic composition needs to be broadened through prebreeding and application of genetic engineering and biotechnological approach (Singh et al. 2007, 2013; Singh 2017b). Serious efforts are needed to breed new varieties with higher nutritional content as higher CO_2 tends to reduce/affect nutritional content adversely. Although, through the legal instrument Protection of Plant Varieties and Farmers' Right (PPV & FR) Act (2001), farmers have been treated at par with formal breeders, they have been facing difficulty in getting their varieties protected legally, particularly in cross-pollinated crops. To take the full advantage of the rich genetic diversity in the form of FVs, an alternative registration and certification system developed by the Food and Agriculture Organization (FAO) known as Quality Declared Seed (QDS) system is advocated to channelise and commercialise FVs using the formal/semi-formal systems (Singh and Agrawal 2018). The integration of both formal and informal seed systems is also a prerequisite for complementing the strengths and weaknesses of the formal and informal seed systems in order to strengthen/improve the seed systems' resiliency for adaptation and mitigation of climate change, ensuring food and nutritional security as well as agro-biodiversity conservation (Singh 2016; Singh et al. 2018).

References

Anonymous (2018) Seed industry in India 2018–2023: market trends, structure, growth, and key players – CAGR to grow at 14.3%. www.researchandmarkets.com/research/nf2gqz/seed_industry_in?w=4

Anonymous (2019) Access to seeds index 2019 – global seed companies – nutrition. http://www.accesstoseeds.org/index/global-seed-companies/key-findings/nutrition/

Atlin GN et al (2017) Rapid breeding and varietal replacement are critical to adaption of cropping systems in the developing world to climate change. Glob Food Sec 12:31–37

Baranski MR (2015) Wide adaptation of green revolution wheat: international roots and the Indian context of a new plant breeding ideal, 1960–1970. Stud Hist Phil Biol Biomed Sci 50:41–50

Beachet HR et al (2019) Combining the effects of increased atmospheric carbon dioxide on protein, iron, and zinc availability and projected climate change on global diets: a modelling study. Lancet Planet Health 3:e307–e317

Brooker RW et al (2015) Improving intercropping: a synthesis of research in agronomy, plant physiology and ecology. New Phytol 206:107–117

Ceccarelli S (2015) Efficiency of plant breeding. Crop Sci 55:87–97

Ceccarelli S (2016) Harnessing the power of evolution in participatory seed breeding, Comments. Available at https://theecologist.org/2016/feb/29/harnessing-power-evolution-participatory-seed-breeding

DAC & FW (2016a) State of Indian agriculture 2015-16. Department of Agriculture, Cooperation & Farmers Welfare, DES, MOA & FW, GOI, New Delhi, p 280

DAC & FW (2016b) All India report on input survey 2011–12. Agriculture Census Division, DES, MOA & FW, GOI, New Delhi, p 521

Dunphy S (2019) Higher temperatures offset the adverse effects of CO_2 on crop nutrition. https://www.europeanscientist.com/en/agriculture/higher-temperatures-offset-the-adverse-effects-of-co2-on-crop-nutrition

Ebi KL, Loladze I (2019) Elevated atmospheric CO_2 concentrations and climate change will affect our food's quality and quantity, vol 3, pp e283–e284. Comment. w.thelancet.com/planetary-health

Eckstein D et al (2017) Global climate risk index 2018. Briefing Paper, Germanwatche. V. Office, Bonn

Economic Survey (2018–2019) Sustainable development and climate change. Ministry of Finance, GOI(2), pp 104–128

FAO and others (2017) The state of food security and nutrition in the world 2017. Building resilience for peace and food security. FAO, Rome, p 132

FAO and others (2018) The state of food security and nutrition in the world; building climate resilience for food security and nutrition. Rome, FAO

Guhakar RT (2015) Local plants for rural food security. In: Lichtfouse E (ed) Sustainable agricultural reviews. Springer International, Cham, pp 141–158

Gutierrez AP (2018) Hybrid Bt cotton: a stranglehold on subsistence farmers in India. Curr Sci 115 (12):2206–2210

Howard PH (2015) Intellectual property and consolidation in the seed industry. Crop Sci 55:1–7

IPBES (2019) UN report: nature's dangerous decline 'unprecedented'; species extinction rates 'accelerating'. Report from the intergovernmental science-policy platform on biodiversity and ecosystem services (IPBES). https://www.un.org/sustainabledevelopment/blog/2019/05/nature-decline-unprecedented-report

Kotak S (2002) Perspectives of private trade. In: Indian cotton trade-past present and future. Silver Jubilee Lecture Series, 26 April 2002. ISCI-CIRCOT, Mumbai, pp 27–44

Kotschi J, Horneburg B (2018) The open source seed licence: a novel approach to safeguarding access to plant germplasm. PLoS Biol 16(10):e3000023

Litrico I, Violle C (2015) Diversity in plant breeding: a new conceptual framework. Trends Plant Sci 20(10):604–613

NAAS (2018) Soil health: new policy initiatives for farmers welfare. Policy Brief No. 3, NAAS, New Delhi, 19 p

NSSO (2016) Income, expenditure, productive assets and indebtedness of agricultural households in India. NSS 70th Round (January–December 2013). GOI, report no 576 (70/33/3), p 1605

Saxena KB et al (2018) Breeding pigeonpea cultivars for intercropping: synthesis and strategies. Breed Sci 68:159–167

Sigaard AS (2019) A study in pink: Gramscian reflections on Bt cotton, the pink bollworm and bio-hegemony in the Warangal District, India. Master thesis in Development, Environment and Cultural Change Centre for Development and the Environment. Universitetet i Oslo, p 142. Available at http://www.duo.uio.no/

Singh RP (2015) Varietal replacement rates among field crops: current status, constraints, impact, challenges and opportunities for the Indian seed industry. Seed Times 7(3, 4):71–89

Singh RP (2016) Local seed system strengthening in marginal environments to mitigate climate change effects *vis-a-vis* ensuring food security in India. Clim Change Environ Sustain 4 (2):158–177

Singh RP (2017a) Seed: an ambassador of science, technology, trade and agrarian revolution. Invited paper for the thematic issue on journey of seed. Seed Times 10(2):16–23

Singh RP (2017b) Role of biotechnology for breeding climate resilient varieties of field crops. Invited paper for the thematic issue on role of biotechnology in crop improvement. Seed Times 10(1):62–72

Singh RP (2018a) Appropriating varietal and seed replacement raters through policies for sustainable agriculture and seed industry. Souvenir Indian seed congress, held at Colombo, Sri Lanka on 5–6th Feb 2018, pp 120–135

Singh RP (2018b) Integration and commercialization of local varieties under sub-optimal environments for food security, promoting sustainable agriculture and agro-biodiversity conservation. MOJ Eco Environ Sci 3(2):65–67

Singh RP, Agrawal RC (2018) Improving efficiency of seed system by appropriating farmer's rights in India through adoption and implementation of policy of quality declared seed schemes in parallel. MOJ Eco Environ Sci 3(6):387–391

Singh RP et al (2007) Influence of high temperature and breeding for heat tolerance in cotton – a review. Adv Agron 93:313–385

Singh KK et al (2009) Pulses in cropping systems. Technical Bulletin, IIPR, Kanpur, p 47

Singh RP et al (2013) Impacts of changing climate and climate variability on seed production and seed industry. Adv Agron 118:49–110

Singh RP et al (2015) Climate change: implications for stakeholders in genetic resources and seed sector. Adv Agron 129:117–180

Singh RP et al (2016) The prevalence, productivity, and protection of traditional varieties vis-à-vis modern varieties in Eastern India: an appraisal. JJDMS 14(2):6955–6970

Singh RP et al (2017) Varietal and seed replacement in the era of climate change, pp 156. ISBN: 978-81-925128-2-6 (published by ICAR-IISS, Mau)

Singh RP et al (2018) Channelizing protected farmer's varieties through semi-formal seed systems for effective utilization and conservation of agro-biodiversity: an overview. Seed Res 46 (2):87–97

Smith MR et al (2018) The impact of rising carbon dioxide levels on crop nutrients and human health. GCAN Policy Note 10. International Food Policy Research Institute, Washington, DC

van Bueren ETL et al (2018) Towards resilience through systems-based plant breeding. A review. Agron Sustain Dev 38:42

Willett W et al (2019) Food in the Anthropocene: the EAT–Lancet Commission on healthy diets from sustainable food systems. Lancet 393:447–492

Farmers' Varieties and Ecosystem Services with Reference to Eastern India 20

R. P. Singh and R. C. Agrawal

Contents

Abstract

Since independence, India puts emphasis on ensuring food security, and consequently, scientific crop breeding programs were initiated initially in the coarse cereals and cereals. Through genetic improvement in these crops, a large number of high-yielding varieties (HYVs) were released for their cultivation by the National Agriculture Research System. However, the adoption and diffusion of HYVs is not uniform, and in different regions and seasons, farmers' varieties (FVs) are still being cultivated and cover significant area. FVs are locally adapted in particular region/environment. Farmers meticulously selected and maintained

R. P. Singh
Birsa Agricultural University, Ranchi, Jharkhand, India

R. C. Agrawal (✉)
Protection of Plant Varieties and Farmers' Rights Authority, Ministry of Agriculture and Farmers Welfare, New Delhi, India

K. Bauddh et al. (eds.), *Ecological and Practical Applications for Sustainable Agriculture*, https://doi.org/10.1007/978-981-15-3372-3_20

the identity since long. Some of the FVs are known to have special nutritional/medicinal/therapeutic value in addition to adaptive traits and have been protected through Protection of Plant Varieties and Farmers' Rights (PPV&FR) Act (2001). Moreover, highest number of legislatively protected rice FVs are from the eastern region recognizing their importance in ecosystem service as well as agro-biodiversity maintenance. The FVs with special nutritional/aroma/medicinal traits with distinct identity and protected under PPV&FR Act should also get higher economic returns if produced through organic agriculture, and for this, eastern region is ideally suited due to its agroecological conditions in addition to their specific adaptation. Finally, FVs that are locally adapted and that have excellent source of traits for abiotic stress tolerance and resource efficiency are ideal candidates to be utilized for developing new varieties and for future crop improvement, for adaptation and mitigation to adverse effects due to climate change, for enhancing food and nutritional security, and for higher return due to premium market price. FVs with PVP certificates need to be integrated into mainstream agriculture like the GI-protected Basmati rice for trade.

Keywords

Farmers' varieties · High-yielding varieties · Adaptation · Adoption · Diffusion · Yield and variability in yield

20.1 Introduction

The seed journey started about 10 millennia ago when our ancestress moved away from hunting and gathering to agriculture for which the domestication of crops was a prerequisite, and in the process, possibly both unconscious and conscious selections were made during adaptations of crop plants. Since the dawn of civilization, farmers are involved in the selection of better performing crop plants at local level and maintained and conserved them. A perfect harmony was maintained through natural and farmer's selection, and different varieties had been part of the expression of art, culture, and indigenous traditional knowledge (ITKs). The initial and important traits for which selections were made had been seed dispersal and dormancy. The rediscovery of law of inheritance laid down the establishment of scientific plant breeding, which, within a relatively short span of past one and a half century, created formal seed system. Crop varieties are developed either through organized formal seed system using well-established scientific principles and systematic steps that were released and notified according to the prevailing regulatory procedures and the varieties known as high-yielding varieties (HYVs) or through the informal seed system (owned and managed by farmers) based upon the indigenous knowledge with respect to the local environments and other desirable traits of interest. The varieties developed by farmers are known as farmers' varieties (FVs). The technologies in the form of HYVs (characterized by their semidwarf nature, photothermal-insensitive behavior, and input responsiveness) along with the assured supply of quality seed of these varieties with other inputs such as irrigation, fertilizers, farm power, and machineries required for intensive agriculture along with policy reforms in the form of input subsidies and minimum support price (MSP) successfully led the establishment of green revolution (GR) in India (Baranski 2015). The GR in India has completed golden jubilee in 2015–2016, and the achievement could be judged by overall food-grain productivity increase by a factor of 3.24 times (from meager 629 to 2042 kg/ha) and overall food-grain production increased by 3.48 times (from 72.35 MMT in 1965–1966 to 251.37 MMT in 2015–2016) with marginal area increase (7% only) from 115.1 to 123.22 Mha during the period of reference in India.

The adoption and diffusion of HYVs in the form of open-pollinated varieties (OPVs)/hybrids in many crops under mostly rainfed conditions has occurred since the early 1980s (Evenson and Gollin 2003; Pender 2008). The very essence of the GR in India is the development and diffusion of a series of modern HYVs in irrigated and favorably rainfed areas and subsequent acceleration in public sector investments in complementary infrastructures in terms of seed, fertilizers, farm machineries, irrigation, and institutions (Hazell 2009). The long-term trend of rising temperatures, declining average precipitation, and increase in extreme precipitation events have been observed, and long-term weather patterns predict that climate change could reduce annual agricultural incomes in the range of 15%–18% on average and up to 20%–25% for rainfed areas (Economic Survey 2017–2018). The very first two sustainable development goals (SDGs) of the United Nations, related to end poverty and hunger completely are to be achieved by 2030 by all the member Nations. Similarly, 12 of the 17 SDGs emphasize on the nutrition-linked indicators, which reflect the increased concerns of the global community. Although, at the national level, India has achieved food security, it varies and is yet to be achieved at the regional level. It is also found that ≈60% of the malnutrition cases are from the states that ranked high in poverty index (Kahlon and Singh 2019).

During GR period and thereafter, the HYVs, however, replaced FVs in the well-endowed areas but couldn't displaced them completely, and the performance of HYVs under low-input and marginal environmental and other harsh conditions was not recorded due to their susceptibility to the prevailing abiotic stresses. Although modern HYVs have been widely adapted, covered large areas, and replaced farmers' varieties (FVs), HYVs couldn't displace them completely. Under unfavorable conditions, however, farmers' varieties (also known as landrace or traditional (TVs)/local varieties (LVs)), being climate resilient and having specific adaptation, performed well, and therefore, the adoption and diffusion of one kind (improved HYVs vs. local/farmers) of variety in a particular season and total absence of another kind in the same region/state or even grow FVs for own consumption, while modern HYVs grow only for market purpose (Singh and Agarwal 2019), are recommended. Farmers grow different kinds of varieties depending upon the season and resource availability. The FVs still cover significant area and virtually grown in all regions and seasons in India. Therefore, highly asymmetrical adoption pattern of HYVs in different seasons in the same region and in different region in the same season does exist (Singh et al. 2016, 2017, 2018a). Moreover, the yield gains obtainable through HYVs over FVs vary from season to season in the same region and in different regions in the same season including the variability in yield, which also varies according to kind of varieties as well as seasons (Singh and Agarwal 2020). The FVs are supposed to have comparatively higher nutritional contents as compared to HYVs, in addition to the specific adaptation and climate resiliency. The present article reviews the contribution of formal and informal seed systems in the form of HYVs and FVs, respectively, and the extent of coverage of both kinds of varieties and relative contributions with special emphasis on rice FVs in the eastern region of India.

20.2 Farmers' Variety Versus Modern Variety

The literature with regard to nomenclature of local varieties, traditional varieties, landraces, folk varieties, farmers' varieties, etc. and these terms are used interchangeably, and therefore, these kinds of varieties need to defined in order to differentiate them.

1. *Farmers' variety*: The term was adopted as generic for all FVs including those that are bred and maintained by active seed selection on farm. The FVs often emerged as farmers' selections from modern and traditional cultivars in market-oriented production systems (Salazar et al. 2006). In India, the PPV&FR Act (2001) defined farmer's variety as a variety that has been traditionally cultivated and evolved by the farmers in their fields or is a wild relative or landrace of a variety about which the farmers possess the common knowledge. The term *traditional varieties* (TVs) refers to local varieties (LVs) also known as landraces, which have never been subjected to a formal plant breeding program. Dennis (1987) defined LVs in rice that have been grown in an area for many years or have been bred or selected from varieties long used in the area, and the author also mentioned that a local variety is a pre-GR variety. The traditional varieties (TVs) are all varieties that are either landraces (have never been improved by breeders) or cultivars that have not been supported by the official channels (Zeven 1998). Sinha (2016) referred TVs as climate resilient, having fixed traits with specific adaptation, which are actively cultivated for at least half a century.
2. *Modern variety*: Most literature on seed supply systems refers to seeds as either "modern varieties" (MVs) or "landraces." "Modern variety" (MV) is understood as a variety that is improved by a formal breeding program (Morris et al. 2003). Heisey and Morris (2002) explained that, conventionally, the term "modern variety" refers to improved varieties developed by scientific plant breeding programs, and in the past, the products of scientific plant breeding programs were commonly referred to as "improved varieties," reflecting the fact that their characteristics have systematically been altered in ways that bring economic benefits to those who grow them, and according to Mahadevappa (2015), these products of crop improvement yielded what are today called "high-yielding varieties" (HYVs). However, there is disagreement about what constitute a MV. High-yielding variety is often used, but Herdt and Capule (1983) had

shown reservation to use it because new varieties may not be high yielding unless a high level of input is used, so it is better to use modern varieties. In case of rice, the term MV refers to the short-statured, stiff-straw, fertilizer-responsive, photo-period-insensitive *indica* rice varieties (Chandler 1982).

20.3 Varietal Development Among Cereal Crops in India

The first All India Coordinated Maize Improvement Project (AICMIP) on maize was established in 1957 to intensify maize improvement in India by having the inter- and intradisciplinary cooperation among various research centers including multidisciplinary and multilocation approach, which served as an efficient mechanism for mobilizing and integrating resources for production-oriented agricultural research. The success of AICMIP led to the establishment of All India Coordinated Crop Improvement Projects (AICCIP) in wheat (1961) and rice (1965). Currently, ICAR has 78 crop-/commodity-/thematic-based All India Coordinated Research Projects (AICRPs) and All India Coordinated Network Project (AINP), and of these, 30 are in Crop Science Division (Chauhan et al. 2016). In India, large number of varieties among cereal and coarse cereal crop grown in different agroecologies, 5-year average crop wise area, total number of varieties, the number of notified varieties after the enforcement of Seed Act (1966), and the average number of varieties in active seed chain have been presented in Table 20.1. The proprietary hybrids by private seed sector are not included in the calculation. The area covered under hybrids in rice, maize, sorghum, and pearl millet, figures obtained based upon the average of availability and requirement of hybrid seed in India since 2012–2013 to 2017–2018 (Agriculture Statics at a Glance 2018).

20.4 Area Coverage by HYVs and FVs

In the country, more than 250 double-cropping systems of primary, secondary, and tertiary importance do exist, while 30 are of primary importance. Moreover, in India, significant area under pulses is covered under intercropping systems (ICSs) where pulses are intercropped with oilseeds, coarse cereals, and commercial crops (Singh et al. 2009).The input survey, conducted by the Agricultural Census Division of "Department of Agricultural Cooperation and Farmers Welfare" (DAC & FW), Ministry of Agriculture and Farmers Welfare's (Government of India), published report in 2016; out of a total of 138.11 million operational holdings, only 39.41% used certified seeds, while 26.96% used seeds of notified varieties (DAC & FW 2016b), which indicated the significant contribution of non-notified and/or FVs. These varieties are being maintained, multiplied, exchanged, and continuously developed by farmers themselves through informal seed system. The extent of coverage of FVs among different crops varied in different regions and seasons based upon the environmental conditions and input availability. The average area (2000–2001 to 2012–2013) under high-yielding varieties (HYVs) and local varieties

Table 20.1 Coverage area, total number of varieties and varieties notified from 1966 onward, and varieties in formal seed chain and farmers' varieties (FVs) with plant variety protection (PVP) certificates in different cereal crops grown in India

S. no.	Group crop	Five-year ave. area (Mha)	Total no. of varieties[a]	No. of varieties notified from 1966 onward[b]	Average no. of varieties in seed chain (2014–2015 to 2018–2019)[c]
1	2	3	4	5	6
(A) Cereal					
1	Rice (total)	43.70	1652	1195	267.2
	Varieties	41.00			257.4
	Hybrids	2.70			9.8[d]
2	Wheat	30.63	545	504	163.2
3	Barley	0.6	124	102	38.8
	Subtotal (including triticale)	74.93	2321	1801	
(B) Coarse cereals					
1	Jowar (total)	5.97	342[e]	298	44.4
	Varieties	3.57			36.4
	Hybrids	2.40			8.0[d]
2	Maize (total)	9.07	629[e]	477	41.0
	Varieties	3.39			17.4
	Hybrids	5.68			23.6[d]
3	Bajra (total)	7.40	318[e]	278	22.4
	Varieties	2.40			12.0
	Hybrids	5.00			10.4[d]
4	Ragi	1.2	132	127	20.2
5	Other small millets (5)	0.73	146	118	12.0
	Subtotal	24.91	1567	1298	
	Grand total	99.84	3888	3099	

[a]http://seednet.gov.in/SMIS/LoginForm.aspx
[b]seednet.gov.in/SeedVarieties/index.aspx accessed on 25/11/2019
[c]Proprietary hybrids not included
[d]No. of public sector hybrids
[e]Includes varieties, hybrids and fodder (VHF)

(FVs) of different cereals and coarse cereal crops among different regions (states) are mentioned in Table 20.2. The coverage areas under different kinds of varieties HYVs and FVs were taken from Agricultural Statistics at a Glance, and only up to 2012–2013 data were available. Rice is grown throughout length and breadth of the

Table 20.2 Average coverage area under HYVs and FVs in cereals in different regions/states of India[a]

Region	State	Area (Mha) (2012–2013)	Average area (2000–2001 to 2012–2013) under HYVs (%)	Average area (2000–2001 to 2012–2013) under FVs (%)	Actual average area under FVs (Mha)
Rice					
Eastern	W. Bengal	5.43	90.17	9.83	0.53
	Odisha	4.03	60.68	39.32	1.58
	Bihar	3.25	54.46	45.54	1.48
	Chhattisgarh	3.78	29.47	70.53	2.66
	Jharkhand[b]	1.77	79.00	21.00	0.37
Northern	Punjab	2.85	98.21	1.79	0.05
Southern	Karnataka	1.27	82.97	17.03	0.21
	Subtotal	22.38 (52.4%)	69.26		6.88 (30.74%)
	All India	42.75			
Wheat					
Eastern	Bihar	2.22	74.05	25.95	0.57
Western	Rajasthan	2.82	81.23	18.77	0.53
	Gujarat	1.05	91.20	8.80	0.09
	Maharashtra	0.59	94.43	5.57	0.03
Central	Madhya Pradesh (MP)	5.30	59.44	40.56	2.14
Northern	Haryana	2.50	96.90	3.10	0.07
	Subtotal	14.48 (48.3%)	76.32		3.43 (23.68%)
	All India	30.00			
Maize					
Eastern	Bihar	0.69	36.47	63.53	0.43
Southern	Karnataka	1.31	98.02	1.98	0.026
Western	Rajasthan	0.99	46.48	53.52	0.53
	Gujarat	0.48	60.15	39.85	0.19
	Subtotal	3.47 (40.0%)	66.10		1.176 (33.9%)
	All India	8.67			

[a]Source: Adapted and modified from Singh et al. (2018a)
Data from Sinha (2016)

country under different ecologies, but the states in which area under HYVs and LVs recorded have been taken into consideration. Rice crop in the eastern region covers significant area in West Bengal (5.43 Mha), Odisha (4.03 Mha), Chhattisgarh (3.78 Mha), Bihar (3.25 Mha), and Jharkhand (1.77 Mha), while in Punjab, it covered 2.85 and in southern state Karnataka, 1.27 Mha. Altogether, seven states covered 22.38 Mha, representing 52.8% of the total area in the country under the crop during 2012–2013/2013–2014. The FVs covered ranged from 1.79% average area under rice in Punjab to Chhattisgarh (70.53%). Unlike rice, wheat is grown predominantly in all regions except southern, and altogether, six states cover 14.48 Mha (48.84% of all India wheat area). The area under FVs in different wheat-growing region varied from 3.1% in Haryana to 40.56% in MP. Among coarse cereals, sorghum, pearl millet, and maize are the dominant crops grown predominantly under rainfed conditions in different regions and seasons of India. Like rice, maize is also grown across the country during different seasons, and four states, namely, Karnataka, Rajasthan, Bihar, and Gujarat, jointly covered 3.47 Mha (34.84%). The area occupied by FVs in maize also varied among different states like in Karnataka (1.98%) to Bihar; the FVs occupied highest average area (63.53%).

Different kinds of varieties being high yielding, hybrids and farmers' varieties are grown under diverse agroecosystems. In the eastern region, in Assam, Bihar, Odisha, and West Bengal, rice is grown 2–4 times in a year. Like rice, maize is also grown under diverse climatic (temperate, tropical, subtropical, and semiarid) conditions along with varying altitudes, while pearl millet is grown under arid and semiarid conditions. These crops are grown around the year in different regions and seasons, and accordingly, to suit specific agroecological condition, different varieties are known to adapt under these diverse conditions in terms of maturity period and yield. Coarse cereals are being grown mainly in rainfed areas (85%). Crop wise sorghum and pearl millet are grown under rainfed conditions, and at national level, >91% area covered by these crops is rainfed, while in maize about 75% area is unirrigated.

20.5 Informal Seed Systems in India

The formal seed sector in India contributes only about 30–35% of the total seed requirement in India, while informal seed sector's contribution is enormously high, mainly in the form of farmer's own seed and commonly known as FSS to the tune of 65–70% (DAC & FW 2016a). Overall, access to quality seed by small and marginal farmers is reported to be only 20% (Roy 2015). Various factors, belonging to institutional, environmental, socioeconomic, and technical categories, are responsible for the occurrence of FVs reported by Singh and Agarwal (2019). India has enacted law known as the Protection of Plant Varieties and Farmers' Rights (PPV&FR) Act (2001) in order to comply the obligations under Trade-Related Intellectual Property Rights (TRIPs) Agreement of the World Trade Organization (WTO). The Protection of Plant Varieties and Farmers' Rights (PPV&FR) Authority of India established during 2005 started accepting applications for the protection of varieties to grant PVP certificates in 2007 under the PPV&FR Act (2001). The PPV&FR Act is a *sui generis* legislation ("of its own kind") and a unique one worldwide because it combines plant breeders' rights with elements of the Article 8 (j) of the United Nations "Convention on Biological Diversity" (CBD) and Article 9 of the FAO's "International Treaty on Plant Genetic Resources for Food and Agriculture" (ITPGRFA), also known as Seed Treaty, wherein farmers' claims as stewards of plant genetic resources (PGR) are enshrined. The PPV&FR Act (2001) under farmers' rights recognizes a farmer as a "*breeder*," "*conservator*," and "*user*" who is entitled to save, use, sow, resow, exchange, and share or sell his farm produce, including seed of a variety protected under this Act, just as he was entitled before the enforcement of this Act, provided that the farmer shall not be entitled to sell branded seeds of a variety protected under this Act. Also, the Indian *sui generis* system of PVP is unique, because under this Act a variety can be registered under either of three categories: new variety, essentially derived varieties, and extant variety. The pace of farmers' variety registration picked up in 2013.

The crop wise status of providing legal protection to varieties developed by either formal seed system (variety developed through organized institution (public sector or private seed industry)) or informal seed system (farmer-managed seed system and where such kind of provisions like formal system do not exist) in the form of PVP certificates under PPV&FR Act (2001) given in Table 20.3. The formal seed sector is working on almost all crops, but generally, the public seed sector is involved in high volume and low value, which is very important from food and nutritional security point of view, while the private sector is involved mainly in vegetable and hybrids in cereals, oilseed, and pulseswith low volume and high value. As can be seen from Table 20.3, a maximum number of farmers' varieties have been protected in the case of food crops that are extremely important for food security (rice) and those FVs that belong to self-pollinated crops, too, and relatively, only a few FVs have been protected in cross-pollinated crops. This could probably be due to the stringent distinctiveness, uniformity, and stability (DUS) requirement.

Table 20.3 Status of legislatively (PPV&FR Act, 2001) protected varieties among cereal and coarse cereal crops by formal and informal seed sectors in India (up to October 15, 2019)[a]

S. no.	Crop	Registered HYVs Formal seed sector	Registered FVs Informal seed sector	Total
Cereals and coarse cereals				
1	Rice	321	1527	1848
2	Wheat	146	23	169
3	Durum wheat	16	1	17
3	Dicoccum wheat	5	–	5
4	Barley	15	–	15
5	Maize	260	6	266
6	Sorghum	130	4	134
7	Pearl millet	127	–	127
8	Finger millet	6	–	6
9	Foxtail millet	1	–	1
10	Little millet	1	–	1
	Total	1028	1561	2589

[a]PPV&FR Act, New Delhi

20.6 Nutritional/Medicinal/Therapeutic Attributes of Farmers' Varieties

Rahman et al. (2006) reported that a large number of farmers' rice varieties with high nutritional and medicinal value are still grown by tribals and small farmers in some eastern states. Similarly, Deb et al. (2015) demonstrated that a considerable number of rice landraces are capable of taking up micronutrients like Fe, Cu, Zn, and Mn. Many of these landraces contain up to 13-fold greater concentration of Fe and Zn than transgenic metal-fortified rice. On comparison, at least 67 folk landraces from their sample contain >20 mg/kg iron. Clearly, farmers' rice varieties have enormously high potential of remedying dietary iron and zinc deficiencies in the country. The list of some of the rice FVs with their nutritional/medicinal/therapeutic properties along with their plant variety protection status is given in Table 20.4.

The increased frequencies of extreme events in the form of heat waves, droughts, floods, submergence, cyclones, coastal flooding, salinity, and increased extreme temperatures are certain to intensify in future. Therefore, serious issues related to climate change, agro-biodiversity loss, and food and nutritional security are continuously haunting the researchers as well as policy planners. Moreover, the central challenge before Indian agriculture is low productivity coupled with high variability in yield and yield gaps. The volatile agricultural growth in India seriously questioned the sustainability of agriculture; in the Economic Survey of India in 2016, it was categorically emphasized that "Indian agriculture, is in a way, a victim of its own success, which over time is posing to be a major threat" (Economic Survey 2015–2016). Also, the global hunger index (GHI) reflects the poor performance of the country with respect to ameliorating the problems of hunger and malnutrition, as it continued to lower its position from 55 (2014) to 80 (2015), to 97 (2016), to 100 (2017), to 103 (2018), to 102 out of 117 countries which ranked in 2019. Moreover, the point of grave concern here is that India's position is below our neighboring countries Pakistan (94), Bangladesh (88), and Sri Lanka (66), sufficient enough to remind the scientific community to develop suitable and affordable technologies in order to confront the situation. The FVs provide better solution in terms of having relatively high nutritional contents in order to overcome the malnutrition problems. Moreover, these varieties could be ideal candidates for the organic farming so that the product could fetch premium market price so that farmers' income be enhanced.

20.7 Protected Rice FVs in Eastern India

Rice in India is grown under extreme diverse agroecological conditions ranging from an altitude of 3 m below sea level to 2100 m above sea level and annual rainfall ranging from 500 to 5000 mm (Sathya 2013). The eastern states and other rainfed areas of the country have not been benefited by the GR, and these regions were bypassed (Bhatt et al. 2016). In this region, rice occupied nearly 55% area of the

Table 20.4 Rice FVs with nutritional/medicinal/therapeutic properties and adaptive traits

S. no.	Traditional rice variety	Nutritional/medicinal/therapeutic properties	References	PVP status[a]
1	Bhadoi	Unpolished brown rice rich in iron and antioxidants	Basu (2018) Rahman et al. (2006)	71 of 2017
2	Kabiraj Sal			–
3	Agniban			47 of 2017
4	Shatia	Rich in iron, antioxidants, fiber, low carbohydrate	Deb (2012) Rahman et al. (2006)	–
5	Kartiksal			–
6	Pichha vari	Enhance milk production in lactating mothers		–
7	Karthigai samba			–
8	Dudhsar			251 of 2013
9	Bhejri			–
10	Kelas	For curing anemia		253 of 2013
11	Bhutmoori			–
12	Paramai-sal	Promote child growth		–
13	Nyavara	For neurotic disorders		–
14	Karhainy	Alleviates paralysis		–
15	Gudna	Treats gastric ailments		–
16	Karanga	Treats dysenteric complaints		–
17	Bora	Treats jaundice		–
18	Pakheru	Traditionally used as a tonic		–
19	Saraiphool			–
20	Karia Gora			–
21	Punai Gora			–
22	Bhama	Consumption of this rice can help in working in fields for a whole day without feeling hungry		–
23	Danigora			–
24	Ramdi			–
25	Muru			–
26	Hindmauri			–
27	Danwar	For safe pregnancy		–
28	Karhani	For breathing and epilepsy		292 of 2018
29	Gunda	For chronic gastric problems		–
30	Dular	Drought tolerant	Sandhu and Kumar (2017)	103 of 2014
31	Kalia			279 of 2013

[a]PPV&FR Act (2018) Compendium of Registered Varieties under PPV&FR Act, 2001, p 388. Available at www.plantauthority.gov.in

country and is grown in different regions and seasons under varied ecologies (upland, lowland, rainfed, irrigated, submergence, and deep water). Also, out the total 11.6 Mha rice-fallow area in the country, 82% lies in the eastern states. Out of total area under hybrid rice ($\approx$2.7 Mha), eastern region covers $\geq$80% of it. The eastern Indo-Gangetic plains (IGPs) has over 20 Mha of rainfed rice area, and about 10 Mha area in the region is drought affected (NAAS 2013). In India, >42% area under rice is rainfed, and 47% of the total rice-growing area is located in rainfed ecosystem (19.3 Mha), which contributes to less than 25% of the total rice production (Ellur et al. 2013). Rainfed upland rice constitute about 6.1 Mha area in India, of which about 4.3 Mha falls under eastern region with very low productivity of less than 1.0 t/ha (Das et al. 2010; Shetty et al. 2013). Nevertheless, about 12 Mha land area is waterlogged and flood-prone, and about 60% of the total flood-prone area in the country lies in Indo-Gangetic basin (NAAS 2017). In India, about 17.4 Mha of rainfed lowland rice are grown each year; of which 5.2 Mha are submergence-prone. Rainfed upland rice constitute about 6 Mha area in India and eastern region, representing 85% of the total upland rice area in the country.

The HYV adoption pattern in rice varies among farmers and locations including the differential pattern of adoption of HYVs during different seasons in the same region (Pandey and Gauchan 2012). The classic examples are in case of rice in Odisha, during summer season, while the area is covered with HYVs; in other seasons such as autumn and winter, local varieties play important role. In Bihar, during Bhadai season, only local varieties are grown (Singh et al. 2016, 2018a). As mentioned earlier, rice in the eastern India is grown under most diverge, adverse, and varied ecologies based upon soil topography and moisture availability. Under unfavorable conditions, FVs play an important role, being climate resilient, and as can be seen from Table 20.5, out of the total 1527 rice FVs that have been legally protected through PPV&FR Act (2001), 1430 (>93%) rice FVs are from the eastern region. The rice FVs with local adaptation are grown in the region under low-input marginal environmental conditions.

Table 20.5 Rice farmers' varieties protected through PPV&FR Act (2001) in different states of eastern India

S. no.	State	Rice average area (Mha)	No. of protected rice FVs (up to October 22, 2019)	Remark
1	Assam	2.4	2	(1) Very early (<100 days) = 3 (2) Early (101–120 days) = 123 (3) Medium (120–140 days) = 599 (4) Late (141–160 days) = 490 (5) Very late (>160 days) = 88 Late and very late FVs are characterized by having photoperiod sensitivity with strong seed dormancy in rainfed lowland areas (Collard et al. 2013; Das 2013)
2	Bihar	3.3	3	
3	Chhattisgarh	3.8	332	
4	Jharkhand	1.8	95	
5	Odisha	3.7	760	
6	Eastern UP[a]	≈3.5	6	
7	West Bengal	5.3	232	
	Subtotal	23.8 (55%)	1430 (93.6%)	
8	Rest of India	19.7	97	
	Grand total	43.5	1527	

[a]Area covered under rice in the Eastern UP

20.8 Popular Rice FVs with Adaptive and Other Desirable Properties Grown in Different Ecologies of Jharkhand

Although the yield capacity of traditional varieties is limited, it is compensated by other appreciable characters such as high nutritional value, good cooking qualities including pleasurable aroma, and sufficient volume of cooked meal with less quantity of raw rice. Clearly, folk rice varieties have enormously high potential of remedying dietary iron and zinc deficiencies in the country, without the need to invest in developing transgenic iron-fortified rice. Several Indian rice folk cultivars are either very nutritious or maintain their distinctive aroma and colors. In Jharkhand, rice farmers' varieties are still grown under different rice ecologies. Upland rice FVs are grown under rainfed conditions in upland topography, which are collectively known as Gora Dhan, but each FV has distinct characteristics; Sinha (2016) has described in detail the characteristics including adaptive and other desirable properties of the popular rice FVs of Jharkhand. The list of some popular rice FVs with adaptive and other desirable properties grown in different ecologies of Jharkhand along with their PVP status is given in Table 20.6.

In Jharkhand, the popular farmers' rice varieties with special adaptive, nutritive, and/or medicinal traits and that fetch higher market prices as compared to their HYV counterparts under cultivation are listed in Table 20.7. Some of the FVs are even popular during special religious or social ceremonies. Moreover, a few FVs even have obtained PVP certificates under PPV&FR Act (2001).

Many farmers grow both HYVs, under high inputs, and FVs for their own domestic use under traditional management. The agricultural integration of HYVs into traditional systems may lead to the genetic integration of modern and traditional varieties. The modification of HYVs by traditional farmers is known as "criolloization" (Lamola and Bertram 1994) or "rustication" (Prain 1993). In this process, HYVs exchange genes with FVs. Valuable features of modern varieties are integrated under farmer management, with desirable features of traditional varieties (Wood and Lenne 1997). In wheat and rice, varieties developed in North and South India, respectively, spread, adopted, and diffused in eastern region, thus reflecting the advantage of domestic spillover. Similarly, hybrid rice adoption has even taken place in rainfed areas of eastern India that were not originally targeted for hybrids (Tripp et al. 2010).

20.9 Farmers' Varieties and Ecosystem Services

Modern HYVs are bred to suit for agricultural intensification (high input–high output) and in turn these varieties do provide high provisioning services in terms of higher yield while Ficiciyan and others (2018) have reported other important services like regulating—resistance against pests and diseases appear to often become lost during breeding for HYVs (Ficiciyan et al. 2018). Similarly, extremely important abiotic stress like drought is mainly responsible for high yield gap as well as high variability in yield, and in this regard, the similar kind of observation was

Table 20.6 Popular rice FVs with adaptive and other desirable properties grown in different ecologies of Jharkhand[a]

S. no.	Farmers' rice variety	Adaptive trait(s)[b]	Nutritional/medicinal/ therapeutic attributes	PVP status[c]
Upland rice ecology				
1	Arsnga	EM, DT	Medicinal value, rice beer, High Satiety Index (HSI)	–
2	Arsunga Gora	EM, DT		–
3	Khating	EM, DT	Nutritious, with HSI value	–
4	Dani Gora	EM, DT	Effective in gastric problem; rich in carbohydrate, protein, and minerals; HSI value	–
5	Lal Gora	EM, DT		–
6	Kala Gora	EM, DT		–
7	Bala Gora	Medium DT	High vitamins B1 and B3 and carbohydrates	–
8	Jaunga	EM, DT	Good for lactating mother	–
9	RaiChuni	EM, DT	Grain is white, hard, sub-transparent, non-glutinous, and non-scented	–
10	Asanloya			–
11	Tanr Jhili			–
12	Rani Kajal			181 of 2013
13	Sathi	EM, DT	Parboiled rice is sweet in taste; stale rice is eaten for 1–2 days	714 of 2014
14	Kannu	EM, DT	Rich in carbohydrate, protein, and minerals	–
Lowland rice ecology				
1	Agin Sar	Good yield even during rain fails Flood tolerant	Rich in minerals and vitamins; starchy water is very helpful to keep fresh and energetic	–
2	Khanika Sar			–
3	Bhorang Sar			–
4	Agni Sal	Sown by broadcasting in June/mid-July (Lewa)	Rich in carbohydrate, protein, and minerals	–
5	Tila Sar (sair)			219 of 2016
6	Barah Sar	Stand upright even after the maturity of grains in the panicles	Hydrated starchy water (Mar) is very helpful to keep fresh and energetic	223 of 2016
7	Ram Sar			–
8	Bhagwan Sar			–
9	Bhojni	Low yield but high straw strength	The red or brown unpolished rice is a healthy food	–
10	Saraikela			–
11	Dahia	DR, DT	Rich in minerals and vitamins Parboiled rice is good in taste	–
12	Dhusri	Sown by broadcasting and transplanted		260 of 2018
13	Rani Kajar			294 of 2016
14	Laldhan	Sown by broadcasting and transplanted; DT Non-shattering	Rich in vitamins and minerals; Laldhan is good for puffed rice/ pressed rice	–
15	Don Karanga			–

(continued)

Table 20.6 (continued)

S. no.	Farmers' rice variety	Adaptive trait(s)[b]	Nutritional/medicinal/therapeutic attributes	PVP status[c]
16	Prasad Bhog	Straw liked by livestock	Good market value; used during special occasions/festivals	–
17	Jhaliar Geanda	Sown by broadcasting and transplanted; non-lodging, saline tolerant DT, FT	Arwa rice used to make different cooking items; parboiled rice is good in taste; used in making local rice drink	–
18	Chhorki			–
19	Kalamdani			112 of 2017
20	Sambalpuria			–
21	Ketki	Sown by broadcasting and transplanted; DT, tolerant to pests and diseases	Good cooking quality/taste. Hydrated starch (Mar) is drunk as food supplement	–
22	Khira Bicha			–
23	Rangi			–
24	Chingmohri			–
25	Sursuria	Minimum water consumption Less requirement of chemical fertilizers	Rich in carbohydrate, protein, and minerals Parboiled rice is good in taste	–
26	Dudhras			–
27	Newair			–
28	Jaya			–
29	Motichur	Long straw yield Good for livestock	Good market value; different food items are made with Arwa rice; used during special religious or social ceremonies	–
30	Nanhia			–
31	Basmari			–
32	Karmusal	Pests and disease tolerant; less water requirement	Starchy water is very helpful to keep fresh and energetic	–
33	Sikki			265 of 2018
34	Hathi Panjar (Hathi Panja)	Higher yield without chemical fertilizers; NS	Preferred Usna rice	–

[a]Adapted and modified from Sinha (2016)
[b]DT, drought tolerant; EM, early maturity; NS, non-shattering; DR, disease resistant; FT, flood tolerant
[c]PPV&FR Act (2018) Compendium of Registered Varieties under PPV&FR Act, 2001, p 388. Available at www.plantauthority.gov.in

recorded by Sandhu and Kumar (2017) by mentioning that the modern high-yielding varieties, although possessing high yield potential, are highly vulnerable to abiotic stresses such as drought, and in the course of post GR breeding, unknowingly, the drought tolerance contributing alleles of traditional cultivars have not been properly retained/maintained in the modern cultivars (Sandhu and Kumar 2017). Nevertheless, the ecosystem's resiliency depends upon the diversity at species level, variety level, and gene level, but wide use of improved HYVs has led to a genetic bottleneck, resulting in the loss of crop, variety, and allele diversity (Peroni and Hanazaki 2002; Tsegaye and Berg 2007). Similarly, the modern HYVs are mainly responsible for declining of provisioning, regulating, and cultural ecosystem services, and this observation was made in none other than the Millennium

Table 20.7 Popular rice FVs with special adaptive with higher economic return grown in Jharkhand[a]

S. no.	Farmers' rice variety	Adaptive trait(s)[b]	Economic and other trait(s)	PVP status[c]
1	Badshah Bhog	DT, good yield even crop damage	Higher market prices than the HYVs; unpolished rice is rich in vitamins B1 and B3	–
2	Chaina Bhog			–
3	Kari Jeera (Sonachur)	High straw yield	Higher market prices than the HYVs; used during special religious or social ceremonies	210 of 2016
4	Kishun Bhog			–
5	Kapoor Bhog			–
6	Megh Jawain	High straw yield	Higher market prices than the HYVs; used during special religious or social ceremonies	–
7	Jeera Jawain			–
8	Tulsimanjar	Tall plant with high straw yield	Higher market prices than the HYVs; Arwa rice used to make pitha/idli/dosa (Chilka roti)	205 of 2016
9	Baans Phul			–
10	Shaha Jeera	Stand upright even after the maturity of grains in the panicles	Higher market prices than the HYVs; easily digestible, good for sick persons	–
11	Shyam Jira			367 of 2016
12	Raisdhan	Sown by transplanting and broadcasting	Good market price Good for pressed rice	270 of 2018
13	Jabakusum			–

[a]Adapted and modified from Sinha (2016)
[b]DT, drought tolerant
[c]PPV&FR Act (2018) Compendium of Registered Varieties under PPV&FR Act, 2001, p 388. Available at www.plantauthority.gov.in

Ecosystem Assessment (MEA 2005). The current process of agricultural intensification system is the single most important threat to biodiversity (Gept 2006). FVs are known to possess traits for local adaptation, stress tolerance, yield stability, and nutrition. Moreover, rice FVs such as "Pokkali" are excellent source of salt tolerance as salt stress adversely affects rice yield in rainfed and irrigated agroecosystems (Dwivedi et al. 2016); the various kinds of ecosystem services provided by HYVs and FVs and their comparative roles are mentioned in Table 20.8.

In rainfed areas and other low-input marginal environments, marginal and small farmers operate in complex, diverse, and risk-prone (CDR) environments with minimum or no external inputs. In such environments, farmer-managed seed systems representing informal seed system are the only means on which resource-constrained farmers rely. FVs are preferred over HYVs by virtue of their climate-resilient nature (Sinha 2016), are locally adapted, and have stress tolerance along with yield stability and seed nutrition (Dwivedi et al. 2016).

Table 20.8 Comparison of HYVs and FVs in relation to ecosystem services

S. no.	Ecosystem service	Performance indicator	High-yielding versus farmers' varieties
1	Provisioning services	Yield	FVs yield equally or higher under harsh conditions (Yadav 2010; Brocke et al. 2014; Li et al. 2012); HYVs exhibit higher yield, but input costs may be also high, even counterbalancing the benefit from higher yields (Li et al. 2012); HYVs outyield FVs under optimal conditions (Kante et al. 2017)
		Resource use efficiency	FVs tend to deliver more stable yields under limited environments (Sangabriel-Conde et al. 2014; Lafitte et al. 1997); FVs are sources of nutrient-use efficiency traits, which are very important for sustaining agriculture (Newton et al. 2010)
		Crop storability	Higher storability of FVs with lower levels of storage losses to insects (Maggs-Kolling and Christiansen 2003; Moreno et al. 2006)
2	Regulating services	Resilience to climate change effects	FVs are often better adapted to drought stress (Annicchiarico 2006; Mazvimbakupa et al. 2015; Munoz-Perea et al. 2007); FVs may be more pest resistant (Olson et al. 2012); FVs are better adapted to local climate conditions (Olson et al. 2012; Fenzi et al. 2017)
		Biological pest and disease control	FVs maintain high levels of resistance against pest and disease (Sánchez-Martín et al. 2017; Patil et al. 2014; Tamiru et al. 2011); FVs are sources of host-plant resistance and abiotic stress tolerance genes (Newton et al. 2010)
		Pollination	Declining pollinators (intensified land use, climate change, alien species, and the spread of pests and pathogens (Kearns et al. 1998; Potts et al. 2010); this has serious implications for human food security and health and ecosystem function (Vanbergen 2013)

(continued)

Table 20.8 (continued)

S. no.	Ecosystem service	Performance indicator	High-yielding versus farmers' varieties
		Biodiversity richness	Up to 75% of plant genetic diversity has been lost due to the rapid expansion of industrial agriculture and large-scale adoption of monoculture farming (Jacques and Jacques 2012)
3	Cultural services	Tradition, cooking quality, nutritional values, taste and color, aesthetic, medicinal and cultural significance	FVs are passed over generations together with recipes (Montes-Hernandez et al. 2005); FVs are sources of phytonutrients with desired micronutrient concentrations that alleviate human aging-related and chronic diseases (Newton et al. 2010)

20.10 Farmers' Varieties for Future Crop Improvement

Climate change is a reality now, and on the basis of long-term data (six decades) on temperature, rainfall, and crop production, a long-term trend of rising temperatures, declining average precipitation, and increase in extreme precipitation events have been observed, and long-term weather patterns imply that climate change could reduce annual agricultural incomes in the range of 15–18% on average and up to 20–25% for rainfed areas (Economic Survey 2017–2018). In India, climate variability has increased both spatially and temporally over the past 50 years (Davis et al. 2019). In the future, the frequencies of extreme events in the form of droughts, floods, salinity, coastal flooding, submergence, and increased temperature extremes are bound to intensify further. Therefore, to adapt and mitigate adverse effects of climate change, the climate resilient as well as resource efficient varieties need to be developed. Farmers' varieties being climate resilient and resource efficient have greater role to play as these varieties do possess the adaptive traits required to breed abiotic stress-tolerant varieties. Moreover, FVs do have comparatively higher nutritional contents in a delicately balanced way, and high-yielding varieties along with higher nutritional contents are the need of the day as micronutrient malnutrition problems have been mainly responsible for the country's poor position in terms of global hunger index, which is even below than our neighboring countries.

By providing novel gene(s) for enhancing nutrient use efficiency and improving the nutrition quality of staple grains and against various abiotic and biotic resistance/tolerance, FVs are known to possess such traits; for example, gene for submergence tolerance in rice has been transferred from a farmer's variety into rice variety "Swarna," which became megastar variety (Swarna sub-1) in eastern India (Singh 2018) covering around 30% area in eastern region and neighboring countries. With respect to rice breeding for grain quality, the popular Basmati rice variety "Karnal Local," which possesses better grain and cooking quality, was a selection from the traditional Basmati rice collection in Karnal district of Haryana, later released as Taraori Basmati in 1996 (Singh et al. 2004). It was widely used parents by rice breeders for grain, cooking, and eating characteristics. The International Rice Research Institute (IRRI), Philippines, used it in as many as 249 crosses (Singh et al. 2018b). Kalanamak, a heritage rice from eastern Uttar Pradesh, is under cultivation since millennia, and its usage as a new variety was released as "Kalanamak 3" (KN3) for cultivation in eastern Uttar Pradesh, India. Subsequently, several semidwarf breeding lines, developed through hybridization or induced mutation, outyielded KN3 by 40% (Chaudhary et al. 2012). A first semidwarf, nonblack husk cultivar "Bauna Kalanamak 102," with comparable cooking quality and aroma to "KN3," has been released for cultivation (Chaudhary et al. 2012; Dwivedi et al. 2019). In wheat, also, the widely adapted variety HD2967 is known to have 25 farmers' varieties in its parentage. Similarly, farmers' sorghum variety Mangalwedha Maldandi (M-35-1) dominates the post-monsoon season sorghum areas in India. Moreover, this FV is least affected by temperature fluctuations and sowing dates, flowering, and maturity. Also, this FV has inherent quality/traits, which save it from insects like shoot fly and aphids and diseases like charcoal and rust. Several varieties were developed at ICRISAT incorporating genes from these FVs.

20.11 Conclusion

Eastern region occupies about 21.85% geographical area and supports 34% of the population of the country, having high population density (616 persons/km^2 compared to 382 persons/km^2 at all India level). Having 84% of the population in rural, the region is recognized with high rural population, and therefore, agriculture is the main occupation for food and livelihood security. The comparatively low irrigated area in the region is 39% as against 45% of the national average effect the production and productivity and responsible for high yield gaps. Adequate and timely supply of critical inputs in terms of quality seed and planting material, fertilizers, and credit and low productivity due to increased frequencies of extreme climate events in the form of floods and droughts are the major factors for lower output and higher vulnerability, which altogether culminate in high poverty level in 69 districts (the maximum number of economically most backward districts) out of 150 at national level (Bhatt et al. 2016). The first three sustainable development goals (SDGs) of the United Nations are related to end poverty (SDG 1), end hunger (SDG 2), and health (SDG 3), which is to be achieved by 2030. In this regard, except West Bengal, all states in the eastern region are lagging behind in these three SDGs, and to overall improve the status, FVs could play an important role in securing food and nutritional security, in addition to performing better ecosystem services. The region is ideally suited for organic agriculture to increase farmer's income. Also, the agro-biodiversity can be maintained in better ways for future crop improvements.

References

Agricultural Statics at a Glance (2018) Published by the Directorate of Economics and Statistics under the Ministry of Agriculture and Farmers Welfare, Department of Agriculture, Cooperation & Farmers Welfare (Government of India)

Annicchiarico P (2006) Diversity, genetic structure, distinctness and agronomic value of Italian lucerne (*Medicago sativa* L.) landraces. Euphytica 148:269–282

Baranski MR (2015) Wide adaptation of green revolution wheat: international roots and the Indian context of a new plant breeding ideal, 1960–1970. Stud Hist Phil Biol Biomed Sci 50:41–50

Basu M (2018) Reviving traditional rice varieties in West Bengal. https://earthjournalism.net/stories/reviving-traditional-rice-varieties-in-west-bengal

Bhatt BP, Mishra JS, Dey A, Singh AK, Kumar S (2016) Policy document second green revolution in eastern India: issues and initiatives. ICAR Research Complex for Eastern Region, Patna, p 50

Brocke KV, Trouche G, Weltzien E, Kondombo-Barro CP, Sidibe A, Zougmore R, Goze E (2014) Helping farmers adapt to climate and cropping system change through increased access to sorghum genetic resources adapted to prevalent sorghum cropping systems in Burkina Faso. Exp Agric 50:284–305

Chandler RF Jr (1982) An adventure in applied science: a history or the international rice research institute. International Rice Research Institute, Los Banos, pp 1–233

Chaudhary RC, Mishra SB, Yadav SK, Ali J (2012) Extinction to distinction: current status of Kalanamak, the heritage rice of eastern Uttar Pradesh and its likely role in farmers' prosperity. LMA Conv J 8:7–14

Chauhan JS, Pal S, Choudhary PR, Singh BB (2016) All India coordinated research projects and values for cultivation and use in field crops in India: genesis, outputs and outcomes. Indian J Agric Res 50(6):501–510

DAC & FW (2016a) State of Indian agriculture 2015–16. Department of Agriculture, Cooperation & Farmers Welfare, Directorate of Economics and Statistics, Ministry of Agriculture & Farmers Welfare, New Delhi, Government of India, 280 p

DAC & FW (2016b) All India report on input survey 2011–12, agriculture census division. Ministry of Agriculture and Farmers Welfare, New Delhi, p 521

Das L, Rath NC, Mishra SK (2010) Appropriateness of improved rice cultivation technologies for upland ecosystem. Indian Res J Ext Educ 10(1):29–32

Davis KF, Chhatre A, Rao ND, Singh D, DeFries R (2019) Sensitivity of grain yields to historical climate variability in India. Environ Res Lett 14:064013. https://doi.org/10.1088/1748-9326/ab22db

Deb D (2012) Folk rice varieties, traditional agricultural knowledge and food security. https://www.researchgate.net/publication/233987520_Folk_Rice_Varieties_Traditional_Agricultural_Knowledge_and_Food_Security

Deb D, Sengupta S, Pradeep T (2015) A profile of heavy metals in rice (*Oryza sativa* ssp. *indica*) landraces. Curr Sci 109(3):407–409

Dennis JV (1987) Farmer management of rice variety diversity in northern Thailand. PhD dissertation. Cornell University, Michigan University Microfilms print, Ann Arbor, p 367

Dwivedi SL, Ceccarelli S, Blair MW, Upadhyaya HD, Are AK, Ortiz R (2016) Landrace germplasm for improving yield and abiotic stress adaptation. Trends Plant Sci 21(1):31–42

Dwivedi S, Goldman I, Ortiz R (2019) Pursuing the potential of heirloom cultivars to improve adaptation, nutritional, and culinary features of food crops. Agronomy 9:441. https://doi.org/10.3390/agronomy9080441

Economic Survey (2015–2016) Agriculture: more from less, Chapter 04, 68–85 pp

Economic Survey (2017–2018) Climate, climate change, and agriculture, Vol(1), Chapter 6, pp 82–101

Ellur RK, Singh AK, Gupta HS (2013) Enhancing rice productivity in India: aspects and prospects. In: Shetty PK, Ayyappan S, Swaminathan MS (eds) Climate change and sustainable food security. ISBN: 978-81-87663-76-8. National Institute of Advanced Studies, Bangalore and Indian Council of Agricultural Research, New Delhi, pp 99–132

Evenson RE, Gollin D (2003) Assessing the impact of the green revolution, 1960 to 2000. Science 300:758–762

Fenzi M, Jarvis DI, Arias Reyes LM, Latournerie Moreno L, Tuxill J (2017) Longitudinal analysis of maize diversity in Yucatan, Mexico: influence of agro-ecological factors on landraces conservation and modern variety introduction. Plant Genet Resour 15:51–63

Ficiciyan A, Loos J, Sievers-Glotzbach S, Tscharntke T (2018) More than yield; ecosystem services of traditional versus modern crop varieties revisited. Sustainability 10:2834. https://doi.org/10.3390/Su10082834

Gept P (2006) Plant genetic resources conservation and utilization: the accomplishments and future of a societal insurance policy. Crop Sci 46:2278–2292

Hazell PBR (2009) The Asian green revolution: IFPRI discussion paper 00911, IFPRI, p 40

Heisey PW, Morris ML (2002) Practical challenges to estimating the benefits of agricultural R&D: the case of plant breeding research. Prepared for presentation at the 2002 annual meeting of the American Agricultural Economics Association (AAEA) Long Beach, California, 28–31 July 2002, pp 1–32

Herdt RW, Capule C (1983) Adoption, spread and production impact of modern rice varieties in Asia. International Rice Research Institute, Los Banos, p 59

Jacques PJ, Jacques JR (2012) Monocropping cultures into ruin: the loss of food varieties and cultural diversity. Sustainability 4:2970–2997

Kahlon LK, Singh R (2019) Tradition knowledge and dynamics of edible plants of primitive tribal group 'paudi bhuyan' with changing demography migration patterns in northern Odisha. Indian J Tradit Knowl 18(1):7–15

Kante M, Rattunde HFW, Leiser WL, Nebié B, Diallo B, Diallo A, Touré AO, Weltzien E, Haussmann BIG (2017) Can tall Guinea-race sorghum hybrids deliver yield advantage to smallholder farmers in west and Central Africa? Crop Sci 57:833–842

Kearns CA, Inouye DW, Waser NM (1998) Endangered mutualisms: the conservation of plant–pollinator interactions. Annu Rev Ecol Syst 29:83–112

Lafitte HR, Edmeades GO, Taba S (1997) Adaptive strategies identified among tropical maize landraces for nitrogen-limited environments. Field Crops Res 49:187–204

Lamola LM, Bertram RB (1994) Experts gather in Mexico to seek new strategies in preserving agrobiodiversity. Diversity 10:15–17

Li J, van Bueren ETL, Jiggins J, Leeuwis C (2012) Farmers' adoption of maize (*Zea mays* L.) hybrids and the persistence of landraces in Southwest China: implications for policy and breeding. Genet Resour Crop Evol 59:1147–1160

Maggs-Kolling GL, Christiansen JL (2003) Variability in Namibian landraces of watermelon (*Citrullus lanatus*). Euphytica 132:251–258

Mahadevappa M (2015) Ever green and never fading seed industry: some issues. Organized by Department of Studies in Biotechnology, University of Mysore, Mysuru, Karnataka & National Academy of biological sciences, Chennai, Tamil Nadu at University of Mysore, Mysuru, Karnataka, 21 August 2015

Mazvimbakupa F, Modi AT, Mabhaudhi T (2015) Seed quality and water use characteristics of maize landraces compared with selected commercial hybrids. Chil J Agric Res 75:13–20

Millennium Ecosystem Assessment (2005) Ecosystems and human well-being: current state and trends: findings of the condition and trends. Island Press, Washington, DC

Montes-Hernandez S, Merrick LC, Eguiarte LE (2005) Maintenance of squash (*Cucurbita* spp.) landrace diversity by farmers' activities in Mexico. Genet Resour Crop Evol 52:697–707

Moreno LL, Tuxill J, Moo EY, Reyes LA, Alejo JC, Jarvis DI (2006) Traditional maize storage methods of Mayan farmers in Yucatan, Mexico: implications for seed selection and crop diversity. Biodivers Conserv 15:1771–1795

Morris M, Mekuria M, Gerpacio R (2003) Impacts of CIMMYT maize breeding research. In: Evenson RE, Gollin D (eds) Crop variety improvement and its effect on productivity. The impact of international agricultural research. The green revolution: mythology and reality. CABI, Wallingford, pp 135–158

Munoz-Perea CG, Allen RG, Westermann DT, Wright JL, Singh SP (2007) Water use efficiency among dry bean landraces and cultivars in drought-stressed and non-stressed environments. Euphytica 155:393–402

NAAS (2013) Water use potential of flood-affected and drought-prone areas of eastern India. Policy paper no. 60, National Academy of Agricultural Sciences, New Delhi, 20 p

NAAS (2017) Abiotic stress management with focus on drought, flood and hailstorm. Policy paper no. 87 National Academy of Agricultural Sciences, New Delhi, 14 p

Newton AC, Akar T, Baresel JP et al (2010) Cereal landraces for sustainable agriculture. A review. Agron Sustain Dev 30:237–269

Olson MB, Morris KS, Mendez VE (2012) Cultivation of maize landraces by small-scale shade coffee farmers in western El Salvador. Agric Syst 111:63–74

Pandey S, Gauchan D (2012) Synthesis of key results and implications In: Pandey S, Gauchan D, Malabayabas M, Bool-Emerick M, Hardy B (eds) Patterns of adoption of improved rice varieties and farm-level impacts in stress-prone rainfed areas in South Asia. International Rice Research Institute, Los Baños, pp 1–18

Patil JV, Reddy PS, Prabhakar Umakanth AV, Gomashe S, Ganapathy KN (2014) History of post-rainy season sorghum research in India and strategies for breaking the yield plateau. Indian J Genet Plant Breed 74:271–285

Pender J (2008) Agricultural technology choices for poor farmers in less-favoured areas of south and East Asia. Occasional Papers, Knowledge for development effectiveness by the International Fund for Agricultural Development (IFAD), p 82

Peroni N, Hanazaki N (2002) Current and lost diversity of cultivated varieties, especially cassava, under Swidden cultivation systems in the Brazilian Atlantic Forest. Agric Ecosyst Environ 92:171–183

Potts SG, Biesmeijer JC, Kremen C et al (2010) Global pollinator declines: trends, impacts and drivers. Trends Ecol Evol 25:345–353

Prain G (1993) Mobilizing local expertise in plant genetic resources research. In: de Boef W, Amanor K, Wellard K, Bebbingtoneds A (eds) Cultivating knowledge: genetic diversity, farmer experimentation and crop research. Intermediate Technology Publications, London, pp 102–110

Rahman S, Sharma MP, Sahai S (2006) Nutritional and medicinal values of some indigenous rice varieties. Indian J Tradit Knowl 5(4):454–458

Roy TN (2015) Review on access of small and marginal farmers to improved seeds in India. Econ Aff 60(4):713–719

Salazar R, Louwaars NP, Visser B (2006) On protecting farmers' new varieties: new approaches to rights on collective innovations in plant genetic resources. CAPRi Working Paper # 45. International Food Policy Research Institute (IFPRI), Washington, DC

Sánchez-Martín J, Rispail N, Flores F, Emeran AA, Sillero JC, Rubiales D, Prats E (2017) Higher rust resistance and similar yield of oat landraces versus cultivars under high temperature and drought. Agron Sustain Dev 37

Sandhu N, Kumar A (2017) Bridging the rice yield gaps under drought: QTLs, genes, and their use in breeding programs. Agronomy 7:27. https://doi.org/10.3390/agronomy7020027

Sangabriel-Conde W, Negrete-Yankelevich S, Eduardo Maldonado-Mendoza I, Trejo-Aguilar D (2014) Native maize landraces from los Tuxtlas, Mexico show varying mycorrhizal dependency for P uptake. Biol Fertil Soils 50:405–414

Sathya A (2013) Are the Indian rice landraces a heritage of biodiversity to reminisce their past or to reinvent for future? Asian Agri-Hist 17(3):221–232

Shetty PK, Hegde MR, Mahadevappa M (2013) Innovations in rice production. National Institute of Advanced Studies, Indian Institute of Science, Bangalore, p 262

Singh RP (2018) Integration and commercialization of local varieties under sub-optimal environments for food security, promoting sustainable agriculture and agro-biodiversity conservation. MOJ Eco Environ Sci 3(2):65–67

Singh RP, Agarwal RC (2019) Farmer's varieties in India-factors affecting their preferential prevalence and the current status of their legal protection. Indian J Agric Sci 89(9):1371–1382

Singh RP, Agarwal RC (2020) Farmer's varieties in India – new insights into adaptation patterns, performance and preferences over modern varieties among different seasons and regions. Indian J Agric Sci

Singh VP, Pratik S, Gopala Krishnan S, Singh AK (2004) Role of Indian Agricultural Research Institute in genetic improvement of rice varieties in India. In: Sharma SD, Rao P (eds) Genetic improvement of rice varieties in India. Today and Tomorrow's, New Delhi, pp 141–187

Singh KK, Ali M, Venkatesh MS (2009) Pulses in cropping systems. Technical Bulletin, IIPR, Kanpur, p 47

Singh RP, Kumar A, Pal SK (2016) The prevalence, productivity and protection of traditional varieties vis-à-vis modern varieties in Eastern India: an appraisal. Jharkhand Journal of Development and Management Studies, Thematic title-Seed biotechnology and Eastern India's "New Green Revolution". Issues Challenges 14(2):6955–6970

Singh RP, Prasad SR, Agarwal DK, Sripathy KV (2017) Varietal adoption and replacement: constraint's, opportunities and impact in India. Indian Council of Agricultural Research-Indian Institute of Seed Science (ICAR-IISS), Mau, p 196. ISBN: 9591-82-764-03

Singh RP, Singh S, Lal SK (2018a) Channelizing protected farmer's varieties through semi-formal seed systems for effective utilization and conservation of agro-biodiversity: an overview. Seed Res 46(2):87–97

Singh VP, Singh AK, Mohapatra T, Krishnan G, Ellur RK (2018b) Pusa Basmati 1121 – a rice variety with exceptional kernel elongation and volume expansion after cooking. Rice 11:19. https://doi.org/10.1186/s12284-018-0213-6

Sinha H (2016) Rediscovering the traditional paddy varieties in Jharkhand: conservation priority in hybrid rice area. J Rural Dev 35(2):285–307

Tamiru A, Bruce TJA, Woodcock CM, Caulfield JC, Midega CAO, Ogol CKPO, Mayon P, Birkett MA, Pickett JA, Khan ZR (2011) Maize landraces recruit egg and larval parasitoids in response to egg deposition by a herbivore. Ecol Lett 14:1075–1083

Tripp R, Hu R, Pal S (2010) Rice seed provision and the evolution of seed markets. In: Pandey S, Byerlee D, Dawe D, Dobermann A, Mohanty S, Rozelle S, Hardy B (eds) Rice in the global economy – strategic research and policy issues for food security. IRRI, Los Baños, pp 213–230

Tsegaye B, Berg T (2007) Utilization of durum wheat landraces in East Shewa, Central Ethiopia: are home uses an incentive for on-farm conservation? Agric Hum Values 24:219–230

Vanbergen AJ (2013) Threats to an ecosystem service: pressures on pollinators. Front Ecol Environ 11(5):251–259

Wood D, Lenne JM (1997) The conservation of agro-biodiversity on-farm: questioning the emerging paradigm. Biodivers Conserv 6(1):109–129

Yadav OP (2010) Drought response of pearl millet landrace-based populations and their crosses with elite composites. Field Crops Res 118:51–56

Zeven AC (1998) Landraces: a review of definitions and classifications. Euphytica 104(2):127–139

Ensuring Food Security by Good Seed Governance: A Case Study from Jharkhand 21

R. P. Singh

Contents

Abstract

Agriculturally, the state of Jharkhand belongs to the category of the complex, diverse, and risk-prone (CDR) region and home to the relatively high level of resource-poor marginal and small farmers (M&SFs). Moreover, being a rainfed state, the delayed onset and early withdrawal of monsoon with asymmetrical distribution, increased incidence of droughts including terminal ones, unseasonal heavy rainfalls, and extreme temperatures increased vulnerability, variability in yield, and yield gaps affecting the food and nutritional security seriously. In addition to the poor adoption and diffusion of new technologies, farmers apply comparatively low external inputs, and therefore, the production and productivity remained low, and unsustainability is the hallmark of the Jharkhand agriculture.

R. P. Singh (✉)
Birsa Agricultural University, Ranchi, Jharkhand, India

K. Bauddh et al. (eds.), *Ecological and Practical Applications for Sustainable Agriculture*, https://doi.org/10.1007/978-981-15-3372-3_21

To increase productivity, ensuring timely supply of quality seed of suitable varieties in adequate quantity at affordable price to farmers is of critical importance. In the state, both seed replacement rates (SRRs) and varietal replacement rates (VRRs) are very poor, thus affecting productivity performance. At national level, the formal seed sector comprising both the public sector and private seed industry is advanced and ranked fifth globally, but at regional level, it varies particularly in low-input marginal environments like states in the eastern region of the country, and Jharkhand typically represents this category. The infrastructure related to seed production, processing, storage, and distribution is in a nascent stage in the state and always depends on the external agencies. With the objective to increase both SRR and VRR in the state, the production of quality seed of the adopted as well as newly released varieties was taken up at massive scale using participatory approach during 2016–2019. In this chapter, the practical experience and difficulties encountered have been presented. However, the country is advanced in formulating various legislations, policies, and well-designed seed regulatory framework in the light of international instruments, agreements, conventions, and other laws for fostering growth of the domestic seed industry and well-being of the Indian farmers. However, at state level, the implementation and execution of the regulatory measures to ensure the timely availability of quality seed of suitable varieties at affordable price at local level requires good governance. In the state, the poor seed governance has converted into poor SRR and VRR in various cereal, coarse cereal, pulses, and oilseed crops including contingent crops of local importance. The whole process also culminated in farmers and seed growers' lost confidence. The major qualitative impact of this failed participatory seed production (PSP) program is reflected by less area coverage, low production, and poor yield virtually among all cereals, pulses, and oilseed crops. Based upon the experiences gained so far, it is learnt that good seed governance along with institutional support is needed at every step of seed production, certification, procurement, processing, storage, and timely distribution to improve seed production and supply chain effectively and efficiently. Without good seed governance, both seed and food security would likely be jeopardized.

Keywords

Participatory seed production · Seed governance · Seed production and supply chain management · Semiformal seed system · Seed and varietal replacement rates

21.1 Introduction

The newly created state Jharkhand came into existence in 2000 as the 28th state carved out from erstwhile Bihar under the Fifth Schedule of the Constitution of India. The majority of state population thrives on agriculture, and scheduled tribe caste is dominating. Although the state agriculture is characterized by having good natural resources in terms of ideal climatic conditions, sufficient rainfall but deficient water availability for irrigation, and ample agro-biodiversity, the agricultural production and productivity is low due to complete dependency on monsoon setting. Also, improper management of natural resources (soil, water, and agro-biodiversity), insufficient infrastructure, and poor postharvest management including processing and market facilities are responsible for low agricultural production which also cause migration. The state is having predominantly marginal and small farmers (M&SFs) with the highest poverty among farm household 45.3% in Jharkhand as compared to 22.5% at national level (Chand 2017). The state is inhabited predominantly by the resource-poor M&SFs being >85% who apply relatively low external inputs (seed, fertilizers, plant protection measures, and irrigation) under the rainfed marginal environmental conditions. The poor adoption of new technologies due to the relatively high level of illiteracy in addition to socioeconomic condition among M&SFs is also considered one of the reasons for low yield. Moreover, the increased frequencies of extreme events in terms of erratic rainfall pattern with highly asymmetrical distribution, intermittent droughts including terminal ones, and temperature extremes particularly at reproductive stage are responsible for the increased vulnerability, variability in yield, and yield gaps. The state of Jharkhand also faces the ill effects of climate change and drought being the predominant one in this plateau region. Climate change is a reality now, and every year is warmer than the previous one. Droughts are regular features of the state, and after every 23 years, there is a drought in the state. In addition, the cyclones also affected crop production particularly post-monsoon season crops. Although the average annual rainfall of the state is higher than the national average, due to its erratic and asymmetrical distribution, more than 80% water runoff, rainfall which generally used to cover approximately 100 rainy days during monsoon season, significantly reduced to ≈70 rainy days due to climate change, and therefore, the rainfall intensity has increased drastically, and the undulated land topography, which does not support natural water harvesting and which is mainly responsible for remaining fallow land during post monsoon rice season, culminated in very low cropping intensity and system productivity (land, crops, and farm animals due to acute shortage of green, dry fodder, etc.) as well.

For agricultural purpose, the state can safely be put in the category of the complex, diverse, and risk-prone (CDR) region, and the resource-poor smallholder farmers including tribal farmers apply low external input (seed, fertilizers, irrigation, plant protection measures, farm power machinery, etc.) which resulted in the low output. Furthermore, high poverty level and monsoon-based agriculture resulted in mono-cropping traditionally which has been the hallmark of Jharkhand agriculture cropping intensity (125 vs. 150) as compared to the national average. For any independent sovereign nation, the self-sufficiency with respect to ensuring food and nutritional security at national and regional level is of paramount importance. The food sovereignty in turn depends on the seed sovereignty which ultimately depends on the genetic resources and their creation, utilization, and conservation. Therefore, seed security is considered a prerequisite to food security, and both of them have common indicators to measure them. At national level, food security has been achieved long before, but at regional level, like eastern India which is also represented by Jharkhand state, the issue of food and nutritional security is still a serious concern, and the state is struggling to achieve on sustainable basis. Seed is key to food security because seed is the first link in the food value chain. In the state, the use of farm-saved seed (FSS) is the rule rather than the exception indicating low seed replacement rate which is directly linked with production and productivity. Moreover, due to very slow adoption and diffusion of the newly introduced varieties in each crop and only a few varieties just dominating the scene, it resulted in poor varietal replacement rate. To ensure food self-sufficiency, therefore, making the state self-sufficient with respect to seed is essential, and proper seed governance is crucial. Galiè (2013) defined seed governance as the formal and informal rules and behaviors that affect rights of access to and control of seed at the international, national, local, and individual levels. Nevertheless, participatory plant breeding (PPB), participatory varietal selection (PVS), and participatory seed production at local level enhance farmer's seed sovereignty (Galiè 2013; Ceccarelli 2016). In Jharkhand state, to accelerate the seed and varietal replacement rate in order to adapt and mitigate the adverse effects due to climate change, the seed production program through participatory approach was taken up at massive scale during 2016–2019. The experience gained in the form of constraints faced at various levels is presented in this chapter in the light of seed governance to make the state self-sufficient with regard to seed and food.

21.2 Requirement and Availability of Cereals, Pulses, and Oilseed in Jharkhand

As mentioned earlier, Jharkhand state is insecure with respect to food grains and oilseed. Since its origin, the state is struggling to become self-sufficient but could not achieve it on sustainable basis. The progressive status of the requirement and availability of cereals, pulses, and oilseed is mentioned in Table 21.1. Since 2001–2002, in the case of cereals, it has been the only state that is able to meet its need although not on sustainable basis, and year-to-year high fluctuations are recorded with respect to crop area coverage, production, and productivity. For pulses, the productivity is higher as compared to the national average, but due to devotion of larger area to cereals, particularly rice, the state is deficient in meeting its requirement; however, in the case of pulses, the gap between demand and supply is shrinking progressively. In the case of oilseed, the situation is worst, and a huge gap exists between demand and supply. Nevertheless, at national level also, the demand for edible oils is only met through import which causes heavy burden on national exchequer. However, in the case of Jharkhand, the oilseed productivity is very poor as compared to all India levels.

21.2.1 Operational Holdings in Jharkhand

In the state, 2.73 million farm holdings occupy 3.18 million ha of cultivable land with average farm size of 1.17 ha. For Jharkhand state, more than 84% of marginal

Table 21.1 The requirement and availability of food grains and oilseeds in Jharkhand

Year	Population (in million)	Cereals[a]			Pulses[b]			Fat (oil)[c]			
		Demand (MT)[d]	Avail. (MT)	Excess/ deficit (MT)	Demand (MT)	Avail. (MT)	Excess/ deficit (MT)	Oil demand (MT)	Oilseed demand (MT)	Avail. (MT)	Excess/ deficit (MT)
2001–2002	26.9	4.12	3.06	−1.06	0.39	0.17	−0.22	0.22	0.88	0.05	−0.83
2006–2007	30.0	4.60	3.39	−1.21	0.44	0.26	−0.18	0.24	0.96	0.07	−0.89
2011–2012	33.0	5.10	5.32	0.22	0.48	0.28	−0.20	0.26	1.04	0.16	−0.88
2016–2017	35.6 (Approx)	5.46	5.86	0.40	0.52	0.48	−0.04	0.29	1.16	0.26	−0.90

Calculated on the basis of minimum standard balanced diet by the Indian Council of Medical Research (ICMR): [a]cereals = @ 0.42 kg per capita/day; [b]pulses = @ 0.04 kg per capita/day; [c]oil = @0.022 kg per capita/day; [d]million tons

Table 21.2 No. of operational holdings in different categories and average holding size in Jharkhand state

S.N.	Farmer's category	No. of operational holdings	Area (ha)	Average holding size (ha)
1	Marginal (<1 ha)	1,877,072	776,743.65	0.41
2	Small (1–2 ha)	426,994	587,241.32	1.38
	Marginal and small (up to 2 ha)	**2,304,066 (84.33%)**	**1,363,985 (42.8%)**	**0.59**
3	Small medium (2–4 ha)	281,104	773,077.18	2.75
4	Medium (4–10 ha)	127,269	745,294.66	5.86
5	Large (>10 ha)	19,756	302,578.96	15.32
	Total	**2,732,195**	**318,4935.8**	**1.17**

Source: Agricultural Statistics at a Glance (2018)

and small farmers (M&SFs) cover less than 43% cultivable area as can be seen from Table 21.2. The major chunk belongs to marginal farmers (68.7%) with average farm size of one acre, while small farmers (15.6%) have average farm size of nearly three and a half acres. These M&SFs are resource-poor and socioeconomically backward with relatively high level of illiteracy making them unable to adopt new technologies and thus apply little external input such as improved seed, fertilizers and other agrochemicals, irrigation, and farm power. Nevertheless, these M&SFs are the major producer as well as consumers of food grains, and therefore, to increase production and productivity on sustainable basis is extremely important in making them and the state self-sufficient with respect to food and nutritional requirements. Also, to reduce poverty and vulnerability by mitigating the adverse climate change effects, these M&SFs need to be uplifted by assured and timely supply of required inputs to this marginalized group in particular.

21.2.2 Performance of Field Crops in Jharkhand State

The area production and productivity of cereals, pulses, and oilseed crops since 2001–2002 in the state is given in Table 21.3. In the case of cereals, it was during the past 18 years (from 2001–2002 to 2018–2019). Similarly, for pulses, *eight times* area coverage, *seven times* production, and *ten times* productivity are found to be lower as compared to the previous year. For oilseeds, *three times* area coverage as well as production and *eight times* productivity are found to be lower as compared to the previous year. This phenomena of fluctuating area coverage, low production, and poor productivity can easily be explained by the *unsustainability of Jharkhand agriculture*. The highest cereal yield was obtained during 2011–2012 (2.89 t/ha), while in the case of pulse and oilseed, it was recorded in 2012–2013 for both pulses (1.17 t/ha) and oilseed (0.8 t/ha). Also, in rainfed and marginal low-input environments, the livestock sector becomes extremely important for the livelihood security, and the deficiency of dry and green fodder also adversely affects the farm

Table 21.3 Area, production, and productivity of cereals, pulses, and oilseeds in Jharkhand

		Cereals			Pulses			Oilseed		
S.N.	Year	Area (M/ha)	Prod. (MT)	Yield (T/ha)	Area (M/ha)	Prod. (MT)	Yield (T/ha)	Area (M/ha)	Prod. (MT)	Yield (T/ha)
1	2001–2002	1.73	3.06	1.77	0.21	0.17	0.81	0.08	0.05	0.63
2	2002–2003	**1.61**[a]	**2.46**	**1.53**	0.24	**0.15**	**0.63**	0.09	0.06	0.67
3	2003–2004	1.63	2.74	1.68	0.30	0.17	**0.57**	0.10	**0.04**	**0.40**
4	2004–2005	**1.53**	**2.29**	**1.50**	**0.29**	0.19	0.66	**0.09**	0.05	0.56
5	2005–2006	1.59	**1.87**	**1.18**	0.29	**0.12**	**0.41**	0.09	0.05	0.56
6	2006–2007	1.95	3.39	1.74	0.38	0.26	0.68	0.15	0.07	**0.47**
7	2007–2008	1.97	3.81	1.93	0.39	0.30	0.77	**0.12**	0.07	0.58
8	2008–2009	1.99	3.86	1.94	**0.37**	**0.28**	**0.76**	0.13	0.07	**0.54**
9	2009–2010	**1.24**	**1.84**	**1.48**	**0.30**	**0.22**	**0.73**	0.14	0.07	**0.50**
10	2010–2011	**1.03**	**1.46**	**1.42**	0.41	0.27	**0.66**	0.19	0.09	**0.47**
11	2011–2012	1.84	5.32	*2.89*	**0.34**	0.28	0.82	0.23	0.16	0.70
12	2012–2013	**1.83**	**4.76**	**2.60**	0.59	0.69	*1.17*	0.25	0.20	*0.80*[b]
13	2013–2014	**1.69**	**4.52**	2.67	**0.57**	**0.58**	**1.02**	0.28	**0.18**	**0.64**
14	2014–2015	1.94	5.15	**2.65**	0.59	0.59	**1.00**	0.28	0.19	0.68
15	2015–2016	2.04	**3.14**	**1.54**	**0.55**	**0.50**	**0.91**	0.28	0.19	0.68
16	2016–2017	2.22	5.86	2.64	0.81	0.84	1.04	0.36	0.26	0.72
17	2017–2018	2.25	6.20	2.76	**0.79**	0.87	1.10	0.42	0.30	**0.71**
18	2018–2019	**1.95**	**3.70**	**1.90**	**0.74**	**0.73**	**0.99**	**0.33**	**0.23**	**0.70**

Source: Kharif workshop 2019, Directorate of Agriculture, Deptt. of Agriculture, Animal Husbandry and Cooperative, GOJ
[a]Bold represents less area/production/productivity as compared to the previous year to check the sustainability
[b]Italic indicates the highest productivity

animal productivity particularly in bad years due to prevailing drought conditions which has become a regular phenomenon.

21.3 Indian Seed Sector

The Indian seed industry is represented by public and private seed sectors. In India, presently, the public sector comprises one national-level corporation, namely, National Seed Corporation (NSC), 15 State Seed Corporations (SSCs), 22 Seed Certification Agencies (SCAs), 2 Central Seed Testing, and 122 State Seed Testing Laboratories (3 are ISTA accredited, and 20 have ISTA membership) to provide the required support to serve the seed industry and farmers. India is known to have 15 out of 20 agro-climates of the world, and climatically, the country is divided into 6 ecological, 20 agro-ecological, 60 sub-agro ecological regions, and 127 agro-climatic zones under six ecosystems. India is also bestowed with 22 agro-biodiversity hot spots distributed over 7 agro-geographical zones. The country also possesses 46 of the 60 soil types in the world and holds the second largest agricultural land in the world. India's climatic conditions offer an ideal environment for the production of virtually all kinds of crops as more than 52% of India's land is cultivable against the global average of 11%, which are plus points. Nevertheless, the country is rich in human resource capital, and ample man power is there to fulfill the need of different categories (executives, plant breeders, management personnel, traders, seed dealers, farmers, seed producers, skilled labors, etc.) of the seed industry. There are currently two seed associations, namely, the National Seed Association of India (NSAI) and the Federation of Seed Industry of India (FSII). India is also among the top ten genetically modified (GM) crop-cultivating countries, with insect-resistant Bt cotton covering a land area of >11 Mha.

The Indian seed industry is ranked fifth globally and quite mature with huge infrastructure in terms of seed production, processing, storage, and trade. It deals with a range of products, namely, open-pollinated varieties (OPVs), hybrids, and genetically modified (GM) seed. Contribution of the Indian private seed sector amounted to be >40% of total seed production in India. It encompasses over 500 private seed companies, and more than two dozen of them have close collaboration with transnational seed corporation (TNCs) and many of them with their own breeding programs (Dasgupta and Roy 2011). The Indian seed market reached a value of US$ 3.6 billion in 2017, exhibiting a compound annual growth rate (CAGR) of around 17% during 2010–2017 and further expected to grow at a CAGR of 14.3% during 2018–2023, reaching a value of more than US$ 8 billion by 2023 (Anonymous 2018). The formal seed sector in India accounts for about 30–35% of the total seed distributed in country, while the informal seed sector comprising mainly farm-saved seed (FSS) accounts for the remaining 65–70% (DAC & FW 2016a). With regard to the use of quality seed, out of a total of 138.11 million operational holdings, only 39.41% used certified seeds, while 26.96% used seeds of notified varieties (DAC & FW 2016b). In Jharkhand, the use of farm-saved seed is much more and is the established rule rather than the exception particularly in the case of local and contingent crops like horsegram, linseed, lentil, niger, finger millet, and other small millets. Since improved seed at sowing time is not available in sufficient quantities even for resowing due to droughts, in that case, also farm-saved seed and seed are obtained through informal means (community seed bank, village seed bank, seed exchange mechanisms, etc.), which is the only means to raise a crop. However, in the case of rice and maize, significant area is diverted under hybrids, and in the case of hybrid rice (4–5 lakh ha) and maize, the whole area under hybrids is covered by proprietary hybrids.

21.3.1 Seed Policies

Formally, the seed production and distribution for field crops in the country started with the green revolution (GR) in the form of the improved seed of high-yielding varieties (HYVs) as the seed being the basic input, while other inputs in terms of fertilizers, plant protection measures, irrigation, and farm power machineries create favorable conditions to realize the genetic potential embodied in the seed to get higher yield. Altogether, all these technologies known as green revolution (GR) technologies ultimately established the process of agricultural intensification. Initially, public seed sector shouldered the responsibilities to produce and distribute the quality seed of HYVs and dealt with crops requiring higher seed rate, and later on after the realization of the positive impact of hybrid vigor or heterosis on yield, private seed sector started its activities in crops like maize, sorghum, and pearl millet requiring low seed rate with high economic returns. This gave major boost for the involvement of private seed sector in field crops, and presently, the private seed sector deals with a range of products. Like in any other sectors, there are several issues to deal with through regulatory framework, legislations policies, schemes, programs, to resolve them appropriately. Also, the obligations in the forms of international treaties, conventions, agreements, developmental goals, and intellectual property management need to be in compliance with seed sector too. Therefore, to serve farming communities better and to increase the production and productivity in order to ensure food and nutritional security at local level, the seed governance issues need to be resolved within the ambit of prevailing regulatory framework to manage seed chain efficiently and effectively. For the benefit of farmers and to facilitate the seed industry by making it more competitive, several measures in the forms of policies and legislations were implemented, and important ones are mentioned below:

1. *Seed Act (1966) and Seed Rules 1969*
2. *National Seed Project Phase-I (1977–1978), Phase-II (1978–1979), and Phase-III (1990–1991), strengthened the seed system infrastructure in the country*
3. *Seed Review Team-SRT (1968)*
4. *National Commission on Agriculture's Seed Group (1972)*
5. *Seed Control Order (1983;1994), regulating seed dealers*

6. *Environment Protection Act (1986) for India's biotechnology regulatory framework and Rules for the Manufacture, Use/Import/Export and Storage of Hazardous Microorganisms/Genetically Engineered Organisms or Cells, 1989 under EP Act, 1986*
7. *Technology Mission on Oilseeds & Pulses (TMOP) in 1986 now called The Integrated Scheme of Oilseeds, Pulses, Oil Palm and Maize (ISOPOM)*
8. *Seed Transport Subsidy Scheme (1987)*
9. *New Policy on Seed Development (1988) providing the farmer with the best planting materials available in the world*
10. *New Industrial Policy (1991), to encourage foreign direct investment in the seed industry.*
11. *Seed Bank Scheme (2000)*
12. *National Seed Policy (2002), to safeguard the interest of farmers, promoting growth of domestic seed industry & agro-biodiversity conservation with the major emphasis on to increase seed replacement rates among all field crops*
13. *Plant Quarantine Order (PQO Regulation of Import into India), in January 2004*
14. *The Seeds Bill (2004), which proposes mandatory registration of all varieties, to replace Seed Act (1966), Seed Control Order 1983*
15. *Protection of Plant Varieties and Farmers' Rights Act (2001), which is India's primary legislation based on the principle of sui generis to comply Trade Related Intellectual Property Rights (TRIPS) of the World Trade Organization (WTO) obligations depicting several features protecting breeder's rights as well as protecting the rights of farmers in the light of Article 9 of the FAO's "International Treaty on Plant Genetic Resources for Food and Agriculture" (ITPGRFA), also known as Seed Treaty, India, that has balanced the conflicting obligations to breeders and farmers as represented by the TRIPS and the Convention on Biological Diversity (CBD) in its laws related to the development and use of improved seed*
16. *Formulation of National Seed Plan (2005) to determine the seed replacement rate and estimation of crop wise seed requirement*
17. *National Mission for Sustainable Agriculture (NMSA)-2007, which came into existence under the National Action Plan on Climate Change (NAPCC) focusing on areas critical to agriculture in adapting to climate change with the priority action for the development and promotion of improved technologies including the development of stress resistant crop varieties*
18. *National Food Security Mission (NFSM)-2007*
19. *Rashtriya Krishi Vikas Yojna (RKVY)-2007*
20. *Joining of OECD seed schemes (2008) in order to facilitate seed trade in international market*
21. *Export and import policy to liberalize seed export and planting materials with few exceptions*
22. *New policy on Seed Development/Modified policy (2011) on Seed Sector Provision for import of wheat and rice*

23. *Cotton Seed Price (Control) Order, 2015, for fixation of sale price for cotton seeds to ensure their availability to the farmers at fair, reasonable, and affordable prices*
24. *National Biotechnology Development Strategy 2015–2020 in December, 2015*

The prevailing seed regulatory framework is applicable at state level too, and some of the schemes like National Mission for Sustainable Agriculture, National Food Security Mission, Rashtriya Krishi Vikas Yojna, and Bringing Green Revolution to Eastern India are being operated in Jharkhand, and pulse seed hub programs under NFSM are being executed through State Agricultural University at five different centers. Due to initial stage of the newly commissioned Jharkhand State Agricultural Development Corporation Limited (JSADC Ltd) from the earlier formed Jharkhand State Seed Corporation Limited (JSSC Ltd), Jharkhand State Seed Certification Agency (JSSCA), and the Directorate of Agriculture, the infrastructure is being developed. The deployment of regular officers and other supportive and technical staff to make them fully functional is still awaited in order to make better coordination as well as better governance with fixed accountability.

21.3.2 Seed Governance

The seed governance challenges include the mitigating constraints and creating favorable environment to foster growth in the seed sector and also to formulate adequate and appropriate governance measures to combat the various identified challenges in the seed production and supply chain. Coleman (2008) conceptualized seed governance as an instrumental ensemble of discursive and nondiscursive factors dealing with institutions, regulatory orders, and administrative appropriateness. Similarly, Atalan-Helicke and Mansfield (2012) recognized the fact that seed governance is the product of an open-ended process of strategic elaboration among stakeholders (including rural women who have been shown to play a major and significant role in preserving and creating knowledge about seed—World Bank et al. 2009) involved in seed sector development and trade. Dasgupta and Roy (2011) have identified a number of seed governance issues with regard to seed policies, regulation, seed production, and supply chain management which are presented in Table 21.4.

21.3.3 Participatory Plant Breeding

Participatory plant breeding program also known as the decentralized plant breeding was started during the 1990s in India and elsewhere particularly in rainfed marginal and low-input environments where the high-yielding varieties (HYVs) could not perform better, and their productivity was found to be at par or even inferior to the locally adapted farmers varieties (FVs). The HYVs being uniform and input responsive require higher inputs in terms of high dose of fertilizers and irrigations to

Table 21.4 Major steps in a functional seed production and supply chain at national level

S.N.	Governance issue	Governance inputs	Main governance challenges
1	Seed regulatory framework	Institutional mechanisms for variety release, registration, seed quality control, seed certification, seed labelling, seed import, and export	(1) Evaluating the procedures, costs, and mechanisms for variety release and registration, seed quality, seed certification, seed labelling, seed import, and export (2) Exploring opportunities or flexibility in standards and procedures that allow low-cost and alternative methods (3) Updating seed policy to affect changes in seed legislation, procedures, and resources for implementation
2	Varietal development	Ensuring a steady supply of new crop varieties that farmers can use to replace older varieties	(1) Increasing efficiency and effectiveness of crop breeding toward specific adaptation instead of wide adaptation (2) Adopting the participatory approaches (PPB, PVS, PSP, QDS seed scheme, use of farmer's varieties (FVs), TVs, and other PGRs)—for faster and cost-effective breeding and seed multiplication
3	Varietal registration	Variety registration system is based on the distinctiveness, uniformity, and stability (DUS) plus value for cultivation and use (VCU) criteria	(1) Restricting market access for FVs (2) Exemption from mandatory multilocation yield trials for VCU and DUS can help to local and farmer-led participatory approaches in variety development and acceleration of commercializing FVs
4	Varietal protection	Protection of Plant Varieties and Farmers' Rights Act (2001). India's primary legislation for TRIPS (WTO) compliance. Exhibits several features protecting breeder's rights as well as protecting the farmer's rights by balancing the conflicting obligations to breeders and farmers as represented by TRIPS and the CBD in its laws related to the development and use of improved seed	(1) In the case of cross pollinated crops, FVs face difficult task to pass the DUS test since by nature, FVs are not as uniform as formal seed sector release varieties. Although India has amended through regulation for furtherance of the implementation of the PPV&FR Act 2001, stipulating that for FVs uniformity standard could be released to allow double the number of off-type as otherwise permitted for registration of other categories of varieties under PPV&FR Act (2001)

(continued)

Table 21.4 (continued)

S.N.	Governance issue	Governance inputs	Main governance challenges
			(2) Protecting farmers' rights over their own PGR and seed systems particularly their right for seed saving, etc. (3) Harmonizing national PVP systems with the WTO (TRIPs), CBD, ITPGRFA-the Seed Treaty, UPOV, etc., in order to have uninterrupted access to improved seeds and crop genetic resources
5	Seed quality control	Strengthen institutional capacity through investment in training and monitoring of seed quality	(1) Alternative approaches to seed quality standards and ways to implement them effectively and efficiently (2) Capacity building at local level for quality seed production by imparting trainings in this venture
6	Seed certification	To provide guarantee with respect to varietal genetic identity and genetic purity	(1) Addressing inadequacies of the existing institutional capacity for seed certification for assuring seed quality (2) Review of seed certification standards to increase the seed of varieties through careful maintenance
7	Governance measures to improve seed supply systems	(1) Develop a long-term national policy (2) Upgrade the regulatory framework (seed laws, rules, acts, etc.) (3) Investment in strategic areas of the seed sector: processing and storage, road transportation, and delivery systems	(1) Estimation of import and local production of hybrid seeds and their quality control (2) Facilitating greater private sector involvement (3) Removing trust deficit between public and private sectors (4) Protecting farmers against spurious seed and frauds (5) Eliminating bureaucracy in the administrative procedure to expedite variety release, registration, and seed multiplication processes (6) Capacity building to engage a broad range of stakeholders (farmer groups, private sector, NGOs, community-based organizations, SHGs, etc.) (7) Decentralized with focus on farmers' participation at different stages in the variety

(continued)

Table 21.4 (continued)

S.N.	Governance issue	Governance inputs	Main governance challenges
			development process and create effective institutional mechanisms for mainstreaming farmer-participatory breeding
8	Governance for the GM seed	(1) Streamline procedures for introducing and testing crop varieties developed through transgenic technology within the framework of national seed laws (2) Establishment of the coordination between the agriculture ministry and other government ministries and departments (3) Increasing effectiveness and efficiency of the biosafety regulatory system as well as institutional capacity systems	(1) Eliminating inconsistencies and ambiguities and promoting operational synergy within the prevailing regulatory framework in relation to transgenic crops (2) Removing administrative bottlenecks, inconsistencies in application of rules and procedures, and fast tracking regulatory approval of GM crop varieties through inter-ministerial coordination committee (3) Accreditation of laboratories to gain public confidence including laboratories for seed testing of GM crops (4) Systematic campaign of educating and awareness building around the pros and cons of GM technology as the public perceptions of GM technology are often not based on scientific facts

perform better to give higher yield, and under rainfed marginal and low-input environments, due to input constraints, their performance was found to be far below than expected. Moreover, the HYVs developed using wide adaptation philosophy, and in marginal environments, the different adaptive traits are required for making them resilient to adverse conditions, and consequently, locally or specifically adapted local varieties performed better due to their diversity and resilient nature. Also, farmers prefer different traits in the rainfed marginal and low-input environments; for example, under these conditions, farmers give more weightage to varieties with more straw as compared to short-statured HYV rice to meet the requirement of their livestock which is very important for their livelihood security under harsh conditions (Singh 2012; Singh and Singh 2016).

The participatory plant breeding and/or participatory varietal selection at local level is very important as these approaches are evolutionary, dynamic, and less time-consuming as compared to varieties developed through formal seed system. Moreover, the instant adoption and diffusion of the selected varieties make these approaches more attractive and less costly with relatively high rate of success. Ceccarelli and Grando (2019) reported that the participatory approach of plant breeding and seed production has been adopted in 10 developed and 59 developing countries summing to 69 countries altogether virtually among all self-, cross-, and vegetatively propagated crops. Moreover, this approach was even adopted by various International Agricultural Research Centres (IARCs) under the umbrella of CGIAR, in addition to agricultural universities and NGOs. In Jharkhand, through PVS approach, Kalinga III variety of upland rice in Jharkhand state of Eastern India has been released for its cultivation, while through client-oriented breeding (COB), other upland rice varieties Ashoka 200F (BVD 109) and Ashoka 228 (BVD 110) have also been released. Originally released in Jharkhand, Ashoka 200F is now also recommended in Gujarat, MP, and Rajasthan and Ashoka 228 in MP. Similarly, for rainfed, maize variety BVM-2 for eastern India was released (Witcombe and Yadavendra 2006; Singh et al. 2019).

21.3.4 Participatory Certified Seed Production

Appropriate seed delivery systems is crucial for marginal and small farmers in low-input marginal environments (Bishaw and Turner 2008). Jharkhand, being a rainfed state, always faces the problems of availability of quality seed in adequate quantities at the affordable price of suitable varieties. To avoid dependence on the external agencies for supplying seed of suitable varieties in the state, it is necessary that seed of locally adapted and adopted varieties be produced locally for ensuring the timely availability of most appropriate and climate resilient varieties for ensuring seed and food security in state. Moreover, due to its dependence on monsoon like other rainfed areas of the country likely to hit more by the increased frequencies of extreme events due to climate change, to mitigate the negative effects due to climate change contingent planning is of prime importance for which assured supply of even second batch of quality seed is a prerequisite for reducing vulnerability, reducing variability in yield, and ultimately bridging yield gaps. Participatory seed production program was started by different Krishi Vigyan Kendras (KVKs) of the university in 2011–2012 and was continued till 2015–2016. The range of activities taken up by the different KVKs in order to create awareness with respect to use of quality seed for crop production program (Table 21.5) along with entrepreneurship development by imparting trainings/seed days/field days BAU under the aegis of ICAR Seed Project, Tribal Sub Plan (TSP), not only helped in dissemination of technologies at untapped far-flung tribal areas of Jharkhand but also helped in enhancement of their income and development of huge human resource in the seed sector through the establishment of seed villages (Singh et al. 2015a). A massive seed production program through participatory approach was taken in 2016 as per the directives given by the chief secretary, Jharkhand Government. In the process, various stakeholders such as Department of Agriculture, Jharkhand State Agriculture Development Corporation Ltd, Jharkhand State Seed Certification Agency, State agricultural University, and its KVKs in addition to a large number of farmers in each

Table 21.5 Participatory certified seed production (*Quintals*) in farmers' field—from 2013–2014 to 2017–2018

S.N.	Crop	2013–2014	2014–2015	2015–2016	2016–2017	2017–2018
Monsoon season						
1	Rice	12,550	7550	12,223	90,323	164,393
2	Finger millet				270	240
3	Pigeonpea	240	100	30	2045	1218
4	Greengram				215	67
5	Blackgram				352	485
6	Horsegram				260	
7	Sesame			3	46	83
8	Groundnut				100	2009
	Subtotal	**12,790**	**7650**	**12,256**	**93,611**	**168,496**
Post-monsoon (Rabi) season						
9	Wheat	1050	275		8577	472
10	Chickpea			94	7997	314
11	Lentil				1389	121
12	Field pea				20	91
13	R&M		75	80	5159	882
14	Linseed				205	137
	Subtotal	**1050**	**350**	**174**	**23,347**	**2016**
	G. Total	**13,840**	**8000**	**12,430**	**116,958**	**170,513**

district were involved in seed production activities. The progressive status of seed production through participatory approach before and after 2016–2017 is given in Table 21.5. As can be seen, the quantity of quality seed produced along with the number of crops in which seed production activities were extended increased, and virtually, each and every crop was included under this program including contingent crops.

21.3.5 Constraints Experienced

The main constraints encountered during 2016–2018 participatory seed production program have been related to the seed certification, procurement, and the lifting of the produced seed from the seed growers and an inordinate delay in making payment followed by cumbersome seed certification process and inadequate logistic facilities in terms of collection, processing, and proper seed storage facilities. No availability of technically qualified and experienced seed certification personnel, their lackluster approach, non-coordination among officials from different stakeholders such as the Department of Agriculture (DOA-GOJ), Jharkhand State Seed Corporation Limited (JSSC Ltd) created in 2012–2013 and later restructured as Jharkhand State Agricultural Development Corporation Limited (JSADC Ltd) as major inputs supplier agency, and Jharkhand State Seed Certification Agency (JSSC Agency) further complicated and exaggerated problems, while administratively, adoption of "ad hoc" approach, no advance indent given to the university with regard to basic generation seed, adoption and enforcement of different seed pricing purchase policy from the seed growers, and high level of bureaucratic and hierarchical approach. All these factors culminated in the institutional failures, and seed growers sold the seed as grain and suffered huge loss economically and also lost confidence to move on further in the same venture. The failure of participatory seed production may be attributed to sudden decision of launching massive seed village program without proper planning, the absence of a "seed production, procurement, processing, and marketing module," the lack of resource inventory consisting an adequate number of seed processing machines and transport system, and administrative lapses and poor seed governance by the officials from the line department and State Department of Agriculture/Jharkhand State Agriculture Development Corporation/Jharkhand State Seed Certification Agency.

21.3.6 Impact of Participatory Seed Production Program Failures

The seed production program using participatory approach was taken up on such a massive scale with the sole objective to increase the seed replacement rate in various field crops instantly as SRR is known to have positive and strong association with yield. Moreover, as mentioned earlier, the production and distribution of quality seed using participatory approach have two other ancillary objectives, namely, to reduce the use of farm-saved seed in a phased manner and also to develop the seed entrepreneurships among rural youth including farm women. The quality seed was produced at local level, but poor seed governance ultimately reflected through poor seed replacement rate of major field crops as presented in Table 21.6. The SRR decreased significantly in most of crops since 2014–2015, and this program was a major failure.

21.3.7 Varietal Replacement Rate (VRRs)

Varietal development, release, notification, and their introduction into seed chain is a continuous process through which the old varieties get replaced by the new one in order to take the advantage of superior genotypes (varieties) with improved yield, quality, and/or tolerance to the prevailing abiotic and biotic stress(es). Through increased VRRs, the investment made is also realized by getting higher yield and/or better quality. The VRRs in general are relatively low, and only a few widely adapted varieties cover large area in each crop (Singh 2015; Singh et al. 2017). Moreover, in some crops including important ones like rice, in some of the South Asian regions like eastern India, the VRRs are reported to be at par or even poor with the sub-Saharan region of the African continent (Singh et al. 2020). To mitigate the climate change adverse effects, the seed production program of climate-resilient new varieties was also taken up in order to realize the positive impact of their genetic

Table 21.6 Seed replacement rate status of different crops grown in Jharkhand

Crop	Norms	2010–2011	2011–2012	2012–2013	2013–2014	2014–2015	2015–2016	2016–2017	2017–2018	2018–2019
Rice	33	25.1	17.1	22.4	15.65	15.41	10.74	7.54	7.25	8.17
Maize	50	14.3	7.3	13.6	1.15	13.65	1.4	0.96	1.46	1.74
Wheat	33	27.6	31.9	41.2	8.34	22.26	20.20	17.70	16.22	16.99
Lentil	33	13.9	6.9	6.3	4.50	2.77	2.56	0.77	7.93	4.24
Chickpea	33	2.4	1.3	2.9	0.73	2.48	0.00	0.18	4.13	9.53
Pigeonpea	50	11.7	23.6	21.6	3.26	6.70	0.00	0.93	0.80	0.09
Greengram	33	–	21.2	19.0	7.03	0.00	2.22	7.15	1.78	10.24
Blackgram	33	–	6.9	5.1	2.96	10.01	14.06	37.46	13.24	36.33
Field Pea	33	17.4	9.4	10.2	–	–	–	–	–	–
Horsegram	33	23.7	7.6	8.8	–	–	–	–	–	–
R/mustard	50	72.1	20.5	34.4	2.43	15.59	2.78	0.03	1.30	1.55
Niger	50	–	–	39.3	–	–	–	–	–	–
Groundnut	33	–	–	–	–	–	–	35.51	58.60	34.38
Soybean	33	–	–	–	–	–	–	12.57	12.44	3.05

Source: Zonal seed review meeting for Rabi 2011–2012 (North-East Zone dt. 12.9.2011), Rabi 2012–2013 dt. 14.9.12, MOA-GOI, Deptt. of Agriculture & Cane Development, Jharkhand 2015–2016 to 2018–2019 (Directorate of Agriculture, GOJ)

Table 21.7 Participatory seed production of new varieties during 2016–2019

S.N.	Crop	Variety	Year	2016–2017	2017–2018	2018–2019
Monsoon (Kharif) season						
1	Rice	IR 64 (Drt.1)	2015	60	5175	90
		Sahbhagi	2011	29,173	38,938	9121
		R. Mansuri	2007		9370	600
		Abhishek	2007	7311	5100	3200
2	Pigeonpea	NDA-2	2008	60	564	756
		IPA 203	2012	10	200	612
3	Blackgram	Azad-3	2006		71	122
4	Greengram	IPM-2-3	2009	60	61	81
5	Groundnut	Kadri 6	2005	100	1113	
6	Sesame	GT-3	2009		23	23
		Savitri	2008		60	
		RT 351	2011	36		
	Subtotal			**36,810**	**60,676**	**14,605**
Post-monsoon (Rabi season)						
1	Wheat	HD 2967	2011	6375	–	
2	Lentil	PL 8	2010		121	
3	Chickpea	JAKI 9218	2006	7587	314	
		GNG1581	2008	125		
4	Field pea	Aman	2010		91	
5	R&M	Pusa-30	2013	3254	–	
		Pusa 26	2011	44	–	
		Pusa 28	2012		250	
6	Linseed	Azad Alsi-2	2011	155		
	Subtotal			**17,540**	**776**	
	G. Total			**54,350**	**61,452**	**14,605**

potential in the form of higher yield in addition to the tolerance to abiotic condition particularly drought (Table 21.7), but due to nonlifting and nondistribution of seed, the varietal replacement rate also could not be improved. Among all cereals, pulses, and oilseed crops, new varieties suitable for the state were introduced, and their seed production through participatory approach was taken up. But their impact could not be realized due to the lack of coordination and poor seed governance.

The poor seed governance has converted into poor seed and varietal replacement rates (SRRs and VRRs) in various cereal, coarse cereals, pulses, and oilseed crops including contingent crops of local importance (finger millet, horsegram, sesame, niger, linseed, etc., Table 21.5) The whole process also culminated in farmers and seed growers' low confidence. The major qualitative impact of this failed PSP program reflected in 2017–2018 the less area coverage in pulses and poor yield in oilseeds that were recorded in spite of good monsoon, while in 2018–2019 less area coverage, low production, and poor yield virtually among all cereals, pulses, and oilseed crops were recorded (Table 21.3); of course, unfavorable weather condition also played a significant role.

21.3.8 Seed Governance Measures

The experience gained along with the constraints encountered in using participatory seed production approach in Jharkhand state has been discussed including how the negligence and non-seriousness from either of the state stakeholders involved in seed certification, procurement, processing, storage, and distribution lead to the institutional failure. The concepts of seed governance and seed system resilience have started gaining momentum in the recent past. Rietberg et al. (2014) stressed upon urgent need to link those concepts to real-life situations and empirical data, in order to understand the relations between governance of seed systems, seed system resilience, and seed security in order to ensure food and nutritional security at local level. In the following sections, the practical difficulties and experience gained in seed production and distributions have been discussed and the critical issues presented in Table 21.8.

In Jharkhand, farmers' varieties of rice also cover 21% area (Sinha 2016). To improve the quality of FVs is also very important to enhance production and productivity. In the case of rice, 372 varieties from farmers and processing plants were found to be infected with rice bunt pathogen. Moreover, in rice, the problem of seed discoloration was also observed which was caused by the infection due to a number of *mycoflora* (*Helminthosporium oryzae*, species of *Drechslera*, *Fusarium*, *Curvularia*, *Alternaria*, etc.) and ultimately resulted in drastic reduction of seed germination below the seed certification standards (Singh and Agrawal 2018). Seed production of farmers' varieties (FVs) using an alternative registration and certification system (semiformal) developed by the Food and Agriculture organization of the United Nations known as "Quality Declared Seed (QDS)" system is advocated to channelize and commercialize FVs using formal system (Singh et al. 2013, 2015b). Like rice, seed production of HYVs and FVs of other crops may also be taken using participatory approach at local level.

Ceccarelli and Grando (2019) reported the restricted institutional support for participatory approach related to participatory plant breeding, participatory varietal selection, and participatory seed production activities; also in the present case dealing with participatory plant breeding and participatory seed production (PSP) program during 2016–2018, the similar kind of experience encountered along with poor seed governance and almost no institutional support were found to be major problems which confirm the difficulties encountered herewith (Singh et al. 2019). The failure program of PSP has led to poor seed and varietal replacement rates, reduced area coverage, and low crop production along with poor yield, and ultimately the seed insecurity jeopardized the food and nutritional security at local level. The experience gained in the form of severe constraints, lack of institutional support, and lack of coordination among seed regulatory agencies including other stakeholders from line department along with poor seed governance can be

Table 21.8 Governance issues for seed production and supply chain at state level (Jharkhand)

S.N.	Issues	Governance inputs	Main governance challenges
1	Varietal development	(1) Varietal mismatch and heavy dependence on spillover benefit (Singh 2015; Singh et al. 2017, 2019), etc., mainly responsible for poor performance of MVs in low-input marginal rainfed environment (Kulkarni 2013; Shetty et al. 2013) (2) Poor adoption of MVs of upland rice due to their unstable and unprofitable nature as compared with irrigated transplanted rice (Pandey et al. 2015; Witcombe et al. 2009) and FVs being preferred over HYVs due to their tolerance to abiotic and biotic stresses (Singh and Agarwal 2020) and FVs which are grown under rainfed conditions in upland topography collectively known as Gora Dhan (Sinha 2016) (3) Development of varieties suitable for existing cropping systems to ease adoption and diffusion	(1) Development and deployment of varieties with specific adaptation (2) Uneconomical seed production of upland rice varieties and therefore involvement of producer farmers' organizations (PFOs), NGOs, SHGs, and women farmer entrepreneurs needed (3) Development and deployment of varieties among crops of local importance (small millets, horsegram, niger, traditional root, tuber crops, etc.) (4) Private seed sector's poor interest in OPVs while emphasizing on the hybrids among cereal and coarse cereals
2	Varietal registration, induction in seed chain	(1) Huge gap between varieties released and their induction in seed chain (Gautam 2013; Singh et al. 2017) causing serious problems of the domination of only a few varieties having wide adaptation in each crop (Singh 2015, 2018) (2) Nonavailability of seed pertinent to newly released varieties (Patnaik 2013; Singh et al. 2019) (3) No registration of varieties/ hybrids for the state	(1) Diverse varieties with respect to maturity and tolerance to various abiotic and biotic stress of local importance need to be inducted into seed chain (2) Marketing of varieties/ hybrids which have not been recommended for the state (3) Prebreeding to introduce new germplasm is absolutely necessary with the objective to enhance specific adaptation in marginal environments
3	Increasing seed and varietal replacement rates	(1) Poor seed and varietal replacement rates are considered a twin perennial bottlenecks that Indian seed sector is ailing (DSR 2014; Singh et al. 2017; Singh and Agarwal 2019) (2) VRR and SRR are necessary	(1) Participatory plant breeding approaches including varietal selection approaches (PPB & PVS) are needed to increase the adoption, diffusion, and faster varietal replacement at local level in the complex, diverse, and risk-prone (CDR) and

(continued)

Table 21.8 (continued)

S.N.	Issues	Governance inputs	Main governance challenges
		for enhancing productivity, and either of these two though has individual effect but cannot make great strides in productivity stressing the need of proper combinations of both VRR and SRR as the impact of these two is synergistic and multiplicative (Singh 2018) (3) Slow varietal replacement rates due to poor genetic gain (Witcombe et al. 1998; Nagarajan et al. 2008; Walker et al. 2015; Muralidharan et al. 2019) (4) Poor seed replacement in crops with low seed multiplication rate but higher seeding rate per unit area such as upland rice, soybean, groundnut, chickpea, and fieldpea (Singh et al. 2017). (5) Major emphasis placed by the formal seed sector on the production of large quantities of seed and the marketing of just a few varieties with wide adaptation (Salazar et al. 2007; Louwaars and De Boef 2012; Singh et al. 2017) (6) Prevalence of breeding philosophy of wide adaptation (Baranski 2015) (7) Preference of less-vulnerable, climate-resilient, and locally adapted FVs, in spite of full adoption of MVs (Raghu et al. 2015) and Less/nil use of MVs for domestic and household consumption (Raghu et al. 2015)	low-input marginal areas is advocated (Pope 2013) (2) Modification and improvement of modern varieties (MVs) by traditional farmers known as "criolloization" (Lamola and Bertram 1994) or "rustication" (Prain 1993) used widely for appears to be widespread (3) Addressing to résolve issues in variety testing (Ceccarelli 2015); introduction of innovative designs in place of obsolete experimental designs and statistical analysis, with the attempts to capture spatial variability and correlation between the plot errors (Singh et al. 2003; Ceccarelli 2012); or increasing the number of locations by the use of partial replication (Cullis et al. 2006) (4) Increasing awareness among farmers by conducting trails at farmer's field as higher rate of illiteracy is still prevailing (~30%) among M&S farmers
5	Seed quality control	(1) Nonavailability of trained and qualified persons in seed certification (2) Lack of coordination among officials of Jharkhand State Agriculture Development Corporation Limited (JSADC Ltd), Jharkhand State Seed Certification Agency (JSSCA),	(1) Advance indent is prerequisite for the desired quantity of FS from Directorate of Agriculture (DOA) and/or Jharkhand State Agriculture Development Corporation Limited (JSADC Ltd) (2) JSSCA should work on recruiting qualified persons on

(continued)

Table 21.8 (continued)

S.N.	Issues	Governance inputs	Main governance challenges
		Directorate of Agriculture (DOA), etc. (3) Cumbersome registration process causing harassment to seed growers	regular basis (3) Minimum 25% of the estimated cost of the advance indent need to be deposited to the university
6	Administrative Issues	(1) Poor implementation of the State Seed Roll Plan (2) Nonlifting/partial lifting/ delay in lifting of the seed produced by the farmers (3) Inordinate delay in making payment to seed growers caused disenchantment among farmers seed producers (4) Poor time and resource management	(1) JSADC Ltd should take lead in strengthening of required infrastructure (2) JSADC Ltd should take lead in strengthening of backward and forward linkage (3) An MOU is required between (a) JSADC Ltd/DOA and University and (b) JSADC Ltd/DOA and farmer seed producers
7	Governance measures to improve seed supply systems	(1) Assessment of seed requirement based upon increasing SRRs including contingent planning (2) Greater attention need to be paid for seed production among crops with higher seed rate and poor seed multiplication ratio (SMR) such as chickpea, groundnut, field pea, soybean, wheat, upland rice, and potato (Singh 2013) (3) Creation of adequate infrastructural facilities and logistic support required for seed production, processing, storage, and distribution (4) Improving seed multiplication ratio (SMR)	(1) Confidence building measures to bring back the seed growers into fold (2) Improving seed conversion ratio (BS → FS → CS) (3) Skill development for entrepreneurs in the field of seed production, processing, storage, and marketing (4) Proper time management since all activities related to seed production, processing, storage, and distribution are time bound (5) Maintaining strong linkage among all stakeholders

considered serious while formulating and implementing such kind of programs in order to remove difficulties and strengthening seed systems at local level.

References

Anonymous (2018) Seed industry in India 2018–2023: market trends, structure, growth, and key players-CAGR to Grow at 14.3%. ResearchAndMarkets.com. https://www.researchandmarkets.com/research/nf2gqz/seed_industry_in?w=4

Atalan-Helicke N, Mansfield B (2012) Seed governance at the intersection of multiple global and nation-state priorities: modernizing seeds in Turkey. Glob Environ Polit 12(4):125–146. https://doi.org/10.1162/GLEP_a_00290

Baranski MR (2015) Wide adaptation of green revolution wheat: international roots and the Indian context of a new plant breeding ideal, 1960–1970. Stud Hist Phil Biol Biomed Sci 50:41–50

Bishaw Z, Turner M (2008) Linking participatory plant breeding to the seed supply system. Euphytica 163(1):31–44

Ceccarelli S (2012) Plant breeding with farmers: a technical manual. ICARDA, Aleppo

Ceccarelli S (2015) Efficiency of plant breeding. Crop Sci 55:87–97

Ceccarelli S (2016) Harnessing the power of evolution in participatory seed breeding, comments. https://theecologist.org/201M6/feb/29/harnessing-power-evolution-participatory-seed-breeding

Ceccarelli S, Grando S (2019) Participatory plant breeding. Who did it, who does it and where? Exp Agric:1–11. https://doi.org/10.1017/S0014479719000127

Chand R (2017) Doubling farmers' income. Rational, strategy, prospects and action plan. NITI Policy Paper No. 1/2017. National Institution for Transforming India (NITI Aayog), Govt. of India, New Delhi, p 40

Coleman M (2008) US immigration law and its geographies of social control: lessons from homosexual exclusion during the cold war. Environ Plann D Soc Space 26(6):1096–1114

Cullis BR, Smith AB, Coombes NE (2006) On the design of early generation variety trials with correlated data. J Agric Biol Environ Stat. https://doi.org/10.1198/108571106X154443

DAC & FW (2016a) State of Indian agriculture 2015–16. Department of Agriculture, Cooperation & Farmers Welfare, Directorate of Economics and Statistics, Ministry of Agriculture & Farmers Welfare, Govt. of India, New Delhi, p 280

DAC & FW (2016b) All India report on input survey 2011–12, agriculture census division. Ministry of Agriculture and Farmers Welfare, New Delhi, p 521

Dasgupta S, Roy I (2011) Chapter 5. Agricultural governance: seed supply system and plant genetic resource management. In: Good Agricultural Governance: a resource guide focused on smallholder crop production. Rap Publication 2011/18. Food and Agriculture Organization (FAO) Regional Office for Asia and the Pacific Bangkok, Thailand, pp 47–60

DSR (2014) ICAR seed project-Seed production in agricultural crops. Annual Report 2013–2014. ICAR-DSR, Mau, p 262

Galiè A (2013) Governance of seed and food security through participatory plant breeding: empirical evidence and gender analysis from Syria. Nat Res Forum 37:31–42

Gautam M (2013) Strategies for quick spread of new varieties among the farmers. In: Advancement in agriculture through Quality Seeds: a Souvenir Brought out on the Occasion of 6th National Seed Congress Held at Lucknow from 12–14 Sept, 2013. Ministry of Agriculture and Farmers Welfare, Govt. of India, Lucknow, pp 49–53

Kulkarni SD (2013) Post production agriculture and their role in making the agriculture sustainable. In: Shetty PK, Ayyappan S, Swaminathan MS (eds) Climate change and sustainable food security. National Institute of Advanced Studies; Indian Council of Agricultural Research, Bangalore and New Delhi, pp 162–171

Lamola LM, Bertram RB (1994) Experts gather in Mexico to seek new strategies in preserving agrobiodiversity. Diversity 10:15–17

Louwaars NP, De Boef WS (2012) Integrated seed sector development in Africa: a conceptual framework for creating coherence between practices, programs, and policies. J Crop Improv. https://doi.org/10.1080/15427528.2011.611277

Muralidharan K, Prasad CGSV, Rao CS, Siddiq EA (2019) Genetic gain for yield in rice breeding and rice production in India to meet with the demand from increased human population. Curr Sci 116(4):544–560

Nagarajan S, Yadava SP, Singh AK (2008) Farmers' variety in the context of protection of Plant Varieties and Farmers' Right Act., 2001. Curr Sci 94(6):709–713

Pandey S, Velasco ML, Yamano TS (2015) Scientific strength in rice improvement programmes, varietal outputs and adoption of improved varieties in South Asia. In: Walker TS, Alwang J

(eds) Crop improvement, adoption, and impact of improved varieties in food crops in Sub-Saharan Africa. CGIAR and CABI, Wallingford, pp 239–264
Patnaik BB (2013) Quality seeds—contribution of national seeds corporation. National Seeds Corporation Ltd. Golden Jubilee Souvenir 2012–13, pp 14–18
Pope HAN (2013) Participatory crop improvement: the challenges of and opportunities for institutionalisation in the Indian public research sector. PhD thesis, University of Sussex
Prain G (1993) Mobilizing local expertise in plant genetic resources research. In: de Boef W, Amanor K, Wellard K, Bebbington A (eds) Cultivating knowledge: genetic diversity, farmer experimentation and crop research. Intermediate Technology, London, pp 102–110
Raghu PT, Erenstein O, Bober C, Krishna VV (2015) Adoption and outcomes of hybrid maize in the marginal areas of India. Q J Int Agric 54:189–214
Rietberg P, Gevers H, Hospes O (2014) Seed Governance. From seed aid to seed system security in fragile areas. Wageningen University, Cordaid, P.O. Box 16440, The Hague, p 50
Salazar R, Louwaars N, Visser B (2007) Protecting farmers' new varieties: new approaches to rights on collective innovations in plant genetic resources. World Dev 35(9):1015–1528
Shetty PK, Hegde MR, Mahadevappa M (2013) Innovations in rice production. National Institute of Advanced Studies, Indian Institute of Science Campus, Bangalore-560012, p 262
Singh RP (2012) Enhancing food and seed security through participatory approach in the Eastern region to mitigate the climate change impact. J Res (BAU) 24:1–12
Singh RP (2013) Issues and strategies to correct missing links in seed sector of India. J Res (BAU) 25(1):1–15
Singh RP (2015) Varietal replacement rates among field crops: current status, constraints, impact, challenges and opportunities for the Indian seed industry. Seed Times (National Seed Association of India) 7(3&4):71–89
Singh RP (2018) Integration and commercialization of local varieties under sub-optimal environments for food security, promoting sustainable agriculture and agro-biodiversity conservation. MOJ Eco Environ Sci 3(2):65–67
Singh RP, Agarwal RC (2019) Farmer's varieties in India—factors affecting their preferential prevalence and the current status of their legal protection. Indian J Agric Sci 89(9):1371–1382
Singh RP, Agarwal RC (2020) Farmer's varieties in India—new insights into adaptation patterns, performance and preferences over modern varieties among different seasons and regions. Indian J Agric Sci (In press)
Singh RP, Agrawal RC (2018) Improving efficiency of seed system by appropriating farmer's rights in India through adoption and implementation of policy of quality declared seed schemes in parallel. MOJ Eco Environ Sci 3(6):387–391
Singh RP, Singh S (2016) Optimizing seed replacement rates in Jharkhand: present scenario, challenges and opportunities. Jharkhand J Dev Manage Stud 14(2):6987–7007
Singh M, Malhotra RS, Ceccarelli S, Sarker A, Grando S, Erskine W (2003) Spatial variability models to improve dryland field trials. Exp Agric 39:151–160
Singh RP, Singh 'Dron' DK, Chattopadhyay S (2013) A glimpse of success stories. Business Planning & Development (BPD) Unit, Birsa Agricultural University, Ranchi, p 20
Singh RP, Kumar A, Ranjan R, Singh H, Kumar M, Singh RK, Dwivedi AK, Singh S (2015a) Participatory seed production with tribal farmers in Jharkhand—a success story. In: Decade of ICAR seed Project: retrospect and Prospects. Rajendra Prasad S, Udaya Bhaskar K, Umesh R. Kamble, Ramesh K.V., Radhika C., S. Natrajan, Sripathy K.V., Rajiv K. Singh, Govind Pal. ICAR-Directorate of Seed Research, Mau, UP, pp 80–85. ISBN: 978-81-925128-6-6
Singh RP, Prasad PVV, Reddy KR (2015b) Climate change: implications for stakeholders in genetic resources and seed sector. In: Advances in agronomy, vol 129. Academic, New York, pp 117–180
Singh RP, Prasad SR, Agarwal DK, Sripathy KV (2017) Varietal and seed replacement in the era of climate change. Indian Council of Agricultural Research-Indian Institute of Seed Science (ICAR-IISS), Mau, UP, p 196. ISBN 9591-82-764-03

Singh RP, Kumar R, Kumar A, Singh S, Mishra A, Singh RK, Singh SK, Jerai MC, Das LK, Jha S, Ram B, Singh RK, Kumar P, Dwivedi AK, Jha AK, Ekka AB, Kumar V, Singh UK, Kumar S, Kandeyang S, Mahto CS, Singh S, Kumar SB, Kumar M (2019) Stakeholder's repulsive approach leading to distorted seed chain: lesson from the case study of participatory seed production in Jharkhand state. International conference on "food security through Agriculture & Allied Sciences" (May 27–29, 2019); Journal of Pharmacognosy and Phytochemistry. Special Issue 5:450–458

Singh RP, Kumar SPJ, Chintagunta AD, Agarwal DK, Kureel RS (2020) Varietal replacement rate: prospects and challenges for global food security. Glob Food Secur. https://doi.org/10.1016/j.gfs.2019.100324

Sinha H (2016) Rediscovering the traditional paddy varieties in Jharkhand: conservation priority in hybrid rice area. J Rural Dev 35(2):285–307

Walker TS, Alwang J, Alene A, Ndjuenga J, Labarta R, Yizgezu Y, Diangne A, Andrade R, Murthoni Andriatsitona R, De Groote H, Mauch K, Yirga C, Simotowe F, Katungi E, Jogo W, Jaleta M, Pandey S, Charyulu Kumara D (2015) Varietal adoption, outcomes and impact. In: Walker TS, Alwang J (eds) Crop improvement, adoption, and impacts of improved varieties in food crops in Sub Saharan Africa. CGIAR and CABI, Wallingford, pp 388–405

Witcombe JR, Yadavendra JP (2006) Cultivating partnerships: better choices for Rainfed farming. Published by the GVT, Kribhco Bhavan, A-8-10, Sector 1, Noida, UP, p 18

Witcombe JR, Packwood AGB, Virk DS (1998) The extent and rate of adoption of modern cultivars in India. In: Witcombe JR, Virk DS, Farrington J (eds) Seeds of choice: making most of the new varieties for small farmers. IBH; Intermediate Technology Group, Oxford and London, pp 53–68

Witcombe J, Devkota K, Virk D, Rawal K, Prasad S, Kumar V, Joshi K (2009) Client oriented breeding and seed supply. In: Scoones I, Thompson J (eds) Farmer first revisited-innovation for agricultural research and development. Practical Action

World Bank, FAO, IFAD (2009) Gender in agriculture sourcebook. The World Bank, Washington, DC

The manufacturer's authorised representative in the EU is Springer Nature Customer Service Centre GmbH, Europaplatz 3, 69115 Heidelberg, Germany. If you have any concerns regarding our products, please contact ProductSafety@springernature.com

Printed and bound by CPI Group (UK) Ltd, Croydon, CR0 4YY
30/09/2026
02225248-0017